T0224553

Affinity, That Elusive Dream

Transformations: Studies in the History of Science and Technology
Jed Buchwald, general editor

Affinity, That Elusive Dream

A Genealogy of the Chemical Revolution

Mi Gyung Kim

The MIT Press
Cambridge, Massachusetts
London, England

First MIT Press paperback edition, 2008

© 2003 Massachusetts Institute of Technology

Set in Sabon by SNP Best-set Typesetter Ltd., Hong Kong.

Library of Congress Cataloging-in-Publication Data

Kim, Mi Gyung.
Affinity, that elusive dream : a genealogy of the chemical revolution / Mi Gyung Kim.
p.   cm. — (Transformations)
Includes bibliographical references and index.
ISBN 978-0-262-11273-4 (hc. : alk. paper), 978-0-262-61223-4 (pb)
1. Chemical affinity—History. I. Title. II. Transformations (MIT Press)

QD505.5 .K47   2003
541.3'9—dc21

2002029574

in memory of my grandmother Bong Ak Kim

Contents

Acknowledgments

This project has had a long pre-history since my intellectual adolescence in graduate school at the University of California, Los Angeles. I was introduced to Wilhelm Ostwald and the concept of chemical affinity in a graduate seminar taught by Professor M. Norton Wise, who supervised my dissertation on the development of affinity chemistry in the nineteenth century. I am truly grateful for his advice, interest, and concern, and above all for the intellectual environment and stimulus he provided during my graduate school years. Professor Robert G. Frank spent countless hours in individual studies, helping me decipher difficult historical works and German texts. He persevered through the chapters of my dissertation while dealing with a death in his family. Without his kindness, finishing the dissertation in time would have been impossible. Professor Robert S. Westman taught an excellent class in historiography which prepared me for further explorations in methodology. I am deeply indebted to these superb teachers for my training in the history of science. The department of history at UCLA provided an exciting intellectual atmosphere which broadened my horizon as a historian as well as sharpening my capacity for critical thinking. I have particularly fond memories of the Cultural History Group.

Various institutions supported my nomadic existence during the subsequent Wanderjahre—the Verbund für Wissenschaftsgeschichte in Berlin, the Center for the Critical Analysis of Contemporary Culture at Rutgers University, the program in History and Philosophy of Science at Northwestern University, the Edelstein Center at Hebrew University, Jerusalem, and the Beckman Center for the History of Chemistry in Philadelphia (now the Chemical Heritage Foundation). While the projects I was working on at these institutions have not yet come to fruition, they laid the foundation for my approach to the Chemical Revolution. I wish to thank various people involved in running these institutions. Professor George Levine at the CCACC cultivated a truly unique intellectual environment in which I thrived. I also spent some time in Korea, teaching at Seoul National University. I would like to thank Professor Yung Sik Kim and the bright, eager graduate students, particularly Ji Yung Yu, for this exceptional opportunity to develop a more comprehensive perspective on the history of science. Chapter 6 of this book is drawn from the seminar I taught there on the Chemical Revolution.

I began to work on eighteenth-century chemistry in earnest after I received an appointment at North Carolina State University. The library resources at the three research universities in the Research Triangle made the project appealing. I wish to thank the staff at the Duke University Libraries, especially those in the history section of the medical library, for accommodating my requests for large quantities of photocopies. The staff at the interlibrary loan office of the North Carolina State University delivered a large number of books and articles in a short period. I have also drawn on the material I gathered at the Smith Collection of the University of Pennsylvania and at the Edelstein Collection of Hebrew University. Various grants from North Carolina State University allowed me to make yearly visits to the Parisian archives. I owe special thanks to Mm. Greffe, Mm. Mines, Mr. Le Roy, and especially Mm. Pouret at the archive of the Académie des sciences and to Mm. Molitor and Mr. Lopes at the history section of the École de médecine for friendly service and smiles. I also wish to thank the staff at the library of the Museum nationale d'histoire naturelle, the Bibliothèque nationale, and the École polytechnique. Dr. Martine Chauney-Bouillot kindly accommodated my last-minute effort to consult the material at the Dijon Academy archive. Dr. Patrice Bret provided me with information on the Dijon Academy archive. A grant from the National Science Foundation (SES 9985886) funded the final phase of research and writing.

My debt to the existing scholarship should be obvious throughout the text. I am thinking particularly of Hélène Metzger, Rhoda Rappaport,

W. A. Smeaton, Henry Guerlac, C. E. Perrin, Michelle Sadoun-Goupil, and Frederic L. Holmes. Without their painstaking and meticulous work on French chemistry, this kind of interpretative project would have been impossible. I am also indebted to the works of Bernadette Bensaude-Vincent and Jan Golinski for historiographical guidance. In understanding the French academic scene, the works of Pierre Bourdieu have been indispensable, although I found it impossible to cite him for specific points. Feminist epistemologies, especially Donna Haraway's "Situated Knowledges," have shaped the way in which I discern the voices of historical actors in their performative contexts.

Less visible but no less important has been the support of various mentors, colleagues, and friends. Professors Dorinda Outram, Simon Schaffer, and M. Norton Wise have offered unwavering support that sustained my career and motivation. Professor Seymour Mauskopf read various raw drafts. Without his trust in me and his enthusiasm for this project, this book would not have come to light. Various colleagues and friends also supplied valuable comments. I would like to thank Professors Marco Beretta, Trevor H. Levere, Alice Stroup, and Mary Terrall for their thoughtful reading. Thanks are also due to the participants in seminars at the Center for History and Philosophy of Science at Johns Hopkins University and at UCLA at which I presented the outline of this book. My colleagues in the history department at NCSU have provided an exceptionally supportive environment for a junior faculty member who had foolishly decided to write a wholly new book for tenure. Edith Sylla proofread parts of earlier versions of the manuscript. Gail O'Brien, Anthony LaVopa, and Jerry and Nelia Surh have sustained me with unfailing friendship, taking care of my son Tony on various occasions.

I am also indebted to Jed Buchwald for encouragement and helpful comments and to Larry Cohen for taking on the manuscript at a relatively early stage. Without their concerted efforts, it would have taken a much longer time to complete this book. Chapters 1 and 2 have been previously published in *Studies in History and Philosophy of Science* and *Science in Context*, respectively. I wish to thank Elsevier Science Ltd. and the Cambridge University Press for their permission to include them here.

The memory of my grandmother, who devoted her entire life to the education of her children and grandchildren, and of Korean students who sacrificed their lives fighting for a democratic government, have kept me going through tough times. My husband Jung-Goo Lee has suffered long periods of separation. Our son Tony has been an indefatigable cheerleader, working as a remarkably efficient copier during the last phase of this project. Their love has kept me grounded.

Affinity, That Elusive Dream

Introduction

I can put my meaning together with letters. Suppose an A connected so closely with B that all sorts of means, even violence, have been made use to separate them, without effect. Then suppose a C in exactly the same position with respect to D. Bring the two pairs into the contact; A will fling himself on D, C on B, without its being possible to say which had first left its first connection, or made the first move toward the second.

—Goethe, *Elective Affinities* (1809)

Chemistry supplied the central metaphor for Goethe's novel *Elective Affinities*. The Captain's explanation of double affinities in the beginning, quoted above, structures the plot: Edward and his first love, Charlotte, find their way to a happy union only after long, ill-conceived marriages thrown in their way. Settled in a country castle, they plan a joyous life together, building a summer house and laying the plans for a park. Into this peaceful scene sewn with blissful love enters a third person, the Captain. His presence gradually separates the couple's spheres of activity so that a fourth person, Ottilie, is called in to keep Charlotte's company. The ensuing passion between Edward and Ottilie and a somewhat measured, thoughtful love between the Captain and Charlotte lead the story to its conclusion: Edward and Ottilie lie together in the vault, while the Captain and Charlotte remain among the living with uncertain possibilities for a permanent union. Just as complete union of an acid and an alkali is impossible except in the form of a precipitate, Edward and Ottilie filter out of life to the place of an eternal union, the vault, leaving the Captain and Charlotte in a more mobile bond in solution, life.

Goethe's use of the analogy between chemical and human relationships is explicit.[1] Chemical substances unite "like friends and acquaintances" or stay as "strangers side by side," depending on their natural differences. Just as a third substance could displace one of the two substances in union, the introduction of an "alien element" or a "third person" could put the existing relationship between two persons "off balance" or invert the "whole moral condition." Two pairs could exchange partners. A whole array of moral lessons could be drawn from the behavior of chemical substances and their human analogues: "Whoever strikes at marriage . . . either by word or act," Goethe warns, "undermines the foundation of all moral society." Just as chemical combinations result from the immutable forces of nature, however, a true love binds human beings forever, beyond death. Against such forces, stern moral warnings sound hollow.

Elective Affinities captures a cultural moment when chemistry increasingly occupied the mind and the leisure time of the European public.[2] Images of Faust and Frankenstein that still haunt us today derive from the early nineteenth century. Chemistry was then a vigorous, powerful, voguish science that made its way into the salons and drawing rooms of the upper bourgeoisie. Chemists seemed to wield magic in manipulating the powers of nature to practical and amusing ends. Chemical demonstrations for the lay public, portable chemical laboratories, hydrogen balloons, and electrochemical experiments all secured the place of chemistry in public culture and imagination. It is not surprising, then, that Goethe chose this fashionable science to frame one of his novels. Nevertheless, we might pause at the choice of elective affinity as the central metaphor. Our historical memory does not accord a central place to chemical affinity in early nineteenth-century chemistry.[3] There exists a considerable distance, therefore, between the cultural memory Goethe enacted and our historical memory.[4]

Our historical amnesia can be traced to Lavoisier's omission of chemical affinity in his *Traité élémentaire de chimie* (1789). Lavoisier regarded it as the "transcendental part" of chemistry, a subject that would require a more sophisticated treatment than he could provide in a purposely elementary text. Although he valued affinity as the frontier of chemical investigation and theory, historians focusing on Lavoisier as the father of modern chemistry have overlooked its importance and concentrated on the Daltonian succession or stoichiometry. To restore chemical affinity to its proper place in the historiography, we must broaden our perspective and bring to life a spectrum of less famous chemists who plied their art in many different (and nowadays unfamiliar) milieu. In particular, chemical teachers played an important role in inventing a coherent scientific tradition. We also must take a more comprehensive look at the development of chemistry as a scientific discipline than has been usual. Chemistry became "an authentic discipline" with prestige and public visibility well before the onset of the Chemical Revolution, nearly a century earlier than modern physics.[5]

In the seventeenth century, chemistry already possessed a stable material culture of the laboratory, a strong didactic tradition, and the discourse of Chemical Philosophy (which provided it with theological and cosmological justification). The primary locales of chemical practice—the laboratory, the classroom, and the public arena—were sufficiently well established for chemistry to acquire a unique designation as a field of practice and inquiry.[6] Chemists did perhaps trail mathematicians and natural philosophers in asserting the cognitive authority of their discipline as an organized field of knowledge, which is to say its "philosophical" character. Their alchemical and Paracelsian linguistic heritage came under increasing attack from natural philosophers seeking to domesticate this rich empirical field in order to refurbish their systems of philosophical knowledge. This was precisely Robert Boyle's aim in promoting chemistry as the "key" to "true" natural philosophy. Boyle endeavored to dissociate the laboratory practice of chemistry from the discourse of Chemical Philosophy and to forge a philosophical chemistry [chemia philosophica] that would translate the laboratory culture of chemistry for a learned audience. The institutionalization of philosophical chemistry required a social evolution of chemists from the "sooty empirics" who prepared medicaments in the basement into learned savants who could converse on equal terms with the scholarly world.

The articulation of philosophical chemistry, as opposed to Chemical Philosophy, is a notable feature of eighteenth-century chemistry.[7] It allowed chemists to assume the social identity of scholars and to publicize their discipline as one of the enlightened sciences that contributed to forging a true representation of nature. For this reason, historians have characterized the Chemical Revolution in terms of the conservation principle, the nomenclature reform, and the epistemological changes that made chemistry more palatable to the philosophical and literary tastes of the Enlightenment public.[8] As diverse as these interpretations may be, they nevertheless share a common premise that the theoretical part of chemistry was "extrinsic" to its laboratory culture.[9] Chemistry, it is assumed, required a new philosophical language, one less hermetical and more in tune with natural philosophy, rather than a disciplined practice, to become a "science." Much like Boyle's contemporary philosophers and the Enlightenment public, historians have succumbed to the lure of intelligible language in judging pre-Lavoisian chemistry as an art without an overarching theoretical system. Even if it could be "described as a discipline, as a set of investigative and interpretive practices, as a livelihood, and as a tradition of texts," Arthur Donovan argues, it lacked the methodological rigor to become a "science."[10]

What constitutes a "science" or a legitimate system of knowledge depends, however, on the criteria specific to each historical period. Eighteenth-century philosophical chemists did not regard apothecaries' practice as philosophical knowledge for their lack of intelligible language. In turn, we might fault these philosophical chemists for their very dependence on philosophical, speculative, and non-specialized language. From the perspective of modern laboratory science, the apothecaries' labor laid the true foundation of chemistry as a science. Eighteenth-century philosophical chemistry was more successful than Paracelsian Chemical Philosophy in its crusade to win for chemistry a respectable place among the other sciences. Nevertheless, it did not guide the practice of apothecaries and metallurgists whose labor molded chemistry as the prototype for modern laboratory sciences.

To trace the evolution of practical chemistry in general terms, we must pay attention to the discourse of *theoretical* chemistry that mediated between chemical philosophy and practice. Whereas chemical philosophy—be it Aristotelian, Paracelsian, Cartesian, or Newtonian—aimed at explaining chemical phenomena as a part of the general order of the universe, chemical theories catered to the demands of laboratory practice. Whereas chemical philosophy fulfilled important functions in the social legitimation and pedagogy of chemistry, chemical theories organized chemists' laboratory experience by mapping the expanding chemical territory. Central to pre-Lavoisian French chemistry were the two theory domains that developed out of the practices in pharmacy and metallurgy. In contrast to specific theories such as the phlogiston or the oxygen theory, which were useful in making sense of a limited range of phenomena, the theory domains of composition and affinity reflected broader yet essential practical strategies on which chemists relied in their laboratory work. Simply put, chemists needed to know what quantities of various substances should be mixed together to make successful products, which substances reacted together, and which did not. In more concrete terms, the apothecaries' trade depended on making an exact mixture of "simples" to produce a medically salient "composition,"[11] while metallurgists had to sort out various metals by way of their differential dissolutions in acids and in fire. Chemists worked out these practical mandates in their laboratory experience through concrete experimental systems[12]: the distillation of vegetables and the selective dissolution of metals in acids. They also expanded these systems into the general theory domains of composition and affinity through their didactic practice. In other words, chemists generalized their experience with concrete experimental systems into a set of theoretical concerns that would organize chemical territory and lay out new investigative paths as well as establishing an interface with the reigning philosophical systems. As a nexus of thought and practice, these theory domains defined the identity of chemical practitioners. By acknowledging them as meaningful categories of discussion, we can recognize that procedural changes in chemical practice, rather than philosophical fashions, regulated the advancement of chemistry. As the analytic tools of chemistry became more and more refined while philosophical fashions kept changing, chemists reformulated these two theory domains to accommodate the changes. By delineating the domain of chemical theory, I analyze pre-Lavoisian chemical discourse into the three layers of philosophy, theory, and practice. The vitality of French chemical tradition depended on the interactive stabilization of these domains of discourse and reality:

philosophy	principles	attractions
theory	composition	affinity
practice	substances	operations

The articulation and stabilization of theory domains in French chemistry owed much to the dialectic between the didactic tradition at the Jardin du roi and the research tradition at the Acadèmie royale des sciences. Established in 1635 for the supplemental education of medical students, the Jardin harbored from the beginning notable Huguenot Paracelsians who advocated chemical medicine and had strong ties to the Protestant medico-pharmaceutical community. Through their triple functions as public lecturers, private laboratory teachers, and textbook writers, these Protestant chemical teachers nurtured a strong didactic tradition unmatched in other European countries. In their struggle to legitimate chemistry as the theoretical counterpart of "vulgar pharmacy" and as a worthy companion of the physicians' art, they articulated broad theory domains of composition and affinity that encompassed the general concerns of chemical practitioners. These theory domains were fully enunciated through intensive research by a group of academicians. Although the Acadèmie as a royal institution did not encourage theoretical research, the dynastic structure and mechanism of succession within this elite institution made it possible for certain academicians to consolidate theoretical investigative programs at exceptional moments and to pursue them systematically.

The first such theoretical moment came when a cluster of chemists around Wilhelm Homberg—Nicolas and Louis Lemery and Etienne-François Geoffroy—stabilized the theory domains of composition and affinity during the first two decades of the eighteenth century. Ostensibly Cartesian in their public, corpuscular language, Nicolas Lemery and Homberg brought to the foreground of chemical investigation the long-standing concern in French didactic tradition with the composition of bodies, expressed in the analytic/philosophical ideal of five principles. However, their principalist approach to chemical composition modeled on distillation method was soon undermined by the success of solution methods. While Louis Lemery shared the corpuscular modeling of chemical actions with his father and Homberg, he announced the death of the five principles by questioning the legitimacy of distillation methods in determining chemical composition. Geoffroy brought the full theoretical potential of solution methods into view with his affinity table. Affinities, rather than principles, were to determine the composition of bodies. In short, French academicians established in the early eighteenth century two contrasting approaches to the problem of composition: principles and affinity.

The second theoretical moment coincided roughly with the period conventionally known as the Chemical Revolution, when Louis Bernard Guyton de Morveau and Antoine-Laurent Lavoisier sought to rationalize chemistry through a new level of linguistic and quantitative discipline. While these two lawyers-turned-chemists tended to be more concerned with the trans-disciplinary rather than the disciplinary appeal of chemistry, each picked up a different strand of the French theoretical tradition—affinity and principles, respectively. Through Guyton's effort, affinity emerged as the frontier of mathematical and quantitative chemistry, while Lavoisier introduced algebraic and quantitative precision to the problem of composition. Their rapprochement during the 1780s provided French chemistry with a quantitative outlook and a new nomenclature. More important, their collaboration amidst the Arsenal Group produced a more sophisticated, dynamic conception of body contingent on the interplay of heat and affinities. The constitution, rather than the composition, of bodies became the enduring vision of French elite chemistry, although it failed to attract broader support from other European chemists.[13]

These two theoretical moments, roughly half a century apart, provide the backbone of this book. During these moments, French academicians drew on their native didactic tradition to define consciously the central theory domains of chemistry, elaborated them with the most current procedural and philosophical trends, and laid out a further course of investigation. This development of theoretical structures moved the rather backward French chemistry to the forefront of the European chemical community by the end of the eighteenth century. The Chemical Revolution centered around Lavoisier and the French Academy may accordingly be understood as a theoretical revolution that refurbished and offered a new vision for the practice of French elite chemistry. Although it reached a broader chemical audience through the nomenclature reform, it did not change radically the way in which the majority of European chemists practiced their art. Not the studies of constitution but stoichiometry defined the investigative frontier of European chemistry after Lavoisier's revolution. Not the "inheritors" of Lavoisier but Berzelius led this fundamental yet "quiet" revolution in chemical practice that yielded chemical atomism.[14] The Chemical Revolution as we know it was indeed a "French" (even a Parisian) affair, writ large by the participants' rhetoric and by our historiographical tradition.

This book's approach can perhaps be described as a genealogy of the Chemical Revolution. Genealogy as a common-sense term serves well here because it delineates the continuity of methods, problem domains, theory domains, and tacit knowledge in French chemistry without prescribing a deterministic causality. The family tradition sets many parameters for the child's development and identity, that is, without determining his or her personality, motivations, thoughts, behaviors, or career. Similarly, the institutional genealogies of Parisian chemists at the two elite institutions of the Jardin and the Académie contributed significantly to the coherence and vitality of French theoretical chemistry by providing relatively stable contexts of performance and reflection. Successive teachers at the Jardin did not simply recite their predecessor's lectures. They had to reenact them for each generation of students, which offered opportunities for invention and re-integration. Each generation of academicians built on the works of previous generations, creating an institutional memory that was both stable and amenable to innovation. Without the active collective memory of these elite institutions that spilled over into the public sphere, it would be difficult to explain how the different approaches to composition—principles and affinity—were passed down to the authors of the Chemical Revolution. They were a part of a living tradition rather than the ossified set of definitions often found in modern textbooks.

Despite the focus here on theoretical chemistry, a genealogical approach differs from the history-of-ideas tradition in historiography in that ideas do not determine the course of scientific development but instead become one element in a spectrum of resources upon which scientists may draw. When ideas, instruments, and methods become a part of written literature, they also become a part of collective memory which could be reenacted at different historical moments.[15] Central to this story of the Chemical Revolution are the methods or techniques—material, social, and literary—that contribute to the stabilization of a scientific tradition, yet are amenable to innovation.[16] A genealogical approach seeks to configure the contingent stability of a scientific tradition—a stability that depends upon the bringing together of many different elements at a given time—by delineating the ensemble of techniques that sustain the tradition.[17] Indeed, a significant change in any part of this ensemble could produce something as radical as an epistemological rupture, even the sort of incommensurability that is associated with scientific revolutions.[18]

Relationships between techniques for controlling nature and for representing it form an important part of the story. New analytic techniques played a crucial role in transforming the theoretical discourse of chemistry and in shaping the two theoretical moments in French chemistry. The infusion of solution methods into pharmaceutical practice led to the first theoretical moment by dissolving the traditional liaison between the distillational analysis of vegetables and the five principles of bodies. During the second theoretical moment, chemists had to integrate the physical concept of heat generated by thermometric practice with the chemical concept of phlogiston derived from experiments with ordinary flame and with the burning glass. They could create a coherent structure for chemical theory only by stabilizing a network of methods and techniques that allowed them to control nature in a reliable manner. The theoretical structure of chemistry depended critically on the stability of its analytic methods.

A genealogical perspective can sensitize the historian to the authority and power relations that shaped the collective research style,[19] the uses of particular philosophies and theories, the rise and fall of a research paradigm,[20] and the local translations of foreign authorities. Owing in part to the small size of the Academy, its representative chemists exercised strong control over the content and the style of its research tradition. Although their research interests often reflected the changing social niches of chemistry, their personal styles and exercise of symbolic power left a lasting legacy through the institutional memory embodied in the *histoire* and the *mémoires* of the Academy. The dynastic mechanism of intellectual succession and the relations of symbolic power modulated to a significant degree the research tradition of the Academy.[21] The relationships of competitive rivalry, master-protégé, and representative-subordinate pair often played a role in producing representations of nature. The master chemists screened foreign authorities carefully to bolster their own symbolic status. The differential receptions of Newton, Stahl, and Boerhaave in France are quite instructive in this regard. Whereas natural philosophers such as Boyle, Newton, and Boerhaave provided precious symbolic capital on which master chemists could draw to address their non-chemist colleagues and more general audiences, chemical authorities such as Stahl offered concrete knowledge that could enhance their expert status and leadership within the chemical community.

Changing techniques of linguistic control and communication also played a significant role in shaping French chemistry.[22] Previously relegated to the vocational domains of pharmacy and medicine, chemistry began to compete as a rational entertainment with other sciences toward the end of the seventeenth century. Nicolas Lemery drew a fashionable audience (including ladies) to his lectures by utilizing the corpuscular language. Boerhaave set an excellent example for a new kind of literary/philosophical chemist—the savante-chimiste—with his penchant for classical literature and his network of correspondence. He provided chemists with a philosophically viable discourse that became a part of the conversational culture of Parisian salons and thus helped chemistry become a vital part of the Enlightenment public sphere.[23] Chemists could leave behind a century-long struggle to gain acceptance in the Parisian medical community and look forward to a brighter future in which chemistry participated as an equal partner to physics in constructing a rational representation of nature. Caught between the commerce of goods and that of words, the Enlightenment chemists provide an ideal subject to bring together the duality of Habermas's public sphere as a bourgeois social space and as a peripherally noble cultural space.[24] Despite the constant deployment of the rhetoric of utility, the raison d'être of chemistry in the Enlightenment was no longer limited to its medical and industrial applications.[25] On the surface, the clamor for a rational discourse of chemistry among the articulate, literary audience made it necessary for chemical teachers such as Rouelle and Macquer to strike a balance between establishing the independent status of chemistry and its place in the overall scheme of scientific knowledge.[26] Newtonian philosophy became the new public language of chemistry for this reason.

At a deeper level, the participation of a broader public in chemical debates wove the techniques of persuasion and argumentation into the fabric of chemical theories. Debates over chemical theory were often conducted by non-vocational chemists who sifted through the existing chemical literature to craft a consistent representation and ontology of nature.

This book is no more than an interpretive essay designed to open up new venues of historical investigation.[27] Each chapter, more or less confined to a chronological period, is structured around a set of questions, which eventually should be answered by much more nuanced "thick descriptions."[27] Chapter 1 establishes the importance of didactic tradition in setting the theoretical agenda for French chemistry. Although Owen Hannaway has argued that chemistry emerged as an "integral and distinctive discipline around the turn of the seventeenth century" thanks to the formal codification of its practices in textbooks, historians have not taken up his cue in delineating the origins of the Chemical Revolution.[28] I argue that the didactic tradition played a crucial role in stabilizing the social and cognitive space of chemistry in Paris by incubating a theoretical discourse that could mediate between chemists' laboratory practice and the reigning philosophical paradigms. In their struggle to establish a social niche in the Parisian medical scene controlled by the Faculté de Médecine, French Paracelsians with strong ties to Montpellier and Protestant medical community carved out a domain of chemical theory between the domains of chemical philosophy and practice. They stabilized an analytic/philosophical ideal of five principles by incorporating the philosophical premise of quality-bearing principles into their analytic practice of distillation.[29] Soon, however, changing philosophical paradigms (from the Aristotelian to the Cartesian) and analytic methods (from distillation to solution methods) began to undermine the doctrine of five principles. In this context, Boyle's critique of chemists' principles and elements served as a negative heuristic to dissociate the analytic component of the doctrine from the philosophical component. Nicolas Lemery defined the analytic ideal of chemical principles as distinct from philosophers' natural principles or elements. Nevertheless, the assumption that chemical composition should be understood in terms of property-bearing principles remained in chemical thought throughout the eighteenth century, resurfacing in Lavoisier's and Berthollet's theories of acidity.

Chapter 2 focuses on the works of Wilhelm Homberg, who articulated a sophisticated program of corpuscular chemistry within the Académie des sciences around the turn of the eighteenth century. As a natural philosopher/chemist with a strong instrumentalist bent in the Boylean fashion, he approached the problem of chemical composition with a much more systematic perspective than other Paracelsian chemists trained in pharmacy. While the previous generation of chemists in the Academy lacked intellectual prestige despite their expert knowledge, Homberg brought chemistry closer to the reigning Cartesian discourse by rehabilitating the theory domain of chemical composition. His philosophical outlook on the analytic practice of chemistry and his new, powerful furnace (the burning glass) produced a somewhat fractured discourse in which one can easily trace several layers of meaning and reality he had to negotiate. Because of his binary identity and perspective, Homberg's discourse on chemical principles illustrate clearly the three layers of chemical language—philosophy, theory, and practice—that remained implicit in many chemical writings throughout the eighteenth century.[30] Although his decades-long effort to bring the doctrine of five principles closer to philosophers' ideal of natural elements and to chemists' ideal of analytic principles did not reach a satisfactory conclusion, a detailed analysis of his language will help us to understand how the practical problem of composition became a central problem in chemical theory.

Chapter 3 concerns the stabilization of the theory domain of affinity, which also fostered an affinity approach to composition. Although the selective dissolution and displacement of metals in acids had been a long-standing problem in metallurgy, the infusion of solution methods into pharmaceutical practice made it a pressing theoretical concern by the turn of the eighteenth century. Louis Lemery and Etienne-François Geoffroy worked out the theoretical implication of solution chemistry through mechanistic imagery and tabular representation, respectively. Despite the divergent approaches, their combined work brought into the open the inadequacy of distillation methods for determining the theoretical composition of bodies and hastened the demise of the five principles. As a consequence, the quest for the "principles of mixts" lost its paradigmatic status to characterizing the "compound" [composé]. Geoffroy's 1718 table des rapports visualized the composition of middle salts from acids and alkalis, embodying an empirically determined view of, or an affinity approach to, composition. That is, affinity became a more comprehensive investigative program with tacit directives for studying the composition of bodies. The chemistry of salts as embodied in the affinity table became the main line of chemical inquiry in the eighteenth century,[31] constituting something like a "normal science" of chemistry. Stahl's works were deeply implicated in this development, much earlier than what historians have identified as the French Stahlist period of the 1760s.[32]

Chapter 4 discusses the reconstruction of French theoretical discourse in the mid-eighteenth century. The demise of the five principles shattered the existing structure at the very time when the articulate audience of the High Enlightenment demanded a more rational and universal language of chemistry. It was in this milieu that foreign authorities such as Herman Boerhaave, Georg Ernst Stahl, and Isaac Newton received serious attention. Boerhaave offered a methodical discourse based on elaborate accounts of chemical and physical experiments. His scholarly treatment of chemical art attracted a large following among the medical students and literary public.

Guillaume-François Rouelle, a popular lecturer at the Jardin, drew on Boerhaave to address Parisian audience, but he did not subscribe to Boerhaave's vision of physical chemistry. Instead, he enshrined Stahl as a chemical authority who offered a wealth of new operations and a philosophical, principalist reflection on the issue of composition. Pierre-Joseph Macquer, who wrote a popular textbook to replace Nicolas Lemery's, also extolled Stahl's genius that ushered in modern chemistry. Newton's works had little impact on chemical practice, but his authority as the new master of natural philosophy prompted chemical teachers to change the public language of chemistry from Cartesian to Newtonian. The presence of Newton in the public sphere[33] also exercised a selective pressure on the existing theory domains of chemistry. Affinity as a kind of attraction received a larger share of chemists' attention as a result, particularly in Macquer's writings. Despite their shared interest in salts and their endorsement of Stahl's authority, Rouelle and Macquer advocated somewhat divergent theoretical structures of chemistry, each favoring the principalist Stahlian approach and the affinity-oriented Newtonian discourse, respectively. These simultaneous shifts in analytic methods, chemical doctrines, and philosophical systems transformed the French didactic tradition during the High Enlightenment. New recruits to chemistry had to deal with the discipline's fractured authority, discourse, and identity.

Chapter 5 deals with a variety of Newtonian languages that plagued the domain of affinity in the years leading up to the Chemical Revolution, particularly in the provincial academies. Although Macquer established affinity as the theoretical frontier of chemistry, Parisian academicians tended to work on more concrete industrial and medical problems. In contrast, non-vocational chemists in the periphery took up his cue and pursued chemical affinity more systematically. The most prominent figure in this regard was Guyton de Morveau, a Dijon lawyer who became the first author of the French nomenclature reform in 1787. His *Digressions* (1772), known mostly for the dissertation on phlogiston, contained in addition a comprehensive program for affinity chemistry.[34] Guyton's reputation grew rapidly through his contributions to the *Encyclopédie*, which brought him the title of the "first chemist of France." His correspondents included Torbern Bergman, who produced the most elaborate affinity table of the time in 1775, and Richard Kirwan, who sought to quantify affinities in the early 1780s. These three chemists constituted the foremost chemical authorities in the European Republic of Science in the mid-1780s, immediately before the public phase of the Chemical Revolution. Their prominence illuminates the extent to which Enlightenment practice of chemistry required reading and writing as much as laboratory work.

Chapter 6 contains a selective discussion of Lavoisier's investigative paths to highlight his theoretical inclinations. I have taken care to chronicle his thoughts on composition and affinity with equal attention to his theoretical outlook and to his experimental program. Lavoisier's earlier thoughts on elements and hydrometric measurements reveal his identity as an amateur *physicien* with a strong instrumentalist bent. He wished to chart a trans-disciplinary approach to the truth of nature that would result in a universal theoretical structure by conducting systematic measurements with a variety of meters—barometer, thermometer, and hydrometer. Lavoisier's principalist approach to composition yielded more immediate and tangible fruit thanks to the development of pneumatic chemistry. Once Lavoisier emerged from the thicket of pneumatic chemistry with the oxygen theory, however, his attention was drawn to the question of affinity and heat, although the task of consolidating and implementing his new system did not allow him time to pursue it systematically. Lavoisier's move toward affinity chemistry is dealt with in the following chapter.

Chapter 7 characterizes the second theoretical moment in French elite chemistry, which took shape during the 1780s when the Chemical Revolution entered a collaborative phase. Although phlogiston defined the polemical focus of the revolution, it did not encapsulate the entire vision of its four authors—Guyton, Lavoisier, Berthollet, and Fourcroy. Nor did the oxygen theory provide the revolution's exclusive rallying point. Throughout the period, affinity remained at the theoretical frontier of French chemistry. In the first volume of the *Encyclopédie methodique* (1786), Guyton announced the onset of a theoretical moment, again expounding his dissolution model of chemical action based on attraction. Lavoisier's investigation of heat with the calorimeter was supposed to provide a reliable measure of chemical attraction. Nomenclature reform was an integral part of Guyton's vision to which Lavoisier's oxygen theory became relevant only because acids constituted the core subject of, or the "key" to, chemistry. By focusing on affinity chemistry, therefore, we can discern multiple voices that shaped the Chemical Revolution. A gradual rapprochement of Guyton and Lavoisier, rather than Guyton's "conversion" to Lavoisier's chemistry, is an important feature in my genealogy of the Chemical Revolution. The most important product of their collaboration was a new conception of chemical bodies as determined by heat and affinities. The "constitution" in Berthollet's terminology, rather than the "composition," of bodies became the leading theoretical quest for French elite chemists.

Chapter 8 traces the evolution of Berthollet as an affinity chemist in order to understand the ways in which the Chemical Revolution succeeded and failed on its own terms. Berthollet's research interest was tuned to the chemistry of salts from early on, perhaps because of his training in the more conventional track of medicine. The theories of acidity, alkalinity, and causticity defined his initial investigative paths. He acquired an intimate knowledge of a broad range of chemical actions rationalized in the language of affinities. Lavoisier's antiphlogistic campaign never encompassed the entire scope of Berthollet's chemical experience. Rather, Kirwan's experiments and Guyton's thoughts on affinities shaped his investigative paths. The conversational culture of the Arsenal Group did have a strong effect on Berthollet, however, as can be discerned in his dynamic conception of the constitution bodies. Already visible in his lectures at the short-lived école normale during 1795, it was expounded in greater detail in his *Essai de statique chimique* (1803), a fitting finale to his history of philosophical chemistry as the unfolding of the affinity program. Though Berthollet's *Essai* may have represented the success of Lavoisier's and Guyton's physicalist program within French elite chemistry, it failed to guide other European chemists in the immediate aftermath of the Chemical Revolution. It was only in the middle of the nineteenth century, after chemists stabilized chemical species through painstaking stoichiometric investigations, that they turned to "Berthollet's problem" and to the quantitative determination of affinities.[35]

1

The Space of Chemical Theory

Chemists began the seventeenth century with a stable material culture inherited from the alchemical, pharmaceutical, and metallurgical practices. The art of chemistry also acquired "sufficient coherency and identity to sustain a vigorous and ongoing didactic tradition," earning it the "recognition as a unique discipline and activity."[1] The consolidation of chemical didactics took place, however, in the context of a highly polemical confrontation between the traditional Galenists and Paracelsians. Andreas Libavius, a gymnasium teacher and humanist, composed the prototype of seventeenth-century chemical textbooks with his *Alchemia* (1597).[2] Wary of the potential dangers Paracelsian chemistry posed to one's health, social position, and scholarly reputation, he produced a methodical discourse of chemistry for teaching.[3] Drawing on Libavius's text, French Paracelsians developed a strong didactic tradition unmatched in other European countries.[4] More specifically, it was the Huguenot physicians and apothecaries around the royal court who carved out a social niche for the practice of chemical medicine in Paris and, along with it, an institutional space for chemical didactics around the Jardin du roi.[5] Many of them had been educated in Protestant cities and had ties to the medical community of Montpellier, whose Faculté was tangled in an interminable feud over the use of chemical remedies with the Paris Faculté.[6] In their struggle to gain acceptance in the Parisian medical community, which was dominated by Galenic medicine, these Protestant chemical teachers forged a unique tradition that emphasized the medical utility of chemistry, or concrete pharmaceutical operations, rather than Chemical Philosophy. Despite their advocacy of chemical remedies, they were on the whole critical followers of Paracelsus.[7]

The practical orientation of French Paracelsians changed the discourse of Chemical Philosophy significantly. While the discussion of Paracelsian principles and Aristotelian elements often took the center stage in the theoretical part of their textbooks, they sought to merge these philosophical ideals with the analytic practice of distillation. This effort to match the philosophical imperative to the reality of chemical analysis shaped a new space of chemical theory—that of composition—and the doctrine of five principles. Just as the traditional apothecaries fussed over the proportion of medicinal "simples" in making an effective "composition," French iatrochemists sought to elucidate the composition of chemical remedies to vindicate their efficacy. Throughout the seventeenth century, they claimed that their theoretical knowledge of composition would help them prepare better medicaments than the "vulgar" pharmacists. According to Jean Béguin, chemistry aimed at preparing medicaments more agreeable to the taste, more salubrious to the body, and less dangerous in their operation. The art of chemistry differed from "vulgar pharmacy," which could not prepare with perfection medicaments of similar virtue.[8] Christophe Glaser argued in the same vein that the "noble art" of chemistry "does furnish us with the more effectual Medicines for the more inveterate and obstinate Affections, and often supply the failings and deficiencies of those of the vulgar *Pharmacy*."[9] By the end of the seventeenth century, Nicolas Lemery identified chemists' prowess in preparing better medicaments with their knowledge of the analytic—that is, distillational—composition of bodies.[10] Compositional analysis, or efforts to discern the "principles of mixts," shaped a positive identity of the chemist, separate from the identity of the "vulgar" apothecary.

The war over chemical remedies concluded in the 1660s in favor of chemists after Louis XIV was cured of typhoid fever with a concoction of antimony. With the creation of the Académie royale des sciences in 1666, Parisian elite chemists acquired another strong institutional basis for practicing their art as an integral part of medicine. However, the infusion of solution methods into pharmaceutical practice during the second half of the seventeenth century began to undermine the alliance between distillation methods and the Paracelsian/Aristotelian five principles. Van Helmont's critique of fire analysis reinvigorated the existing criticisms of distillational analysis, prompting chemists to seek more fundamental methods of decomposition for the "true" principles of bodies. Robert Boyle attacked the very premise of chemical analysis by negating the existence of principles. While his search for ultimate particles proved no less elusive than chemists' pursuit of principles, his corpuscular chemistry struck a resonant chord with the leading members of the Académie, who pursued a unified vision of sciences. Boyle's critique functioned as a powerful negative heuristic, driving a wedge between the analytic ideal of principles and the philosophical ideal of elements. Nicolas Lemery distinguished the analytic, "chemical principles" from the philosophical, "natural principles." His success as a popular teacher and textbook writer depended on his ability to negotiate the change in analytic methods from distillation to solution methods and the change in philosophical language from the Aristotelian to the Cartesian.

This chapter deals, then, with the nascent space of chemical theory in Parisian elite chemistry. Although Hélène Metzger did an excellent job of tracing various definitions of principles and elements in French didactic discourse, she was interested primarily in the philosophical integrity of chemical discourse.[11] In the following discussion, I will analyze these definitions for their relevance to chemists' laboratory reality as well as for their rhetorical efficacy in the institutional context of chemistry. The space of chemical theory was shaped not only by the philosophical strength of the concepts such as principles and elements but also by their performativity in the laboratory and in public lectures. In other words, Parisian chemical teachers adapted various philosophical languages to their laboratory and social reality. Their efforts stabilized composition as a major theory domain of chemistry and shaped the analytic ideal of chemical elements and principles, which was later endorsed by Lavoisier. In addition to configuring the cognitive and social space of chemical theory in the seventeenth century, therefore, this chapter points to the genealogy of the definitions of elements and principles running down to Boyle and Lavoisier through the medium of French didactic discourse. This continuity in didactic tradition ensured the survival of the "principalist" approach to chemical composition—that the principles contained in bodies determined their properties—well into the nineteenth century. The didactic discourse functioned as a center of intergenerational communication and collective memory.

1.1   The Jardin du roi

Paracelsian medicine found its way into the French court as early as 1560s, inducing vigorous opposition from the established medical community. The Faculté de médecine, which had a monopoly over the practice of medicine in Paris, banned the use of antimony in 1566 and brought charges against Roch le Baillif, a professed Paracelsian and médecin ordinaire du roi, before the parlement of Paris in 1578.[12] Despite the open hostility of the Faculté, Paracelsian physicians began to take root after the ex-Huguenot Henry IV's entry into Paris in 1593, establishing a niche for chemical medicine in a Parisian medical community dominated by the Galenic medicine of the Faculté. The appointment of Jean Ribit (ca. 1546–1605) as premier médecin du roi in 1594 induced further appointments of chemical physicians—including Joseph Duchesne (Quercetanus) (ca. 1544–1609) and Turquet de Mayerne (1573–1655), who also functioned as secret agents in Henry IV's diplomacy with Protestant countries. Ribit had been born near Geneva, had studied medicine in Turin, and had traveled extensively throughout Europe before his attachment to the court. While criticizing Paracelsian "sectaries" for their exaggerated claims, he evaluated carefully the merits of chemical medicine.[13] Du Chesne had learned chemistry in Germany and had taken a medical degree at Basel before coming to Paris in 1593. A long-time advocate of Paracelsian medicine, he published two theoretical treatises in its defense in 1603 and 1604.[14] Although he offered a conciliatory note toward the established medical community by portraying spagyric medicine as a supplement to traditional medicine, his books touched off an acrimonious debate over Paracelsian medicine and chemical remedies.[15] In 1597, Ribit also secured the position of médecin ordinaire for Mayerne, a Swiss Protestant who had studied at Heidelberg and had taken medical degrees at Basel and Montpellier.[16] A major rival of the Paris Faculté, the Montpellier Faculté accommodated Protestant students and chemical remedies. Montpellier, a favorite resort town for English gentry, also fostered an active community of apothecaries.[17] The doctrinal and religious heresy of the Montpellier Faculté only fueled the Paris Faculté's hostility toward their rival. The Paris Faculté did not acknowledge the rights of the Montpellier graduates to practice medicine in Paris, with the exception of royal physicians.[18] Mayerne defended chemical medicine in his lectures, which drew serious opposition from the Faculté. He left for England and served as a physician to James I and then to Charles I.

It was these chemical physicians at Henry IV's court, especially Ribit and Mayerne, who helped Jean Béguin establish his lecture course in Paris (ca. 1604–1620). Along with his immensely popular textbook *Tyrocinium Chymicum* (1610), Béguin's lectures inaugurated a distinct tradition of French didactic chemistry.[19] In the reactionary milieu of Marie de Médeci's regency after Henry IV's assassination in 1610, Béguin moved further away from speculative Paracelsianism by focusing on the preparation of chemical remedies. The next generation of chemical physicians to Louis XIII, Jean Hérouard and Charles Bouvard, helped Guy de la Brosse (1586–1641), whose father had been a physician at the court of Henry IV, to establish a botanical garden.[20] A royal edict was issued in 1635 for a Jardin royal des plantes médicinales—later called the Jardin du roi—to supplement the education of medical students.[21] La Brosse felt that the Faculté, content with bookish philosophy, ignored the potential benefits of studying the vegetable kingdom and chemistry. A botanist and a critical follower of Paracelsian medicine, he also objected to the exclusive use of minerals and sought to complement them with vegetables.[22] His efforts might account for Lefebvre's mid-century turn toward vegetable remedies. In an explosive milieu of renewed confrontation around Théophraste Renaudot (a protégé of Richelieu), La Brosse maneuvered carefully to secure an institutional space for chemical medicine.[23]

Despite strenuous opposition from the Faculté, which sought to protect its guild rights over Parisian medicine, the Jardin came to provide a secure institutional space for teaching chemical medicine through its public lectures.[24] The lectures at the Jardin were designed initially for botany, divided into two parts, l'extérieur des plantes (taught by the intendant and a sous-démonstrateur) and l'intérieur des plantes (taught by three demonstrators). Two positions of demonstrator were soon transformed, however, into lectureships in chemistry and anatomy. In 1647, William Davidson, a Scot who had been lecturing in Paris from the 1620s, became the indendant and the first demonstrator of chemistry.[25] He was succeeded in 1651 by Nicaise Lefebvre (ca. 1610–1669), son of a Protestant apothecary in the Huguenot center of Sedan. Lefebvre, educated at the Calvinist academy of Sedan, had apprenticed to his father and to another physician at the academy.[26] When Lefebvre left for England with Charles II in 1660, the position fell to Christophe Glaser (1615–1678), an apothecary from Basel. He taught until 1672, when an unfortunate involvement in a famous poisoning scandal forced him to leave Paris.[27] Moyse Charas (1619–1698) occupied the position briefly before Guy-Crescent Fagon was appointed as the intendant. (See appendix.)

The Protestant chemical teachers in Paris cultivated a strong didactic tradition through their public lectures, private laboratory courses, and textbooks. They carved out an institutional space and a cognitive authority for chemistry as a distinct form of knowledge with its own method of inquiry and principles. By extolling the virtue of chemical remedies, they advocated chemistry as an integral part of medical training and practice. This medico-pharmaceutical use provided chemistry with an institutional foothold in France, where the mining industry and allied metallurgical chemistry lagged behind those of Germany and Switzerland.[28] While emphasizing the value of concrete chemical operations in producing medically salient substances, the Parisian chemical teachers endeavored to transform chemistry from an art to a philosophical knowledge by crafting a theoretical discourse of elements and principles. To this end, they sifted through a variety of philosophies, ancient and modern, and tailored them to the practice of distillation analysis. Their efforts to bridge the gap between various philosophies and chemical operations shaped the mediating space of chemical theory and the institutional space of iatrochemistry. Consciously posing the question whether "chemistry ought to be called Art or Science," Lefebvre differentiated three species of chemistry in his *Cours de chimie* (1660): chemical philosophy, iatrochemistry, and pharmaceutical chemistry. Chemical philosophy aimed at "the knowledge of the nature of the Heavens and Starres, the source and original of the Elements, the cause of Meteors, original of Minerals, and the way by which Plants and Animals are propagated; having not in her power to frame or make any one of all these things, but being sufficiently pleased with entertaining her discourse and reason upon the causes of so many various effects." Iatrochemistry or medicinal chemistry had the goal of operation guided by contemplative and scientifical chemistry, "for as the art of Physick, consists in two parts, Theory and practice; the former being but a Clue and help to lead unto the other." Pharmaceutical chemistry had "for its end only operation, belonging to the Apothecary's profession, who is to direct his work by the Precepts and Orders of Iatrochymists." A successful physician had to possess all three species of chemistry. Chemistry could thus be defined as a "practical or operative science," since it attained the characteristics of an art as operations as well as those of a science as contemplative knowledge.[29]

Lefebvre's definition of iatrochemistry projected French Paracelsians as a new kind of physicians who shook off the bookish pretensions of the Faculté while guiding apothecaries in their operations through theoretical reflection. They were critical of the speculative tendencies in Paracelsianism as well as of the philosophical pretensions in traditional medical training. Most of them emphasized the experimental foundation of chemical medicine, devoting the bulk of their textbooks to a set of practical procedures for the preparation of chemical and vegetable remedies. Their social aspirations came through, however, in the theoretical and transparently rhetorical parts of the textbooks, where they gave the definition, the object, and the goal of chemistry in general terms. The changing institutional and rhetorical contexts are pertinent, therefore, to our understanding of French didactic discourse. Although a detailed contextual history of seventeenth-century French chemistry is beyond the scope of this book, I will point to the rhetorical dimension of the theoretical discourse in the hope that further studies will illuminate its sociocultural dimensions in more concrete terms.

1.2 Principles and Elements

Chemists inherited from the sixteenth century an odd mix of the Aristotelian "elements" and the Paracelsian "principles." In attacking the scholastic philosophy,[30] Paracelsus (1493–1541) had not abandoned the four elements, but he had introduced the tria prima of Salt, Sulphur, and Mercury without clarifying the relationship between the two systems.[31] The French Paracelsians who taught at the Jardin were, on the whole, critical followers of Paracelsus. Even as they defended the use of chemical remedies, they tried to cleanse chemistry of the Paracelsian excess of mysticism and to shape it into a respectable philosophy. To this end, they utilized the reigning Aristotelian philosophy.[32] Such an endeavor was perhaps necessary in the polemical context of acrimonious debates with the established Galenist physicians of the Faculté. The crisis in French intellectual life of the 1620s also fostered censorship of a variety of naturalist thoughts, including alchemical and Paracelsian varieties.[33] Although Parisian chemical teachers devoted the bulk of their textbooks to concrete pharmaceutical preparations, they mixed into their theoretical discussion the Aristotelian "elements" and the Paracelsian "principles."[34] At the same time, they increasingly associated the philosophically laden terms of elements and principles with their analytic practice, fusing them with the products of distillation. By the middle of the seventeenth century, they succeeded in stabilizing the five "elements or principles" as the ultimate chemical constituents of all bodies, matching them to the five categories of distillation products. They created thereby an analytic/philosophical ideal of chemical elements and principles that performed on the one hand the rhetorical function of legitimating chemical medicine and on the other hand the theoretical function of rationalizing the prevailing mode of chemical analysis.

Joseph Du Chesne utilized in his philosophical works both the Aristotelian elements and the Paracelsian principles. In *Le Grand miroir du monde* (1587), on the authority of Moses, he designated water and earth as the two "productive elements" to which God gave the light in Genesis. In addition, the three principles of salt, sulphur and mercury were found in all things and were related to the two productive elements.[35] Chemists knew "how to separate heterogeneous or dissimilar bodies, and to join the homogeneous or similar ones" by virtue of fire. The knowledge of principles and elements set them apart from the "operators & artisans," or mere empirics, who "amuse themselves by making various fires, by amassing pots and bottles, by wasting their coal, and by making poor use of their time."[36] In other words, the discourse of elements and principles served the rhetorical function of asserting chemists' superiority over common apothecaries. In his polemical work *De priscorum philosophorum*(1603), Du Chesne recounted a similar story of Genesis, but further speculated on the relationship between the four elements and the spagyric principles. Excluding fire for the reason that Moses did not mention it in Genesis, he accepted water, earth, and air as the three passive elements and mercury, sulphur, and salt as the three active principles. These three principles existed in all bodies of nature, even in the four Aristotelian elements. Du Chesne sometimes utilized all four elements, pairing salt with earth, sulphur with fire, and mercury with water and air.[37] On the authority of the Bible and experience, he attempted to thwart the rhetoric of the "dogmatists"—the Paris Faculté—who relied on ancient textual authorities to defend their turf.[38]

Jean Béguin probably arrived in Paris in the midst of the disputes between Galenists and Paracelsians. Under Ribit's protection, he set up a pharmaceutical laboratory and offered private instruction on spagyric remedies. He also defended their utility and necessity in his textbook *Tyrocinium chymicum* (1610).[39] The rhetorical mission of his text is clear. On the one hand, he deplored the hypocrisy of the Faculté in pretending to uphold the oracles of Hippocrates, while ignoring the real beneficence of spagyric remedies. On the other hand, he shunned the "imposters" who pretended to know chemistry but abused the art through their incompetence and fraudulent uses. To remedy the bad situation created by these two camps and to defend chemistry's "innocence," he had resolved to offer the public free of charge the knowledge he had acquired "with much pain and expense" in preparing more agreeable and more effective chemical remedies.[40] True to his purpose, Béguin allocated only a small portion of the book to general remarks, leaving the bulk of the book to practical operations of mineral substances.

Béguin's rhetorical intent comes through more clearly in his definition of chemistry, his defense of chemical remedies, and his use of Aristotle's authority to establish the unique principles of chemistry. Chemistry was "an art that teaches to dissolve natural mixt bodies[41] and to coagulate them when dissolved, in order to make more agreeable, salubrious, and safe medicaments." Chemistry differed from other sciences in that it was a theoretical and practical science, while all the other sciences were either theoretical or practical. Unlike physics, chemistry did not aim at contemplation and knowledge of mixt bodies, but sought to operate or to make all kinds of useful substances which placed it among the practical sciences. The goal of chemistry was to prepare medicaments more agreeable to the taste, more salubrious to the body, and less dangerous in their operation. Therein lay the difference of this art from "vulgar pharmacy," which prepared medicaments neither with such perfection nor with similar virtue. Béguin defended chemical remedies explicitly by addressing every objection raised against their uses. He argued that chemical remedies were not poisonous or contrary to human nature, that they were salubrious despite their acrid origins from metals and minerals, that they were not dangerous despite the acrimoniousness they acquired through fire, and that their preparation with fire did not destroy the virtues of the mixts. Although Béguin endorsed the condemnation of Paracelsus and his sect by the Faculté, he did not shun the use of chemical remedies, since they offered indispensable aids to a good doctor.[42]

Aristotle's authority secured the status of chemistry as a science with its own method of inquiry and principles. According to Aristotle, Béguin emphasized, every science had its own principles that constituted its object. Such a philosophically sanctioned inquiry for chemistry meant that it resolved mixt bodies into the three spagyric principles. In other words, Béguin justified the analytic premise of Paracelsian chemistry with the Aristotelian philosophy. Chemistry differed from physics[43] and from medicine in its method of inquiry. It possessed "principles proper & intrinsic, formally constitutive of its object." Even if they were studying the same body, physiciens examined its motion and rest, physicians the causes of health and disease, and chemists its resolution and coagulation. Through their art, chemists could reveal the internal virtues of the body and render them useful. Chemists had to proceed in all their examinations, theories, and operations by the three principles of mercury, sulphur, and salt. Otherwise, their knowledge and artifice would be without foundation and outside chemistry's principles. According to Aristotle, all things were resolved into what they were composed of. Matter and form were the natural principles of physics, four elements were the principles of medicine, and three spagyric principles were the principles of chemistry. Béguin utilized Aristotle's authority, then, to legitimate chemistry as philosophical knowledge superior to the merely empirical operations of "vulgar pharmacy."

The three principles Béguin stipulated differed somewhat from Paracelsus's tria prima, however, in their material characteristics. Although Béguin defined them in terms of general qualities,[44] they appeared in chemical analysis as concrete substances. In burning wood, for example, one obtained an aqueous, non-flammable vapor (which represented mercury), an inflammable vapor (which could be changed into oil, and which represented sulphur), and ash (from which one could easily extract a recognizable salt).[45] Therefore, there were as many mercuries, sulphurs, and salts as there were objects of analysis. Although the principles extracted from natural bodies were as numerous as the different kinds of extractions, experiments reduced them to the three "first principles," which did not depend on others yet on which all others depended: mercury, sulphur, and salt. Béguin's notion of "first principles" hinted at the last point of analysis, beyond which the virtues of mixts would be destroyed:

Chemist by this same means has discovered his principles Mercury, Sulphur, & Salt, seeing from experiment that Chemical & artificial resolution could well arrive at these three principles. It would stop at these principles and not go beyond it. Otherwise, it could destroy totally the virtue of the resolved body. But then this would no longer be chemical resolution which always has to conserve the virtues of mixts & stops at the principles which sustain them, in order not to go beyond its goal, which is to dissolve & coagulate the mixt without losing anything of its internal virtues.[46]

None of these principles appeared alone, however, as a simple substance. The fact that chemists could not make complete extractions and produce these principles in pure state rendered Béguin's theory somewhat uncertain. His three principles remained as yet ideal substances.

The introductory part of the *Tyrocinium* also discussed the main operations and instruments of chemistry, the taxonomy of which changed little over the course of the seventeenth century. Chemical operations consisted of solution and coagulation. Béguin listed two main methods of solution: calcination (solution in fire) and extraction (solution in menstruum). Calcination was "a reduction of mixt into calx, which was called by Geber pulverization of compound made by fire." Calx consisted of a subtle powder, principally of minerals, made by the dissipation of the humidity that linked their parts. Extraction was "a species of solution, by which the most subtle parts of mixt body are separated from grosser ones." General extraction comprised various procedures such as sublimation, distillation, digestion, putrefaction, circulation, and fermentation. "Special extraction" referred to a procedure in which the most subtle and noble parts of mixts were extracted by some menstruum and then made into a syrup through evaporation or distillation. Coagulation was an operation by which liquid substances became solid by the privation of their humidity. Béguin allocated a separate chapter to instruments such as furnaces, vases, and utensils of chemistry.

Chemists' claim to the internal composition of bodies became a focus of contention during the period of intellectual crisis in Paris.[47] In 1624, Antoine de Villon (a professor of peripatetic philosophy at the University of Paris) and Etienne de Clave (a chemical physician) attempted to post fourteen theses refuting the Aristotelian and the Paracelsian doctrines. They advocated the "true" natural principles—earth, water, salt, sulphur or oil, and mercury or acid spirit—which could be neither generated nor corrupted. All bodies in nature were supposedly produced by the "union & mixtion" of these material principles in different quantities. Although the authorities quickly suppressed their anti-Aristotelian, atomistic proposal, the backlash it induced from a range of physicians and philosophers testified to the entrenched Aristotelian orthodoxy of the time. Their critics played down their truth claim by problematizing the analytic methods. Jean Pagès, a physician, pointed out the inadequacy of "spagyric resolution" in isolating the said principles in pure form.[48] While refuting Robert Fludd's claims to truth and certainty through alchemical experience, Mersenne called for a determination of the proper elements of bodies, preferably with quantitative precision.[49]

Etienne de Clave later produced a more systematic treatise in which he pointed out the shortcomings of the Paracelsian and Aristotelian philosophies on analytic grounds. He argued that a natural or physical principle had to be a simple and homogeneous body and therefore could not be compounded. Ideally, the "elements of nature" should be discovered through "the resolution of mixts into their principles," which the Aristotelian theory did not prescribe. Furthermore, such resolution had to be "an exact & exquisite purification & the last or final reduction of diverse substances . . . into their homogeneity & elementary simplicity," but the Paracelsian practice stopped at the level of gross separation.[50] De Clave directed his polemical force, therefore, to securing the empirical foundation of the doctrine of principles and elements. He adopted five "principles or elements" based on the chemical experience of fire analysis. Of the four Aristotelian elements, he excluded fire and air: Fire was not a material principle but a cluster of phenomena. Air was indeed a simple body, but it did not enter into the composition of other bodies. Only water and earth, along with the three spagyric principles of mercury, sulphur and salt, qualified as "elements or first principles" that entered into the composition of all mixts. De Clave made the compositional qualification of elements explicit:

We say therefore that there are five simple bodies that we call elements, not because they are simple, otherwise the sky and the air would be elements, but only because they compose all mixts as we have said above.[51]

De Clave's five elements, seemingly qualitative to the modern reader, referred to several classes of distillation products: Fire resolved mixts "so happily as it does into five elements or first principles which were actually included and hidden in them, that is, water or phlegm, spirit or mercury, sulphur or oil, salt and earth."[52] Water was an insipid, "congealable element" that served as the vehicle of all others. Spirit or mercury was a "fermentable element, always acid, the most penetrating and the heaviest of all bodies." Oil or sulphur was an "inflammable element or principle quite subtle and the least heavy of all." Salt was a "coagulable, caustic, and viscous element which bound others together in the mixt." Earth was a "discontinuous, or friable principle or element with some bitterness."[53] In this manner De Clave grounded the long-standing doctrine of principles and elements on the chemical experience of distillation analysis. Only experiments could teach chemists the number and the properties of these elements. At the same time, De Clave presumed that these "homogeneous substances, elements, principles or simple bodies" were incorruptible and could not be converted one into another. Principles or elements were complete and simple bodies that formed diverse mixts according to various proportions. De Clave thus introduced what was, as Metzger observed, a "strangely modern" definition of chemical elements:

The element is a simple body which actually enters into the mixture [mixtion] of compound bodies [corps composés] and to which they can be finally resolved.[54]

De Clave's definition thoroughly fused the philosophical requirement that elements should be the building blocks of the universe with the act of chemical analysis. Since chemists were not able to isolate these five elements in pure state, however, his definition of elements remained an analytic/philosophical ideal or a mirage yet to be attained in practice.[55] Nonetheless, this ideal indicates that distillation became the representative method of chemical analysis by the middle of the seventeenth century. Chemists invented increasingly sophisticated furnaces to accommodate their complex art; they even made them public knowledge.[56] The link between the five principles and the distillation products soon became a stable part of French didactic discourse. The representative status of distillational analysis derived from the importance of vegetable analysis in pharmaceutical practice, since mineral substances seldom yielded all five principles. In other words, the distillation of vegetables provided an exemplary system for chemical theorizing in the middle of the seventeenth century, and this granted a paradigmatic status to the pursuit of the "principles of mixts."

A shift to vegetable analysis was noticeable in the textbook of Davidson's successor at the Jardin, Nicaise Lefebvre. In the preface to his Cours de chimie (1660), Lefebvre offered a broad justification of chemistry by emphasizing its utility over the course of human history. Chemistry was the "Art and Knowledge of Nature itself; that it is by her means we examine the Principles, out of which natural bodies do consist and are compounded; and by her are discovered unto us the causes and sources of their generations and corruptions, and of all the changes and alterations to which they are lyable." Chemistry had been a useful art. It had helped mankind from its infancy with knowledge of minerals and metallurgy and it had provided the ancient sages with "the occasions and true motives of reasoning upon natural things." The Bible and a host of ancient authorities (including Hippocrates and Aristotle) testified to this fact, despite the contention of "his sectators of this Age, to oppose Chymistry." Chemistry, the "true Key of Nature," allowed the "expert Artist" to discover Nature's "hidden beauty" and true remedies.

Lefebvre acknowledged Paracelsus, van Helmont, and Glauber as the "most ingenious and exact Artists" who provided the "Beacons and Lights" for the theory and the practice of chemistry.[57]

Lefebvre defended the use of chemical remedies strenuously. His defense offers a glimpse of the polemical context in which the teachers of chemical medicine operated. The running controversy with the Faculté made it necessary for them to pay attention to how the remedies were made, to their sources, and to their taste. In addition to the broad theoretical justification he offered in the preface, Lefebvre included "An Apology in the behalf of Chymical Remedies":

> I thought it necessary to vindicate the Professors of this noble Art, from the calumnies and aspersions which ignorant persons do cast upon it, before I come to describe the Preparations of those Remedies which are used by true Physicians; to arm with Reasons and defensive Arguments the Lovers of this Science, against the weakness of their Adversaries. I say, that these despisers of Chymists, and haters of Chymistry, are ignorant; because not only unacquainted with the true Preparation and Effects of these Remedies; but moreover ignorant in the knowledge of Nature and its Operations, which only can be discovered by those that work upon natural Products, and do exactly and curiously Anatomize all the parts which they contain in particular.[58]

Lefebvre accused Galenists of being ignorant of the very doctrines of Hippocrates and Galen they supposedly defended. They were "only Physicians by name," being acquainted only with some "University Writings" which persuaded them that "Physick is nothing else but an art of discerning heat and cold." In order to settle the controversy between "Chymists and Galenists," Lefebvre appealed to "Peripatetick Philosophy, provided it be seasoned with the noble knowledge both of Galenical and Chymical Physick" and guided by "Reason which is the Touch-stone by which the truth or falsity of all Learning is discovered." Galenists had three objections to chemical remedies: that they required the use of fire, that they were derived from minerals, and that they were too violent. Lefebvre did not see much harm in the use of fire, insofar as cooks used it all the time to prepare food. As for the origins of the remedies, Lefebvre contended that the "greatest and best part of the choysest Chymical Remedies are taken from the family of Vegetables and Animals." In fact, he devoted the bulk of his book to the preparation of vegetable remedies, and he advertised it for the "most abstruse and secret particulars in the matter of Vegetables."[59] He insisted, however, that even the remedies derived from minerals were not necessarily "venomous, or contrary to the nature of Man's Body." Galen and contemporary Galenists used minerals with little preparation and could not "correct the vices and violence" of the mineral remedies they used "with that requisite nicety and perfection." Chemists, in contrast, "prepared, corrected, and devested them from the malignity and venom they did contain, by the separating of Purity from Impurity." The strongest defense of chemical remedies derived from their effectiveness. They offered "the true weapons, wherewith a Physician must arm himself to conquer and extirpate the most stubborn and rebellious Diseases; and even such as are held incurable by the ordinary Remedies of *Galenical* Physick."[60]

Lefebvre's text is full of meandering thoughts and does not fit the mold Béguin provided. He accepted the common Paracelsian precept that there existed a "universal spirit" which was "a substance voyd and divested of all Corporeity" and the "primary and sole substance of all things." The "universal spirit" had a threefold denomination of sulphur, mercury and salt. It entered the composition of mixts only in these "specificated or embodyed" forms of sulphur, mercury, and salt, imparting specific properties. Mixt bodies were, therefore, fine or volatile, liquid or solid, pure or impure, and dissolved or coagulated "according as this spirit contained more or less Salt, Sulphur, and Mercury."[61] "Let us then conclude," wrote Lefebvre, "that this radical and fundamental substance of all things, is truly and really one in its essence, but hath a threefold denomination; for in respect of its natural heat and fire, it is called *Sulphur*; in respect of its moysture, which is the food and aliment of this fire, *Mercury*; and finally, in respect of the radical drought, which is, as it were, the knot and cement of the fire and moysture, it is called *Salt*. . . ."[62]

Despite his extensive speculation on the original composition of mixts, Lefebvre conceded that chemists found, upon actual resolution of mixts, the five principles or elements. Homogeneity served as the singular criterion for their selection; "after that the Artist hath performed the Chymical resolution of bodies, he doth finde last of all five kinde of substances, which Chymistry admits for the principles and Elements of natural bodies, whereupon are layd the grounds of its Doctrine, because in these five substances is found no heterogeneity; these are, the Phlegmatick or waterish part, the Spirit or Mercury, the Sulphur or Oyl, the Salt, and the Earth."[63] Lefebvre thus endorsed the idea of five principles linked to the distillation products. The importance given to vegetable remedies explains in part why distillation became the standard method of chemical analysis by the middle of the century, legitimating the five principles as the chemical constituents of all bodies.

The vacillating meanings of elements and principles were not lost to Lefebvre, however. He acknowledged that some named water and earth as elements and the other three as principles. He objected to such usage since the latter three did not meet the Aristotelian definition of principles that they should not be made from other things or from each other. Instead, they should be called "elements," since they were "the last thing to be found, after the resolution of the Compound." Elements were by definition the substances that composed mixts originally and to which mixts were ultimately resolved. Such a fine distinction ran counter to the commonsense usage, however, in which "principles" meant the components of mixt bodies and "elements" the substance of the entire universe. "But because the Elements are considered in two ways, either as they are parts which do constitute the Universe, or as they only compound Mixt Bodies; to accommodate ourselves to the ordinary way of expression, we shall attribute unto them the name of Principles, in regard they are constitutive parts of the Compound; and shall reserve the appellation of Elements, for those great and vast Bodyes, which are the general Matrixes of natural things."[64]

Christophe Glaser (1615–1678), who succeeded Lefebvre at the Jardin, paid more attention to practical operations. His popular textbook *Traité de la chymie* (1663)[65] was organized in much the same way as Béguin's, except for longer sections on vegetables and animals. Glaser defined chemistry as a "scientifick Art, by which one learns to dissolve bodies, and draw from them the different substances of their composition, and how to unite them again, and exalt them to an higher perfection." A "Noble Art," chemistry was "the key which alone can unlock to all Naturalists the door of Nature's secrets; by reducing things to their first principles; by giving of them new forms; and by imitating Nature in all its productions and Physical alterations." Chemistry furnished "more effectual medicines for the more inveterate and obstinate Affections" and thereby often complemented "the failings and deficiencies of those of the vulgar *Pharmacy*." Without the aid of this "noble and excellent Art," the apothecaries did not know how to "make their compositions like true Artists," or "how to preserve the principal vertue of their ingredients, and separate the pure from the impure, and heterogeneous in natural commixtures." Without chemistry, physicians would be at a loss to prepare effective medicaments.[66]

Glaser also endorsed five "first principles" to which chemists reduced animal, vegetable, and mineral substances by fire. Mercury, sulphur, and salt were active principles; water or phlegm and earth were passive. He made the relevance of these principles to the distillation products explicit:

These names are bestowed upon them for the likeness they have to common *Mercury*, *Sulphur* and *Salt*, and the Elementary water and earth: the *Mercury* appears to us in the resolution of bodies, in form of a most aerial subtil liquor; the *Sulphur* is apparent to our smell and tast, by which we distinguish it from the insipid and inodorous plegm, which sometimes ascends with it; and it appears to us in the form of a penetrating inflamable Oil; the *Salt* remains joyn'd to the body of the earth till it be extracted by elevation.[67]

According to Glaser, these "first principles" also possess certain qualities that indicate their presence in the mixt body. The most active principle, Mercury or the Spirit, is "a subtil, light, and penetrating substance, which gives life and motion to bodies, makes them grow and vegetate." Because of its activity, "those Mixts in which this substance predominates, are not very durable." The "second Active Principle" or sulphur is a "subtil, penetrating, and inflamable" substance that appears as a subtle or heavy oil in the distillation process. It is the cause of "Beauty or Deformity in Animals, of the different smells and colours in Vegetables, and of the ductibility and malleability of Metals," it "preserves bodies from corruption," and it "sweetens the *acrimony* of Salts and Spirits." Salt, extracted from the earthly residue, is the "cause of the different tasts of things," and it "renders those bodies where it abounds, so durable, that they are almost incorruptible." Water or phlegm makes the bodies that contain it easily corruptible. It dissolves salt and incorporates it with spirit and sulphur, bringing the activity of spirit and the inflammability of oil under control. Earth prevents salt from dissolving in water.

In the seventeenth century, then, chemical teachers at the Jardin sought to institutionalize chemical medicine, mediating between the bookish culture of the Faculté and the laboratory culture of the apothecaries. In the process, they tailored the Aristotelian and Paracelsian philosophies to the analytic practice of distillation to stabilize the doctrine of five principles. By the middle of the century, despite their fluctuating commitments to various discursive traditions, they had stabilized five "principles or elements" as the chemical constituents of all mixts. The stability of five principles in chemical theory owed much to the chemists' mainstay analytic method: distillation.[68] A skilled chemist who knew how to regulate the intensity of fire usually was able to extract five groups of substances from a variety of natural bodies (mostly vegetables): spirits, oils, aqueous liquors (phlegms and water), earths, and salts. Spirits were the volatile products of distillation that rose up during the process and were collected after cooling. Unlike oils, they did not "swim in" water (that is, they mixed with it), and they did not burn easily. Unlike phlegms, they were clear liquids. Oils and aqueous liquors were easy to differentiate. Oils "swam" in water and were inflammable. Phlegms were thick, insipid liquors that mixed with water. Through repeated dissolution, chemists were also able to treat the earthly residue that remained at the bottom of the distillation vessel to obtain the lixivial salt or fixed alkali and an earth.

Although the distillation process regulated one set of chemical terminology, it did not provide an exhaustive criterion of classification. Chemists had another, simultaneously operating system of classification that was based on the projected qualities of the five principles—mercury, sulphur, water, earth, and salt. The defining qualities of these principles derived from the earlier Aristotelian and Paracelsian philosophies. As chemists freely acknowledged, these philosophically defined principles were never isolated in pure form. Since chemists usually obtained a variety of mixtures from distillation, the correlation between the two sets of naming schemes was contingent on the judgment of the individual chemist. Often a chemist opted to discuss the philosophically defined five principles in the theoretical part of a textbook and to ignore them in the practical part. The five "principles or elements," with their uncertain ties to the distillation products, necessarily remained a mirage. Suspended between the analytic and the philosophical gaze, the analytic/philosophical ideal of elements and principles conformed to neither. Chemists in the seventeenth century employed two parallel systems of classification simultaneously, one based on the morphology of distillation products and the other drawn from the philosophical systems of Aristotle and Paracelsus. Between these two systems mapped by the analytic and the philosophical gaze, respectively, there remained a considerable gray area, which the individual chemist had to negotiate in a haphazard manner. It was precisely this state of affairs that rendered French didactic discourse vulnerable to the criticisms of Robert Boyle, the "sceptical chymist."

1.3   *The Sceptical Chymist*

In England, chemical remedies became a part of established medical practice much more quietly than in France. No controversy comparable to the Parisian debacle broke out over the use of chemical remedies themselves.[69] It was only during the political unrest of the Civil War and the Interregnum that Paracelsian and Helmontian philosophies began to make waves as a source of radical thought.[70] Although these chemical philosophies and worldviews were adapted to diverse religious and ideological opinions, their frequently subversive political association became a serious concern for the ideological warriors of the Restoration who wished to tame the new discourse for a stable political order.[71] The line of battle was drawn not over chemical remedies, but over the discourse of chemical philosophy in all its varieties—alchemical, Paracelsian, Helmontian.

In this context, Robert Boyle, as the representative natural philosopher of the Royal Society, had every reason to tread cautiously in advocating chemistry as the key to the "New philosophy."[72] While he appreciated chemists' labor and skill in harnessing nature, Boyle could not endorse the discourse of Chemical Philosophy, which was open to wayward political interpretations. Instead, he wished to construct a philosophical chemistry based on sound experiments and solid reasoning thereupon.[73] In this endeavor, he found it necessary to reform chemists' discourse of principles and elements. On the one hand, it had to conform to chemists' analytic practice. On the other hand, it had to acquire a level of respectability and legitimacy. Boyle introduced Helmontian methods of solution analysis for the former and the language of corpuscular philosophy for the latter.

The circulation of van Helmont's works transformed the English Paracelsian tradition.[74] Drawing on the existing criticism of fire analysis, van Helmont insisted that the principles separated out by the distillation method did not exist in the bodies before the analysis. In the *Ortus medicinae* (1648), he criticized the doctrine of tria prima, arguing that fire analysis did not simply extract the pre-existent principles, but that it created new ones. Instead, he proposed water and semina as the principles of mixts, presumably based on the solution method of analysis via the alkahest, or the universal solvent. Boyle became acquainted with van Helmont's works as early as 1648, thanks to the rising interest in them among the members of the Hartlib circle.[75] Along with George Starkey, he searched for the elusive alkahest during the early 1650s.[76] Moving in 1655 to Oxford, the royalist stronghold during the Interregnum, Boyle came into contact with a different group of scientists, composed mainly of mathematicians. Aside from lacking intellectual prestige, chemistry had acquired a subversive political resonance among this group.[77] Addressing these mathematicians, Boyle tried to provide chemistry with a more "solid" philosophical foundation, or to create a new "chemia philosophica" in the language of mechanical philosophy.[78] He argued that the iatrochemical definition of elements and principles could not be realized in practice, nor could it provide the foundation for a rigorously philosophical chemistry. By interpreting the theoretical part of French didactic discourse literally and perhaps more seriously than the authors had intended, Boyle drummed out the inadequacy of chemical principles in linking the practice of chemical analysis to the philosophical system. While it is not clear how much water Boyle's criticisms held among chemical practitioners, his voice held sway among the natural philosophers and mathematicians who sought to integrate chemistry into a more orderly structure of philosophical discourse. Ironically, the Parisian chemical teachers who tried to distance themselves from the "vulgar" chemists through their theoretical discourse were themselves labeled as

such by Boyle.[79] Clearly, Boyle perceived that medical utility and borrowed philosophical words could not legitimate chemistry in Restoration England.

In *Certain Physiological Essays* (1661), Boyle recalls how he came
to propose "the desirableness of a good intelligence betwixt the
Corpuscularian Philosophers and the Chymists."[80] There were many
"learned men" who had acquired such an ill opinion of chemical art
from "the illiterateness, the arrogance and the impostures" of chemists
that they were "apt to repine, when they see any person, capable of
succeeding in the study of solid philosophy, addict himself to an art
they judge so much below a philosopher, and so unserviceable to him."
Having been subjected to such concerns himself, Boyle thought it insufficient to demonstrate the utility of chemistry in producing medicaments
and other useful substances for trades. It was "requisite" to show some
"specimens" which would illustrate that "chemical experiments might
be very assistant even to the speculative naturalist in his contemplation
and inquiries." In his effort to execute such examples, however, he had
been hindered by the lack of time, by his ignorance of "the intire system
of either the Atomical, or the Cartesian, or any other whether new or
revived philosophy," and by the division between Epicureans and
Cartesians. Nonetheless, he decided to publish an example, ignoring
the fine points of the Atomical or Cartesian systems. Instead, he would
focus on the common aspects of these systems "in opposition to the
Peripatetic and other vulgar doctrines," explaining the phenomena
"intelligibly" by grounding them on the "little bodies various figured and
moved." Because both atomism and Cartesianism explained phenomena
by "corpuscles, or minute bodies," he could lump them together as
the "corpuscular" philosophy. Insofar as they referred to mechanical
engines, they could also be called the "mechanical hypothesis or
philosophy."

Through his example of saltpeter, Boyle wished to "beget a good
understanding betwixt the chymists and the mechanical philosophers,
who have hitherto been too little acquainted with one another's learning." On the one hand, mechanical philosophers looked upon the
spagyrists as "sooty empirics" or as "a company of mere and irrational
operators" whose experiments might be useful to apothecaries but were

useless to a philosopher. Boyle hoped to "do no unseasonable piece of service to the corpuscular philosophers, by illustrating some of their notions with sensible experiments" and by showing that chemical experiments could be explained without recourse to forms, qualities, elements, and principles. Chemists, on the other hand, regarded corpuscularians as "empty and extravagant speculators, who pretend to explicate the great book of nature, without having so much as looked upon the chiefest and the difficultest part of it; namely the phaenomena." Boyle contended, however, that some of the "hermetic opinions," such as the transmutation of bodies, would be better supported by the corpuscular philosophy than by the existing doctrine of elements and principles. He sought to mediate a "confederacy" between chemists and corpuscularians that would "conduce to the advancement of natural philosophy."[81]

Boyle took a keen interest in the general principles of chemistry. He criticized the authors of chemical textbooks for conveying a "confused" sense of principles unfit for a philosophical discourse. He charged that they were prone to write "systems" comprising "an entire body of physiology," which gave rise to "not a few inconveniences." First of all, not being able to cover the entire realm by their own effort, textbook writers were forced to repeat what others had said. Second, they gave these textbooks "specious and promising titles and comprehensive method of these systems," intended to "persuade unwary readers, that all the parts of natural philosophy have been already sufficiently explicated." Third, their textbooks tended to suppress or bury the individual "excellent notions or experiments" to which the author could truly lay a claim and thus injured his reputation.[82] Particularly problematic was the doctrine of principles and elements commonly stated in the textbooks. Even if one could grant chemists the privilege of guarding their "grand" secrets such as the preparation of the Elixir or the Philosopher's Stone, one could not condone their equivocal way of writing when they conversed on the general principles of natural philosophy:

...I fear that the chief Reason why Chemists have written so obscurely of their Principles, may be, That not having Clear and Distinct Notions of them themselves, they cannot write otherwise then Confusedly of what they but Confusedly Apprehend: ... But though much may be said to Excuse the Chymists when they write Darkly, and Aenigmatically, about the Preparation of their Elixir, and Some few other grand Arcana, the divulging of which they may upon Ground Plausible enough esteem unfit; yet when they pretend to teach the General Principles of Natural philosophers, this Equivocal Way of Writing is not to be endur'd. . . . And if the matter of the Philosophers Stone, and the manner of preparing it, be such Mysteries as they would have the World believe them, they may Write Intelligibly and Clearly of the Principles of mixt Bodies in General, without discovering what they call the Great Work.[83]

In *The Sceptical Chymist* (1661), Boyle played out the role of a philosophical chemist casting doubt on the validity of the spagyric tria prima and of the Peripatetic elements.[84] He picked up where Etienne de Clave had left off, criticizing both the Aristotelian and the spagyric doctrine for their lack of analytic rigor.[85] Boyle's solution differed radically from De Clave's, however. Instead of distillation methods, he espoused solution methods. Although he was even less enamored of the Aristotelian doctrine and knew that Aristotelian elements and spagyric principles had slightly different connotations,[86] Boyle defined "elements" and "principles" as equivalent terms referring to "those primitive and simple Bodies of which the mixt ones are said to be composed, and into which they are ultimately resolved."[87] This definition, often endorsed by French textbook authors, perturbed Boyle deeply because it assumed the permanence of these principles or elements throughout natural or chemical changes. Utilizing the long-standing criticism of fire analysis strengthened by van Helmont, Boyle argued that fire did not simply sort out the pre-existing principles from mixts, but altered them. There was no guarantee, therefore, that the products of fire were the principles that constituted the mixts in the first place. Furthermore, the products of fire did not meet the basic criterion of elements or principles that they should be pure substances. Boyle devoted the bulk of *The Sceptical Chymist* to four general considerations related to the fire analysis:

Since, in the first place, it may justly be doubted whether or no the Fire be, as Chymists suppose it, the genuine and Universal Resolver of mixt Bodies;

Since we may doubt, in the next place, whether or no all the Distinct Substances that may be obtain'd from a mixt body by the Fire were pre-existent there in the forms in which they were separated from it;

Since also, though we should grant the substances separable from the mixt Bodies by the fire to have been their component Ingredients, yet the Number of such substances does not appear the same in all mixt Bodies; some of them being Resolvable into more differing substances than three, and Others not being Resoluble into so many as three.

And Since, Lastly, those very substances that are thus separated are not for the most part pure and Elementary bodies, but new kinds of mixts.[88]

In order to shape chemistry into a more "solid" philosophy, Boyle proposed a corpuscular ontology for chemists' principles and elements. There existed but one "Universal Matter," which was divided into particles of various sizes and shapes. These minute particles could form a variety of masses or clusters, which would resist dissipation into their original particles and thus correspond to what chemists termed "principles."[89] Boyle did not believe, however, that there could exist only three or four such clusters.[90] Multiplication of these intermediate clusters (subject, it was hoped, to empirical confirmation) could serve as a bridge between the few simple bodies, philosophically posed, and the numerous varieties of chemical substances obtained in practice.[91] Opposing the atomistic conception that chemical analysis by fire merely separated out the particles of elements, which retained their properties throughout, Boyle contended that chemical fire might alter the existing clusters of particles to form new clusters. He thus proposed a more general notion of "Mistion":

I consider that it very often happens that the small parts of Bodies cohere together but by immediate contact and Rest; and that however, there are few Bodies whose minute parts stick so close together, to what cause soever their Combination be ascrib'd, but that it is possible to meet with some other Body, whose small Parts may get between them, and so dis-joyn them; or may be fitted to cohere more strongly with some of them, then those some do with the rest; or at least may be combin'd so closely with them, as that neither the Fire, nor the other usual Instruments of Chymical Anatomies will separate them. These things being promis'd, I will not peremptorily deny, but that there may be some Clusters of Particles, wherein the Particles are so minute, and the Coherence so strict, or both, that when Bodies of Differing Denominations, and consisting of such durable Clusters, happen to be mingl'd, though the Compound Body made up of them may be very Differing from either of the Ingredients, yet each of the little Masses of Clusters may so retain its own Nature, as to be again separable, such as it was before.[92]

Boyle made it clear that chemists' principles were not permanent, but transmutable. As Kuhn has pointed out, Boyle's "dynamic atomism" entailed transmutation reactions of "not simply gold from lead, but anything from almost anything." Boyle exploited the polemical utility of transmutation reaction relentlessly to refute the permanence of chemists' principles.[93] Only then could he validate the usefulness of the corpuscu-lan philosophy:

> ... if you do acquiesce in what hath already been done, you will, I presume, think it no mean confirmation of the corpuscularian principles and hypotheses: for if, contrary to the opinion, that is so much in request among the generality of modern physicians and other learned men, that the elements themselves are transmuted into one another, and those simple and primitive bodies, which nature is presumed to have intended to be the stable and permanent ingredient of the bodies she compounds here below, may be artificially destroyed, and (without the intervention of a seminal and plastick power) generated or pro-duced: if, I say, this may be done, and that by such slight means, why may we not think, that the changes and metamorphoses, that happen in other bodies, which are acknowledged by the moderns to be far more liable to alterations, may proceed from the local motion of the minute or insensible parts of matter, and the changes of texture, that may be consequent, thereunto.[94]

Boyle's use of the mechanical philosophy for chemistry has been a con-tentious issue. The mechanical philosophy did not provide an adequate explanation for chemical phenomena. Boyle accepted it gradually and to a limited degree.[95] Contrary to Boas's claim that Boyle's move to Oxford entailed a conversion to the mechanical philosophy, Clericuzio argues that Boyle continued to work out the chemical program he had formu-lated during the earlier period. By tracing carefully Boyle's evolution as a chemist and his views on composition and corpuscular philosophy, Clericuzio shows that Boyle's interest in chemical composition and the particulate theories of chemical action had developed in tandem with his Helmontian search for the alkahest.[96] Newman and Principe have further shown Boyle's deep immersion in alchemical literature and practice. His esteem for the alchemical adepti and their prized secrets grew with his increasing knowledge of chemistry.[97] It is entirely possible that Boyle designed his "corpuscular" philosophy as a hybrid of the alchemical particulate theories and the mechanical philosophy. The two different orders of particles, linked by elaborate yet fragile connections, could meet the demands of chemical practice and systematic philosophy simultaneously.[98]

Given the strength of these new works, one could reasonably argue that Boyle used the mechanical philosophy for the purposes of social legitimation. Chemistry was practiced in Boyle's time by men of lower social status, although there existed a measure of social stratification between physicians, on the one hand, and apothecaries and metallurgists, on the other. As much as Boyle appreciated chemistry as the "key" to the "New Philosophy" or experimental philosophy, he probably perceived the need to dress up chemical labor in more respectable and intelligible terms.[99] If he appreciated the potential utility of chemists' or alchemists' skills, instruments, and labor in shaping a new experimental philosophy, the mechanical philosophy offered him a convenient "vehicle in elevating the social and intellectual status of chemistry and its practitioners, hopefully to that of natural philosophy and natural philosophers."[100] He conducted a highly visible campaign to rid chemistry and alchemy of their mystical past and their obscure languages.[101] Such pruning was necessary to make chemistry a legitimate part of natural philosophy. He carefully distinguished the positive contributions chemists made through their labor from the dark, mysterious Hermetic philosophy, which he deemed below the philosophical standard.[102] If he practiced alchemy in private, or in communication with a close circle of friends, Boyle did not endorse it publicly.[103] Thomas Sprat echoed Boyle's sentiments in the *History of the Royal Society of London*. While declaring that "from their [chemists'] labor, the true philosophy is likely to receive the noblest Improvements," Sprat pointedly excluded alchemists from the ranks of chemists who had contributed positively to the development of "modern" experimental culture.[104] Dissociated from its mystical past and its radical political associations, chemistry could help reform natural philosophy to fashion a new, experimental philosophy. This, in turn, would provide an empirical foundation for the mechanical philosophy and for a rational Anglican theology.[105]

It is not strange in this light that Boyle took issue with the doctrine of chemical principles and elements given in French textbooks. They offered not only the most public discourse of chemistry but also the most clearly stated body of knowledge on the practical operations of chemistry. In a way, it offered a model for the philosophical chemistry Boyle wished to found, combining systematic philosophy with the analytic, experimental knowledge. The fact that Boyle took these French textbooks seriously is itself a measure of their success in the seventeenth-century culture of chemistry. To Boyle's disappointment, however, they merely handed out a set of definitions based on the old, shaky, thin philosophical languages which were not heeded to in the practical part. Boyle called on the text-book writers to fashion a philosophical language that would conform to practical operations. He wished "to take those Artists off their excessive Confidence in their principles and to make them a little more Philosophy with their Art."[106]

Boyle pursued not a chemical philosophy, but a philosophical and theoretical chemistry. He did not wish to extend chemical experience into a worldview, but sought to turn chemistry into a respectable branch of philosophy by rendering linguistic precision to its practice. He needed not simply a logically coherent corpuscular ontology of chemical principles, but one that could match chemical experience. He thus sought to bring the popular doctrine of five principles in line with the corpuscular ontology. He knew that the doctrine generalized the distillation experience of animal and vegetable substances. Even so, however, it did not "follow, that these five Distinct substances were simple and primogeneal bodies, so pre-existent in the Concrete that the fire does but take them asunder."[107] To prove his point that these five principles were not permanent, Boyle planned a four-part appendix to *The Sceptical Chymist*, to consist of "The Producibleness of chymical Principles; the Uncertainty of the vulgar Analyses made by distillations; The various Effects of the Fire according to the differing ways of employing it; and Doubts whether there be any Elements, or material Principles of mixt bodies, one or more, in the sense vulgarly received." Only the first part appeared as an appendix to the second edition (1680).[108]

In *Producibleness*, Boyle expressed the hope that a serious examination of chemical principles and their relationship to the distillation products might "do some service to the inquirers into *the material principles of things*; by obliging the chymists, at least, to reform their doctrine about them, and build it more cautiously, and that upon a larger, as well as more solid foundation of natural history." To this end, he directed sharp criticism to the speculative or "notional" part of chemistry, while expressing warm admiration for chemists' skills and instruments in elucidating the material composition of bodies. As an experimental philosopher/chemist, Boyle made explicit what many chemists took for granted in their practice, in effect translating the esoteric chemical terms for non-chemists. While expressing bewilderment and frustration over chemists' indiscriminate uses of the terms, therefore, Boyle actually provides us with a more lucid characterization of the distillation products. As modern historians, we occupy a symmetric position with the natural philosophers vis-à-vis seventeenth-century chemical discourse. While Boyle no doubt exaggerated the illegibility of chemical discourse with regard to the contemporary chemists, natural philosophers probably experienced a similar frustration in their efforts to assimilate chemistry into their philosophical systems.

Salt was the first principle of the tria prima. Because of its considerable activity, it was deemed to have precedence over the other two principles. Chemists attributed two common properties to salts: "that it is easily dissoluble in water, and that it affects the palate with a sapor, whether good or evil." Acid salts had a sharp and piercing taste, dissolved coral and diverse stones, and calcined lead and several other minerals. They were produced naturally when wine degenerated into sour vinegar. Many woods also produced acid spirits upon distillation. Volatile salts had the smell and the taste of urine. The distillation of soot produced a white volatile salt that, "in smell, taste, and diverse operations, appear to be of great affinity with those of human blood, or urine." Fixed salts were alkalis "which seem to have an antipathy with acid ones, by making a conflict with them, and exercising divers operations contrary to theirs." Boyle valued van Helmont's opinion that fire produced fixed salt by uniting volatile salt to sulphur.[109]

For Boyle, spirits constituted the most confusing category of distillation products. Although they had clear morphological characteristics, they consisted of at least three families: acid salts, volatile alkalis, and vinous or inflammable spirits. Vinous spirits (probably alcohols and ethers), subtle and penetrating, were neither acid nor alkali. The overlapping categories of salts and spirits indicate to us that chemists employed two sets of classification simultaneously, one based on the external characteristics and the other on the projected internal quality of bodies. "As for what the chymists call *spirit*," wrote Boyle, "they apply the name to so many differing things, that this various and ambiguous use of the word seems to me no mean proof, that they have no clear and settled notion of the thing. Most of them are indeed, wont, in the general, to give the name of spirit to any distilled volatile liquor, that is not insipid, as is phlegm, or inflammable, as oil."[110]

The "constituent character" of sulphur was inflammability. There existed at least three kinds of sulphurs, however, "manifestly differing in consistence, texture, or both": oils, inflammable spirits, and "consistent sulphurs" such as common sulphur. The imprecision in terminology was just as evident here as in the case of spirits. What chemists normally called sulphur, or common sulphur, was a mineral body. The two qualities chemists attributed to phlegm or water were "its appearing to them insipid, and its being of a volatile and fugitive nature."[111] Earth seemed "the most simple, elementary, and unchangeable" principle; it did not dissolve in water, did not affect the taste, and did not fly away from the body in combustion.[112]

The "darkness and ambiguity" chemists allowed themselves were most pronounced in their writings about mercury. Chemists seemed to agree, Boyle observed, that chemical mercury or spirit was a volatile liquor, "not inflammable like oil or sulphur, nor yet insipid like phlegm." It had to comprise, then, a wide range of different substances, such as "acid spirits, those of niter and vitriol; urinous, as those of blood, hartshorn, &c. and anonymous ones, as those of guaiacum, honey, raisins, &c."[113] In this sense, "mercury" and "spirit" had identical meanings. Running mercury, or quicksilver, belonged to an entirely different category, however. Many chemists believed that they could obtain running mercury from minerals, particularly from metallic bodies. After examining many experiments in which chemists reportedly obtained quicksilver from other metals, Boyle proceeded to address the main theoretical question: "whether or no the mercury of metals and minerals be principles pre-existent in them, and only extracted from them." While chemists answered this question "resolutely . . . in the affirmative," Boyle was "inclined to think that the mercuries obtained from metals do not clearly appear to have been pre-existent in them, and only separated from them by the artist."[114] Furthermore, in Raymund Lull's *Clavicula* Boyle found a process that turned a metal into mercury. Boyle staunchly opposed the permanence of chemical principles.

Boyle isolated the distinctive properties of chemists' principles to devise a corpuscular explanation. For example, the two agreed-upon qualities of salts were solubility in water and a distinct taste. Boyle proposed a corpuscular explanation accordingly:

> ... there appears no absurdity in conceiving, that by the action of the fire or other fit agents, small portions of matter may be so broken into minute parts, and these fragments may be so shaped and connected, as, when they are duely associated, to compose a body capable of being dissolved in water, and of affecting the organs of taste.[115]

It is difficult to assess Boyle's influence on contemporary chemists because they drew more extensively on the available alchemical, metallurgical, and pharmaceutical literature than Boyle did. Many of them explicitly rejected Boyle's corpuscular program. In contrast, Boyle's efforts at creating a philosophically coherent discourse of chemistry based on positive experimentation opened the door for other natural philosophers to take a serious interest in chemistry. Boyle also became a significant role model for the philosophical chemists of the eighteenth century, a new species of chemists who, in shaping their investigative agenda, relied as much on printed literature as on the available material culture of apothecaries and metallurgists.

The differential reception of Boyle's works among vocational chemists and among natural philosophers can be gauged in the setting of the Académie royale des sciences. While chemists rejected Boyle's corpuscular program, natural philosophers and mathematicians regarded it favorably. The language of corpuscular chemistry also won some favor in the context of public lectures, if Nicolas Lemery's success is any indication. Thus, while the vision of corpuscular chemistry did not alter significantly the material culture of pharmacy or metallurgy, it functioned as the legitimating discourse of chemical practice in the emerging public sphere of the early Enlightenment.

1.4 The Académie royale des sciences

The creation of the Académie royale des sciences in 1666 inaugurated a new era for Parisian chemistry. Chemistry occupied a significant place in the plans for the new Academy. This can be discerned from the proposal for a Compagnie des Sciences et des Arts by Thévenot, Auzout, and Petit, which envisioned a chemical laboratory.[116] Charles Perrault's competing proposal for an Académie Générale also included chemistry in the philosophy section, along with botany, anatomy, and experimental physics. The medal struck for the inaugural meeting of the Academy (December 22, 1666) depicted on one side the effigy of the King and on the other side the goddess Minerva surrounded by symbols representing astronomy, anatomy, and chemistry.[117] Although foreign mathematicians such as Huygens and Cassini enjoyed the most prestige, chemists received generous allocations of salary, space, and laboratory funding to conduct long series of collaborative investigations.[118] Chemical laboratory was also a part of the plans for the new observatory intended as the physical site of Academy's activities.[119] If the completed observatory did not live up to the expectations,[120] the chemists of the Academy had at their disposal a well-equipped laboratory in the king's library, along with the living quarters for the chief chemist.[121] To publicize an experimentalist orientation, an imaginary portrayal of Louis XIV's visit to the Academy depicted chemical vessels along with skeletons and astronomical instruments.[122]

Chemistry owed its recognition to its relevance to pharmacy and medicine, rather than to its intellectual prestige. Chemists had not figured prominently in any of the scientific circles from which the leading members of the Academy were recruited.[123] The first chemists of the Academy, Samuel Cottereau Du Clos (1598–1685) and Claude Bourdelin (1621–1699), came from the royal households. Du Clos was a Huguenot physician to Louis XIV who had participated in preparing the 1658 Geneva edition of Paracelsus's works.[124] Bourdelin had been an apothecary to the household of Gaston, Duc d'Orléans, uncle to Louis XIV.[125] Not much is known about the background or the education of these two chemists. They encountered a new social and intellectual environment in the Academy, whose members with backgrounds in law or in Parisian scientific salons shaped the philosophical and literary fashions.[126] While offering a prestigious social space in which these Paracelsian chemists could legitimately practice chemical medicine, therefore, the Academy also forced them to interact with other natural philosophers and mathematicians who favored Boyle's mechanistic program. There is no indication that Du Clos and Bourdelin succeeded in negotiating this intellectual distance from their learned colleagues. Instead, they attempted to prove their worth through empirical work.

As the senior and more zealous member, Du Clos initially set the agenda for chemistry in the Academy. Well informed about contemporary developments in chemical analysis, he offered a vision of solution chemistry substantially different in philosophical orientation from Boyle's. Just over a week after the inaugural meeting, apparently following on an earlier resolution by the "Compagnie," Du Clos announced a research program seeking the "true principles of Mixts."[127] Du Clos shared the contemporary concern over the validity of fire analysis, most strongly expressed by van Helmont and Boyle. His endorsement of water as the only material principle of all mixts, his interest in alkahest and archeus, and his critique of distillation methods indicate his familiarity with the Helmontian literature.[128] More clearly than Boyle, however, Du Clos sought to implement the analytic innovations of Helmontian chemistry in practice. He knew that the use of solution methods in chemical analysis demanded more stringent criteria for discerning chemical principles.

Du Clos's vision of analytic chemistry carried on Helmontian ideas that had been circulating since the mid 1650s. Du Clos noted that chemical analysis produced five different substances: phlegm, spirit, oil, salt, and earth. Spirit (or mercury), oil (or sulphur), and salt constituted the essential virtues of mixts, while phlegm and earth were purely material substances without virtue. Chemists recommended the first three as the "true essential principles of physical mixts," because they were the principles of the grand elixir in Hermetic philosophy. Du Clos insisted, however, that these three substances were neither primary nor simple and therefore could not be called principles. Those who proposed mercury, salt and sulphur as the "principles or primary constitutive pieces" of natural mixts did so because they employed chemical analysis less exact than an "extreme resolution." In order to resolve natural mixts into their primary components, Du Clos argued, it was necessary to perform a more precise and exact analysis, or an extreme resolution. This involved the use of alkahest, a "dissolutive liquid, capable of penetrating and dissolving radically all mixt bodies." It signified "all spirit" because it supposedly reduced bodies completely into volatile liquids without leaving any earth behind. Since this "mysterious and secret" solvent was not well known, one had to employ a number of other specific solvents, such as alkalis and acid spirits, for the extreme resolution of mixts. These solvents would further resolve sulphurs into water, salt, and earth, mercury into salt and phlegm, and salt into phlegmatic water and earth. Sulphur, salt, and mercury could not be regarded as principles, then, because they were neither simple nor primary. An extreme resolution of natural mixts produced only water, verifying its status as the only primary matter. An impalpable and spiritual "efficient" transformed this primary corporeal matter into mercury, salt, and sulphur. This efficient or alterative spirit had two manifestations, one hot and igneous and the other cold and aerial. They could form only imperfect mixtures, which gave rise to the different qualities of mixts. Perfect mixtures required the participation of the vivifying soul, or archeus.

Du Clos continued to pursue his analytic vision while conducting a broad range of investigations. In 1667, he presented papers on saltpeter, on the augmentation of weight through the addition of fire, and on the analysis of wine, salts, saltpeter, mineral waters, etc., displaying a vast array of knowledge and experience. To judge from the number of records in the procès-verbaux, he was the most active member of the Academy during its first three years.[129] In 1668, while awaiting completion of the Academy's chemical laboratory, Du Clos again discussed the importance of solution analysis. He stipulated that the principal means of chemical analysis were fire, air, and dissolutive liquids. Fire separated the parts of mixts through heat. The "heat of fire" excited movements in mobile matters according to the degree of their mobility. Different parts of bodies receiving different movements could be separated and form a new compound. With flame and combustion, fire could directly separate the parts, but it could not separate fixed or non-combustible matters. Nor could it separate the "constitutive parts" of bodies. One had to use "resolutive menstruum" for this level of analysis. Other menstrua—corrosive or extractive—only attenuated compact matters or divided them into their "integrant parts."[130] Undoubtedly, Du Clos used solution methods in his extensive analysis of mineral waters, which became an enduring research agenda for the Academy.[131] Although his vast analytic programs did not yield immediate results, they created an impetus for systematic, organized research toward a theoretical end.

If Du Clos wished to continue his quest for an effective solution method, studying the alkahest from Paracelsus's and van Helmont's writings, he was diverted from that endeavor to evaluate Boyle's program. Continuing the pattern of active exchange with English scientists that had existed before the establishment of the Academy, the academicians paid close attention to the contemporary debates in science, and particularly to the activities of the Royal Society of London.[132] Boyle and his works figured prominently in this exchange. Huygens visited England in 1661 and 1663 when he was elected a Fellow of the Royal Society. He kept up an active correspondence with Moray, the president of the Royal Society, who sent him many of Boyle's books, including *The Sceptical Chymist* and *Certain Physiological Essays*.[133] The astronomer Auzout also became a Fellow of the Royal Society in May, 1666, and he maintained an active correspondence with Oldenburg.[134] Upon Auzout's request, Du Clos undertook to examine Boyle's *Certain Physiological Essays*, presenting page-by-page reports to the Academy between September 1668 and February 1669. In his reports, Du Clos criticized Boyle's corpuscular philosophy as applied to chemistry and his lack of requisite chemical knowledge and skill. His reports opened up a prolonged debate within the Academy, with Perrault and Mariotte leaning toward a corpuscular understanding of chemical composition.[135] Although Du Clos concluded by 1669 that it was not necessary or even possible to reduce chemistry to the figures and the movements of particles as Boyle had wished, he could not silence the other academicians who were more positively inclined toward Boyle's mechanistic program for chemistry, or at least toward a chemical discourse more in tune with natural philosophy. Permanent secretary Du Hamel, summarizing Du Clos's conclusion, characterized the spirit of chemistry as confused and hidden, in contrast to the spirit of physics, which was distinct, simple, and clear.[136]

Du Clos's vision of solution chemistry as the ultimate method of analysis also suffered a significant blow with the take-off of the Academy's larger project on the natural history of plants. His proposal on the principles of mixts had been followed by Claude Perrault's proposal on the natural history of plants, which included chemical analysis as a way of studying the causes of the medical "virtues" of vegetables.[137] Nearly a year and half later, Du Clos further elaborated the role of chemical analysis in the natural history of plants.[138] With the approval of this proposal by the "Compagnie," the ambitious project of analyzing all known plants for their medical "virtues" began. Bourdelin faithfully carried out distillation of numerous plants on and off until 1699, when he died. Du Clos's voice began to wane noticeably in the mid 1670s, however. After the appointment of Perrault's protégé Denis Dodart in 1671, Du Clos lost control over the natural history project. Young, energetic, and meticulous, Dodart succeeded in publishing *Mémoire pour servir à l'histoire des plantes* (1676), in which he defended the method of distillation analysis despite increasing criticism and frustration over its validity in discerning the composition and the medical virtues of plants.[139] Du Clos's vision, which leaned toward solution methods as the superior means of determining more intimate composition of mixts, was ignored in Bourdelin's protracted analysis of plants. The persecution of Protestants in the 1680s made it increasingly difficult for Du Clos to maintain his grace within the royal institution of the Academy.[140] Although French chemists adopted many techniques of solution analysis, especially the color indicators, Du Clos's vision of solution chemistry found breathing room within the Academy only after Bourdelin's death in 1699. Simon Boulduc (1652–1729), who joined the Academy in 1694, began to criticize the distillation method openly and to advocate the method of extraction.[141]

1.5   The Analytic Ideal

To judge from Boyle's and Du Clos's works, solution methods gained a level of currency during the second half of the seventeenth century, undermining the monopoly of distillation methods in determining the composition of bodies. The changing landscape of chemical analysis and theory resulting from the use of solution methods can also be discerned in Nicolas Lemery's popular textbook *Cours de chimie* (1675).[142] Born and apprenticed in Rouen, Lemery spent some time in Paris as a boarding student of Christophe Glaser before embarking on a tour of

France.[143] He stayed nearly three years at Montpellier, taking courses in pharmacy and anatomy at the Faculté and teaching in the private laboratory of Henri Verchant. In 1672 he returned to Paris, where he acquired the patronage of the Prince of Condé and settled down as a master apothecary. Lemery's private courses attracted a large and fashionable Parisian audience, including women, which accounts in part for the use of corpuscular language and imagery in his lectures.[144] Corpuscular philosophy lent intelligibility and entertainment value to chemical lectures, legitimating the art of "sooty empirics" in the public sphere. The fresh outlook Lemery introduced to the Parisian didactic tradition can be attributed in part to his education in the provinces, particularly at Montpellier, among the Protestant apothecaries.[145] Although he retained much of the format and the content of Glaser's textbook,[146] Lemery clarified many of the chemical terms, introducing corpuscular imagery to explain the acid-alkali reactions. Bernard le Bovier de Fontenelle, a noted Cartesian popularizer of science, eulogized Lemery as "the first who dissipated the natural or attributed darkness of chemistry, who reduced it to the clearest and simplest ideas, who abolished the useless barbarism of its language, who promised instead only what he could and what he knew to be capable of executing." Following Fontenelle, Hélène Metzger has credited Lemery with reforming the French didactic tradition with the mechanical philosophy.[147] Although the use of corpuscular language did not significantly alter the content of the chemistry he taught, it probably helped Lemery to address his fashionable Parisian audience. He displaced chemistry from the apothecaries' shop to the emerging public sphere of demonstration lectures.[148]

Lemery's textbook indicates a significant transformation in French chemistry resulting from the use of solution methods both in vegetable and in mineral chemistry. Though it is difficult to document these changes step by step, it is possible to gauge the extent of such a change by comparing Lefebvre's and Lemery's textbooks.

Lefebvre used "extract" loosely, often to refer to the products of distillation. In the sections on the "manner of extracting the Principles of Honey" and on "how to extract the Oyl of Wax," for example, he described an ordinary distillation procedure in which he placed the substance in a closed vessel and subjected it to varying degrees of fire.[149] In preparing the "extract or Bloud" of Great Comfrey, however, its roots were beaten, sprinkled with white wine, mixed with bread crumbs, and imbued again with wine before the mixture was placed in a closed vessel for digestion in a vaporous bath. The liquid expressed from this mixture was then subjected to another period of digestion for "a more exact purification." Afterward, Lefebvre evaporated about two-thirds of the liquid to obtain the "true balsamick extract" of the roots, or the "Bloud of the Great Comfrey."[150] In the section on "How to extract out of Juniper-berries what is more profitable and useful in them for Chymical Pharmacy," the wood and the berries went through the distillation process for an "extract" with "some extraordinary vertue."[151] In Lefebvre's few indiscriminate uses of "extract," it refers mostly to the product of distillation with or without prior digestion. Such usage conformed to Béguin's conception of distillation as a method of extraction.

Lemery's *Cours de chymie* illustrates changes that had been brought about in part by the use of solution methods. In discussing vegetable chemistry, he clearly distinguished the "extracts" as substances obtained with the use of solvents, as distinct from those obtained by means of distillation. "*Extracts* of all *Vegetables* are made after the same manner," Lemery stated, by digesting the substance in various kinds of menstruum (often with mild heat), then expressing the liquid and evaporating or distilling the humidity. Water and the spirit of wine were the most common solvents. The demand for a gentler method of extraction derived from the need to prepare more efficacious medicaments. The same plant that produced an excellent remedy through the method of extraction often became ineffective when reduced to its "principles" through the distillation method. Extracts of vegetables often exhibited medical virtues that were destroyed in the distillation process. Therefore, Lemery suggested, the medical virtue of vegetable remedies was not contained in any particular principle of the vegetable, but in "a different mixture of Principles as is requisite to produce certain Fermentations in our Bodies":

There was a great contest among Chymists, heretofore, in which of the Principles it is that the Purgative virtue of many Remedies doth consist. Some maintained it to be in the Salt, others in the Sulphur, and others again in the Mercury. But when every part had very diligently separated each their Principle, and came to try it, they found after all that none of them was Purgative; which hath persuaded many of them to think that this Purgative was of so subtle and penetrating a nature, that glass it self was not able to keep it from being lost in the tryal.

For my part I cannot grant any such imperceptible Purgative, and I rather am apt to believe that the Purgative virtue of a Mixt consists in nothing else but such a different mixture of Principles as is requisite to produce certain Fermentations in our Bodies. So that when once we separate the Sulphur, Mercury, or Salt, the position of parts, or proportion of Principles being changed, there remains no longer any Purgative effect, because the Principles being separated can no more produce that Fermentation which they did while they were united, and united together some kind of way that Art is ignorant how to imitate.[152]

Lemery also brought mineral chemistry back to the center of the French didactic tradition. He offered extensive descriptions of the procedures of purification for each metal and its dissolution in acids. The amount of space allocated to mineral chemistry (particularly that of metals) in his textbook is quite remarkable. The bulk of Lemery's book is devoted to the discussion of minerals (240 pages). There are relatively short sections on vegetables (80 pages) and on animals (20 pages), and there is a brief discussion of generalities (18 pages). In marked contrast, most of the practical section of Lefebvre's textbook is devoted to vegetable chemistry. Aside from the theoretical section (part I; 71 pages) and a general discussion of chemical operations (part II, book I; 30 pages), Lefebvre's book deals mostly with various preparations of vegetables (part II, book II; 212 pages). Most of these procedures used distillation methods to procure medically salient components.

Most important, Lemery identified metallic dissolutions and displacements as "one of the most difficult [questions] to resolve well, of any in Natural Philosophy,"[153] musing on their quantitative proportions and their speed of action. It was well known to chemists that metals dissolved selectively in acids and that alkali salts could precipitate metals from such solutions. For example, oil of tartar or spirit of sal ammoniac could precipitate gold from its solution in regal water. Lemery, recognizing an important theoretical question in such phenomena, tried to give a corpuscularian answer:

I do suppose that when the *Aqua Regalis* hath acted upon Gold, so as to dissolve it, the points or edges that enabled it to do so, are fixed in the particles of Gold. But seeing that these little bodies are very hard, and consequently hard to penetrate, these points do enter but very superficially, yet far enough to suspend the particles of Gold and hinder them from precipitation. Wherefore if you would add never so much Gold more, when these points have seized upon as much as they are able to joyn with, they cannot possibly dissolve one grain more; and it is this suspension that renders the particles of Gold imperceptible. But now if you add some body that by its motion and figure is able to engage the Acids enough to break them, the particles of Gold being left at liberty will precipitate by their own weights. And this is what I conceive the Oyl of Tartar, and Volatile Alkali Spirits are able to do. They are impregnated with very Active Salts, which finding bodies at rest presently fail to move them, and by the quickness of their motion do shake them so violently, as to break the points by which they were suspended; and this occasions the Ebullition which presently happens when these Spirits are poured upon the Dissolution.[154]

Whereas the body of Lemery's *Cours* reflects the influx of solution methods into pharmaceutical practice, the introductory part of the book maintains the conventional form, revealing only a trace of these changes. True to the long-standing French didactic tradition, Lemery devoted only a short section to general discussion (9 pages out of 323), leaving the bulk of the book to practical operations. He defined chemistry as "an Art that teaches how to separate the different substances which are found in Mixt Bodies." By "Mixt Body" he meant "those things that naturally grow and encrease, such as Minerals, Vegetables, and Animals." The "First Principle" Lemery admitted for the composition of Mixts, as did Lefebvre, was a "Universal Spirit, which being diffused over all the World, produces different things according to the different Matrixes, or Pores of the Earth in which it settles." Since this principle was admittedly "a little Metaphysical, and [fell] not under our senses," he saw it fit to establish some "sensible" principles that could be isolated in chemical analysis:

Whereas the Chymists in making the Analysis of Mixt Bodys have met with five sorts of Substances, they therefore concluded that there were five Principles of Natural things, Water, Spirit, Oil, Salt, and Earth. Of these Five, Three of them are Active, the Spirit, Oil, and Salt, and two passive, Water and Earth.[155]

Lemery made the procedural roots of the five principles clear. From his discussion of each principle one can easily infer the concrete types of substances obtained in the distillation of vegetables and animals. "The Spirit which is called Mercury" was a "subtile, piercing, light substance" that appeared when chemists made "the Anatomy of a Mixt Body." Like other principles, it could not be drawn pure. When it was "involv'd in a little Oil," it appeared in the form of "volatile spirit," such as spirit of wine. When it was "detained by some salts, which check its Volatility," it appeared as "fixed spirit," like the acid spirits of vitriol, alum, and salt. Because of its greater capacity for motion, spirit tended to corrupt the bodies containing it. Water, "which is called Phlegm," emerged in the distillation process before fixed spirits or after volatile spirits. After the spirits came "the Oil which is called Sulphur by reason of its inflammability." This "sweet, subtile, unctuous substance" was always drawn impure, mixed with spirits or salts. Sulphur caused the diversity of colors and smells. Salt remained "disguised in the Earth" after the other principles were extracted. One extracted it by "pouring Water upon the Earth to imbibe its Salt; then filtring the dissolution, and evaporating all the moisture." It was a "fixt, incombustible substance, that gives Bodies their consistence, and preserves them from corruption." What remained after all this treatment was, of course, earth.[156]

The morphology of distillation products did not provide an exclusive criterion for classification, however. Lemery used another, overlapping system of classification that was based on the representative qualities of substances—mercury, sulphur, water, earth and salt—inherited from the earlier Aristotelian and Paracelsian traditions. Since chemists often obtained a variety of mixtures through distillation, the correlation between the two sets of naming schemes was not clear cut; it was often contingent on the judgment of the individual chemist. The tension caused by these two simultaneously operating schemes applied to an expanding repertoire of analytic procedures is palpable in Lemery's text. Having described "salt" as a fixed substance embedded in earth, Lemery immediately lists three different kinds of salts: fixed, volatile, and essential. In this case, "salt" referred to a group of substances that shared a set of common characteristics, regardless of the form in which each emerged in the distillation process:

There are three differences of Salt, as Fixt, Volatile, and Essential. The Fixt Salt is that which remains, after the Volatile Principles are separated: The Volatil is that which easily riseth, as the Salt of Animals: And Essential Salt is that which is obtained from the Juyce of Plants by Crystallization. This last is between the Fixt and Volatil.[157]

Furthermore, there were three kinds of spirits: "the Spirit of Animals, the Burning spirit of Vegetables, and the Acid spirit." The first (e.g., spirits of hart's horn) was a volatile salt; the second (e.g., spirit of Wine, spirit of Juniper) was an oil; the third (spirits of Vinegar, Tartar, and

Vitriol) was an Acid Essential salt, dissolved and put in fusion by fire. These different types of salts, spirits, and oils created considerable confusion over the classification of distillation products.

Lemery knew the limits of the distillation method in extracting the said five principles, particularly in the case of minerals. Some minerals produced barely two, and others (such as gold and silver) produced none, frustrating chemists' attempts to find the salts, sulphurs, and mercuries of these metals. Whereas Boyle had charged on the basis this observation that chemists' doctrine of principles was flawed, Lemery used a corpuscular reasoning to account for such difficulty:

The five principles are easily found in Animals and Vegetables, but not so easily in Minerals. Nay there are some Minerals, out of which you cannot possibly draw so much as two, nor make any separation at all (as Gold and Silver) whatsoever they talk, who search with so much pains for the Salts, Sulphurs, and Mercuries of these Metals. I can believe, that all the Principles do indeed enter into the composition of these Bodies, but it does not follow that they must remain in their former condition, or can be drawn as they were before; for it may be these substances which are called Principles are so strictly involved one within another, as to suffer no separation any other way than by breaking their Figure. Now it is by reason of their Figure that they are called Salts, Sulphurs, and Spirits: For example, if you mix an Acid Spirit with the Salt of Tartar, or some other Alkali, the edges of the Acid will so insinuate into the Pores of the Salt, that if by distillation you would separate the Acid spirit again from the Salt, you'l never be able to effect it, because the edges of these Spirits are so far destroyed or changed, that they no longer preserve their former Figure.[158]

Lemery's defense of the five principles and his corpuscular account of the acid-alkali reactions (quoted above) raise the possibility that he was aware of Boyle's critique of chemists' principles and of the acid-alkali theory. As a leading Protestant apothecary employed in the royal household, Lemery should have been acquainted with other Protestant chemists in Paris.[159] He returned to Paris in 1672, soon after Du Clos's critique of Boyle caused a stir within the Academy. He also knew Wilhelm Homberg, who had worked briefly in Boyle's laboratory.[160] It seems quite likely that Lemery incorporated Boyle's critique of chemical principles into the French didactic tradition, creating a unique blend of chemical philosophy. His limited yet flexible use of the corpuscular imagery illustrates well the fluidity of French didactic discourse in accommodating the current philosophical fashion.

Lemery also articulated a theory of "universal salt" that functioned as a viable heuristic for years to come. He believed that there was "one chief [salt], of which all the rest are compounded, and do conceive it to be made of an Acid Liquor sliding through the veins of the Earth, which doth insensibly insinuate and incorporate in the Pores of stones." After a long fermentation and concoction, it was formed into a fossil salt, called "Gemma [rock salt]." The "long fermentation and concoction which is made in the stone, serves to digest, and perfectly unite the Acid with the stony parts, for the making of Salt." As the acid liquor traveled through the earth, it produced with other earths different kinds of salts such as vitriols and alums. Such a conjecture was "the more likely to be true, because from the mixture of Acids with some Alkali matter we always draw a substance very like unto Salt." In other words, Lemery utilized the common chemical experience of acid-alkali reactions in producing crystallized salts to account for the original formation and the subsequent distribution of all mineral salts in nature. Vegetables produced two kinds of salt, one volatile and essential and the other fixed and alkali. The volatile salt originated from the "universal salt, which being spread over all the earth, is attracted by the spirits of the Plant, and volatilised, and so made that which is called Essential Salt." Fixed salt was incorporated in a great deal of earth and had to be drawn by calcinating the plant. It acted with acids with a great effervescence or ebullition "until all the pores have been sufficiently opened by the Acid points to make way for the igneous particles to fly out, that enter'd into it during the Calcination." Animals also yielded two kinds of salts.

In the subsequent editions of his book, Lemery's discussion of salts expanded considerably. Although he maintained the theory of universal salt, he sought to clarify the effects of fire in the production of vegetable salts. He supposed three kinds of salts drawn from vegetables—"an Acid salt called Essential, a *Volatile*, and a *Fixt Salt*." The acid or essential salt was prepared by expressing the juice from a plant and setting it in a cool place to form crystals. Lemery regarded the acid salt as the "*true Salt* that was in the *Plant*, because the means that are used in drawing it are *Natural*, and such as cannot change its nature." The same could not be said of the other two salts, because "the violent fire that are used about them make impressions of another nature, and their effects are very different, so that the fire seems to alter and disguise them." The second, volatile salt of plant was drawn from fermented seeds or fruits. Fermentation set the particles of plants at work and disposed them to an easier separation, causing the salt of the plants to unite with oil, which rendered it volatile. This union with oil made distillation necessary in separating the volatile salt from fermented fruits and seeds. Lemery took it for granted that "fire has chang'd, or else added something to this salt," since it effervesced with acid (thus accounting for the name "volatile alkali"). No alkali existed in the plant before the fire analysis[161]:

> The *Chymists* will needs have this *Volatile Alkali* to be in the *Plant*, just the same as when it is drawn; that is to say, they make this a different *species* of *salt*, lying hid under the *acid*, until it is laid open by the force of fire. But this opinion is founded on no credible experience, for *Anatomize* the *Plant* how you think fit, without using fire, and you shall never find any other but an *acid salt*. Doubtless it will be said, that all other ways of dissecting *Plants* even into their *salts*, prove too weak without the assistance of this grand dissolvent fire. But if we consider impartially how fire does act, we shall be forced to acknowledge that it rather destroys, and confounds the greatest part of the bodies it opens, and does not leave them in the *natural* state they were in before, and especially when it is driven with that force which is necessary to draw this *salt*.[162]

Rather than multiply the species of salts, Lemery granted only one species of salt in plants. Volatile alkali salt was the acid essential salt, dragging some oil and calcined matter. These extra substances changed the nature of the acid salt by "breaking the *Saline* points, and rendring them *Porous*, so that any *acid* liquor being cast upon it, enters into the Pores, and violently divides the parts, whence follows the *Effervescency*."[163] Similar reasoning applied to the fixed salt of plants, which retained the earthly part after distillation. While some chemists argued that there was an alkali hidden in all bodies that fermented with acids, Lemery thought that these terrestrious bodies themselves were alkalis. There existed in plants "only one species of salt," which gave rise to other salts through the action of fire.[164]

In addition to defining acids and alkalis according to origin and process of preparation, Lemery considered their nature in corpuscular terms in order to remove the obscurity surrounding the subject. He affirmed that the "*acidity* of any liquor does consist in keen particles of *salts*, put in motion" because "the nature of a thing so obscure as that of salt, can't better be explicated, than by admitting to its parts such figures as are answerable to the effects it produces." Aside from the fact that an acid pricked the tongue, it also crystallized into edges. In addition, "all Dissolutions of different things, caused by acid liquors, do assume this figure in their *Crystallization*." From the fact that alkalis produced an ebullition with acids, it was reasonable to conjecture that "an *Alkali* is a terrestrious and solid matter, whose *pores* are figured after such a manner that the *acid* points entring into them do strike and divide whatsoever opposes their motion." The strength of the ebullition between an alkali and an acid depended on the solidity of the parts of the alkali that put up more or less resistance to the acid parts. There were "as many different Alkali's, as there are bodies that have different pores," yielding fermentations with differing strength. There had to be a "due proportion between the acid points, and the pores of the Alkali."[165]

Lemery's corpuscular definition, albeit speculative, provided an ontology of acids and alkalis independent of their manner of preparation. As such, it could cross the boundaries of different analytic methods and serve as a universal definition. While the proliferation of analytic procedures employing acids and alkalis increasingly removed them from the procedural roots of distillation, chemists could deploy the corpuscular ontology to transplant them to a new analytic environment. The protean points of acids and pores of alkalis made such transposition possible. The corpuscular language also helped Lemery to unpack esoteric chemical practice for the consumption of fashionable Parisians who were drawn to his popular lectures.

Although Lemery's corpuscular language differed substantially from Boyle's, the two chemists seem to have interacted at some level. The increasing persecution of Protestants in the 1680s caused a great deal of disturbance in Lemery's life. Contemplating exile to England, he wrote to Boyle in 1682, offering a copy of the fourth edition of his *Cours de chimie*. Thus, Lemery knew of Boyle's reputation and work by 1682 at the latest.[166] He left for England in 1683, but stayed only for a short period. It is not known whether he met Boyle while there.[167] Afterward, he took a doctorate at Caen and converted to Catholicism after Louis

XIV's revocation of the Edict of Nantes in 1685. Although the king restored Lemery's privileges as a master apothecary, he had to fight the Paris guild of apothecaries to regain his right to establish a shop. Despite these difficulties, his *Cours de chimie* enjoyed a long life, going through more than thirty editions. Subsequent editions show an evolving program of French analytic chemistry. In the fifth edition, published in 1683, Lemery differentiated "Chymical principles" as the limit of current analysis from the "Natural principles" stipulated in various philosophical systems. He put his trust in the "chymical principles" as a more practicable guide to the "true" principles of nature:

> The word *Principle* in *Chymistry* must not be understood in too nice a sense: for the substances which are so called, are only *Principles* in respect of us, and as we can advance no farther in the division of bodies; but we well know that they may be still divided into abundance of other parts which may more justly claim, in propriety of speech, the name of *Principles*: wherefore such substances are to be understood by *Chymical Principles*, as are separated and divided, so far as we are capable of doing it by our weak imperfect powers. And because *Chymistry* is an Art that demonstrates what it does, it receives for fundamental only such things as are palpable and demonstrable. It is in truth a great advantage to us, that we have *Principles* so sensible as they are, and whereof we can have so reasonable an assurance. The fond conceits of other *Philosophers*, concerning *Natural principles*, do only puff up the mind with *grand Idea's*, but they prove or demonstrate nothing. And this is the reason that going to discover their *Principles*, we find some of them do frame one System, and others another. But if we would come as near as may be to the *true Principles of nature*, we cannot take a more certain course than that of *Chemistry*, which will serve us as a Ladder to them; and this division of substances, though it may seem a little gross, will give us a very great *Idea of Nature*, and the figure of the first small particles which have entered into the composition of mixt bodies.[168]

Lemery's dissociation of "chemical principle" from "natural principle" signaled a breakdown of the long association between the Aristotelian and Paracelsian philosophies, on the one hand, and the distillation analysis of plants, on the other, in understanding chemical composition. As Bourdelin's analysis of plants at the Academy failed to deliver on the promise, distillation analysis as the only legitimate method of determining chemical composition met with increasing resistance. An explicit denunciation of the distillation method had to wait until after Bourdelin's death in 1699. By that time, Lemery had joined the Academy and, with his long-time acquaintance Wilhelm Homberg, had begun to articulate a full-blown program of corpuscular chemistry that appropriated the prestige of the new Cartesian paradigm. The intellectual fashion of Cartesianism in the French Academy, led by Fontenelle, probably strengthened this pragmatic liaison between chemistry and the corpuscular philosophy. Homberg, in pursuing the seemingly unattainable Boylean agenda of bridging the gap between the analytic ideal of chemical principles and the actual products of chemical analysis, stabilized the theory domain of chemical composition in the new philosophical language.

2

A Theoretical Moment

The doctrine of five principles as the analytic/philosophical ideal of chemical composition served two functions in the seventeenth century. On the one hand, its philosophical and theoretical orientation served to differentiate chemists from the "vulgar" apothecaries. On the other hand, its analytic component set chemists apart from the bookish physicians of the Faculté. In other words, the doctrine helped French Paracelsians to create a mediating social and cognitive space between those of apothecaries and physicians and to forge a positive identity for chemistry and chemists. From about the middle of the century, however, it had to face serious challenges. First, the philosophical paradigm began to shift from the Aristotelian to the Cartesian. The mechanistic understanding of matter did not easily accommodate the idea of permanently stable principles or elements. Moreover, chemists' claim to the internal composition of bodies was severely contested by natural philosophers who wished to tame this rich material culture to build a new, empirically grounded philosophy of nature. These contestations entailed close scrutiny of the analytic practice of chemistry. Robert Boyle, as an experimental philosopher/chemist, criticized chemists' principles harshly, urging them to look for material principles (preferably corpuscular). He envisioned a philosophical chemistry based on positive experimentation and expressed in a language conducive to the mechanical philosophy. Second, the legitimacy of distillation methods was undermined by the increasing use of solution methods in pharmaceutical practice. While Boyle took advantage of the changing philosophical and analytic paradigms to subsume chemical doctrines to the mechanical philosophy, French iatrochemists responded to these changes in a different manner.

Du Clos rejected completely the mechanistic approach to chemistry, while endorsing solution analysis as the legitimate method for elucidating chemical composition. His vision of solution chemistry did not find a congenial environment in the Académie, however, in part because of Bourdelin's and Dodart's insistence on distillation methods. Nicolas Lemery simply dissociated the analytic ideal of chemical principles from the philosophical ideal of elements, endorsing the former. At the same time, he introduced corpuscular explanations of chemical actions, adapting chemical theory and practice to the current philosophical fashion. These efforts did not find a clear point of convergence, however, leaving the doctrine of five principles in serious jeopardy.

The reorganization of the Academy in 1699, after the noted Cartesian popularizer Fontenelle was appointed permanent secretary, created a unique environment in which chemical composition received serious attention from the elite chemists Wilhelm Homberg, Nicolas Lemery, Etienne-François Geoffroy, and Louis Lemery.[1] They mixed creatively the divergent visions of French iatrochemistry and Boyle's philosophical chemistry to re-stabilize the theory domain of composition and to invent a new domain of affinity. Nicolas Lemery, the most experienced of these chemists, provided a foundation on which others would build more elaborate edifices. Homberg, who had wide-ranging interests in experimental philosophy and in the chemistry of metals, assumed a leading role in articulating a sophisticated version of corpuscular chemistry that resonated with Fontenelle's vision of unified science. As an experimental philosopher looking into chemistry, he approached chemical practice with a much more critical and systematic perspective than other Parisian chemists trained in pharmacy or medicine, trying to streamline chemical discourse in tune with the mechanical philosophy. He wished to "perfect" chemistry, or to make it more philosophical, by clarifying its connections with general physics.[2]

Homberg explored the issue of chemical composition systematically in a series of *Essais de chimie* presented to the Academy between 1702 and 1709. His approach combined various traditions then known and defies simple characterization. On the one hand, he sought to rehabilitate the doctrine of five principles, by separating them out in actual analysis with the burning glass, a new and more powerful furnace. He thus defined "chemical principles" as the "simplest matters to which a mixt is reduced by chemical analysis," as Lemery had done, and tried to implement it in the classification of more complex substances. On the other hand, he doubted the permanence of these principles and endeavored to reduce them to ultimate particles. Homberg's dual outlook implied a hierarchical composition of chemical bodies, although he could not specify the relationship between the two levels of composition. On the whole, he approached chemical problems with the inclinations of an experimental philosopher who wished to introduce instrumental precision to chemical practice and systematic reasoning to chemical discourse.

The burning glass played a crucial role in Homberg's efforts to resolve the crisis in chemical theory. Homberg was less innovative in chemical operations than in physical methods. While he doubted the validity of the distillation method, he did not modify the procedures,[3] but relied on the areometer to measure the quantity of principles. Nor did he dispute the legitimacy of the doctrine of five principles. Instead, he relied on the burning glass to confirm their existence in a manner compatible with the corpuscular philosophy. In the face of mounting opposition to distillation analysis, he thus reasserted the principalist outlook and the supremacy of fire analysis. The burning glass helped him to conceptualize pure fire as a tangible, corpuscular substance that made up other bodies and participated in chemical actions. The new instrument did not simply reinforce the existing theoretical structure, however. It also allowed Homberg to tap into the long-standing conception of fire as the universal solvent that resolved bodies to their ultimate constituents. This projected capacity of the burning glass in effecting the ultimate dissolution of bodies promised to surpass the analytic efficacy of solution methods and to provide a common instrumental foundation for the distillation and solution methods. Although Homberg did not succeed in this endeavor, Louis Lemery further articulated this ontology of fire as the ultimate solvent, establishing it as an enduring legacy in French chemistry.

Homberg consolidated a distinctly theoretical moment in French elite chemistry. The existing theoretical structure of chemistry—the doctrine of five principles—faced a serious threat from changes in analytic methods and philosophical languages. The chemists of the Paris

Academy had to reassess the efficacy of distillation methods in the laboratory and the legitimacy of the Aristotelian/Paracelsian elements and principles as a public language. Through careful reevaluation and realignment of the available instruments, methods, theories, and philosophies, the academicians re-stabilized composition as a main theory domain of chemistry at the cross-section of various chemical philosophies and practices. They wielded corpuscular imagery and language freely while fleshing out the problem of chemical composition in their investigation. The notion of stable, unchanging principles reflected an essential practical strategy on which chemists relied in their daily practice. However unphilosophical it may have seemed to Boyle, chemists could not meaningfully conduct their analyses without subscribing to the persistence of chemical units. Without this presupposition, chemists' experience would be too fleeting and too transient to constitute a domain of reality. The analytic ideal of chemical principles and elements was a projection based on chemists' experience with the substances stabilized in practice. Boyle accepted this practical mandate when he acknowledged the relative stability of a cluster of particles that could pass for chemical principles.[4]

Homberg's *Essais* open for us a complex chemical reality that came into existence in the early eighteenth century at the boundary between chemistry and experimental philosophy. Homberg should be recognized as a pioneer of the French physico-chemical tradition. His instrumentalist approach to the problem of chemical composition set an important precedent for Lavoisier. His decade-long struggle to tame the different dynamics of chemical philosophy, theory, and practice and to build a viable chemical system offers us a privileged window on the unstable, complex chemical reality a natural philosopher/chemist had to negotiate. The *Essais* encoded a mature corpuscular program centered on the problem of composition. As such, they constitute an important missing link in the historiography between Robert Boyle and the Chemical Revolution.[5]

2.1 The Reorganization of the Académie

The fortunes of the Paris Academy fluctuated somewhat with the changing ministerial protectors and with Louis XIV's wars. After Colbert's death in 1683, Louvois (1639–1691) reduced financial support for the Academy which, combined with the departure of its Protestant members such as Huygens, demoralized its members. The chemistry section did not fare well under Louvois's direction. Increasing persecution of the Protestants made it difficult for Du Clos to maintain his enthusiasm, while failing health prevented Bourdelin from working in the Academy's laboratory. Pontchartrain (1643–1727), who took over a smaller, much discouraged company in 1691, tried to revive it despite the financial difficulties caused by Louis's wars. He formalized the position of president and appointed his nephew, Abbé Jean Paul Bignon (1662–1743), who became a strong advocate of the Academy. Bignon immediately enlisted Tournefort and Homberg, who resumed the languishing project of the natural history of plants. [6]

Wilhelm Homberg (1652–1715), a Dutch Protestant by birth and a law student at Leipzig and Jena, had traveled extensively before his appointment to the Academy. His itinerary and acquaintances included Magdeburg (Otto von Guericke), Padua (studying anatomy), Bologna (chemistry), Rome (mathematics, mechanics, and astronomy), France (Nicolas Lemery), and England (Boyle). After taking the degree of doctor of medicine at Wittenberg, he migrated through the mines of Saxony, Bohemia, and Hungary and was employed briefly in the laboratory of chemistry established by the king of Sweden. He floated into Paris in 1680, hoping to gain Colbert's patronage. Perhaps to this end, he converted to Catholicism in 1682 and became a French subject in 1683, but he left for Rome after Colbert's death. His return in 1691 presumably set off a productive period in French chemistry. In a eulogy, Fontenelle praised Homberg for "the wealth of knowledge and the excellent instruments" he possessed (phosphorus,[7] pneumatic machine, microscope, and so on), placing him first even among the academicians.[8] Relative to the previous generation of chemists in the Academy, Homberg practiced chemistry in a manner more accessible to his colleagues in other fields, utilizing the gadgets of experimental philosophy as well as the corpuscular language.

Under Bignon's care, the Academy was set on a course of expansion. Fontenelle (1657–1757), a member of the Académie française since 1691, became permanent secretary of the Académie des Sciences in 1697. He had won literary fame with two popular expositions of Cartesian cosmology: *Entretiens sur la pluralité des mondes* (1686) and *Digression sur les anciens et les modernes* (1688).[9] During the years separating these two publications, he attended the gatherings of the "Norman group" at Varignon's residence, where he met Du Hamel, Du Verney, de la Hire, Tournefort, and other members of the Academy. He apparently maintained a serious interest in science, studying botany with Tournefort, medicine with Homberg, chemistry with Lemery, and anatomy with Du Verney.[10] Fontenelle's vision for a rational, mathematized science can be discerned in his introduction to the *Histoire de l'Académie*:

The sterile sort of natural science, which had not progressed for several centuries, has been abandoned; the reign of words and terms is over; we want things; we establish principles that we can understand and pursue; and it is thus that we progress. Authority has ceased to have more weight than reason. What had been accepted without contradiction because it was accepted traditionally, is now examined and often rejected. As we consult nature, rather than the ancients, about natural phenomena, nature makes herself more easily accessible, and coerced often enough by the new experiments we are undertaking to investigate her, she grants us knowledge of some of her secrets. Mathematics, on the other hand, has not made less considerable progress. Those sciences which are linked with physics have advanced with it, and pure mathematics is today more fertile, more universal, more sublime, and, so to speak, more intellectual than it has ever been. To the extent that the sciences have broadened their scope, their methods have become simpler and easier. Furthermore, for some time mathematics has not only yielded an infinity of its own kind of truths, but has in addition produced in minds quite generally a habit of exactitude and precision even more precious than all these truths.[11]

Although Fontenelle criticized dogmatic systematizers and espoused the empirical foundation of scientific knowledge, he was committed to Cartesian cosmology, to the use of mathematics in natural sciences, and to the unity of nature as prescribed in its laws. During his long tenure as permanent secretary of the Academy until 1740, Fontenelle nurtured a vision of unified science. His skillful mix of mechanical/mathematical rationality devoid of teleology with an empirical orientation helped to induce a rather different research imperative in French elite chemistry, as can be seen in Homberg's theoretical works of the next decade. Fontenelle also created a collective memory and shared cultural heritage for French scientists and the European literary public through his *Histoire* and *éloges*. The *Histoire* summarized the achievements of academicians in a language accessible to the lay audience, thereby establishing an institutional memory that was shared by French scientists and the literary public. The *éloges* projected a lofty image of scientists in the pursuit of truth and public good through their diligent labor,[12] thus helping to transform scientists into cultural icons of the Enlightenment.

The reorganization of the Academy in 1699 added Homberg's longtime acquaintance Nicolas Lemery to the chemistry section as an associé, along with two élèves: Homberg's protégé Etienne-François Geoffroy and Simon Boulduc's son Gilles-François.[13] By this time, Nicolas Lemery had completed two pharmaceutical treatises: *Pharmacopée universelle* (1697) and *Traité des drogues simples* (1698). His extensive knowledge of pharmaceutical operations, combined with teaching experience, provided the new cohort of academicians with a solid reference point for their individual and collective research efforts during the next decade. Upon Bourdelin's death, Lemery was promoted to pensionnaire, with Geoffroy filling the position of associé. The proposals submitted to the Academy in 1699 by this newly constituted group do not indicate particularly novel orientations in their research. Nicolas Lemery proposed to study the reactions of antimony systematically by all the known methods of analysis, "by solvents, by distillations, and by calcinations," and some medical preparations of mercury.[14] Geoffroy, an heir to a prosperous pharmaceutical dynasty, proposed to work on essential salts. He also initiated a correspondence with Hans Sloan at the Royal Society so as to inform the Academy of curious and particular experiments that did not appear in the *Philosophical Transactions*. Gilles Boulduc wished to study the stones formed in human body with different solvents. The existing members continued their previous works, Bourdelin on the analysis of plants, Homberg on glass and invisible ink, and Simon Boulduc on purgatives.[15] Louis Lemery, who joined the Academy a year later than his father, first as Tournefort's élève in botany section, then as his father's in chemistry section in 1702, began to work on antiscorbutic plants. Within a few years, however, the dominant style and language of chemistry in the Academy began to change. Homberg assumed a leading role in this transformation by articulating a sophisticated version of corpuscular chemistry that resonated with Fontenelle's vision of unified science.

2.2   The Analysis of Plants

Homberg's lectures to the Academy soon after his arrival covered a wide range of topics in experimental philosophy and chemistry of metals. Phosphorus, the air-pump, and antimony figured prominently as his long-term interests. Abbé Bignon directed him, however, to the central research program of the previous generation of chemists: the chemical composition of plants.[16] Homberg began with Dodart's three-volume summary, carefully chewing over Dodart's methods and Bourdelin's data.[17] Indications are that he was not as familiar with plant analysis as with metallic chemistry; it took him a long time to prepare the initial report, and even longer to produce experimental results worth mentioning. In his initial report to the Academy, Homberg chose his words carefully, perhaps so as not to offend his senior colleague Bourdelin. The decorum of the Academy would have demanded such courtesy. He praised the "quantity" of plants analyzed and the "exactitude" of observations contained in the record. Nevertheless, the great "uniformity" found in plant composition forced him to seek the individual characteristics of plants in the "different combinations" of the same matters. He stipulated that only the varying quantities of these matters could account for the individual differences among plants. The fact that plants shared the same ingredients (albeit in diverse proportions) and the fact that plants of opposite effects yielded similar analytic products provided "convincing proof that one would not be able to judge fully the [medical] effects of plants by analysis." The value of distillation analysis for medicine was very much in doubt. Homberg tried to put a positive spin on this disappointing conclusion, noting that some of the experiments contained "quite considerable discoveries" that supported some contested truths, such as the existence of volatile salts in plants and the diversity of the lixivial salts found in plants. He also hoped that the products of distillation analysis might still have some utility, and he proposed to take a closer look at each of them. He had begun to work with oil of tartar (a fetid, stinking oil that came out toward the end of distillation), extracting an oil of lesser smell and some medical value.[18]

Homberg's subsequent presentations to the Academy provide no indication that he continued to work on the subject of plant analysis. In view of the quantity and the intractability of the data Bourdelin had compiled, it was probably difficult for him to come up with something conclusive. Nearly three years elapsed before he could report on some experiments related to the composition of fetid oils. He found "a great difference of taste, odor, & consistency in oils." Since he obtained much more oil through distillation even after all the oil had been squeezed out of plants, Homberg concluded that oil was a "thickening of saline aqueous & terrestrial parts." In other words, fire united the parts of plants that were disposed to form oil. He also observed that young seeds of plants had a large quantity of phlegm, little oil, and much fixed salt, while mature seeds had little phlegm, much oil, and very little fixed salt. Since the successive distillation of plants also yielded less oil and more phlegm, Homberg concluded that the phlegm in young seeds eventually composed the oil found in mature seeds by uniting with the salt and the earthly matter. Art separated this oil into the same "simple matters" nature had formed.[19] Fetid oils did not lend themselves easily to experimental control or to theoretical understanding, it seems, since it took another five years for Homberg to extract more oil with better smell by treating the fetid oil of plants with a variety of acid spirits, a procedure he deemed useful for perfumers when they distilled the oil of aromatic plants.[20] By that time, he had given careful consideration to the limitations of distillation methods and to the utility of solution methods.

Homberg was obviously concerned with the validity of distillation methods. He should have been familiar with Boyle's criticism of fire analysis. Denis Dodart, who had to cope with such criticism in publishing the mémoire on plants and was a close associate, could also have given him an earful.[21] Homberg alluded to the contention by many physiciens that the chemical principles drawn from plants did not exist naturally and that the "violent action of fire" created them. He sought to silence such criticism with an interesting approach that combined his interest in experimental philosophy and in chemistry. Having observed that "spiritous" liquids evaporated more quickly in vacuum, he evaporated liquids in vacuum for a "distillation without heat" to compare them with the products of ordinary distillation. Placing eau de vie in the distillation apparatus enclosed in the air-pump,[22] he found that the liquid gathered in the recipient was much stronger than the liquid that remained in the distillation vessel. The spirituous part was seemingly separated from the phlegmatic part without the "heat of fire," providing a convincing proof for the existence of chemical principles.[23]

Despite his defense of distillation methods, Homberg did not believe in the permanence of the principles extracted by these methods. He reported some "extraordinary" effects that raised the possibility of the transformation of plant salts. When he distilled lixivial salt with spirit of wine, for example, this supposedly "fixed" salt became a quite volatile liquid at mild heat. He interpreted that the lixivial salt, penetrated and divided by sulphurous spirits of wine, was "volatilized."[24] When he returned to the subject in the next year, however, he found other procedures in which lixivial salt retained its solid form. He concluded, therefore, that fixed salts and spirit of wine changed into one another depending on the different manner of operation.[25] Homberg also believed that the urinous salt of plants could be transformed into something else, although it had not yet been shown. At the least, it had to pass from plants to animals, since all urinous salts of plants and animals resembled one another.[26] In contrast to the fixed and urinous salts, Homberg believed that the acid salt of plants derived from the earth, all formed as a mineral salt.[27] These three salts extracted from plants through distillation—fixed, urinous, and acid—had to compose the essential salt, which was extracted from plants without fire or fermentation.

Homberg's keen interest in the transmutation of chemical principles, combined with his support of the five principles, suggests that he implicitly believed in a hierarchical composition of bodies. He did not question the validity of distillation methods, despite the mounting criticism of its effectiveness in extracting valuable medicaments. Nor did he challenge the doctrine of five principles. He tackled subjects of current interest with physical instruments, such as the areometer and the air-pump, without questioning the existing structure of chemical theory. He probably regarded distillation as a means of approximate analysis that separated out "chemical" principles. This did not preclude his search for the ultimate composition of bodies at the corpuscular level.

2.3   Acids and Alkalis

While examining the distillation products of plants, Homberg also took stock of the development in solution methods. He shared Nicolas Lemery's notion of universal salt—i.e., that all acids contained the same salt and that fire modified this salt to produce alkalis. Nevertheless, the expanding uses of acids and alkalis in chemical manipulations called for a more precise definition and understanding of their actions. Beginning in 1695, Homberg scrutinized various aspects of acid-alkali reactions seeking linguistic and quantitative precision. Using the areometer, he determined the composition of acids and alkalis quantitatively and interpreted their properties accordingly.[28] In other words, he maintained the principalist approach to chemical composition and action. This did not produce a satisfactory explanation for the selective action of acids and alkalis.

One can easily detect Homberg's preference for physical rather than chemical properties in his investigation of acids and alkalis. In 1695, he began to work on the composition of acids using the areometer.[29] He contended that acid spirits were nothing but volatile salts dissolved in aqueous liquor or phlegm. They consisted of "what is properly acid, which can be reduced to a concrete salt, & a certain proportionate quantity of phlegm, which gives fluidity to this salt and produces effects quite opposed to the effects of the volatile salts of vegetables and animals."[30] Homberg used the areometer to measure the quantity of these acid salts more precisely, separate from the phlegm in which they were embedded.[31] By 1699, he made a new areometer with a very thin neck and measured the specific weights of acid spirits, compiling a table of their weights at extreme temperatures to calibrate for their variation with temperature. He then mixed these spirits with an ounce of salt of tartar, dried the product thoroughly, and measured the weight increase. Taking this weight increase as the "true quantity" of the "pure acid salt," he calculated its proportion in a given acid spirit. The quantity of the acid salt contained in each acid spirit determined its "degree of force." For example, an ounce of regal water made of spirit of niter and sal ammoniac dissolved twice as much gold as an ounce of spirit of salt. In order to explain this phenomenon, one used to invoke the differential strengths

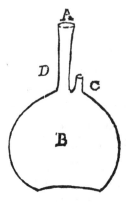

Figure 2.1
Homberg's areometer, described as follows (Homberg, "Observations sur la quantité exacte des sels volatiles acides contenus dans les differens esprits acides," *Memoires*, 1699, 44–51, at 46–47):

The construction consists of a glass vessel similar to a small matrass with a neck AD so narrow that a drop of water occupies the space of five to six lines. On the side of this neck, there opens from the belly B of the vessel a small tube C of the same capacity as the neck & of the length of about six lines parallel to the neck AD. This small tube serves to give an outlet for the air which is in the vessel as one fills it with the liquid; the reason why the neck is so narrow is because by this one can recognize more easily the true volume of the liquid which enters into the vessel. One makes a mark D on its neck AD to know up to which point it has to be filled. It is good to make the extremity A a little wider in order to pour in the liquid more easily.

The usage of this areometer is to fill it with an acid spirit up to the mark on the neck, to weigh it subsequently with a good assay balance [trébuchet] & to compare the weight of this spirit to that of another spirit. One would then know quite exactly how one weighs more than the other: since one drop of water occupies the space of five to six lines in the neck of this areometer, if one pours the spirit by one line, more or less, the error would be only one fifth or one sixth of the drop, which is quite small. Nevertheless, this will be quite sensible in the areometer & quite easy to correct by adding a little of liquid if it is less, or by tapping on the neck with the finger if it's too much, which will leave a little of the liquid by the tip of the small tube.

of the points of these two acids; one was easily blunted, while the other was more durable, and so on. Instead of such explanations with a "weak appearance of truth," Homberg argued, one could now attribute it to the different quantities of the acid salt contained in these acid spirits. The differential reactivities of acids depended on their acid salt content. In these calculations, Homberg assumed that all acid spirits contained the same acid salt.[32] His method set a clear precedent for the later determinations of saturation capacities by Kirwan and Lavoisier.

If Homberg wished to pin the differential strengths of acids on the respective quantities of the same acid salt contained in them, the facts spoke otherwise. For example, salt of tartar in its saturation seemed to retain one-eighth more of the acids of vegetables than of the acids of minerals. The acid salt contained in vegetable acids seemed weaker, that is, than that contained in mineral acids. Homberg attributed this difference to the movements of fermentation and distillation the parts of vegetable salt had to suffer in the plants, which divided them into much smaller parts than those of mineral acids. Although the acid salts of plants and mineral acids exhibited the same nature, he argued, they differed in the size of their parts. This would explain why salt of tartar absorbed more of the plant salt, why the same quantity of the acid salt would drag much more aqueous liquid in distilled vinegar, and why vegetable acids were more volatile than mineral acids.

When Homberg deployed his new areometer to estimate the "degrees of forces" of various alkalis, he discovered again that the quantity of the acid salt did not determine the strength or reactivity of acid spirits.[33] Moreover, all acids were "not of the same nature, because some dissolve certain bodies other acids do not dissolve." One usually divided acids into two categories, regal waters and strong waters whose main components were spirit of niter and spirit of salt, respectively. Regal water dissolved gold, but not silver; strong water dissolved silver, but not gold. In order to estimate the "degrees of forces" of alkalis, Homberg made spirit of niter and spirit of salt with known degrees of forces to react with a number of earthly alkalis. He found a considerable difference in the "dissolving forces" of spirit of niter and of spirit of salt. The former usually dissolved the same alkali more than the latter, sometimes more than twice as much. Homberg attributed the difference in part to the quantity of the acid salt contained in each acid spirit, but he could not fully explain the selectivity in acid-alkali reactions. Other examples of selective action could easily be multiplied. He conjectured that diverse "configurations" of the points of these acids were responsible for such selectivity, but he could not solve this puzzle in a satisfactory manner. In short, Homberg's principalist approach to composition and reactivity ran counter to laboratory experience. The selective dissolution of metals in acids remained an unsolved puzzle.

In 1701, Homberg attempted to demarcate the proper domain of acid-alkali reactions, expressing some reservation about the indiscriminate application of the acid-alkali theory, as Boyle had done earlier. He criticized that the system of acids and alkalis was rendered too general when one identified it with the appearance of fermentation, effervescence, or ebullition. Not only was it a common error to confound these three classes of phenomena; in addition, some of these phenomena were caused by something other than the mutual action of acids and alkalis. For example, mixing an extremely pure and dry acid spirit with an essential oil of aromatic plant produced a most perfect effervescence, accompanied by a great flame. Camphor (the most inflammable of all resins),[34] when dissolved in oil of Canelle, caught fire with spirit of niter, although the oil on its own did not do so. Since these sulphurous matters, not alkalis, produced such violent effects with acids, Homberg concluded that the system of acids and alkalis did not apply to these experiments.[35]

Throughout the first decade of his tenure at the Académie, Homberg characterized the actions of acids in terms of the quantity of the acid salt they contained. This principalist assumption did not help him to explain the selective dissolution of metals in acids. He had to resort to a speculative corpuscular ontology to explain this category of chemical action. The fact that Homberg took serious interest in these phenomena bespeaks the importance they held among his colleagues. Fontenelle applauded Homberg's work as an effort, as yet limited, to "predict" the changes corresponding to different chemical suppositions. It would bring chemistry closer to modern geometry, in which a universal formula applied to all cases:

If one could reduce chemistry, & in general physics to a few universal formula which contain all possible cases, as one reduces the most sublime Questions of modern Geometry, one would be in a state of predicting the changes which would respond to the different suppositions one would like to make, & often one would see quite easily the changes in the suppositions produce quite great variations in the effects. But physics is too vast, & too little known, at least until now, & experiment alone teaches us what is the power of circumstances in varying the phenomena.[36]

Homberg's presentations to the Academy indicate that he started out as a versatile experimental philosopher with some knowledge of metallurgical chemistry. Initially, he invested more time in experimenting with the objects of curiosity, such as the air-pump and phosphorus. Gradually, he immersed himself in the main problem domain of chemistry practiced at the Academy: the composition of plants. This led him to examine the theories and techniques of chemical analysis. Though he proved quite innovative in applying the instruments of experimental philosophy (such as the areometer and the air-pump) to chemical problems, he accepted the basic theoretical premise of distillation analysis: that the principles thus obtained should somehow determine the properties of mixts. While experimenting with acids and alkalis, he did not take them seriously enough to perceive that they posed a threat to the philosophical, principalist approach to chemical composition.

2.4   Chemical Analysis and the Domains of Reality

The tides began to change quickly after Bourdelin's death on October 14, 1699. Within three months, Simon Boulduc pronounced to the Academy the inadequacy of distillation methods in obtaining medically salient ingredients.[37] Through an analysis of ipecacuanha, a root known to have emetic and purgative virtues, Boulduc showed that the distillation method yielded the usual five principles, but that none had medical value. He asserted that the products of distillation, "which are improperly named principles, retain none of the virtues of the mixt from which they are drawn."[38] As an alternative, Boulduc proposed an "analysis by extraction" that would require digesting the plant in a solvent for a certain period before proceeding with the distillation process. His method of extraction was similar to Béguin's special extraction, but he could utilize the sophisticated knowledge of solvents and their actions that had accumulated over the previous century. Boulduc's challenge made it clear that the ineffectiveness of distillation methods threatened the legitimacy of the five principles as the constituents of all bodies. In the following year, he stepped up his rhetoric, arguing that the "common method" of distillation gave only a "general & superficial knowledge" of mixts.[39] In his long-term project of analyzing all known purgatives, he consistently used water and spirit of wine as the "two great solvents" for the saline and the sulphurous or resinous parts of the plants, respectively, to prepare various "extracts" with strong medical effects.

On June 18, 1701, less than two months after Boulduc's presentation on Jalap,[40] Homberg took up the challenge of answering the most serious question facing the method of distillation: whether the five principles of plants found in chemical analysis were the "true principles" that composed the mixt before the analysis.[41] Not only did chemists reduce a variety of plants quite different in taste, smell, figure and virtues to a similar set of principles through analysis; they also could not recompose the original mixt by combining the principles separated from them. Through a comparative analysis of the same plant with different intensities of fire, Homberg showed that they all yielded the same kinds of principles, albeit with different "degrees of volatility & fixity according to the fermentation & the degrees of fire these mixts have suffered in their analyses." These principles could not recompose the original mixt, because fire had changed its natural arrangement and degrees of volatility and fixity. Fire also dissipated some principles, which were irrevocably lost.[42] The remaining principles were no longer found in the same quantity, quality, or arrangement as they had been in the mixt before the analysis. Homberg thus concluded that great fire was "not so proper to discovering the true principles & virtues of a plant as when by a small heat & some fermentation one aids the natural separation of the principles which compose these simples."[43] Fontenelle summarized Homberg's conclusion in more succinctly corpuscularian terms: The long labor of the academicians analyzing numerous plants had been all but in vain, except to learn the single truth that great fire was not the proper means of obtaining the true principles and virtues of plants.[44]

At this juncture, when the legitimacy of the principles obtained through the distillation method came under serious attack within the Academy, Homberg began to sketch out an ambitious program of analytic chemistry by utilizing a new, powerful furnace—the burning glass. Early in 1702, he announced to the Academy his intention of putting together a textbook, *Elemens de Chimie*, to be followed by a *Cours d'Operations*.[45] Fontenelle applauded Homberg's project as an effort to establish chemistry as a part of physics: Although chemists had already dissipated the "artificial darkness" surrounding the old chemistry, making it more clear and more related to physics, there remained as yet much uncertainty and obscurity. Many connections were yet to be established, which required much skill and inventive ways. Through a long series of works, Homberg had acquired many views which would "perfect" much of chemistry even if they were not new.

The *Elemens*, containing the "principles of modern chemistry," was divided into six chapters, each addressing the subject of the general principles of chemistry, sulphur, salt, mercury, water, and earth.[46] As a natural philosopher looking into chemistry, Homberg scrutinized the structure of chemical theory to strengthen its internal coherence and to render it more intelligible for a broader audience. The corpuscular language and imagery played a crucial role in this overhaul of chemical theory and philosophy. First, it allowed Homberg to establish connections between the results of different analytic methods and thereby to form a uniform terrain of chemical theory. Second, it dressed chemical operations in a philosophical language that was shared by a broader scientific and lay audience. The corpuscular language thus helped enhance the heuristic and cognitive power of chemical knowledge. At the same time, Homberg's attempt to bridge the gap between chemical practice and philosophical language diffracted chemists' terms, making them acquire multiple meanings. In order to understand Homberg's text, then, it becomes necessary to differentiate his language and to discern the multiple domains of reality he negotiated as an experimental philosopher/chemist.[47]

Chemical analysis had to start, of course, at the most complex level of composition, as nature presented it. I will call this the *domain of nature* (N). For chemists, this domain comprised the three kingdoms of minerals, vegetables, and animals. As chemical analysis became more refined, this domain receded further into the background, in part because it was taken for granted and in part because chemists increasingly began their experiments with materials that had been processed. In Homberg's text, this domain is the least visible of the four that I outline here.

Fire, the main instrument of chemical analysis, "reduced" natural objects into various "matters"[48] and "principles." The *domain of accomplished analysis* (A) will thus comprise those substances that were actually separated out by chemical analysis and constituted chemists' laboratory reality. Lemery's and Homberg's "chemical principles" as the last point of analysis existed within this domain. A skilled chemist who knew how to regulate the intensity of fire for a desired outcome usually extracted some spirits, oils, aqueous liquors (phlegms and water), earths, and salts from a variety of natural bodies (mostly plants). All these distillation products could be dubbed "chemical principles" if they could not be decomposed further into simpler ones. They constituted, then, the domain of reality crafted in chemical practice.

But Homberg, as a natural philosopher, was not satisfied with the plethora of distillation products, which did not allow clear-cut classification. He sought to reduce them further to simpler ones common to all natural objects, or to the "true" principles. These principles, not yet isolated, were held responsible for the ultimate composition of natural bodies. With the *domain of projected analysis* (A'), therefore, I would like to designate this world of true principles that Homberg hoped to isolate and were few in number. Homberg's true principles were not merely a projection based on the accumulated analytic experience, however. Their existence had been stipulated in the French didactic tradition long before any empirical evidence became available. Chemical teachers had invoked the Aristotelian four elements and the Paracelsian *tria prima* throughout the seventeenth century without being able to materialize them in their analyses. Following in Boyle's footsteps, Homberg wished to bridge this gap between chemical analysis and the philosophically postulated principles. In doing so, he provided a significant impetus for materializing chemical principles. One could argue, therefore, that these true principles existed in the domain of chemical theory carved out between the domains of chemical practice and chemical philosophy.

In addition to the above three domains, Homberg often immersed himself in a constructed world of "integrant parts," comfortably speaking of the pointed particles of acids and the spongy receptacles of alkalis. In the predominantly Cartesian milieu, such corpuscular language enhanced the heuristic power and the cognitive authority of chemical knowledge, and in this secondary function it shaped Homberg's analytic program. I will call this domain of reality carved out in the boundary region between chemistry and "physics" (i.e., natural philosophy) the *domain of chemical philosophy* (P). This domain was not positively related to the art of chemical experiments; it served to decipher chemical arts in the prevailing language of natural philosophy, be it Aristotelian, Cartesian, or Newtonian. Changing winds of philosophy could radically alter this domain, which was exempt from the standards of experimental proof, without significantly affecting the other three domains.[49] Nicolas Lemery's *Cours de chimie* (1675) thus introduced the "Cartesian" mode of explanation into the predominantly Aristotelian/ Paracelsian discourse of mid-seventeenth-century French chemistry. This Cartesian fashion in chemical philosophy lasted until the 1740s, when it gave way to the Newtonian.

Needless to say, the domain of accomplished analysis grounded chemists' expertise and identity. They used various methods of chemical analysis to negotiate the distances between the different domains of reality, while linking the domain of philosophy to the other three domains through elaborate reasoning and imagination. Each chemist crafted a vision of chemical reality by melding together these four domains in his own way, but each knew enough not to wander too far from the domain of accomplished analysis. Chemists had been negotiating these four domains for some time in their didactic discourse, but they became much more pronounced in the French corpuscular program, particularly in Homberg's *Essais*. Although he vowed to keep his gaze at the level of accomplished analysis, Homberg often wandered into the domains of theory and philosophy. These domains offered him mirages, or visions of a world uniform and consistent in its composition, urging him to push the limits of chemical analysis. Mediated by experiments and imagination, the four domains blended effortlessly into the unique chemical world Homberg inhabited.[50] The stability of this world hinged, for this reason, upon the available methods of chemical analysis and the current philosophical fashion.

2.5   The Principles of Modern Chemistry

Homberg's first *Essai* consisted of the first article on the principles of chemistry and the second one on salts.[51] True to the French didactic tradition, he began the first article by defining chemistry. He reaffirmed the analytic ideal, but added another dimension—that of synthesis:

> I call chemistry the art of reducing composed bodies into their principles by means of fire & of composing new bodies in fire by mixing different matters.

Homberg did not simply define the analytic/synthetic ideal; he also tried to use it in practice, as can be seen in his definition of "principles." To the degree that chemistry was a part of "physics," it shared the principles of physics that pertained to the figure, arrangement, and movement of primary matter. Despite the long discussions concerning these principles, however, it had been impossible to determine anything incontestable on these principles. Since "chemical physics,"[52] which "consists only of experiments and exposition of facts," seeks "only the truth which is certain," it has established another sense of "principles," more material and more sensible, as "the simplest matters to which a mixt is reduced by chemical analyses." Through these material principles, chemistry recognized more directly the bodies subject to its analysis, and explained their operations.

Homberg intended "principle" in the sense pertaining to chemical operations, but the word vacillated between the domain of actually accomplished analysis and the domain of projected analysis. Without a careful tracking of the meaning of "principles," his article falls short of intelligibility. As much as he wished to ground chemistry on facts, he could not let go of the projected domain of experimental possibilities, where the principles would be simpler than chemists could tease out in their analyses. In other words, Homberg found himself in a tortuous position as he performed the analysis of the multifarious phenomenal realm while searching for a set of simple principles that made up all natural bodies. He kept expanding the analytic horizon of chemistry toward this end.

In order to facilitate our understanding of Homberg's project, we should perhaps start by delineating the domain that was least visible in his text: the domain of nature. Chemists manipulated this domain to produce concrete and imaginary chemical substances and alternative yet overlapping domains of reality. The domain of nature consisted of three kingdoms: mineral, vegetable, and animal. The mineral kingdom comprised metals and metallic minerals, fossil salts, stones, and earths. For analyses of metals and metallic minerals, chemists employed mercurification by solution, resuscitative salts, and the burning glass, thereby reducing them to mercury, a sulphurous matter, an earthly matter, and in some cases a saline matter. For analyses of other minerals, chemists used distillation and subsequent lixiviation as the major means of analysis: fossil salts (saltpeter, marine salt, and vitriolic salt) produced much acid, which contained a sulphurous matter, a little fixed salt, and a little earth. Stones produced much earth with a little sulphurous vapor. Earths contained an acid, a little fixed salt, and a little sulphurous matter. Vegetable and animal matters were more uniform in their composition. Chemists analyzed them primarily through distillation and lixiviation. In distillation with prior fermentation, they produced first spirituous liquors and volatile salts, then aqueous liquors, then fetid oils, leaving a residue, which was further reduced through lixiviation to fixed salt and insipid earth. In distillation without fermentation, aqueous liquors preceded volatile salts and spirits. Since they all produced some salt, water, earth, and a sulphurous matter, Homberg lumped vegetable and animal kingdoms together. Nevertheless, plants produced three kinds of salts (acid, urinous, and lixivial salt); animal substances produced only the latter two.

In the projected scheme of things, then, there existed only five principles: salt, sulphur, mercury, water, and earth, whose combinations produced the great variety of bodies in nature. Sulphur was the active principle, since it could either act alone or activate other substances. Earth was the passive principle, since it never acted alone and it only served as the receptacle or matrix of other principles. Salt, water, and mercury were "middle" principles; they acted only when joined to sulphur, which modified them in infinite ways. In practice, chemical analysis produced a multitude of saline matters or salts, sulphurous matters or sulphurs, mercury, water, and earthly matters or earths. Homberg assumed that sulphurs and salts were not isolated in pure state but were joined to a third principle which served as a "vehicle." This tenuous generalization had to be proved through painstaking analyses. In order to build a comprehensive theory of chemical composition and an intelligible chemical systematics, he had to sort out the variety of saline and sulphurous matters produced in the analysis of natural bodies. Earthly matters had not yet been subjected to extensive chemical analysis. Not surprisingly, Homberg's quest in the next decade concentrated on isolating the two principles of salt and sulphur in a pure state. He was keenly aware of the need to correlate the concrete distillation products with the projected simple principles:

[N]   plants, animals, minerals
[A]   salts, sulphurs, earths, phlegms, spirits
[A']   salt, sulphur, earth, water, mercury.

2.6   Salts

The second article of Homberg's first *Essai* dealt with salts—essential salts, fossil salts, and a variety of spirits produced in the distillation process. Homberg regarded essential and fossil salts as mixts because they were easily reduced to other simpler substances. The spirits not yet analyzed into simpler substances had to be considered as *salt principles*. In this definition within the domain of accomplished analysis, salt principle was a "matter soluble in water and one that does not change by fire." Following Lemery's and Charas's classification, Homberg adopted three kinds of salt principles—acid salts, volatile alkali (or urinous) salts and fixed alkali salts:

Salts   [N]   mixts   ⎡ fossil salts
                           ⎣ essential salts
         [A]   principles ⎡ acid salt   (volatile)
                           ⎢ urinous or volatile alkali salt (volatile)
                           ⎣ fixed alkali salt   (fixed).

The relationship between salt principles and mixt salts was not yet clear, although Nicolas Lemery had made a significant contribution to this end with his theory of universal salt. Lemery conjectured that the acid salt traveling through the earth formed different fossil salts by combining with different earths. He thus generalized the chemical experience that the combination of an acid and an alkali produced a more complex salt into a theory of salt formation. The definition of *salted salts* or *middle salts* derived from this understanding. Fontenelle reiterated this common-sense definition in the *Histoire*:

It is a piece of axiom in chemistry that the mixture of acids and alkalis has to produce a movement and an ebullition, which is only the same action by which these two different species of salts are penetrated and intimately united, & that this union has to give birth to a *middle salt* [*Sel moien*] which one calls a *salted salt* [*Sel salé*] such as ordinary salt or Sel ammoniac from which ensues that acids and alkalis cannot be together without first having fought and consequently destroyed.[53]

Homberg used this "axiom in chemistry" to classify salts by the analytic/synthetic procedures. Acids combined with fixed alkali salts or lixivial salts, earthly or metallic alkalis, and metals to produce mixed salts [sels mixtes], also called middle salts [sels moyens], which were "in part fixed, in part volatile."[54] Homberg characterized the production of middle salts as fixation of volatility rather than as neutralization of acidity, diverging somewhat from Lemery's definition of the term. The attention to the physical rather than the chemical characteristics of substances and their reactions set him apart from most chemists. Applying the physical criteria of volatility and fixedness, he placed sal ammoniacs (another commonly accepted "salted salt") in a separate category because they had two volatile components, which rendered them volatile:

middle salts   = acids + fixed alkali
               = acids + earthly or metallic alkalis
               = acids + metals
sal ammoniacs = acids + volatile alkali.

Once the composition of middle salts was elucidated, the next step in analytic reasoning pointed to the composition of acids. Homberg had been working on an exact, quantitative determination of acids in their reactions with alkalis since 1695. Now he offered a principalist conjecture that acids might be composed of a common "pure acid" and a variety of sulphurous matters which differentiated them as well as conferring activity on them. Depending on the kind of sulphurous matters they contained, therefore, acids could be divided into three classes. The first class of acids contained, in addition to the presumed pure acid, animal or vegetable sulphur. Spirit of niter and all acids distilled from plants, fruits, and alcohol belonged to this class. The acids of vitriol, common sulphur, and alum belonged to the second class of acids, which contained bituminous sulphur. The third class of acids, containing a more fixed mineral sulphur, comprised the acids drawn from a variety of marine salts and rock salts:

[A]   [A']
Acid spirits = pure acid + a sulphurous matter

$$\left[\begin{array}{ll} \text{animal or vegetable sulphur} & \text{(first class)} \\ \text{bituminous sulphur} & \text{(second class)} \\ \text{metallic sulphur} & \text{(third class).} \end{array}\right.$$

Homberg classified acids, then, according to the *projected* analysis, which would produce a pure acid and sulphurous matters. Not being able to isolate this "pure" acid, however, he tried to justify his classification through a reference to the other domains of reality. He could rationalize it by examining the natural origins of various acids. He deemed it "reasonable" to say that the first class of acids contained a vegetable or animal sulphur, because all the acids distilled from plants could easily retain a portion of the plant oil or its sulphurous matter. As for spirit of niter, which was extracted from saltpeter, it also made sense, since all saltpeter was derived either from earth soaked with the excrement of animals or from wall scrapings from old basements filled with the sulphurous matter of animals.

Homberg also invoked the domain of philosophy to explain the differential strengths of acids. Nicolas Lemery had already stipulated that the acidity of any substance consisted in its pointed particles, while alkalis were endowed with pores of various shapes and sizes.[55] Such a corpuscular ontology supported his theory of universal salt by making explicit the mechanism through which the acid salt contained in plants could be transformed into alkali salts by the fire used in chemical analysis. Homberg further speculated that vegetable and animal sulphurs, quite light and occupying much space, augmented considerably the volume of the acid points to which they were joined. This made it more difficult for these acids to penetrate compact matter with small pores.

Being light and having much surface, however, they tended to give in more easily to the flame and to act faster than other acids.

The same line of reasoning applied to the reactions of acids and metals. Different sizes of pores explained the differential solubility of metals in acids.[56] Strong waters, with the main component of spirit of niter, dissolved silver and lead without dissolving gold or tin. Regal waters, containing mainly spirit of salt, dissolved gold and tin without dissolving silver or lead. Both waters dissolved iron, copper, and mercury. Such differential actions of these two liquids made sense, Homberg argued, if gold was a quite sulphurous, compact metal with very small pores and silver was a less compact metal with little sulphur and larger pores. Spirit of niter, belonging to the first class of acids, had pointed parts covered with expansive animal or vegetable sulphur, which enlarged their points significantly. These enlarged points of the acid found it difficult to penetrate the small pores of gold and thus could not dissolve it. They would have no problem with filling the larger pores of silver, however. Spirit of common salt belonged to a different class of acids. The metallic sulphurous matter in this case did not enlarge the points of the acid significantly, allowing them to enter the small pores of gold. They were too slim, however, to fill the larger pores of silver.

In Homberg's text, the four domains I have outlined blend into a seamless chemical reality. The stability of this world depended on his ability to weave together the realities created by the different practices of chemical analysis, theory, and philosophy. With a little effort, however, we can make visible the distance Homberg had to negotiate between the analytic/synthetic ideal of chemical principles and the actual level of chemical analysis. His endeavor to bridge this gap betrays him as an outsider to the chemical profession. Most chemists dwelt only briefly on the five principles before going about the business of making new chemicals. They seldom worried about the disparity between their philosophical-looking principles and their mundane practice. When forced to confront such disparity after Boyle's critique, Nicolas Lemery simply distinguished chemical principles from natural principles, eliminating the need to bring them together. Homberg, as a natural philosopher, demanded a more uniform construction of the world. In order to achieve it, he had to impose an order that was not readily discernible in chemical practice.

This divergence of vision between a philosopher/chemist and an apothe-
cary/chemist became more pronounced in the ensuing years with the use
of the burning glass.

2.7   Sulphurs: Burning Glass Chemistry

Sulphur was supposed to be the first topic in Homberg's proposed
Elemens after the general discussion of principles. He had in fact taken
up the analysis of fetid oils as early as 1692.[57] Locating this singularly
active, universal principle among the multitude of concrete chemical substances proved to be a tricky affair, however. Analysis of the oily, sulphurous matters of plants did not yield promising results,[58] which
probably led him to present first the chapter on salts. A "new species of
furnace" came to Homberg's aid at this juncture.[59] His patron Philippe,
duc d'Orleans, had ordered from Germany one of Tschirnhous's burning
glasses, over three feet in diameter. It was set up at the royal palace and
made available to him. Although the burning mirror was not a new
instrument, the older type used concave mirrors which reflected the solar
rays coming from above to focus them from below. The object of investigation had to be placed in an upside down vessel, which made this
instrument a "nearly useless object of curiosity." Tschirnhous's burning
glass, made of convex glass, directly focused the solar rays coming from
above, allowing him to burn bodies on a regularly positioned vessel.
With its extraordinary size, this new instrument was far more powerful
than ordinary chemical furnaces, although the availability of favorable
sunlight limited its use significantly. Fontenelle hoped, echoing Homberg,
that this "new key to enter into the interior composition of bodies" might
lead to a new physics.[60]

With the Tschirnhaus burning glass, Homberg embarked on a series
of experiments to unravel the nature of the sulphur principle. He
expected that the new instrument would reveal the "intimate" composition of bodies at the corpuscular level, beyond the ordinary chemical
analysis of distillation. He was already familiar with Du Clos's experiment with the old burning mirror. It proved to him in an "incontestable"
manner that the "matter of fire" or "igneous particles" entered into the
composition of bodies and augmented their weight. He thus attributed

the differential "vivacity" of effervescence exhibited by quicklime and slaked lime to the presence of "igneous particles" or "the matter of fire" in quicklime.[61] Homberg's reliance on the burning glass as the key to the corpuscular composition of bodies, despite his earlier reservation about the use of great fire in chemical analysis, suggests that he believed in a hierarchical composition of bodies. He wished to resolve with the burning glass the composition of metals at the ultimate, corpuscular level of the "true sulphur principle" and to somehow link such corpuscular composition to the intermediate level of usual distillation products. His project reflected Boyle's concern with the further resolution of chemical principles and elements to their ultimate corpuscles. Unlike Boyle's, however, Homberg's corpuscles still retained their chemical identities. If he took Boyle's corpuscular chemistry seriously, he sought to arrest it at the level that could be matched to chemical manipulations. To anticipate, Homberg's plan would have looked as follows.

[N]	Metals	Common Sulphur	vegetables and animals
[A]	metallic sulphur	bituminous sulphur	inflammable oils or vegetable sulphur
[A']		The sulphur principle or the matter of light.	

Homberg first experimented with perfect metals.[62] He found that gold and silver became "volatile metals" with the fire of the sun, just as imperfect metals did with the regular fire of the furnace. At the focus of the burning glass, gold evaporated into smaller, nearly invisible particles. Placed at various distances from the focus of the burning glass, refined gold produced fumes and violet glass that weighed less than the initial amount. From these observations, Homberg theorized that gold consisted of mercury, a metallic sulphur, and an earth. Mercury, although the heaviest of all three components, escaped in fumes because it was divided into fine particles. Having lost the heaviest part, the remaining glass containing the sulphur and the earth, weighed less. The fire of the burning glass was not different in nature from the fire of ordinary flame, which was "a gross mixture of the oil of wood with the matter of light." Their actions differed in one important aspect, however: To penetrate other bodies, the flame required the presence of air, which, being heavier than the flame, pushed it against the other bodies according to "the laws of the equilibrium of liquids." In contrast, the fire of the sun was a "simple matter, whose parts are infinitely smaller than those of ordinary flame." Being pushed constantly by the sun, it could act on bodies in the absence of air.

As a superior furnace, the burning glass promised to unleash a greater potential of the fire analysis. It would allow chemists to command pure fire, nature's most powerful solvent, instead of looking for the mysterious alkahest. It had already demonstrated that gold and silver did not possess "invincible fixity." It could induce a "great progress in clarifying the principles of chemistry" and even to open the door for "a new physics," as microscopes and pneumatic machines had done in their own times. Fontenelle echoed these hopes in stating that fire was the "universal solvent" which resolved mixts into principles in their corpuscular state. In other words, the burning glass experiments preserved the legitimacy of fire analysis in the context of intensifying scrutiny over distillation methods:

Until now chemistry has employed for the decomposition of bodies no other agent which was more proper than fire. Fire has been its universal solvent, or nearly always the soul of its other solvents, & it has recognized the mixts only in so far as it has known their contexture & develop their principles.[63]

After analyzing metals with the burning glass, Homberg tackled common sulphur, but with conventional techniques.[64] The easy inflammability of sulphur probably precluded use of the burning glass. The choice of common sulphur was not fortuitous; it was a logical step in his pursuit of the "sulphur principle." Of the three sulphurous matters that supposedly composed acid salts, he had already explored metallic and vegetable sulphurs. Bituminous sulphur, the sulphurous matter contained in common sulphur, was next in line. By using an open flame instead of closed heating (such as distillation or sublimation), Homberg separated out the principles of common sulphur—an acid salt, an earth, and a bituminous and inflammable oily matter—in addition to a small quantity of copper[65]:

Common sulphur = an acid + an inflammable oil + an earth
+ trace metal.

The acid of sulphur was precisely the same as that of vitriol and alum, as Homberg had earlier predicted. It produced alum with simple earth, vitriol with an earthly and a metallic matter, and common sulphur with an earthly and a bituminous or inflammable matter. Homberg believed that the second component, a dense, bloody red oil, was the "true sulphur" or the "inflammable part" of common sulphur, carried in the minimal amount of distilled oil which served as a vehicle. The sulphur principle still eluded him, however.[66] The third component, an earth, was extremely fixed (having lost the volatile component of oil) and nearly inalterable. Even when exposed to the burning glass, it only produced fumes without burning.

Homberg delivered the long-anticipated "Treatise on the sulphur principle" to the Academy on April 22, 1705, making his hierarchical view of chemical composition explicit.[67] He acknowledged that until then the "chemical principle" of sulphur was the "oily or fatty matter" sensibly perceived in the analysis of animals, plants, and some minerals. Since chemists could now further decompose it into simpler matters, it could no longer be called a principle. Furthermore, some minerals lacked this oil, indicating that it was not the active principle that the sulphur principle was supposed to be. Analysis of this oil produced much aqueous liquor, an earth, and a salt without giving a candidate substance for the sulphur principle. Homberg concluded, therefore, that the "true sulphur principle," which must have been there to constitute an inflammable oil, was "absolutely lost" in the analysis. Instead of concrete sulphurous matters separated out from a variety of natural substances, he defined the "true sulphur principle" as a completely pure matter that escaped chemists' analyses and senses yet conferred activity on all other bodies. This elusive principle had to be the "matter of light":

All mixts which go through a rigorous or very exact analysis lose, as we have said, the sulphur principle which composed these mixts. The more the artist perseveres to discover it, the less he finds it. We have therefore no positive knowledge of the sulphur principle by means of our analysis, or by the decomposition of mixts. This made me think that one could perhaps discover something in the composition of artificial mixts. In fact, several operations of this nature have given me indications that it is *the matter of light which is our sulphur principle*, the single active principle of all mixts.[68]

The identification of the true sulphur principle with the matter of light defied Homberg's earlier definition of chemical principle as an undecomposed matter. The matter of light would satisfy the criterion that the sulphur principle should be the sole active principle of all bodies, but it was not accessible through the usual methods of chemical analysis. In fact, he had ventured into the domain of projected analysis, relying on a powerful instrument—the burning glass—that supposedly manipulated this principle. He sought to make his project intelligible by providing a plausible account of how the matter of light constantly agitated and how it acted on other bodies. For the first point, he supposed that the matter of light was the smallest of all sensible matters, so that it could pass freely through the pores of all other bodies. When it entered an inflammable oil, it augmented its volume and filled all its pores, transforming it into a flame. The flame was, then, an oily body whose pores were exactly filled with the matter of light. It was more "solid" than any other body. That is, flame was a mixture of the matter of light with an inflammable oil. Furthermore, Homberg supposed that the entire universe was full of the matter of light pushed out continually from the sun and other fixed stars. Because of this, the matter of light was continuously in motion, acting on all porous bodies in its passage. For the second point, to prove that the matter of light could enter other principles and cause qualitative changes in them, Homberg enlisted some "facts." Common mercury, for example, changed upon heating into black, white, or red powder without any addition of sensible matter. Something must have been added to mercury in this operation, Homberg reasoned, since the powders weighed more than the mercury. Whatever was introduced into the mercury also changed its nature by uniting perfectly with it. However, the only matter that could have touched mercury in the operation came through the glass vessel, which could not be traversed by ordinary matter. If one assumed that flame consisted of the matter of light and some combustible matter, the first could travel through the vessel and combine with the mercury, changing its nature and augmenting its weight.

An "incontestable fact" for Homberg's new sulphur principle came from Du Clos's 1667 experiment on the regulus of antimony using the old burning mirror.[69] Homberg repeated the experiment with the regulus of Mars, using the new burning glass.[70] He placed 4 ounces of its powder a foot and a half from the focus of the burning glass. The calcination proceeded, sending out thick, abundant fumes. At the end of the process, the calx weighed 3 gros [1 ounce = 12 gros] more than the regulus. He then placed the calx at the focus of the burning glass, which melted the calx, turning it into an orange glass weighing $3\frac{1}{2}$ ounces. Homberg reasoned that, since the regulus lost $\frac{1}{2}$ ounce in the entire cycle, the lost weight must have escaped as thick fumes in the calcination process. This meant that solar rays contributed to the weight of the calx by $\frac{3}{4}$ ounce [9 gros], since no other matter touched the regulus during the calcination process:

regulus   →   calx   →   melt.
4 ounces $4\frac{1}{4}$ ounces $3\frac{1}{2}$ ounces

Qualitatively and quantitatively, then, solar rays contributed to the weight of the calx, lending credit to Homberg's idea that it must be the sole active principle in the universe, or the sulphur principle. He preferred to call it the "matter of light" rather than the "matter of fire" because "fire" evoked much ambiguity:

Being persuaded that the matter of light is the only one that can penetrate quite liberally all the porous bodies and the only one that acts always, as we have demonstrated in the first part of this article, and persuaded that this matter is capable of being introduced into all other bodies, remaining there and thereby changing their figure, weight, and volume, we believed that no other matter could be our sulphur principle and our single active principle than the matter of light.[71]

Now that the involvement of the matter of light in these processes had been established, all that remained was to figure out exactly how the matter of light acted on bodies to produce the sulphurous matters known to chemists. That was to be the subject of another mémoire.

Homberg's "Treatise on the sulphur principle" alerts us to the difficulties he experienced in implementing the analytic/synthetic ideal of chemical principles. The old alchemical/Paracelsian chemical philosophy dictated that the sulphur principle should be the singularly active principle, which conferred activity on all other bodies. Chemical analysis yielded a range of sulphurs or inflammable oils, which were decomposed further into aqueous matters, earths, and salts, without providing a candidate substance for the sulphur principle. The Boylean natural philosophy demanded that the chemical principles obtained in distillation should be further broken down to little corpuscles. Homberg tried to satisfy these disparate demands of chemical philosophies and practices through the use of a powerful new instrument. The burning glass proved to his satisfaction that the calcination of metals involved the "matter of light." By equating the true sulphur principle with the matter of light, corpuscular in nature, he sought to bring together the old and the new chemical philosophies. The burning glass offered, however, different types of products that could not be matched directly to the products of distillation. Only by postulating that common flame consisted of the matter of light and an inflammable oil could he hope to confirm the existence of the matter of light in other chemical bodies. This involved a long speculation unwarranted by the regular techniques of chemical analysis, as we have seen.

Despite the difficulties, Homberg's pursuit of the "true sulphur principle" established a clear precedent for the later development of the phlogiston theory in France. His sulphur principle contained one of the core tenets of the phlogiston theory: that it eluded detection while conferring activity on other bodies. His student Geoffroy, who had much more extensive training in pharmacy, took the investigation in a more empirical direction, interpreting the sulphur principle as a concrete, oily substance that was involved in the transformations of metals and acids. Geoffroy later identified this material sulphur principle with Stahl's phlogiston on the basis of their shared operational characteristics. Homberg's and Geoffroy's work on the sulphur principle fed into mid-century French phlogistic theories, thereby paving the way for the reception of Stahl's works.

2.8 The Composition of Metals

The difficulties Homberg had to negotiate to stabilize his "true sulphur principle" became evident in the subsequent debates concerning the composition and the transmutation of metals. His outlook diverged significantly from Geoffroy's, although they both accepted the sulphur principle as a uniquely active principle that induced chemical transformations. Soon after Homberg announced the decomposition of common sulphur, Geoffroy "proved" its composition by recomposing it from the three supposed principles: the spirit of sulphur (acid), the inflammable part, and the oil of tartar (earth).[72] To judge from Geoffroy's confident statement, such a method of decomposition and recomposition had become an accepted norm in chemical analysis:

One is never sure, in fact, of having decomposed a Mixt into its true principles until one can recompose it from the same principles. This reestablishment is not always possible. When it is not possible, it does not necessarily work against the analysis of the mixt, but when it is successful, the analysis is demonstrated.[73]

Substituting oil of vitriol for the acid salt and oil of Terebenthine for the inflammable part, Geoffroy went further than simply proving the composition of common sulphur; he proceeded to recompose it from other similar principles:

Common sulphur
    = acid of sulphur + the inflammable part + oil of tartar
    = oil of vitriol   + oil of Terebenthine +   "   .
      [acid salt]   [sulphurous matter]   [earthly matter]

Fixed salts that contained an acid and an earth needed only an inflammable oil to produce sulphur. Some acid salts, such as dried marine salt and fixed niter, did not produce sulphur, however; this indicated that their acids were probably different from that of sulphur, vitriol, and alum. Geoffroy named the acid of sulphur "vitriolic acid" to distinguish it from others.

In the same paper, Geoffroy also cautiously breached another subject of inquiry: the composition of iron. He offered a "conjecture" that iron, like sulphur, might also be a compound of the sulphur principle or an inflammable part, a vitriolic salt, and an earth.[74] Common sulphur and iron would differ then only in the earth they contained, sharing two other common constituents. The facility with which iron was inflamed showed that it was sulphurous. The "rust of iron," or the dissolution of iron by the humidity of air, indicated that iron was saline and vitriolic. In addition, Geoffroy produced a more "philosophical proof" by recomposing iron, the "black, heavy powder attracted by the magnet," from the three named constituents:

Iron = vitriolic salt + inflammable part + earth.

In fact, Geoffroy seems to have been interested in the composition of metals in general. Iron was admittedly "the most imperfect and thus the most alterable and the most common" of all metals. If one could establish its composition for certain, however, it would have immediate implications for the general composition of metals, an outstanding riddle in chemistry, as Fontenelle surmised.[75] If all metals had compositions similar to that of iron, sharing their sulphur principle with common sulphur, metals could be transformed into sulphur. Such transformation between different classes of matters would prove that the sulphur principle was the sole active principle in the universe.

Homberg endorsed Geoffroy's efforts to establish the general composition of metals, but the master and his student were soon engaged in a lengthy debate with Louis Lemery over the composition of iron.[76] In order to prove the composition of the "artificial iron," Geoffroy tried to recompose iron from the sulphur principle, vitriolic salt, and an earth. He had to start with the material free of iron, but he could not find an earth or wooden ash that was completely free of iron. Assuming that neither plants nor wood contained iron, he deduced that every time wood burned it produced iron by mixing its three components.[77] From the distillation of honey which yielded iron grains,[78] Lemery countered that plants contained iron before calcination.[79] In the debate involving Geoffroy, Homberg, and Lemery, the interpretation of experimental results depended critically on the status of the burning glass as a chemical instrument. Disputing Homberg's "new chemistry of which the single furnace is the burning mirror," Louis Lemery insisted, based on his investigation through "ordinary chemistry," that iron was "an oily matter intimately united to an earth." The acid salt often found in iron was only a foreign part; it did not belong to iron's "true substance." That is, the acid salt was not a principle of iron, but a solvent that dissolved iron and destroyed it.[80] Endorsing Geoffroy's synthesis of artificial iron, Homberg concluded nevertheless that iron had a certain quantity of superfluous oily matter or sulphur that was separate from its "truly metallic part." This sulphur, combined with some other matter such as charcoal, made iron inflammable. Not every metal had an inflammable sulphur, however. The sulphurs of gold, copper, and tin were not inflammable. Homberg argued that only the burning glass could detect such a fine difference among the intimate principles of metals.[81]

The composition of metals was an integral part of Homberg's program of analyzing salts and sulphurs. Despite the slow progress of the experiments on metals (attributed to unfavorable weather), he went ahead with a theory of chemical composition, utilizing the new sulphur principle.[82] In 1706 he proposed that various sulphurs which chemists identified as principles of natural bodies were compounds of the sulphur principle or the matter of light and another principle. The union of the matter of light with other principles could be effected directly with the power of the sun or fire, or indirectly through the transfer of the sulphur principle from one substance to another. There were four kinds of sulphurous matters—metallic, vegetable, animal, and bituminous— although vegetable and animal sulphurs belonged to the same class. Except for the metallic sulphur, all sulphurous matters were volatile; this facilitated the transfer of the sulphur principle among them.

Metallic sulphur, a combination of the matter of light and mercury, was in part volatile and in part fixed. When the matter of light was bound to mercury only superficially, through a long digestion, it formed volatile metallic sulphur and produced imperfect metals. When it was attached to mercury by a stronger fire, it penetrated the very "substance of mercury," transformed volatile metallic sulphur into fixed metallic sulphur, and produced gold or silver. Gold had more fixed sulphur than silver, which had just enough to be a metal. Gold and silver had simple compositions, then, each consisting of some mercury and some fixed sulphur. Other metals contained additional matters, such as volatile metallic sulphurs, bituminous sulphurs, earths, and saline matters. Homberg's idea of the composition of gold and silver came close to the alchemists' dream.

The subsequent course of experiments reveals the divergent perspectives of an experimental philosopher/chemist and an apothecary/chemist particularly well. Homberg moved away from the dictates of chemical analysis through the use of the burning glass. Instead of producing simpler substances through a stepwise chemical analysis and synthesis, he began by supposing the simplest principle—the matter of light—and built up more complex substances in his imagination. In contrast, Geoffroy worked more in line with chemical methods.[83] First, his thoughts on the composition of metals differed significantly from Homberg's. Although he adopted Homberg's program as it pertained to the composition of metals, Geoffroy curtailed excessive speculations. In 1707, again defending the composition of artificial iron, he rejected mercury as a principle of iron. He even insinuated that mercury could not enter into any metal, although it ordinarily passed for the base of all metals. Sulphur, acid, and earth sufficed to compose a metal. Their relative doses, strengths of union, and manners of uniting accounted for all the differences among metals. Geoffroy thus sought to generalize his analysis of iron for all metals. Second, while endorsing the possibility of metallic transmutations, Geoffroy tried to effect them at the level of concrete substances. His response in 1707 also entailed that vegetable matters contained the principles of minerals, a premise fully laid out in Homberg's 1706 paper. Geoffroy embraced this seemingly paradoxical conclusion, since it conformed to the "great uniformity of Nature." He also showed by "curious" experiments that such imperfect metals as iron, copper, lead, and tin, when deprived of their sulphur and reduced to an earth, resumed their metallic form when treated with a sulphur, even a vegetable one. For Geoffroy, the sulphur principle was invariably a concrete oily substance separated out in chemical analysis, rather than Homberg's "true sulphur principle." As for gold and silver, their reduction with the burning mirror proved the existence of their sulphur sufficiently. Once they were reduced to an earth or vitrified, however, Geoffroy was not able to restore them to their metallic state with the addition of sulphur. Such revivification of gold and silver would have proved his theory that mercury was not a part of metallic composition and that the artificial production of perfect metals required only knowledge of the earth proper to each.

Having failed to restore vitrified gold and silver to their metallic state, Geoffroy persisted with imperfect metals. In 1708 unfavorable weather again slowed the investigation, allowing Geoffroy to make use of the burning glass on only a few days. He also had to take care that the materials in the "supports" on which the metals were placed at the focus of the burning glass did not partake in the action.[84] In May of 1709, he offered more detailed experiments supporting his thesis that sulphur could restore metallicity to iron, copper, tin, and lead.[85] He argued that each of these four imperfect metals contained an earth that was susceptible to vitrification. Other than the earth, the metals contained a sulphur (an oily substance) that produced opacity, brilliance, and malleability.

The sulphur was the same in all four metals, even in vegetables or animals, as Geoffroy had mentioned in 1707. Now came the proof. In order to vitrify metals, one had to leave them exposed at the focus of the burning glass on a vessel that did not supply fresh oil. One could restore them to the metallic state by placing them on charcoal and melting them with the burning glass. The process proved that charcoal supplied the single principle that the vitrified metals lacked, restoring them to the metallic state. The same process applied to mercury. Geoffroy concluded, therefore, that each of the four imperfect metals was composed of an earth susceptible of vitrification and a sulphur.

In sharp contrast to Geoffroy's meticulously empirical approach, Homberg's treatise on mercury laid out in more speculative detail how the transformation of mercury could produce a variety of metals.[86] Although he no longer believed that mercury was a "chemical principle" or "the substance that cannot be reduced into simpler matters by any analysis," Homberg retained it as such because he was not yet able to analyze it.[87] Even as he endorsed the analytic ideal of chemical principles, however, Homberg ventured deeply into the domains of projected analysis and philosophy. He stipulated that mercury existed in three different states: first as liquid mercury, second as a metal, and third as an earth. The first form was pure mercury, consisting of small, solid, polished balls. The second form was a metal, consisting of mercury balls pierced all the way through by the matter of light, which made thin holes in the balls to lodge themselves. When this metallic form of mercury was subjected to the burning glass, the holes had to pass so much matter of light at once that they became large and confounded. They could no longer hold the matter of light, and they became empty. This third form of mercury was a simple earth, which had only ruined fragments of mercury.

Metallic sulphur was nothing but the matter of light that penetrated the globules of mercury, made thin holes, and attached itself to these globules with its natural glue. Because these holes were so tiny, the matter of light stuck out of the globules of mercury, binding them together. The perfect metal, a combination of pure mercury with the matter of light, was then a mass of mercury, the parts of which were attached together by the matter of light lodged in the holes of the small balls of pure mercury. In order to make the metallic sulphur of gold, the matter of light had to penetrate the balls of mercury much more deeply and abundantly to cover the entire surface of the balls. In making the metallic sulphur of silver, however, it had to penetrate only enough to lodge itself in the hole, and in much less quantity. Since silver had less of the metallic sulphur that connected the balls of mercury, it also had more space or "interstices" between these balls. In other words, silver possessed larger pores than gold, and this caused the difference in their reactions with acids. Since the difference in silver and gold lay only in the quantity of the matter of light that each contained, one could transform silver into gold by subjecting it to fire and infusing it with more matter of light. This was why one could find gold in silver mines. Homberg tried in vain to perform this operation in the laboratory. He postponed discussing imperfect metals.

Despite their differences, Homberg and Geoffroy shared a belief in pervasive transmutation. Geoffroy's experiments had already shown that vitrified metals could recover their metallicity if imbued with a vegetable sulphur drawn from charcoal. Homberg continued the project into the next year to prove that vegetables and minerals shared the same sulphur with metals. The reciprocal processes would provide a firm proof, leaving nothing to be desired. He thus set out to show that a metallic sulphur could pass into a vegetable matter and make an oil as easily as a vegetable sulphur passed into a metallic matter and regenerated a metal. In order to obtain the sulphur of metals, which usually went up in fumes, he placed a mixture of iron and tin at the focus of the burning glass and collected the unusually thick fume on a piece of cotton. He placed this cotton in cold, distilled vinegar which was deprived of its oily part. The vinegar then recovered its oily part and gave, upon distillation, much phlegm and a true oil. Homberg concluded from this that the oil must have come from the cotton, which gathered the sulphur from the metals. To be certain, he repeated the same experiment with spirit of vitriol.[88]

In dealing with the composition of metals, Homberg and Geoffroy illustrated well the differential standpoints of a natural philosopher/chemist and an apothecary/chemist. Homberg approached the problem with a prior conviction that the true sulphur principle must be the active principle which conferred activity on all other bodies. It had to be possible, by transferring this active principle, to transform a metal into another metal, bituminous sulphur into metallic sulphur, and consequently common sulphur into a metal. Multitudes of transmutations among mineral, vegetable, and animal substances would reveal the uniform composition of nature. Exceedingly simple in thought, such a complete analytic control of chemical substances was not obtained in practice. As Homberg's student, Geoffroy obligingly tackled metallic transmutations. Instead of theorizing on the transfer of the true sulphur principle, however, he sought to effect transmutations of concrete sulphurs. In doing so, he curbed much of Homberg's speculation, staying within the bounds of chemical analysis. Though Homberg stated a clear definition of chemical principles, he more often than not overrode it to construct the world of integrant particles. Geoffroy used it in practice.

Although Homberg and Geoffroy shared the "new chemistry" of the burning glass, they differed markedly in their uses of the instrument. Homberg acted more as an impatient theoretician who sought to overhaul the chemical system with a limited use of the burning glass. Geoffroy patiently built his case through extensive experimentation with the instrument, gradually integrating it into chemical practice. While the reception to Geoffroy's theory seems to have been favorable, he did not elaborate it further.[89] In general, he refrained from the type of excessive corpuscular speculation Homberg excelled in. Geoffroy's "sulphur principle" invariably referred to the concrete oily substance contained in bodies. He later identified it with Stahl's phlogiston.[90] Geoffroy's "sulphur principle," the oily substance that restored metallicity to metals, formed the empirical core of the phlogiston theory in the course of the eighteenth century.[91] Rouelle and Macquer, the leading chemical teachers at the middle of the century, coupled it with Homberg's sulphur principle or the matter of fire or light. During the summer of 1772, when Lavoisier began to make preparations for a wide range of burning glass experiments, he also reviewed Homberg's and Geoffroy's papers, tracing the phlogiston theory in France back to Geoffroy's experiments with the burning glass.[92] He had to work hard to dissociate phlogiston and the matter of fire.

2.9   Negotiating the Instrumental Boundaries

In 1708, when Homberg returned to the subject of acids and alkalis, he attempted cautiously to correlate the results of various analytic methods: burning glass chemistry, distillation analysis, and solution chemistry. He wanted to devise a classification of acids and alkalis that would weave together the products of all three analytic methods. He also had to sort out the implications of introducing a new "true" sulphur principle to the classification of salts. He affirmed his earlier position that the difference of salts consisted only in the different sulphur each salt contained. Developing a clear classificatory scheme for sulphurous and saline matters was not easy, since sulphurous matters other than metallic ones contained salt, water, and earth in addition to the sulphur principle. Matching this interpretation derived from the burning glass chemistry to the existing classification of salts based on the distillation methods as well as to the characteristics of acids and alkalis emerging in solution chemistry proved even more challenging. This effort at correlating the substances derived from different analytic methods was crucial, however, in constructing a unified theoretical framework. The corpuscular ontology played an important part in negotiating the instrumental boundaries and stabilizing the fragile links between different analytic products. Once formed, these links would strengthen the corpuscular ontology.

Homberg's efforts to merge the results of burning glass chemistry with those of other analytic methods are evident in his supplement to the article on salt.[93] He sought to devise a consistent classification of acids and alkalis by matching the new salt principle with the products of distillation. He then used this new classificatory scheme to explain the common characteristics of acids and alkalis, such as their differential strengths. He began by defining as an "obvious acid" [Acide manifeste] every substance that produced a sour taste on the tongue and "obvious alkali" [Alkali manifeste] as every substance that received acids with ebullition and effervescence. In studying the acid-alkali reactions, the three fossil salts held a primary status because their combinations with various oily matters produced all the other salts chemists knew of. There were three kinds of simple or fossil salts produced by nature: saltpeter, marine salt, and vitriolic salt. With some effort, chemists decomposed them into an acid matter, an aqueous matter, an earthly matter, and an oily or sulphurous matter:

fossile salt = an acid matter
+ a sulphurous matter ⟶ particular acid
+ an aqueous matter ⟶ acid spirit
+ an earthly matter.

The acid matter was the "pure salt" or the "salt principle," which served as the base of all salts. It was never found pure, however; it was always accompanied by some sulphurous matter. This combination of pure salt with a sulphurous matter made a "particular acid" specific to each fossil salt. Even this particular acid was not visible or palpable until it was lodged in some earthly or aqueous matter. In the fossil salts, it was found lodged in some earthly matters to form crystals as in saltpeter or marine salt. When it was artificially united with an aqueous matter, it became an acid spirit, such as spirit of niter, spirit of salt, or spirit of vitriol. From this new definition of the "salt principle" we can gauge the distance Homberg traveled with his burning glass. In his earlier scheme (1702), acid spirits were salt principles, a definition consistent with the distillation method. Now these spirits were deemed to contain the "salt principle," a sulphurous matter, and an aqueous matter. Although one could argue from this that Homberg's allegiance to the empirical definition of principles had weakened considerably, it is more likely that he believed in pushing the frontier of accomplished analysis closer to the domain of projected analysis. Homberg's analytic scheme would have looked as follows:

N	saltpeter	marine salt	vitriolic salt
A	spirit of niter	spirit of salt	spirit of vitriol
A'		pure salt or salt principle.	

The more "composed" salts of plants and animals could be analyzed in a manner similar to the simple or fossil salts, except that they held a larger proportion of water or earth, which weakened them. The acid spirits, lixivial salts, and essential salts separated from the plants contained different proportions of the acid salt, which determined their solubility. Volatile and fixed alkali salts were produced by the action of fire. The strength or "alkaline force" of fixed alkali depended on the different proportions of the earth and the acid. Volatile alkalis had the same kind of composition as fixed alkalis except that they contained much fetid oil of the plant, which made them volatile.

As for the manner in which acids acted on alkalis, Homberg supposed that acids, consisting of small solid, pointed bodies, swam freely and continuously in an aqueous liquid, being pressed continually by the matter of light. Alkalis had pores specifically designed to lodge the points of acids. The pressure of the matter of light, aided by exterior heat, made the points of acids enter the pores of alkalis, displacing the aerial matter from the pores and thus producing bubbles. Acids acted on other alkaline matters in a similar way, albeit more slowly and at more intense heat. Vegetable acids, being lighter and more rarefied because of the fermentations they suffered in plants, occupied more space in the pores of alkalis and did not penetrate as deeply as mineral acids. When vegetable acids were lodged first in the lixivial salt, therefore, the more solid and heavier acids of fossil salts could push more vigorously into the pores of alkalis, and could enter them by rearranging and compressing the rarefied acids of vegetables. This was what happened when an acid appeared to act as an alkali toward another acid.

Homberg also dealt with "doubtful" acids and alkalis.[94] Volatile alkalis, either of plants or of animals, did not make effervescence or ebullition with all acids. For them to act, it was necessary that their "forces may be proportioned." Distilled vinegar and spirit of urine, for example, had no effect on each other unless one poured a great quantity of vinegar or weakened the spirit of urine substantially. Only when the two reached a desired proportion did they commence ebullition all of a sudden, consuming the last portion of vinegar as if the previous large quantity had no presence. In another instance, a red liquid distilled from plants, which appeared immediately before fetid oil, had marks of both acids and alkalis. That is, it acted as an alkali in making ebullition with spirit of salt, but as an acid in turning the color of Tournesol red. In this red liquid, acid and alkali seemed to be swimming together without penetrating each other.

All these boundary behaviors occurred only for the mixtures of volatile alkalis with vegetable acids, not with mineral acids. Homberg explained this difference in the behavior of vegetable and mineral acids with an elaborate image of pointed acid particles and alkali pores. He acknowledged that plants absorbed from the earth mineral salts that had bundles of acid salts laid one on top of another. In the plants, they were bound intimately together with the sulphurous matters of vegetables in order to pass the tight filters of the organs of plants. The bundles of acid points were resolved into individual points or needles. On the other hand, volatile alkali was a spongy matter capable of compression. In a small amount of liquid, it remained somewhat compressed, which made it difficult for the acid points to penetrate. In a sufficient quantity of water, however, it resumed its natural state, allowing the acids to penetrate it more easily. The particulate shapes of vegetable acids and volatile alkalis accounted for their behavior: All actions of alkalis and acids consisted in that the matter of light pressed one against another so that they could penetrate each other. Since vegetable acids consisted of simple, light, delicate individual acid points, they presented little mass to the matter of light pressing them against the pores of alkali. In a small amount of liquid, these pores were compressed and the delicate acid points glided on the lump mass without being able to penetrate the pores. In a larger quantity of acid, however, the pores resumed their natural state and received the acid points more easily when pressed by the matter of light. The mineral acids, possessing heavier bundles of points, could penetrate the pores even in the compressed state because the pressure from the matter of light was correspondingly greater.

Homberg built on Nicolas Lemery's theory of universal salt to correlate different analytic methods for a consistent definition of acids and alkalis. In traveling through the earth, the universal salt combined with various alkalis to produce fossil salts and was absorbed by vegetables to constitute the acid salt in vegetables. The distillation process produced various alkali salts by driving out the acid salts initially lodged in natural bodies. The strength of alkalis depended on the number and the size of the pores they developed through the action of fire. Lemery's theory of salt combined the conventional definition of salts deriving from the distillation of plants with the emerging understanding of acids and alkalis in the solution chemistry of metals. Lemery linked these two methods with a simple ontology of pointed acid particles and alkaline pores. Homberg further complicated this scheme by introducing the results of his own investigation with the burning glass. Introducing the true sulphur principle as the active constituent of all bodies made it necessary for him to design an elaborate classificatory scheme which went beyond the positive norms of ordinary chemical practice. His corpuscular ontology was correspondingly more complex. Homberg used this elaborate scheme linking three analytic methods to explain the selective dissolution of acids (a problem Nicolas Lemery had identified as one of the most difficult question), offering a corpuscular explanation for the differential strengths of acids and alkalis and for their selectivity. Homberg's approach to these problems reflected his principalist outlook that the principles contained in mixts would explain their reactivities. Though he failed to produce a crystal-clear definition of chemical principles or a satisfactory answer to the selective action of acids, his efforts at elucidating these problem domains point to their central place in chemical theory during the first decade of the eighteenth century.

Coming at a critical juncture when the distillation analysis was under serious attack, Homberg's burning glass chemistry represented his effort to offer an alternative method of analysis to study chemical composition by fire. He failed to provide a generally applicable alternative, however. The instrument was rare, and it did not function under normal circumstances. The excessive fire it occasioned was likely to strengthen the long-standing criticisms leveled against fire analysis. Even if chemists believed the new rhetoric that fire was the ultimate "universal solvent" which "intimately decomposed" the bodies subjected to the burning glass,[95] they could not command it as such without daily use of the instrument. In addition, Homberg's quest for the ultimate particles of matter went beyond chemists' pragmatic concern. As Béguin had elaborated a century earlier, the aim of chemical analysis was to extract principles that possessed medical virtues.

Homberg's intervention consolidated a unique moment in French chemistry by stabilizing composition as a theory domain of chemistry that transcended a particular mode of analysis. While composition had been a main theoretical goal of chemical analysis in the seventeenth century, it was bound to the singular effort at discerning the "principles of mixts" via distillation analysis. As an experimental philosopher/chemist, however, Homberg brought an instrumentalist vision to chemical practice and sought to streamline chemical experience with uniform analysis and measurements. In contrast to the delicate distillation analysis, which depended critically on the operator's skill, the burning glass promised a more uniform analysis of natural bodies to their ultimate level of corpuscular composition. The areometer could reduce the complicated operations of acids and alkalis to the numerical differences of composition. In other words, Homberg dreamed of unpacking the internal composition of bodies in an instrumentally objective and philosophically consistent manner. The burning glass offered the most powerful analytic tool available. In order to realize his dream, however, Homberg had to match the products of burning glass chemistry to the regular products of distillation and solution analysis. The corpuscular philosophy served to link various products of different analytic methods, thereby creating a coherent ontology of nature. At the same time, it translated chemical knowledge into a language fit for public consumption. Homberg's characterization of chemical principles in corpuscular terms forged a discourse that could mediate between esoteric chemical practice and general philosophy. He thus stabilized the analytic ideal of chemical principles as one that transcended the particular method of analysis. By melding it to the current philosophical language, he repositioned the theory domain of composition at the boundary between chemistry and natural philosophy. That the principles of bodies should determine the properties of mixts long remained an unspoken assumption in chemical practice despite the changing methods of analysis.

Various factors helped crystallize the first theoretical moment in French chemistry. The relative decline of the distillation method as the legitimate method of chemical analysis threatened to destabilize the doctrine of five principles. The emergent public sphere of the early Enlightenment called for a rational language of chemistry fit for consumption by the lay public. The Academy offered an interdisciplinary forum in which chemists had to communicate with natural philosophers and mathematicians. The ascendancy of Cartesian philosophy in the Academy as well as in the public sphere made it appealing to codify chemical knowledge in corpuscular terms. The reorganization of this prestigious institution in 1699 brought together a group of distinguished chemists who could debate the issues of chemical theory. Homberg tried to meet these diverse demands on chemical discourse by forging a program of corpuscular chemistry that would coordinate various analytic methods and yield a consistent doctrine of chemical composition. Despite the failure of his program, his works had a lasting effect on subsequent generations of chemists. Louis Lemery took up his work on the sulphur principle and articulated an elaborate discourse of fire that fed into the phlogiston theory of Rouelle and Macquer. Richard Kirwan revived Homberg's method of determining the differential strengths of acids and alkalis by their mutual saturation, establishing an important precedent for the development of stoichiometry in the nineteenth century. Lavoisier followed a path strikingly similar to Homberg's in taking up first the hydrometer and then the burning glass as a kind of universal instrument that would unpack the secret of chemical combination. The corpuscular understanding of chemical bodies became an entrenched, albeit implicit, part of chemical theory.

3
Affinity

The analytic frontier of chemistry at the beginning of the eighteenth century was mapped by solution methods, particularly by the use of acids and alkalis. Traditionally used in metallurgical practices, acids and alkalis made their way into pharmaceutical preparations in the seventeenth century.[1] The infusion of solution methods induced mounting criticism of distillation methods, as can be seen in the mid-century Helmontian literature. Du Clos's call to carry out a program of "extreme resolution" to find the "true principles of mixts" went unheeded in the Academy, however, because Bourdelin and Dodart continued to rely on distillation methods in their analysis of plants. The quest for the alkahest (the all-powerful solvent) proved as elusive as the search for the philosopher's stone,[2] but the solution methods caught on gradually in chemical practice, complementing the traditional analysis of distillation. Boyle's color indicators helped this transition from the fire to the solution analysis by allowing chemists to identify acids and alkalis with more certainty and precision. In his *Experimental History of Colours* (1665), Boyle stipulated that all acids and only acids turned the syrup of violets red. Acids also turned all other blue vegetable indicators red; alkalis turned them green. Neutral substances turned them neither red nor green. With the inclusion in Harris's *Lexicon Technicum* (1704), these tests became available to chemists of all ranks.[3] French chemists, academicians and amateurs alike, employed color tests routinely in their analyses of mineral waters.[4] Homberg regarded the Tournesol test as the most reliable means of detecting acids and alkalis,[5] especially in the case of "doubtful" species which did not offer the distinct mark of acid-alkali reaction in the form of ebullition.[6]

The changing emphasis in analytic method had a radical implication for chemical theory. Although the selective dissolution of metals in acids and the displacement of metallic salts had long been known in metallurgical practice, the reigning status of the distillation method delayed the systematization of these experiences as an integral part of chemical theory. In the quest for the "principles of mixts," metals and minerals did not occupy a central place because they did not yield all five principles. Despite his allegiance to the five principles, Nicolas Lemery acknowledged this analytic reality:

> The five principles are easily found in Animals and Vegetables, but not so easily in Minerals. Nay there are some Minerals, out of which you cannot possibly draw so much as two, nor make any separation at all (as Gold and Silver) whatsoever they talk, who search with so much pains for the Salts, Sulphurs and Mercuries of these Metals. I can believe, that all the Principles do indeed enter into the composition of these Bodies, but it does not follow that they must remain in their former condition, or can be drawn as they were before.[7]

And there were "many minerals from which it is not easy to draw any principle," as Fontenelle casually remarked in 1700.[8] The expanding use of solution methods thus helped relocate metals and minerals from the periphery to the center of chemical theorizing. Chemists began to model their theories on the operations of acids, alkalis and metals and to apply them to the operations involving vegetable and animal substances. In other words, they reversed the direction of analogy in chemical theorizing.[9] Instead of trying to find in minerals the same "principles of mixts" that were distilled from vegetables and animals, chemists reified their object of inquiry as a "compound." Middle salts as the combination of acids and alkalis provided the model for the compound. The focus on these relatively simple substances also enhanced chemists' analytic control of their behavior, setting the stage for a conceptual structure more in tune with their practice. In particular, the selective displacement of metals in acids emerged as the central question in theoretical chemistry. Nicolas Lemery addressed this problem as early as 1675, but without abandoning the doctrine of five principles. Homberg, as we have seen, sought to answer the same problem by measuring the differential strengths of acids and alkalis. Despite his innovative physical methods, he retained the principalist approach to chemical composition and reactivity modeled on vegetable analysis.

Generational change within the Academy brought in new perspectives, however. In 1711, Louis Lemery examined the problem earlier raised by his father in a more systematic manner, thereby establishing the selectivity of chemical substances as a main theory domain of chemistry. His efforts received high praise from Fontenelle, who had been preaching that the "prediction" of chemical actions would help bring chemistry closer to modern geometry.[10] Fontenelle projected that Lemery's work on precipitates would expand the clarity of "modern Philosophy" by bringing chemistry closer to exact physics, away from incomprehensible mysteries. He also took the liberty of discussing the relevance of Lemery's work to the general principles of hydrostatics. Simply put, the particles of metals were sustained in solution insofar as they maintained an "equilibrium of force" with the particles of solvent. The cessation of the equilibrium for some reason caused precipitation. In addition, Lemery systematically exposed such "properly chemical" factors as the "convenance" of the dissolved substances and the choice of proper intermediates. Fontenelle did not think that Lemery explained the preference of alkalis or their "disposition to unite" with acids over metals. His review indicates that the "convenance" or the "disposition to unite" of chemical substances, generally associated with Geoffroy's affinity table of 1718, had already become a central theoretical concern by 1711.[11]

The corpuscular chemistry in the Academy nurtured by Fontenelle began to wane after Nicolas Lemery's and Wilhelm Homberg's deaths in 1715. Etienne-François Geoffroy, a more experienced chemist than Louis Lemery, assumed leadership of the chemistry section and took the solution chemistry in a radically different direction. Instead of musing upon the causes of selectivity, he represented the order of selectivity visually in his 1718 table des rapports. Skillfully exploiting the multiple significations of "rapport" as relationship and as mathematical ratio,[12] Geoffroy inscribed the orderly behavior of chemical substances in an easily accessible form. Ever appreciative of the lawful behavior of nature, Fontenelle declared that "a chemical table is in itself a spectacle agreeable to the mind."[13] Perhaps to distance himself from excessively corpuscular reasoning, Geoffroy did not mention Louis Lemery's contribution to the subject. Traces of shared conversations or debates in the Academy can be discerned only by interpreting Geoffroy's table carefully.

He adopted displacement reactions as the unique method of measuring the "rapports" of bodies as well as the notion of fire as a universal solvent. Both of these assumptions had been considered in detail by Louis Lemery, albeit in corpuscular language. Despite their divergent approaches, Louis Lemery and Etienne-François Geoffroy established chemical affinity or rapports as a new theory domain. Furthermore, they made possible a new conceptualization of chemical composition as contingent upon affinity relations. In the principalist approach, the philosophically posed but distillationally separated principles determined the composition and consequently the properties of chemical bodies, including affinities. In contrast, the affinity approach implied that chemical composition was the result of empirically determined affinities. The composition of middle salts served as a model case. Simply put, Geoffroy's table embodied a new theoretical approach that affinity determined composition, whereas in the principalist approach composition determined affinity. Later generations of French chemists, taught successively by Geoffroy and Lemery at the Jardin, inherited both of these approaches and mixed them creatively according to their needs.

3.1   Fire as a Universal Solvent

Despite his strenuous criticism of burning glass chemistry, Louis Lemery came to embrace Homberg's expository style in later years, apparently with Fontenelle's encouragement. He had a rather different background and commitments than his father. Nicolas Lemery was educated largely within the network of Protestant apothecaries and suffered persecution in the 1680s on account of his religion.[14] Appointed to the Academy at the age of 54, he led the life of a dedicated scientist, focusing his energy on the minute variations of antimony. Despite his extensive knowledge of chemistry, he remained largely in Homberg's shadow. Thanks to Nicolas's conversion, however, Louis was able to attend the Paris Faculté and to circulate among the learned elites of the medical community in the court circles. The membership at the Academy brought more honors and obligations. He taught at the Jardin briefly in 1705, became a physician at the Hôtel Dieu, and was a frequent guest at the Palais de Luxembourg. He nurtured ambition for a higher life than that of a lab-

oratory chemist. Intensely competitive toward his older and more established colleague Geoffroy, he sought quick fame but spent less time in
the laboratory. After he acquired the position of médecin du roi in 1722,
he spent three months a year at the Versailles. As a physician to Princess
de Conti in the 1730s, he stayed regularly at her residence, writing his
memoirs there.[15]

In view of his education and his career moves, it is not surprising that
Louis Lemery favored a discursive rendering of chemical experiments,
akin to Homberg's. Unlike such founding members as Du Clos and
Bourdelin, who had little in common with the other members of the
Academy with backgrounds in Parisian scientific circles, law, or the
Faculté, Homberg fashioned himself as a natural philosopher interested
in rationalizing chemistry. He appreciated the value of corpuscular
philosophy as a public language of chemistry. Notwithstanding his com-
plaint against the "new" chemistry of the burning glass, Louis Lemery
must have admired Homberg's stature as a natural philosopher/chemist.
In the years after the controversy over the composition of iron, he came
to adopt Homberg's style of extensive corpuscular reasoning, evolving
into his true disciple in intellectual style. The first indication of this development is found in his discussion of fire in 1709.[16] As Homberg's burning
glass chemistry reached a tentative closure, Lemery developed a more
articulate discourse on fire or the sulphur principle.

Although Lemery mimicked Homberg's extensive corpuscular reasoning, he did not endorse the principalist approach. Instead, he brought
the radical implications of solution chemistry into the open, re-casting
fire as a universal solvent in an explicit analogy to the operations of water
as a solvent with regard to salts. Lemery's corpuscular language masked
this fundamental shift of emphasis in chemical analysis, facilitating the
subsequent transition in chemical theory. Lemery's discourse on fire set
an important precedent for Boerhaave's *Traité du feu*, which became an
integral part of French discourse on heat, fire, and phlogiston for the
remainder of the eighteenth century.[17] Although Boerhaave explicitly
cited Lemery's articles in his *Elements*, the popularity of his textbook
and the demise of the Cartesian paradigm served to eclipse Lemery's contribution to the subject. More important, Lemery articulated a detailed
mechanism for the fixation of fire in bodies, which contributed to French

phlogiston theories. Rouelle and Macquer later identified phlogiston as fixed fire.

The burning glass played a central role in conceptualizing fire as a chemical substance. Boyle, Du Clos, and other seventeenth-century investigators who employed this extraordinary instrument commonly assumed that fire was a substance that participated in chemical actions and augmented the weight of the bodies. On their authority, Nicolas Lemery also accepted fire as a chemical substance that was involved in chemical actions. Walter Harris, who translated the fifth edition of Lemery's *Cours de chimie* into English, singled out this aspect of Lemery's book:

> He is the first perhaps who has taken such particular notice, what an augmentation of weight is added to many Preparations by the concurrence and incorporation of the *substance of Fire* into their composition, as you may see in the Calcination of *Lead*, p. 107. in the Distillation of *Spirit of Saturn* from the *Salt of Saturn*, p 116. in the Calcination of *Regulus* of Antimony, p 208. and even in the Calcination of *Antimony* by the heat of Sun with a burning glass, p 228. which few instances may possibly lead the way to Inquisitive persons to discover the same augmentation in divers other Preparations.[18]

Homberg also assumed that fire was intricately involved in the composition and action of chemical bodies. He attributed the difference of action between quicklime and slaked lime to the matter of fire contained in quicklime. Despite the common perception that slaked lime, as the ashy residue of quicklime, lost its principal alkali part, Homberg found that acids dissolved equal quantities of both, which indicated to him that quicklime and slaked lime contained equal quantities of alkaline substances. But quicklime experienced a much more prompt and violent effervescence than slaked lime. Homberg attributed this difference in the "vivacity" of effervescence to the presence of the "igneous particles" or "matter of fire" in quicklime.[19] In the principalist approach, the properties of bodies were determined by the kind and the quantity of the principles contained in them. Homberg developed his thoughts on fire further through his experiments with the burning glass. As an "incontestable example" that the matter of fire was introduced to certain bodies, that it remained there for a long time, and that it augmented their weights, Homberg cited the regulus of antimony calcined by the burning mirror.

Since no other substance could have touched the regulus in the process, the weight increase had to derive from the "particles of fire."

Louis Lemery built on Homberg's works, but he reconceptualized fire as a solvent for calcination and combustion processes by drawing an explicit analogy with the action of water as a solvent.[20] He offered a detailed corpuscular reasoning on the function of fire as a solvent and on its fixation in chemical compounds, which provided a clear precedent for Boerhaave, Rouelle, and Guyton de Morveau in conceptualizing chemical action as a process of dissolution. Lemery stated that "the matter of fire is the first & the most powerful solvent of earthly bodies; we have no other agent that penetrates as deeply & disunites the essential substances as perfectly." It could be the true principle of heat and light, as well as of fluidity. In order to function as a solvent, fire had to consist of particles with finite size and shape. It was a "fluid of certain nature" with a set of particular properties that distinguished it from all other fluids. Its properties depended on the figure as well as on the movement of its parts.

Often, fire became imprisoned in solid bodies rather than dissolving them. Two remarkable circumstances ensued. First, the matter of fire enclosed in a solid body increased markedly the weight of the body. Second, it preserved during the entire duration of captivity its particular properties, which became evident when it escaped its prison. All the world knew, Lemery asserted, that several metallic matters, including the regulus of antimony, lead, tin, and even mercury, gained weight when exposed to fire, although they lost much of their proper substance during the operation. Since no other material used in the operation could have penetrated the vessel, this augmentation of weight through calcination had to be caused by fire. Only the matter of fire passed through the walls freely and abundantly, causing a substantial weight gain, sometimes amounting to one-tenth the total weight. Burning glass experiments that did not use wood or charcoal provided a clear proof of this fact: Metallic bodies gained as much weight with the burning glass as when they were exposed to ordinary fire. Because of its unique ability to incite fire without added fuel, then, the burning glass offered a stronger proof for the involvement of fire in chemical actions as a material substance.

The ontology of fire as a chemical substance depended on this manipulative capacity of the burning glass.

In order to strengthen his opinion that fire worked as a solvent despite its imprisonments, Lemery wished to prove that it always conserved its nature. Water, whose fluidity depended on the matter of fire, could likewise be enclosed in an infinity of bodies without losing its fluidity or any other properties. The proof was difficult to come by in the case of metallic calces, however. The matter of fire was so well engaged, hidden, and retained by them that it could not manifest itself by any external sign until it was released. The fire insinuated in the stony or saline bodies through calcination was easier to detect. Since these bodies were much less solid, water sufficed to disunite their parts, to grind them into fine powder, and to set the matter of fire free. The fire thus escaping helped the "disunion" of the parts, and heated the liquid more or less according to its quantity. Charcoal, pure earth from which all salts had been chased out by fire, still heated the water in which it was immersed. The properties of various calcined bodies should be attributed, therefore, to the parts of fire retained in them, which made impressions upon leaving.

Lemery disregarded the objection that the parts of fire only caused a rapid movement and that they lost their movement and effectiveness while passing through the tissue of gross bodies. The matter of fire conserved its movement while confined in solid bodies, he argued, because even the most solid and heavy body was filled only to 1/100,000 of its volume, leaving much room for foreign matters. Even if the matter of fire could not maintain its movements entirely, the property of fire derived not only from its movement, but also from its figure and tenuity. As was well known, water and salt retained their properties in solid state because those properties derived from the figure of their parts.

Acknowledging the difficulty in imagining that a matter so subtle and so active as fire could be arrested in the body, Lemery reasoned that it required a certain quantity of fire to overcome the resistance of a solid body and to rupture the union of its parts. In smaller quantity, the matter of fire had less force and became contained in the solid body. A solid body subjected to fire opened and dilated its pores, which allowed an easy passage for the matter of fire; however, these pores returned to their original state once the fire ceased to operate on the body. The fire introduced could not escape and had to wait for another process of dilatation. Water could free the fire contained in saline and stony bodies easily because the quantity of water used in these cases was much larger than the quantity of imprisoned fire. Besides, water did not work alone. The matter of light always conserved its movement in the interior of these bodies, traveled continually inside its prison, trying to force it open, but could not open it without the exterior means. With the help of water, it would at least contribute to facilitating and effecting the escape.

Lemery clearly distinguished between the fire fixed in bodies and the fire free to pass between bodies in the form of heat. The parts of fire contained in the interior of a body could not be felt on the surface, just as salt was not sensible to the taste until it was sufficiently released from all other bodies. The calcined bodies, which had received an abundant quantity of the matter of fire, did not pass any to the touch of the hand. The vivid impression of heat that bodies made when they had just been removed from fire was due not to the parts of fire that were imprisoned but to those that escaped. Accordingly, there existed two kinds of pores: those sufficiently large to give a free passage to the matter of fire all the time and those that allowed the passage only when dilated by heat. Well before Black articulated the concept of latent heat based on thermometric measurements, then, Lemery offered a detailed mechanism of how fire could become fixed in bodies. Rouelle and Macquer took up this distinction in their articulation of the phlogiston theory in the 1740s.

Lemery adopted Homberg's position that the matter of fire or of light as a pervasive medium could explain the nature of flames. The sun was nothing but a "quite considerable mass of the matter of fire or of light," or "a great flame which did not differ essentially from ours," since they both produced perfectly the same effects. The sun acted either by emanations that broke away or by trails of the matter of light which were pressed onto the bodies. When reunited by the burning glass, the matter of light acted like a violent flame immediately applied to these bodies. This vigorous action indicated that air, much like water, acted as a moderating medium to subdue the violence of luminous rays. The flame consisted, then, of a great quantity of the matter of light that acted as much more vividly as it was more abundant and more reunited. In order for a body to be inflamed, the luminous substance had to exist quite abundantly in the body and to "form a quite robust mass to press on all sides with rigor the matter of light which is found confusedly expanded in the interstices of air" so that the parts of this matter could transmit a sensible impression. Phosphorus provided an excellent example. The matter of light escaping in small parcels would immediately be surrounded by a greater mass of air and become invisible. The matter of light contained in the inflammable bodies should exit in much greater quantity at each instant than that in calcined bodies, either because calcined bodies contained less quantity of the matter of light than oily bodies or because they had more compact tissues that did not permit free exits. This could explain why, when exposed to a small flame, easily inflammable bodies such as paper were sometimes consumed entirely without throwing any flame. The flame was too weak to chase out at once a great quantity of the matter of light contained in these bodies, which escaped successively in small invisible portions, proportioned to the force that procured its deliverance.

Lemery accounted for the different states of bodies by their proportion of the matter of light. As the singular cause of fluidity, the matter of light modified bodies differently according to their natures. Substances such as water melted much more easily than metals. Ice was the natural state of water attained by the "single absence of the matter of light." Consequently, it attained "true fusion," much like metals, when exposed to fire and turned into a liquid. Water differed from metals because it usually had enough matter of light to retain its fluidity. The matter of light formed inflammable bodies by being engaged in certain compounds of salt, earth, and water. When analyzed, these bodies were reduced to salt, earth, water, and a fine substance that passed through the vessels and was always dissipated in great quantity to effect a considerable loss of weight. Salt, earth, and water, together or separately, never became inflammable, and they retarded or impeded the inflammability of other bodies. They served in the composition of inflammable bodies only to contain and to arrest the matter of light, which was truly the matter of flame. It was, then, the matter of flame that escaped the artist in the analysis of inflammable bodies. In other words, Lemery's matter of light became phlogiston or the principle of inflammability in the mid-century didactics of Rouelle and Macquer.

3.2   Displacement Reactions

In 1711, Louis Lemery undertook a systematic reflection of solution chemistry.[21] Although his work focused primarily on metallic precipitates, he offered an overview of displacement reactions. To this end, he set up a general classification of the operations that produced precipitates, building on his father's work.[22] Through this exercise, he charted a theory domain of chemical operations rather than of substances. In particular, he gave prominence to the operations that produced "true precipitates," or displacement reactions, as the model experimental system for solution chemistry. Lemery's systematic reflection on precipitates encoded the selective dissolution of metals in corpuscular language, thereby legitimating this exemplary chemical action as a theory domain in the current philosophical language.

Lemery defined precipitation as an expression chemists used to describe the "fall of a body which had been suspended & dissolved in a liquid from which it has been subsequently disunited." Although Fontenelle construed this as a physical definition based on the principles of hydrostatics, Lemery used it to differentiate "true" metallic precipitates, or the products of displacement reactions, from "false" ones. One could obtain false precipitates, or the matters that lost their initial metallic form and were reduced to a friable and indissoluble mass, in several ways. Calcination (red and violet mercury), incomplete dissolution in acids (antimony in spirit of salt or in regal water), and calcination after dissolution and evaporation (mercury in spirit of niter), all produced such precipitates. True metallic precipitates differed from false ones in that they were directly separated from their dissolution in liquid. As Lemery put it, false precipitates were abandoned by the liquid, while true precipitates abandoned the liquid themselves. True precipitates were made sometimes "naturally" through agitation, but mostly with recourse to the "intermediates" such as alkali salts or other metals. The choice of intermediates depended on the nature of the bodies to be precipitated. Lemery provided an exhaustive discussion for each case. In order to precipitate a resinous matter dissolved in spirit of wine, one could use common water which, by meshing intimately with the spirit, would precipitate the resinous matter. Camphor in spirit of wine could thus be

Table 3.1

Solvent	Dissolved substance	Intermediate
Spirit of wine	Resinous matter	Water
Gross oil	Resinous matter	Evaporation or distillation
Alkalin liquids	Bituminous bodies	Acid
Acids	Metals	Fixed or volatile alkali salts
Acids	Metals	Eau de chaux, marine salt

separated out with water, although it swam to the surface rather than sinking to the bottom. Nonetheless, the process followed the "true laws of precipitations," since it was the natural levity of camphor which raised it to the surface. (See table 3.1.)

Lemery explained the separation of bituminous bodies dissolved in alkaline liquids by acids with vivid imagery. He supposed that the alkali was like a piece of wood with a hole that pierced through its entire thickness. If one pushed through one side of the hole a solid body in the form of a spindle or some other figure that could be contained in this hole by its extremity, penetrating the hole no more than one-third or one-half its length, and then if one pushed through another side of the hole another solid body capable, by its figure and volume, of filling the entire length of the hole, the second body would push out the first body and occupy its place. Common sulphur dissolved in alkaline liquids, for example, only because of the abundant acid it contained which engaged the pores of the alkali salt. Since the acid had too big a volume to penetrate the pores of the alkali well, if one poured on this mixture other acids more liberal and more capable of traversing the entire extent of the pores, they displaced the sulphur, which was then precipitated.

The fourth class of phenomena, the precipitation of metals from their dissolution in acids by fixed and volatile alkali salts, was Louis Lemery's main interest. This was precisely the problem his father had labeled as "one of the most difficult questions to resolve well in philosophy" already in 1675. This type of precipitation consisted yet of two kinds. In the first, the metal dissolved in acid was precipitated as subtle powder by an alkali. In the second, the metallic solution turned, upon the mixture of alkali, into a thick, viscous coagulum. Lemery reasoned that in the first case the acid was not united closely to the metal and abandoned it easily upon the approach of the alkali. That is, acids were "engaged more easily and more deeply" with some metals than others. Metals such as gold and silver dissolved only in certain acids, while iron and copper dissolved in nearly all acids. Acids abandoned silver for copper, not vice versa. Copper precipitated silver from its solution in nitrous acid. All these facts indicated that acids held iron and copper more closely than other metals, making it more difficult for them to leave. In the second case, mostly with copper and iron, the acid was strongly engaged in the pores of the metal. Linked to the metal on one of its extremities, the acid united with an alkali salt on the other extremity.

The difference between these two types of metallic precipitates did not derive from the nature of acids and alkalis employed, Lemery argued, but from the "proper nature of the metal" or its "particular disposition to let go or to retain the acid."[23] The acid solutions of silver more often than not abandoned this metal for copper, not vice versa. "But why," Lemery asked, "do these same acids clothed with the parts of silver abandon this metal for copper or for an alkali salt! Why do they not conserve one & the other! *What is the force that causes them to make this exchange*! How is it made or rather how can it oblige silver to cede to copper or to an alkali salt the acids of which it was in possession."[24] Lemery offered an admittedly "rough" mechanism for such selectivity. If one pressed a stick furnished with a metallic ball at one extremity into a hole by the other extremity, the ball would receive a considerable shock when it reached the hole. If the ball was bound to the stick strongly enough to resist this shock, it would remain attached to the stick but would not enter the hole. If the ball became detached from the stick because of the shock, it would go its own way afterward. This was indeed a "faithful image," Lemery reasoned, of what happened in the case of metallic precipitations. If an acid engaged with a metallic body at one of its extremities entered into the pores of an alkali salt by its another extremity, the violent shock it received would shake off the metal that was not strongly attached to it. If the metal withstood the shock, however, it would form a compound of an acid, a metal, and an alkali salt.

Lemery's discussion of the actions of regal water reveals the changing conception of chemical composition and reaction advanced through the chemistry of metals. Chemists knew that spirit of salt dissolved gold but not silver. Spirit of niter had the opposite effect. According to the principalist conception of chemical composition and reaction, regal water should dissolve gold by its spirit of salt and silver by its spirit of niter. At least, it should dissolve first gold, then silver, as Homberg had reported in his 1706 memoir. Instead, ordinary regal water dissolved only gold. From these facts, Lemery concluded that these two acids had to be "united intimately together" in regal water, or were altered by the "reciprocal union" of their parts. In other words, regal water seemed to be a "compound" in which the two acid spirits formed an "intimate union," allowing one to be absorbed by the other. How could an acid with pointed parts absorb another acid, also with pointed parts? Were they of different sizes? Lemery supposed, as his father and Homberg had, that all acid parts had the same size and figure. They were never perfectly pure, however, and were usually embedded in other sulphurous and earthly bodies. Different concrete salts provided the acids with different effects. One could easily conceive, therefore, how one acid could absorb another. If one acid was accompanied by a grosser and spongier sulphurous matter and the other, with a more subtle sulphur, the subtle sulphur would penetrate the grosser one, joining at the same time the two acids. Lemery proceeded to prove that spirit of niter was the free and less enveloped one and that spirit of salt was clothed by a grosser sulphur, thus absorbing spirit of niter. He questioned Homberg's earlier claim that gold had smaller pores than silver. Lemery's intricate interpretive maneuvers illustrate well why the principalist approach to chemical composition and reaction failed to explain displacement reactions, particularly those of metallic salts.

Lemery followed up his general reflection on solution chemistry with a much more detailed study on the precipitates of mercury.[25] This was a subject in which Homberg had maintained a keen interest, although Lemery listed the abnormal color pattern of mercury precipitates as the main reason for his interest. When one precipitated a metal from its dissolution in various acids with an absorbent salt, the precipitate acquired a variety of colors depending on the circumstances. Silver, lead, and tin, which gave no color to their acid solutions when they were pure, mostly produced white precipitates. Gold gave its own color to the acid solution as well as to the precipitate. Copper gave blue. Iron gave different colors to different acid solutions and their corresponding precipitates. Even in the case of iron, the color of the precipitate was the same as that of the acid solution from which it was produced, regardless of the absorbent salt used for the procedure. Mercury offered a unique exception to this rule. Although it produced limpid solutions with acids, the introduction of absorbent salts gave all of a sudden a vivid color, which offered a "quite agreeable chemical spectacle." Not only did the color originate from two equally clear and limpid liquids (acid solutions and absorbent salt); the same acid solution gave different colors depending on the absorbent salt used. Furthermore, the color fainted if one introduced more acid, reappearing with the addition of more absorbent salt.

Arguing that the general theory of color taught in physics could not explain the colors of mercury precipitates, Lemery attributed them to the different quantity of the "parts of fire" contained in the precipitates. This correlation made sense because the brightness of color, ranging from red to white, depended on the strength of the alkaline salt used as an intermediate. The alkaline property of fixed salts, or their capacity to receive acids, derived from the fire of calcination that chased the acids from their pores. Containing more parts of fire than of acids, these pores became more susceptible of receiving acids. This explained why the parts of fire in fixed alkaline salts would quit them to enter mercury. When these alkali salts were placed in the solution of mercury in acids, mercury released acids which took the place of the parts of fire in fixed salts, and the parts of fire thus released took the place of acids in mercury. That is, mercury and fixed salts effected "a kind of exchange of acids & the parts of fire which take mutually the place of one and the other." Lemery's attribution of color to the parts of fire probably made it easier for the later generations of French chemists to accept Stahl's designation of the phlogiston as the principle of color.

Lemery's endless speculation on the mechanism of calcination and combustion probably rendered his work less quotable for the later generations of chemists, especially in the anti-Cartesian climate after the middle of the century. He made a crucial contribution to French chemical theory, however, by providing a philosophically coherent discourse for the operational chemistry of salts. Through his generalized discourse on fire, he detached Homberg's sulphur principle from its alchemical connotation as the active principle. His discourse on fire, coupled with Geoffroy's empirical work on the sulphur principle, provided the core of the later phlogiston theories. He identified fire as the substance that conferred inflammability and color, two main attributes of Stahl's phlogiston, and provided a detailed account of how fire became fixed in solid bodies. Since Lemery taught at the Jardin from 1730, it is not surprising that Rouelle and Macquer identified Stahl's phlogiston with the matter of fire. We should also note from his papers that red calx of mercury was a well-characterized substance for more than 50 years before it came to play a crucial role in the episode of the Chemical Revolution. Not only did Lemery clearly distinguish between the white and red precipitates in terms of their preparation; he also noted the weight change upon calcination, attributing it to the matter of fire. He conducted an extensive investigation on mercury precipitates using a variety of absorbent salts as intermediates.[26] At the least, Lemery's exhaustive discussion indicates to us the degree to which metallic precipitates held the attention of contemporary chemists. By characterizing fire as a universal solvent and displacement reactions as a model chemical system, he also laid the conceptual foundation of Geoffroy's affinity table. The analogy between fire and water as solvents made possible the conceptualization of chemical action as dissolution, later adopted by Rouelle and Guyton. In other words, Louis Lemery effected a reversal of analogy in chemical theorizing by elaborating an extensive discourse of solution chemistry. His works provided many crucial features for French chemical theory in the second half of the eighteenth century.

3.3   Differential Solubility

The increasing sophistication in solution chemistry can be measured by Louis Lemery's attempt in 1716 to account for the differential solubility of salts in water.[27] His close scrutiny of the action of water on various salts hints at a mounting concern over the exact function of water in chemical analysis. Just as chemists had been anxious to figure out the role of fire when the distillation method became their standard method of analysis, they became increasingly concerned about the involvement of water in solution analysis. Arguing that the dissolution of bodies in liquids did not belong to the ordinary case of hydrostatics, Lemery articulated a common mechanism for the precipitation process. A solvent, owing to its great agitation despite the surface calmness, reduced the dissolving body into fine dust, which it then surrounded and lifted. Should the small particles of the body reunite to form a larger particle, the solvent could no longer suspend it in the liquid and precipitated it. Despite this common mechanism, however, water seemed to show "quite curious differences" in dissolving a variety of salts, probably because of the different natures of the salts. First, some salts took more time than others to dissolve. This could be explained by the different compactness of salts which gave more or less access to the water particles. When the water could act only on the exterior of the body, the dissolution had to be slower than when it could penetrate the interstices between the particles of the body and work on the interior as well. Second, the same portion of water dissolved different quantities of salts no matter how long one allowed the process to continue. Salt of tartar and saltpeter showed the "most remarkable & most singular" differences: Saltpeter required four times as much water in volume as salt of tartar did.

As before, Lemery delved into a detailed modeling to explain such differences. He supposed that the "integrant part" of salt of tartar had a figure that did not permit a compact union. The parts had much interval between them and touched only at a few places without being able to "contract a more perfect union."[28] Owing to this weak union of its integrant parts, water could separate them more quickly and suspend them in solution in greater quantity. On the contrary, the integrant parts of saltpeter had a surface that allowed a strong union to one another and consequently had less interval between them. The smaller interstices caused water to take more time to dissolve it. Water served as a kind of "intermediate" between the particles of saltpeter to prevent them from being reunited. Without this "interposed water," the particles would quickly reunite and precipitate. Upon evaporation, therefore, saltpeter produced crystals, while salt of tartar only yielded a friable and porous powder that melted in the atmospheric air. This proved to Lemery that salt of tartar truly had a weaker "disposition" to be united. He insisted on the different dispositions of bodies as an irreducible factor in chemical action and accounted for other phenomena in these terms. The difficulty in separating certain bodies was due to the disposition of its parts to unite strongly.

Another interesting phenomenon involved the dissolution of multiple salts. Chemists knew that a solution that held a maximum amount of one kind of salt could further dissolve a different kind of salt, after which it again admitted further quantities of the first salt. Although some proposed that such successive dissolution of different salts had to do with more efficient filling of the interstices, Lemery proved otherwise with a "decisive" experiment. He showed that the volume of the water was higher after he introduced a quantity of the second salt. This indicated that the added salt occupied additional space it required independently of water. Instead of more efficient packing, therefore, Lemery proposed a different explanation for successive dissolution. The water charged with the first salt did not exhaust its "movement or force of dissolving," but continued to dissolve new parts of the same salt. The new parts could not sustain themselves in the solution and were precipitated. By the same mechanism, water was capable of dissolving the second salt. The question then became how the second salt utilized the water that was already occupied as an intermediate or vehicle of the first salt without itself uniting with the first salt. Also, how did the water dissolve further quantity of the first salt after the second salt was introduced? Lemery supposed that the integrant parts of the first salt had a "particular disposition" to join one another, but not the integrant parts of the second salt. The parts of different salts did not have the same "convenance" as the parts of the same salt. The parts of second salt could, therefore, utilize the water already employed as an intermediate to the first salt without uniting with the parts of the first salt. Since the same water particle served as an intermediate to the first and the second salt, the two salts served as mutual barrier to the parts of the same salt. That is, "since the introduction of the second salt, there is found greater interval than before between the parts of the first." This increased distance enabled further dissolution of the first salt.

Lemery further supported his theory with an experiment. He made separate solutions of saltpeter and common salt, mixed them together and evaporated the resulting solution. He obtained two distinctly different crystals, as if the solutions were separately evaporated. He thus concluded that different salts had no disposition to be united. He provided much more extensive "experimental proof" of this conclusion in his later studies, suggesting the dual functions of water as a solvent, first in suspending the dissolved bodies and second in acting as a barrier to the parts of the same salt. While all salts needed the same quantity of water for their suspension, they differed greatly in their requirement of the barrier water. The greater the "disposition to unite" of their parts, the farther they had to be kept apart, which required a greater quantity of water acting as a barrier.[29]

3.4 The Death of Five Principles

Lemery's theoretical reflection on solution chemistry was a response as much to the widespread discontent with the distillation methods as to the advancement of solution methods. In 1719, he addressed directly the limitations of "ordinary chemistry."[30] Although the problem of fire analysis had been discussed for over half a century by then, the advances made with the burning glass and with the solution method made Lemery's reflections more timely. He began his report by presenting an "imagery" or a "representation" of what was happening in the "ordinary analyses of plants and animals." He supposed two edifices having the same exterior form but different materials and different arrangements of parts. If in order to discover their materials and arrangements one destroyed these edifices, using an active and violent agent, they would be reduced to dust through the agent's force and natural vivacity of movement. In the resulting chaos, everything would be left not only confounded but also considerably altered. The dust resulting from the demolition of one edifice would be similar to the dust from the other edifice. It would not be possible to distinguish the nature and the diversity of the materials that had entered into the composition of each edifice. From this, one could only conclude that the two edifices had been built with the same materials. Fire in ordinary analysis likewise reduced the substances subjected to its action to a kind of dust either through the trouble, confusion, and derangement caused by its swift action or by the new parts it carried and introduced into the mixt. The substances that inhabited the mixt naturally formed new compounds. For this reason, and because of certain disguises that fire imparted on the different parts of plants and animals, two plants with opposite natural virtues, one quite salutary and the other poisonous, seemed to resemble each other in their distillation products. Aside from their qualities, one would be thus inclined to believe that the two plants were the same.

Lemery's reflection on fire analysis had direct implication for the notion of principles. He did not believe that principles, properly called, would be subject to alteration by fire. By a circular reasoning, the substances that experienced alteration did not deserve the name "principles." In other words, he believed that there existed proper, unalterable principles, but that fire changed their arrangements in the mixts. Taking the futility of distillational analysis as a "certain & incontestable" fact learned through experience, Lemery sought to devise a more exact and convenient mode of analysis that would lead to a more certain knowledge of composition. He assumed that the "natural state" of salts in plants and animals was in the form of middle salts. Many animal matters contained a large quantity of sal ammoniac, a natural salt made of an acid and a volatile alkali salt, while plants contained saltpeter in which an acid was arrested by a fixed alkali salt. From the character, the state, and the "natural composition" of these salts, he deduced what became of them when they passed through fire. For a long time, chemists judged from distillation analysis that animal matters contained only volatile alkali salt and no acid. Lemery believed that the most common natural salt in animal matters was in the form of sal ammoniac. Although it was composed of two volatile components, they were separated during the distillation process, which yielded only a volatile alkali salt. He had to explain what became of the acid part, how it was separated from the volatile alkali salt, and why they were not elevated together. When sal ammoniac was found alone, fire could elevate it as a whole. When it was in the mixt, however, it was "intimately united" to the earthly part of the mixt which hindered the action of fire. Fire thus removed the most volatile part, volatile alkali salt, leaving the acid to be even more deeply

engaged with the earthly part. The ordinary analysis of animal substances thus led to two erroneous conclusions: that there was no acid in
animal matters and that volatile salt existed as such in animal matters.
For the saltpeter contained in vegetable matters, ordinary analysis
yielded a contrary result: Fire removed the acid first, leaving the fixed
alkali behind which, exposed to fire for a long time, suffered significant
alterations. The acids that emerged from the distillation analysis of plants
differed in their volatility and responded differently to fire.[31]

In reviewing Lemery's extensive work, Fontenelle praised his skill, long
labor, and great number of reflections. Lemery had overcome a great difficulty of the subject with "decisive experiments" and offered a "general
theory of the volatility of urinous salts."[32] Lemery's contribution held
deeper meaning for chemical practitioners. Aside from the harsh sentence
on distillation method, his work offered a systematic reflection on the
nature of saltpeter and sal ammoniac. Both were middle salts formed of
an acid and an alkali. The first was united to a fixed alkali, the second
to a volatile alkali. Utilizing this characterization of middle salts, an
assumption derived from solution chemistry, Lemery explained all the
variations one experienced in the distillation analysis of vegetables and
animals. In other words, he extended an understanding of the composition of salts derived from solution chemistry to the salts prepared by the
distillation method. Such a move constituted an important reversal in the
direction of analogy: Nicolas Lemery had sought to explain the behavior of acids and alkalis in terms of the way they were prepared by the
distillation method. The tables had been turned.

Lemery's work was the last step in the transition from the distillation
to the solution methods as the proper analytic method in elucidating
the theoretical composition of bodies. Deeply disappointed by the inconclusive results of the decades-long investigation of plant composition,
French academicians explored new analytic methods. Homberg's
burning glass promised to reveal directly the involvement of fire in chemical operations, but the instrument proved to be too exclusive in clien-
tele and too limited in its application. Solution methods, tested and
expanded over the decades, proved more realistic and more productive
in improving the traditional methods of chemical analysis. At every
step, however, we find these chemists musing upon the implications of

employing new analytic methods. They sought to match the conventional definitions of salts drawn from the distillation method to the new compounds of sulphur from burning glass chemistry as well as to the acids and alkalis which dominated the solution chemistry. Consolidating the products of different analytic methods into a coherent list of stable chemical substances was not a straightforward matter. These efforts at correlating various definitions of salts from different analytic practices were necessary to stabilize middle salts as the objects of theoretical inquiry. The corpuscular imagery and language offered some help in making connections between the products of different analytic methods, but they were necessarily speculative and beyond empirical confirmation. However, as time went on it became increasingly clear that distillation could no longer reign as the only means of discovering the composition of bodies. More to the point, the search for the "principles of mixts" via distillation method failed to deliver on its promise of theoretical understanding and practical utility.

3.5   The *Table des Rapports*

Louis Lemery's lengthy works on the differential actions of metallic precipitates indicate that the problem, metallurgical in origin, acquired a theoretical urgency within the Académie. It received a radically different treatment, however, in the hands of Etienne-François Geoffroy (1672–1731). Born into a prosperous Parisian pharmaceutical dynasty, Geoffroy was groomed from early on for the family business. His father, Matthieu-François, cultivated many political and scientific connections.[33] Leading scientists such as Cassini, Homberg, and Du Verny attended regular conférences at Geoffroy's household. Etienne-François also completed a year's apprenticeship with the Montpellier apothecary Pierre Sanche before he became a maître apothicaire in Paris in 1694. Accompanying French ambassador Comte de Tallard to London in 1698, he befriended Hans Sloan and was elected a Fellow of the Royal Society.[34] On his return, Geoffroy entered the Academy as Homberg's élève, no doubt thanks to the connections made through his father's conférences. Afterward, however, he chose to become a physician and gave up the rights to his father's business to his younger brother, Claude-Joseph.

Figure 3.1
Etienne-François Geoffroy. Courtesy of Académie des Sciences archive.

Although this move resulted in a significant financial loss, Etienne-François had a comfortable income with which to pursue the life of a chemical teacher and a physician. Graduating from the Faculté in 1702, he began to lecture at the Jardin in 1707 and at the Collège in 1712.[35] Geoffroy's motivation for and commitments to chemistry differed considerably from Louis Lemery's. He had a thorough education in contemporary pharmaceutical practice, including a year of apprenticeship at Montpellier. His dedication to the apothecary trade can be discerned from his most significant publication, *Tractatus de materia medica*.[36] No doubt he utilized this knowledge in his practice of medicine. His advice was sought in all parts of France.[37]

If Geoffroy appreciated the importance of solution chemistry and shared the concern over its theoretical foundation, he did not care much for the corpuscular ontology his colleagues employed for the purpose. It was only after the deaths of Nicolas Lemery (June 19, 1715) and Homberg (September 24, 1715), however, that Geoffroy began to deliver his lectures on affinities.[38] It took him another three years to present a written memoir to the Academy, accompanied by the *Table des differents rapports observés entre differentes substances*.[39] Instead of speculating on the corpuscular mechanism of metallic displacement reactions, Geoffroy summarized these reactions in a tabular form, which allowed easy access to the wisdom accumulated through chemical practice. By visualizing the composition of middle salts as a consequence of the affinities of acids and alkalis, the table also generalized the dualistic composition derived from the practice of solution chemistry. Geoffroy presented the affinity approach to composition much more effectively with a table than Louis Lemery ever could with his detailed corpuscular imagery. By doing so, he stabilized chemical affinity as a more comprehensive theory domain than composition.

Geoffroy began his memoire with a simple statement that chemical substances exhibited certain preferences in reactions[40]: "One observes in chemistry," he wrote, "certain rapports between different bodies which allow them to unite easily with one another. These rapports have their degrees and their laws. One observes their different degrees in that among several confounded matters with some disposition to unite together, one unites constantly with a certain other in preference to all others."[41] Although the cause of such a "disposition to unite" or convenance had been a subject of extensive speculation by Louis Lemery, Geoffroy did not mention his colleague's name or contribution to the subject. Instead, he put forward the "laws of these rapports": If among the substances that had a "disposition to unite together" two were found united (A, B) and subsequently mixed with another (C), either the third substance (C) would join one of the substances (A) and shake the other (B) loose or it would not join either of the substances originally in combination (A or B). If C joined A, one could conclude with sufficient probability that C had "more rapport of union or disposition to unite" with A than B did. One could thus predict the actions among these three substances:

Every time when two substances that have some disposition to join each other are found united, if there follows a third which has more rapport with one of the two, one is united with it by letting go of the other.[42]

Geoffroy cautioned that he was not yet able to generalize the above proposition, "not having been able to examine all the possible combinations to assure myself that one will find nothing to the contrary." Nevertheless, such a proposition would find a great scope of application in chemistry where one encountered "at every step" the effects of this rapport. It was the "key" to all the hidden movements that ensued from the mixture of bodies.

The "order" of this rapport was little known, however. Geoffroy thus gathered the experiments and observations of other chemists, as well as his own, to prepare a table in which one could see "at a glance" the different rapports of the "principal matters one is accustomed to work on in chemistry." Such an arrangement would prove quite useful by allowing the beginners to form a "just idea" of the rapports of substances in a short period of time, while providing more experienced chemists with a "method" to decipher complicated operations:

With this table, those who begin to learn chemistry will form in a short time a just idea of the rapports different substances have with one another & chemists will find an easy method for discovering what happens in several of their operations difficult to unravel & what has to result from the mixtures they make from different mixed bodies.[43]

Geoffroy, who had been teaching at the Jardin since 1707, substituting for G.-C. Fagon, was appointed to the post in 1712. He also became professor of pharmacy and medicine at the Collège de France in the same year. He was clearly aware of the pedagogical value of his table. As he later stated in his defense of the table, it would require an "entire course of chemistry" to present the "detail of the operations that serve as the foundation of the table."[44]

The principle for the construction of the table (shown here as figure 3.2) was simple. The top row of the table comprised different substances often employed in chemistry, numbering sixteen in all. Below them were arranged by different columns a number of substances that reacted with the top substance in the decreasing order of rapports. For example, the first column was headed by acid spirits, which was followed by fixed alkali salt, volatile alkali salt, absorbent earth, and metallic substances. That column represented the chemical experience that fixed alkali salt reacted most favorably with acid spirits and would displace all the substances listed below it from their existing combination with acid spirits. Likewise, volatile alkali salt would displace absorbent earth and metallic substances from their combinations with acid spirits. It would not, however, displace fixed alkali from its combination with an acid spirit.

Not as transparent, in part because Geoffroy did not state it explicitly, was the principle of inclusion in the table. How did he choose the substances listed at the top row of the table? F. L. Holmes has pointed out that the left half of the table, dominated by acids and alkalis, reflected Homberg's understanding of middle salts. That is, the substances listed in the top row would produce a variety of middle salts with the substances listed below each of them.[45] Whereas the first column was headed by acid spirits, columns 2–4 were headed by three mineral acids—the acid of marine salt, nitrous acid, and vitriolic acid—to show that metallic substances did not have an equal "convenance" with these acids. Columns 5–8 showed the reactions of a variety of alkalis—absorbent earth, fixed alkali salt, volatile alkali salt, metallic substances—with individual mineral acids. The remaining eight columns were headed by sulphur, mercury, five metals and semi-metals (lead, copper, silver, iron, regulus of antimony), and water. Ursula Klein has noted that this right half of the table derived largely from the age-old metallurgical practices which dealt with metallic sulphides (column 9), amalgams (column 10), and alloys (columns 11–15). Metallic sulphides were used in the sepa-

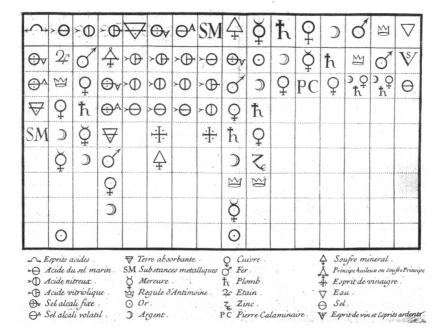

Figure 3.2
Geoffroy's Table des rapports, 1718. From Geoffroy, "Des différentés Rapports observés en Chymie entre différentes substances," *Mémoires*, 1718, 256–269.

ration of gold-silver alloys; Amalgams were formed in the process of extracting gold and silver from their ores.[46]

The table had two intended uses: to "discover" what went on in the mixtures of several bodies and to "predict" what had to result from them. Geoffroy offered with his table a practical solution to Fontenelle's dream of bringing chemistry closer to modern geometry. The language of "rapport" deliberately invoked the mathematical meaning of "ratio," in contrast to that of "affinity," which carried somewhat ambiguous connotations of kinship and analogy. Geoffroy illustrated the predictive function of the table with the preparation of corrosive sublimate (a problem earlier dealt with by Nicolas Lemery[47]), contending that it was a kind of ordinary operation whose "theory" was little known. The ordinary method of preparation involved three salts. Vitriol calcined to red, marine salt, and mercurial nitrous salt formed by the dissolution of mercury in spirit of niter. As soon as these three substances were mixed exactly, one sensed the odor of the spirit of niter, which was elevated in yellow vapor. If one distilled this mixture in a retort, an acid spirit emerged at moderate fire, which was mostly spirit of niter mixed with some spirit of salt. Subsequently, with stronger fire, a white crystalline saline mass rose, leaving at the bottom a reddish mass from which one could obtain a white salt and a red metallic earth. As Geoffroy explained this operation, vitriol was a middle salt composed of vitriolic acid and iron, marine salt was a middle salt of the acid of marine salt and an absorbent earth, and mercurial nitrous salt was a middle salt of spirit of niter and mercury:

vitriol = vitriolic acid + iron
marine salt = acid of marine salt + absorbent earth
mercurial nitrous salt = spirit of niter + mercury.

One could predict the outcome of the mixing by considering the rapports of the six substances that composed the three mixts. There existed three acids (acid of marine salt, vitriolic acid, and nitrous acid) and three bases (absorbent earth and two metallic substances). Geoffroy read from the first column of the affinity table that acids in general had more rapport with absorbent earth than with iron or mercury and from the fifth column that vitriolic acid in particular had more rapport with absorbent earth than other acids. The first product should then be a compound of vitriolic acid and absorbent earth. The acid of marine salt, thus freed, had a greater "convenance" with metals than spirit of niter and "force[d] this acid to abandon mercury." They formed a mercurial saline concretion, sublimated mercury, which was sufficiently volatile and sublimated above the vessel. The spirit of niter, freed in turn, found nothing else to take up and evaporated as yellowish vapor, leaving as a residue a metallic earth containing iron or Saffron of Mars. The nitrous acid was capable of taking up iron, but was driven out by fire. Though it seemed completely superfluous in the net equation, it served several practical purposes.

Geoffroy insisted that the same "theory" could explain another process of making corrosive sublimate by a solution method: One poured the solution of marine salt in water over the solution of quicksilver in spirit of niter. The white precipitate resulting from the mixture was sublimated to a white saline compact mass which was corrosive sublimate.

The clear liquid yielded saltpeter upon drying. The juxtaposition of these two methods—distillation vs. precipitation[48]—relayed Geoffroy's hope that the laws of rapports could mediate between the different chemical procedures to shape a uniform terrain of chemical theory. By going through a variety of processes that produced corrosive sublimates, Geoffroy illustrated that the same order of rapports could explain all the varieties of chemical operations in a consistent manner. The "theory" of the process was the same, no matter what the variations in practice were. Through the example of corrosive sublimates, Geoffroy thus presented concisely and powerfully the descriptive, predictive, and explanatory functions of his table of rapports.

Geoffroy also left open the possibility that the table could be refined with further accumulation of facts. He admitted that some of his analyses were not "perfectly exact and precise," but that they remained within a tolerable limit of precision. That is, the table was "exact and precise" only to the level of the available techniques of chemical analysis. In Fontenelle's words: "The more chemistry becomes perfect, the more the table of M. Geoffroy will become perfect as well, either by a great number of substances that it contains or by the arrangement and the exactitude of rapports."[49] Geoffroy's table offered, then, a synoptic representation that organized the current sum of chemical experience but was flexible enough to accommodate further experiments. This provisional aspect of the table made it a successful theory for the laboratory culture of chemistry: It provided a framework within which chemists could conduct further investigations with the goal of perfecting it. In addition to its functions of summary description, prediction, and explanation, the table allowed dissent within a comprehensive framework. Already in 1820, we find Geoffroy defending his table in its particulars.[50] All three objections—the first two from his brother Claude-Joseph and the third from Caspar Neumann, a Prussian chemist—took issue with the prescriptive function of the table. As a novelty, Fontenelle surmised, Geoffroy's table of "different rapports or different affinities" attracted criticism, and it required much clarification because of the "natural difficulty of the subject."[51]

The first objection took issue with the first column of Geoffroy's table, which stipulated that absorbent earths had less rapport with acids than alkalis. Lime, an absorbent earth, routinely combined with the acid of sal ammoniac, however, displacing urinous alkali salt in the process. Geoffroy replied that this observation did not destroy the "order of rapports" established in the table because lime did not share the characteristics of an insipid and porous absorbent earth. Rather, it had much convenance with fixed alkali salt, exhibiting many of its characteristics. Fontenelle argued that the objection would be "plainly resolved" if lime contained a fixed alkali salt, but Geoffroy could not resolve the issue decidedly.

The second objection also concerned the first column of the table. Claude-Joseph Geoffroy argued that metallic substances, having less rapport with acids than any of the alkalis, should not displace an alkali from the "salted salt or compound salt" [sels salés ou composés] of acid and alkali such as sal ammoniac. Nevertheless, one often observed that in mixing sal ammoniac with iron filings, a little of urinous spirit was elevated at the beginning of the operation. The process seemed to indicate that metallic substance could act on the salt to absorb its acid, releasing its alkali. Geoffroy responded that, although the observation was true, the volatile alkali was produced in a small quantity and only under certain circumstances. He thus reasoned that the volatile salt came from the impurities contained in the iron used in the experiment. Iron filings, when exposed to the humidity of air, were reduced to iron rust, which gave a volatile salt at a moderate heat. If one used new iron filings and sublimated the mixture, not allowing enough time for the filings to turn into rust, no volatile salt arose.

The third objection came from Caspar Neuman (1683–1737), a royal apothecary of the Prussian king, who had recently visited Paris.[52] When one distilled a mixture of three parts of minium (lead calx) and one part of sal ammoniac, one drew a good quantity of urinous spirit, which was quite volatile, penetrating, and caustic. This seemed to indicate that metallic substances sometimes had a greater rapport with acids than with volatile alkali salts. Geoffroy concurred that not only minium but also other metallic calces produced the same effect, absorbing the marine salt contained in sal ammoniac and releasing the urinous alkali salt. They did so, however, by the fixed alkali salt they contained, not by the metal. He asserted that these experiments concluded nothing against the order of rapports. In fact, minium or lead calx was a quite different kind of compound from lead, having lost the "oily principle" that constituted the metal.

One could agree with Fontenelle that Geoffroy failed to give a definitive answer to the objections made against his table, leaving much uncertainty. Geoffroy's answers give us a valuable insight, however, into how the table was utilized as a principle of classification which sharpened the boundaries of chemical species. Addressing the first objection, for example, he had to distinguish clearly between lime and quicklime, identifying one as a substance analogous to fixed alkali salt and the other as an absorbent earth. Without the table, these two cognate substances had been lumped together as absorbent earths. In responding to the second objection, Geoffroy first defined "metallic substances" as "fixed metal [fix metaux] separated from their mines, & in their state of purity under their form of metal, as well as semi-metals such as antimony, bismuth, well purified zinc, & other mineral substances which partake of metal." Other "compositions or metallic preparations" that changed the nature of metal (salts, vitriols, calces, etc.) did not qualify for the same designation. Geoffroy also had to distinguish between the freshly made iron filings and rust. A similar strategy had to be introduced, differentiating metals from their calces, to address the third objection. This kind of debate over how one should identify and classify a distinct chemical substance would have helped to refine the analytic techniques and ultimately to further revise the table.[53]

3.6 The Cognitive Status of the Affinity Table

The affinity table has received much attention from the historians of chemistry as an orderly representation of the seemingly chaotic activity of pre-Lavoisian chemistry. Nevertheless, we are not yet certain of its cognitive status. Was it merely a collection of facts, or did it provide a comprehensive framework, a conceptual structure, or a theoretical system? This musing over the function and the cognitive status of affinity tables reveals the inadequacy of our historiographical distinction between scientific theory and fact. Initially, historians of chemistry characterized the affinity table as a mere collection of facts devoid of theoretical aspirations. In his pioneering work, Alistair Duncan argued that this was exactly what was wanted. Despite its utility in predicting chemical actions, the affinity table was meant only "to provide a convenient summary of a large body of experimental results and not to assume any theories or principles or causes a priori."[54] In a similar vein, Smeaton has refuted I. Bernard Cohen's claim that the table was inspired by Newton.[55] Lissa Roberts has also stated that the table utilized the "rhetoric of theoretical neutrality."[56] Although Duncan and Roberts are quite sympathetic to eighteenth-century chemists' non-theoretical project, one cannot but notice that their (and our common-sense) notion of a theory is quite different from Geoffroy's. For Geoffroy, a theory meant a reasonable explanation of a group of chemical operations that could be applied across the boundaries of different analytic methods. His notion of a theory did not relate back to a system of philosophy that could provide a set of causes and thereby claim a correspondence to nature. He intentionally muted Louis Lemery's corpuscular reasoning on the cause of chemical reactivity. Insofar as eighteenth-century chemistry did not share the aims and methods of natural philosophy, historians have judged that it was a non-theoretical, merely empirical, and theory-neutral "art" rather than a science. The affinity table did not "represent" nature in a philosophically coherent way and therefore was not a theoretical system.

The discrepancy between our notion of a theory and the historical natives' notion invites us to think about what "theory" should mean in what Evelyn Fox Keller calls the "epistemological culture" of chemistry.[57] Keller cites the two definitions of "theory" given in *Webster's Dictionary*: "the analysis of a set of facts in their relation to one another" and "the general or abstract principles of a body of fact." She argues that the first definition is much more pertinent to such "practical" sciences as medicine and molecular biology. As Keller states in regard to molecular biology, "both the interpretation of experimental data and the design of new experiments depend on extensive and sophisticated theoretical analysis of the possible relationship which can be brought into consistency (or inconsistency) with the data at hand." This is exactly the sense in which Geoffroy used "theory."

Recent studies have brought out more forcefully and cogently the theoretical and the philosophical relevance of the affinity table. F. L. Holmes has identified salts as the main subject of theoretical investigation in eighteenth-century chemistry and placed the affinity table in the midst of it. The affinity table, according to Holmes, was "a nodal point in the continuing evolution of a pragmatic chemistry of operations oriented around the concept of middle salts." He points out, therefore, that the left half of the table listed three mineral acids and four different kinds of alkalis that reacted with them to form middle salts. He argues that "Homberg's classification of middle salts and Geoffroy's table provided conceptual frameworks to organize and clarify knowledge of chemical substances and operations that had been accumulating over the preceding century."[58] I would like to strengthen his interpretation by adding that the classification "salted salts" had been in use long before Homberg reified them as "middle salts." Geoffroy's table was also preceded by Louis Lemery's reflections on metallic displacements, which introduced the language of "disposition" and "convenance" to chemical theory. These facts imply that a broader-based ongoing investigation of salts in pharmaceutical practice provided the background for Louis Lemery's and Geoffroy's concern with the actions of metals and salts.

Ursula Klein has argued in a more philosophical vein that the modern "concept of chemical compound" provided the "underlying framework for Geoffroy's table." The concept involves three facets in Klein's analysis: reference to pure chemical substances, their conservation throughout different transformation, and the presumption of lawful connections among them exhibited in the operations of analysis and synthesis. The chemical compound, according to Klein, is thus a "conceptual structure comprising the notions of chemical composition, chemical analysis and chemical synthesis, as well as chemical affinity."[59] Klein's concept of chemical compound is not entirely new: Kuhn, in denying Boyle's modernity, argued that modern chemical atomism included commitments to "a belief in the endurance of elements in their compounds" and the "recognition of analysis and synthesis as fundamental tools of the working chemist."[60] In other words, Klein has traced the origin of modern chemical atomism to Geoffroy's affinity table. Although not all features of modern chemical atomism were clearly present in Geoffroy's table, Klein's thesis in a modified, historicized version can help us understand this important period.

A significant yet implicit assumption Geoffroy used in choosing the "principal matters" for his table was that they were stable substances with their own chemical identities. Because chemists could recover these substances after a series of transformations, they assumed that these substances retained their identities throughout their reactions. Needless to mention, such a notion of stability derived from chemists' manipulative capacity in following through these transformations with various means of identifying the substances. As Klein points out, chemists' experience with the "reversible" actions of acids, alkalis, and metals was crucial in this regard. The centrality of displacement reactions, as opposed to destructive decompositions, makes sense in this context. Despite Geoffroy's clear endorsement of displacement reactions as the means of determining the order of rapports, however, the actual practice involved the use of distillation methods on many occasions. Geoffroy took for granted the distillation of three components in making corrosive sublimates, as we have seen. In other words, even the operations that required the use of fire or distillation methods were conceptualized as displacement reactions. The right half of the table did not even list substances that formed middle salts through displacement reactions, but it represented the preference of each metal for making alloys. Klein has thus concluded that the requirement of displacement reactions did not apply to this part of the table.[61] It is possible to interpret the table in a consistent manner, however, if we see the right half as the actions of metals in the solution of fire (an agent traditionally known for destructive rather than dissociative power) and the left half as containing their actions in aqueous solution. Louis Lemery had taken significant trouble to reconceptualize fire as a solvent, as we have seen. Under this interpretation, one could apply Geoffroy's premise of displacement reactions as the unique method of determining affinities to both sides of the table. In the decades after the publication of Geoffroy's table, chemists strengthened the conceptualization of fire as a solvent or an instrument rather than as a property-bearing principle. It was Louis Lemery who provided this necessary reconceptualization of fire. His theoretical reflections on solution chemistry contributed much to establishing affinity as a theory domain more comprehensive than that of composition, although Geoffroy's table visualized these conceptions more effectively. Their efforts allowed chemists to characterize composition as a result of affinities, rather than encompassing affinities as a facet of composition. Their combined works eclipsed the paradigm of seeking the "principles of mixts" via distillation to institute the chemistry of "compounds" made of stable and interchangeable substances.

Geoffroy's table reorganized the existing fund of chemical knowledge, then, by centralizing a particular category of operations: metals in acid-alkali solution. The table embodied an operational classification that divided the majority of chemical substances into three classes: acids, alkalis or alkaline substances, and middle salts. In presenting a table organized implicitly around the acid-alkali reactions of metals, Geoffroy promoted successfully a central exemplar that could organize the rest of chemists' experience. In short, Geoffroy's table extended the basic premise of solution chemistry to the conventional practice of distillation: that chemistry analyzed bodies into their stable components that exchanged partners through displacement reactions. The model bodies for such behavior were middle salts. By visualizing middle salts as combinations of acids and alkalis, the table accomplished more than prescribing the lawful behavior of chemical bodies. It also stated an ideal that chemical composition was determined by rapports or affinities. In other words, Geoffroy's table articulated through visualization an affinity approach to chemical composition. In the principalist tradition, the principles of mixts were supposed to determine their behavior. According to this logic, composition should determine affinity, although affinity was not a central question in the distillational analysis of composition. In the affinity approach, affinity determined composition. Needless to say, these two opposite approaches called for different notions of chemical composition—philosophical vs. operational. These two notions remained in tension throughout the century.

To appreciate the theoretical relevance of the affinity table, or to characterize it as a theoretical system uniquely suited to the epistemological culture of chemistry, we must pay attention to the interlocking features of the table. The affinity table embodied in Geoffroy's vision a collection of "theories" that explained a variety of concrete chemical operations in a consistent manner and charted a uniform terrain of chemical theory. Chemists' ability to conjure up these theories from the table depended on the lawful behavior of the rapports of the substances listed there. Their rapports were in turn determined by displacement reactions in which these substances retained their identities. On the one hand, then, chemists' manipulative capacities in identifying the substances and in effecting their transformations provided the necessary precondition for the determination of rapports. As we have seen, many displacement reactions on the right half of the table could not be determined with any "exactitude or precision." On the other hand, such shortcomings in manipulative capacities did not stop Geoffroy from generalizing displacement reactions as the model for all other determinations of rapports. It was precisely this move for generalization beyond the available manipulative capacities that enabled the affinity table to function as a theoretical system in tune with Fontenelle's vision of predictive science. In the decades after the publication of the table, there developed a widespread reconceptualization of fire as an instrument or as a solvent. This validated the affinity table as a universally applicable theoretical system. At the same time, middle salts were construed as the natural state of existence for the salts in natural bodies. The composition of these salts in their natural state could then be elucidated by determining their rapports. These related investigations strengthened the theoretical agenda of the affinity table.

3.7 Phlogiston

Geoffroy's table included in the fourth column the "oily or sulphur principle," which was changed to "phlogiston" in subsequent affinity tables. This entry of course invites historians' curiosity because of the notoriety phlogiston acquired during the Chemical Revolution.[62] As we saw in chapter 2, Geoffroy conducted a series of transmutation experiments studying the composition of sulphur and metals. He included the sulphur principle in the table on the basis of these experiments. Though he could not isolate, touch, taste, or smell the sulphur principle, it still possessed a well-defined identity through its actions. The inclusion of the sulphur principle in the table thus reveals better than the choice of any other substance the operational, rather than the ontological, criteria of stability and identity Geoffroy used in the selection of his "principal matters."

The later identification of this principle with phlogiston, sanctioned by Geoffroy himself, also derived from their procedural correspondence.

To understand the staying power of phlogiston in chemical practice throughout the eighteenth century, we must unpack its operational identity.[63] In his 1720 defense of the affinity table, Geoffroy inserted a detailed discussion of a problem earlier proposed by Stahl in a letter to Neumann: "When one has saturated & crystallized vitriolic acid with salt of tartar, find a means of separating this acid from the fixed salt in a moment of time & in the palm of a hand."[64] It seemed like a rather difficult puzzle, Geoffroy mused, since salt of tartar was the strongest alkali and should have "the most intimate rapport" with vitriolic acid, the strongest acid. It was unlikely that any other acid or alkali could break their union, as Stahl himself had acknowledged in his *Traité De Zymotechnia* or *De formentatione*.[65] Geoffroy thought that the solution to this problem would prove why the "oily or sulphurous principle" should occupy the place right below vitriolic acid in the column headed by it. In other words, Geoffroy identified his "oily or sulphurous principle" with Stahl's phlogiston on the basis of their operational characteristics.

Geoffroy proposed two solutions to Stahl's problem. Although Stahl had suggested them in several places in his works, they were not easily accessible to those who had not studied the writings of this "habile Chimiste" with care. Geoffroy had given the "idea" of one solution in the table of rapports by placing the "oily principle, or as M. Stahl calls it, the phlogistic principle, the inflammable principle, or the principle of inflammability" above the alkali salt. Phlogiston was the only substance that could displace salt of tartar from its union with vitriolic acid. In one process, Geoffroy melted vitriolated tartar (salt of tartar saturated by vitriolic acid) on a crucible with a little salt of tartar to facilitate the fusion and threw in some inflammable matter such as powder of wood, crushed charcoal, or even some oily or resinous fatty matter. The mixture was consumed in a great flame, producing copious fumes. When the flame subsided to a subtle, bluish flame and a penetrating acid odor began to circulate, the fire was withdrawn. One found on the crucible a reddish saline sulphurous matter similar to hepar sulphuris, a mixture of salt of tartar and sulphur melted together. By dissolving this mixture in a sufficient quantity of water, filtering it, and pouring it over some distilled vinegar or other weak acid spirit, one obtained sulphur. Geoffroy interpreted that in this operation the oily principle contained in the inflammable matter, rarefied and placed in movement by the element of fire, was insinuated between the two salts of vitriolated tartar. Since it had more rapport with vitriolic acid than with alkali salt, it was united to the former quite tightly and consequently was detached from the latter. There resulted from the union of the oily principle and vitriolic acid a "compound" that was common mineral sulphur. The salt of tartar, abandoning vitriolic acid, remained united with this new sulphur in the form of hepar sulphuris. Although the oily principle had a little more "affinity" with vitriolic acid than with the fixed alkali salt, it conserved a rather tight rapport with the alkali salt. Thus, together these three substances formed hepar sulphuris. The sulphur that had by itself no "disposition to unite with the parts of water" dissolved quite easily with the help of the alkali salt since this salt had a nearly equal "affinity" with water and oil and therefore could unite quite easily with one without abandoning the other. When one dissolved the hepar sulphuris in water, the particles of the salt of tartar, although separated from vitriolic acid, remained united to the "bituminous molecules of sulphur." To facilitate this separation, one had to pour some acid on this liquid. Since the alkali salt had a "much more considerable rapport" with acids than with bituminous parts, it joined with the former and abandoned the latter:

Vitriolated tartar   + inflammable matter

(vitriolic acid + salt of tartar)   (the oily principle + · · ·)

$\longrightarrow$ hepar sulphuris

(salt of tartar + sulphur)

[= vitriolic acid + the oily principle]

$\xrightarrow{\text{acid}}$ (salt of tartar + acid) + sulphur.

Stahl had stipulated a number of conditions: The process had to be a simple, well-known one made in a quite exact manner without heat or fire. It had to be a useful process that could produce pure vitriolic acid. In order to fulfill all these conditions, Geoffroy suggested placing some hepar sulphuris resolved in liquid on the palm and pouring some drops of distilled vinegar, which made an instant precipitation of sulphur. This last step produced vitriolic acid that was quite pure because it was now released from all the metallic or mineral parts mixed in the oil or in the spirit of vitriol distilled in the ordinary manner.

In another process, also discussed by Stahl, one soaked some linen in a strong wash of alkali salts and then exposed them to the vapor of inflamed sulphur. The acid vapor of sulphur, incorporated in the alkali salt, covered the linen with a saline powder. One placed this linen in a sufficient quantity of water to dissolve the salt and evaporated the humidity at mild heat until the salt crystallized in the form of very fine tassels or needles, which was vitriolated tartar. The particles of vitriolic acid in this operation were extremely rarefied by "the element of fire, or as M. Stahl pretends it, by the inflammable principle," whereas the acid in the ordinary preparation of vitriolated tartar was extremely concentrated. One could verify that this salt was indeed ordinary vitriolated tartar, except for the rarefaction of its acid, from a singular remark made by Stahl: If, after having dissolved this salt in water, one holds the dissolution for several months in a glass bottle filled no more than to one-third, there would form little by little through slow evaporation of water the crystals of vitriolated tartar on the surface. They would precipitate as "the matter of fire or the principle of inflammability" dissipated into air, permitting the acid particles to approach one another. In order to separate the acid from this vitriolated tartar, one had to throw it into oil or spirit of vitriol, spirit of niter, or spirit of salt. These acids, which had a stronger affinity with the alkali salt than with rarefied vitriolic acid, attached themselves quite easily to salt of tartar and released the volatile acid. Being liberated, this volatile acid dissipated as a vapor similar to the vapor exhaled from burning sulphur. In fact, by throwing this new vitriolated tartar in some acid spirits one could sense instantly a strong acid odor of sulphur. One could even withdraw volatile vitriolic acid spirit by mixing vitriolated tartar with oil of vitriol in a vessel at moderate fire:

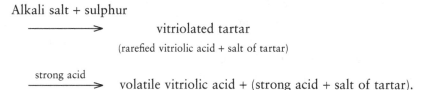

Alkali salt + sulphur

———————⟶   vitriolated tartar

(rarefied vitriolic acid + salt of tartar)

strong acid
———————⟶   volatile vitriolic acid + (strong acid + salt of tartar).

In order to fulfill all the conditions stipulated by Stahl, one could make this operation in the palm of one's hand by dissolving vitriolated tartar in water, taking the solution in one hand, and pouring over it a few drops of spirit of vitriol. The volatile vitriolic acid was separated and elevated into the air; the salt of tartar remained in solution, joined to fixed vitriolic acid.

Geoffroy identified his "oily or sulphurous principle" with Stahl's "phlogiston" or "principle of inflammability," then, by following every minute detail of the procedure. In the operations Geoffroy cited, "the oily principle" referred to a substance contained in the inflammable matters such as wood or charcoal that had the strongest affinity with vitriolic acid. Mineral sulphur was a "compound" of this principle and vitriolic acid, which accounted for its inflammability. If we go back further to his investigation of metals with the burning glass, Geoffroy's oily principle was the "sulphur principle" that regenerated metallic calces. Although he could not isolate it in concrete form, he could attribute a distinct chemical identity to this substance because it retained its peculiar property of inflammability and its affinity for vitriolic acid throughout various transformations. At the middle of the century, these two properties defined the empirical core of French phlogiston theories. By identifying his "sulphur principle" or "oily principle" with Stahl's "phlogiston" or the "matter of fire," however, Geoffroy made it possible for others to correlate it with Homberg's and Louis Lemery's "sulphur principle" or the "matter of fire or of light." Therefore, later generations of French chemists inherited Geoffroy's empirical confirmation of Stahl's phlogiston based on the reactions of sulphur and vitriolic acid along with Homberg's more general characterization of the "sulphur principle" as the matter of fire and light and Louis Lemery's detailed speculation as to exactly how this matter of fire might become fixed in bodies. As a consequence, phlogiston acquired a long, checkered history in France. By the middle of the century, it was identified ontologically as the fixed matter of fire and operationally as a substance that shared the transformations of Stahl's second earth.

Contrary to the common historiographical assumption that French Stahlism began with Rouelle's lectures, it was Geoffroy who introduced Stahl's works into France. He held Stahl in high esteem, and he studied individual operations contained in Stahl's books carefully, making sense of them in terms of the affinities between acids and alkalis. In turn, Stahl seems to have followed the French development, although he invariably acknowledged Becher and Kunckel as his sources. In his treatise on salt, published in 1723, Stahl included an extensive discussion of the "different degrees of force of acids." He followed the order of affinities in the first column of Geoffroy's table to explain the "different degrees of the activity of acids on different substances they dissolve" with concrete examples.[66] Since Geoffroy was the leading chemist of the Academy by this time, it would make sense if other chemists followed his suit and mined Stahl's treatises for curious and useful operations. This assessment is borne out in the works of Boulduc, Duhamel, and Grosse during the next two decades. Gilles Boulduc, in his sophisticated analysis of mineral waters, cited Stahl's *Dissertation de Acidulis & Thermis, Specimen Beccherianum*, and *Traité des Sels* as background to his work. Boulduc's 1726 reference to the *Traité des sels*, which had been published only in German, suggests that it could have been translated into French well before its publication in France in 1771. A sudden surge of interest in a variety of middle salts is noticeable after the publication of Stahl's work on salts. Du Hamel and Grosse, who examined a variety of middle salts systematically in the 1730s, constantly referred to the "illustrious M. Stahl."[67] Through his leadership, Geoffroy rendered French elite chemistry susceptible to a rather different chemical system that was at once more academic and more tuned to practical operations. As Stahl put it, in order to acquire the chemical art, "its Scientifical Elements must be well understood; and its Operations personally view'd, and manually perform'd: whence its two parts, of Theory and Practice."[68] The overlapping developments in the chemistry of salts indicate, as Holmes has argued, "a complex interplay of parallel, partially independent development superimposed on a common core of shared procedures, empirical knowledge, and concepts; and an incomplete, somewhat erratic communication of influences back and forth between these two national specialist communities."[69]

3.8 Laboratory and Industry

The importance of salts derived from their uses in daily life, especially in pharmaceutical preparations. Their natural sources and their artificial production mattered a great deal to the contemporary chemists.[70] Once they had a secure understanding of the artificial salts made in the laboratory, therefore, academicians began to deploy their theoretical knowledge in a range of practical spheres, including industrial production of salts, analysis of mineral waters, and preparation of agreeable medicaments. The investigation of salts in the Academy throughout the eighteenth century was conducted mostly within these practical realms, although these inquiries occasionally yielded insights of theoretical value. Already in 1716, we find Louis Lemery and Claude-Joseph Geoffroy engaged in a debate over how sal ammoniac was manufactured in the Levant and in Egypt. Since sal ammoniac was one of the most common drugs, Fontenelle surmised, it was astonishing that they did not know exactly where it came from or how it was made.[71] Chemists knew that it was a urinous volatile salt penetrated by an acid. Geoffroy reported in detail on various procedures for making sal ammoniac, seeking to refine and to generalize the process. He mixed human urine (which contained a variety of salted salts or middle salts loosely united) with a new salted salt whose acid would combine with the alkali of the urine. He believed that sal ammoniac made with marine salt closely resembled sal ammoniac made in the Levant. This invited the thought that mineral acids and urines of other animals could be used for the "artificial composition" of sal ammoniac. Geoffroy believed that all sal ammoniacs—natural, laboratory, and industrial—had the same composition regardless of where they were produced. Lemery objected to Geoffroy's assessment of the procedures used in the Levant, arguing that the sublimation process used in the laboratory would not be feasible on an industrial scale. He suggested instead that the sal ammoniac from the Levant was made by simple evaporation of some natural material. A letter from the Cairo consulate in 1719 and another from P. Sicard, a Jesuit missionary in Egypt, cleared up all incertitude. Fontenelle proclaimed that Geoffroy had "divined the true formation of sal armoniac."[72] When commerce with Egypt was interrupted by the plague at Marseilles, the French had to get sal ammoniac from the Netherlands. They obtained pains de Sucre from Oriental Indies. Claude-Joseph Geoffroy again applied his "idea of sublimation" to this new kind of sal ammoniac as well as to the "salt of England" that had become fashionable. His work clarified many ambiguities concerning the production and the composition of manufactured sal ammoniacs. "The French genius," Fontenelle proclaimed, "is manifested even in chemistry."[73]

In 1717, nearly simultaneous with Geoffroy's research on sal ammoniac, Louis Lemery began to investigate niter, an important ingredient in the production of gunpowder in the form of saltpeter.[74] Unlike other fossil salts, niter or saltpeter was not mined; it was gathered from stables, cow sheds, and other places where animal deposits had accumulated. The saltpeter from the Oriental Indies was much better, though, and was believed to be found in great abundance naturally. Since the earth from which one drew niter seemed to be replenished with it after a certain period if exposed to fresh, humid air, such as that of a cave, some believed that the primary source of niter was air.[75] Lemery refuted this "system of aerial niter," arguing that neither the air nor the earth was the source of niter. He defined the terms exactly. Niter was an acid salt different from that of common salt, vitriol, or sulphur. When it was engaged in fixed alkali, it formed a composed salt or salted salt [sal composé ou salé], which was saltpeter. When engaged in a volatile alkali, it became nitrous sal ammoniac. This was the main difference between the form of saltpeter derived from plants and the form derived from animals. From the vegetable and animal sources, one obtained "natural" and "artificial" saltpeter, respectively. Plants contained saltpeter that was all formed, as in the case of the saltpeter of Indies. The "artificial" saltpeter made in Europe came from nitrous sal ammoniac drawn from earth filled with animal deposits. Since animals were nourished by plants, and vice versa, Lemery surmised that saltpeter in plants became nitrous sal ammoniac in animals and vice versa. He explained this "double metamorphosis" by supposing that the "niter principle" was always the same, but that it was attached to a "matrix" that became more fixed in plants and more volatile in animals. In other words, Lemery utilized in studying niter a theoretical understanding of middle salts that undercut the differences between the distillation products of vegetable and animal analysis. Against the conventional wisdom that animal matters did not contain any acid, he argued that they also contained the niter principle, which was lodged in them in a different manner. He illustrated by this how chemists' knowledge of composition could chart and control the three kingdoms of nature. Clause-Joseph Geoffroy's and Louis Lemery's works proved the relevance of chemists' laboratory work to the practical sphere, even if the invention of concrete industrial processes did not immediately follow from them.[76]

The Academy had maintained a strong interest in the analysis of mineral waters from its inception, as can be seen in Du Clos's works. The quality or the curative value of mineral waters was attributed to the salts and minerals dissolved in them. Already from the middle of the seventeenth century, chemists sought to replicate the curative virtue of mineral waters by identifying these salts. In 1708 the academician Tournefort supplied recipes for six different artificial mineral waters (using niter, cream of tartar, quicklime, sal ammoniac, spirit of sulphur, etc.).[77] In view of the long-standing tradition of mineral water analysis in the Academy, it comes as no surprise that Gilles-François Boulduc sought to apply the evolving knowledge of salts to this subject.[78] What is remarkable is the degree of sophistication in his analytic reasoning. Several features of Boulduc's work deserve particular attention.

First, Boulduc sought to reconstruct the salts that were present in the natural state of mineral waters from their distillation products. To circumvent the criticism that "salts are the productions of art with recourse to fire, & according to their expression the *creatures of fire*," he had to devise long, laborious procedures. Even so, he confirmed his analysis in the end by solution methods. He precipitated all the different salts by successively applying new portions of spirit of wine that possessed, according to Geoffroy's table, a higher rapport with water than salts. The fact that Boulduc felt compelled to design a proof based on displacement reactions to persuade others indicates that solution methods were regarded as a more natural, less destructive process.

Second, Boulduc used tight experimental reasoning to fret out the principles of middle salts. Notable in this guesswork was his consideration of the rapports between various acids and alkalis, including a notion of "double exchange" between the pairs of acids and alkalis belonging to different middle salts. In order to prove the saline nature of a new middle salt or selenite that he found in the waters, he invoked general traits of middle salts that one could "change middle salts into different compounds, transport one of their principles to another body, decompose them, and through this, be assured of which principles they are combined."[79] In other words, the basic premise of Geoffroy's table was very much at work in Boulduc's analysis. Anticipating the objection that selenite did not dissolve in water and that it thus lacked the essential quality as a salt, Boulduc discounted solubility as the necessary criterion for determining the saline nature of a substance, citing Stahl and Kunckel as authorities. Although he quickly softened his position by adding that the insolubility was only apparent, and was caused by excessive earth, he clearly compromised the conventional boundaries of salts by utilizing a new, compositional definition. He characterized selenite as a middle salt because it was composed of vitriolic acid and excess earth.

Third, Boulduc's careful analysis yielded the important discovery that Glauber's salt, long known only in the laboratory, existed in nature. The next year, he identified two other natural salts as Glauber's salt because they were composed of vitriolic acid and the base of marine salt. These discoveries weakened Glauber's original opinion that his salt was not found all formed in nature[80] and helped close the gap between the natural and the artificial salts. In the ensuing years, Boulduc continued to apply his techniques to a variety of subjects in order to refute the assumption that salts were the artificial products of fire. He hoped to discover a mineral alkali, or a natural and fossil alkali that existed before the fire analysis. Boulduc's careful work won esteem among the academicians. When Boulduc took up the analysis of another mineral water, Fontenelle praised the "precise and exact" nature of his work:

One has seen in 1726 with what care M. Boulduc has examined the new waters of Passy in order to separate out all the different matters they contained, which demands much more work, & more ingenious & finer work than one would ordinarily believe. For one is willingly content with some light & superficial proofs carried out in quite a short time.[81]

After the publication of Stahl's *Ausfürliche Betrachtung und zulänglicher Beweiss von den Saltzen* (1723), a manuscript translation of which was apparently in circulation as the *Traité des sels*, there was a resurgence of interest in salts. Many reports read within the Academy involved the subject of middle salts in one way or another. Although it would be impossible to discuss all the problem domains associated with salts during this period, I would like to point out a couple of continuing trends. First, Stahl's works were read with respect and care by most of the chemists in the Academy. Second, most discussions on salts impinged on industrial or medico-pharmaceutical interests. Chemists strove to narrow the gap between the natural and the artificial salts. In 1728, Claude-Joseph Geoffroy undertook an exhaustive study of vitriols and alum in the manner of his earlier study of sal ammoniac. He wished to make artificial green and blue vitriols, to "imitate perfectly these two natural productions."[82] The next year, he reported on the means of rectifying vinegar, trying to put one of Stahl's methods into practice. He also measured the "degrees of force" of various vinegars using Homberg's method.[83] Lemery examined borax in detail in 1728 and 1729.[84] Claude-Joseph Geoffroy continued this work in 1732, suggesting that borax was not a middle salt but a true natural alkali salt that differed from ordinary alkalis derived from plants.[85] Bourdelin reported on lixivial salts.[86] Boulduc reported work on corrosive sublimate in 1730 and work on two newly discovered salts of great medical utility in 1731.[87] In addition to the investigations by its own members, the Academy received various reports related to the subject of salt. In 1728, Amand reported a method of revivifying strong water after it was used to precipitate gold. Petit sought to explain the precipitation of marine salt in the manufacture of saltpeter.[88]

Although the practical investigation of salts had a heyday during the 1730s, the theoretical treatment lagged behind. Aside from the implicit uses by Stahl and Boulduc, the only indication that Geoffroy's table was not completely forgotten until 1749 comes from Jean Grosse, a German chemist Geoffroy had inducted into the Academy in 1731. He prepared a revised table of 19 columns for the use of his students, apparently in 1730.[89] Soon, he embarked on an extensive investigation of tartar, utilizing the table. Tartar was an essential salt used in medicine. For internal uses, one had to find ways to make it more soluble. The traditional method was to dissolve tartar with lixivial alkali salt. Le Févre proposed to use borax.[90] Grosse was following this trail to see if any other alkaline earth could do the same job, while Du Hamel was looking for a way to purify crude tartar and to render it as good as the one from Montpellier. Since they were using the same materials and the same operations, Du Hamel and Grosse joined their efforts to tackle the problem of making tartar soluble.[91] In the process of testing other alkali salts, they took a more theoretical turn and began to search for a common factor in all the substances that dissolved tartar. They ruled out both phlogiston and the matter of fire individually, which indicates that phlogiston and the matter of fire were not yet closely identified with each other. They also excluded common alkali salt and began to consider earths as responsible.[92] Experimenting further on a variety of alkaline earths that also dissolved tartar, Du Hamel and Grosse began to reflect on the "different alkaline bases of salts & on a comparison one could make between the earthly & saline alkalis," although their immediate task was a comparison of various soluble tartars for their "different degrees of solubility." From the fact that even a weak acid could easily "regenerate" cream of tartar (insoluble tartar) from soluble tartars, they concluded that soluble tartars contained two bases. One, the "natural base" contained in cream of tartar, was not easily drawn out by weak acids. The other was an added portion of alkaline earth or salt which was not intimately united to the acid of tartar. Even so, soluble tartars did not "abandon their new base with equal facility." Du Hamel and Grosse ranked soluble tartars in the order of their precipitation with acids, attributing this difference to the "different degrees of alkalinity." They characterized alkaline earths as substances with distinct identities rather than as substances devoid of acidity. In other words, they laid the foundation for a positive identification of alkali salts. Their works led to "a general acceptance by mid-century that the active base of all the alkaline earths was the same."[93]

Du Hamel and Grosse's work strengthened the conception of middle salt as a combination of an acid with its alkaline base by providing a positive identity to the latter. Du Hamel continued this line of inquiry, focusing on the earthly part of middle salts. In his exhaustive investigation of sal ammoniac, which continued Geoffroy's and Lemery's earlier research, Du Hamel focused on the earthly rather than the acid parts.[94] In 1736, he attempted to isolate the base of marine salt, free of all acids. While a perfect knowledge of salts required decomposition and recomposition, marine salt escaped such a complete analytic control. He succeeded in identifying the base through painstaking research, only to find that it had already been mentioned in Stahl's *Specimen Beckerianum*. He complained bitterly of the abbreviated manner in which this "sçavant chimiste" wrote so that nobody could comprehend it. He had been forced to do the work over. To his regret, his work had lost the merit of novelty.[95]

The flurry of investigations on salts tapered off somewhat after the mid 1730s with the generational shift in the Academy. Following R. A. Réaumur's (1683–1757) pitch for the industrial application of science, the Academy's chemistry sessions were dominated by Jean Hellot (1685–1766) and others reporting on dye, porcelain, alloys, etc. By this time, however, salts constituted a well-defined area of inquiry, something akin to a normal science. French chemists worked on them from time to time, expanding the scope and developing increasingly more sophisticated theories. Louis Lemery took up the subject of vitriol again in 1735, this time much more systematically, and the subjects of vitriols and alums in 1736.[96] Despite the relatively low visibility of salts in the Academy, it is quite likely that chemists continued to work on salts in their practical spheres, particularly in pharmaceutical practices. By the mid 1740s, the proliferation of middle salts called for a new classification. Claude-Joseph Geoffroy envisioned a systematic revision of the affinity table through collaborative efforts,[97] while Guillaume-François Rouelle proposed a classification based on crystal forms.[98] It was in this context that Geoffroy's table experienced a dramatic revival. By this time, however, the deaths of C. F. Du Fay (1739), F. P. Petit (1741), G. F. Boulduc (1742), L. Lemery (1743), and J. Grosse (1744) had left Parisian elite chemistry under the leadership of Réaumur, who curbed the dominance of pharmaceutical chemistry in the Academy.[99] The porcelain, textile, dye, and gunpowder industries offered elite chemists more lucrative careers, freedom from the obligations of a personal physician, and better opportunities for social mobility. While the graduates of the Faculté continued to thrive in this new social space of chemistry, as can be seen in the careers of Macquer and Berthollet, the apothecaries found it increasingly difficult to gain admission to the Academy or to assert control over the discipline of chemistry.

The elite chemists of the Academy helped foster a strong tradition of theoretical chemistry in France through their methodical work. Previously dependent on the medico-pharmaceutical culture of Montpellier and other Protestant cities as a source of analytic innovation, the elite chemists of the Academy now articulated new investigative programs of theoretical and practical importance. They adopted solution methods as the legitimate mode of analysis in determining chemical composition, thereby refurbishing the existing theory domain of composition. They cultivated metallic dissolutions in acids as the model system for chemical theorizing, which gave rise to the new theory domain of affinity. In parallel with these shifts in analytic practice and theory domains, they worked out a sophisticated vision of corpuscular chemistry. By inventing a discursive ontology of chemical substances, they linked together the products of different analytic methods to prepare a level ground for uniform chemical theory. They also utilized corpuscular language to unpack chemical phenomena for public consumption. In other words, the chemists of the Paris Academy transformed chemistry from a medically useful art to a recognizable field of philosophical knowledge. At the same time, they reinvented themselves as savants in the service of the state. Boyle's vision of philosophical chemistry had come of an age in the French Academy, making chemistry palatable to the articulate audience of the Enlightenment public sphere.

4

Chemistry in the Public Sphere

In mid-eighteenth-century France, an eccentric figure of violent chemical opinions and mild temper was drawing the attention of the Parisian intellectual elite. Guillaume-François Rouelle (1703–1770), an inspiring yet "extremely absent-minded and easily distracted" demonstrator of chemistry at the Jardin du roi, was staging chemical spectacles for a diverse audience of prospective apothecaries, physicians, physicists, philosophes, and literary talents.[1]

Rouelle performed invaluable services for French chemistry during his tenure at the Jardin between 1743 and 1768. France was slowly emerging from the devastation of war with active government investment in industrial development. In addition to the entrenched, prosperous business of apothecaries, other social niches began to open up for chemists in mining, porcelain, dyes, soap, candles, etc.[2] Through his private and public lectures, Rouelle taught a whole generation of French chemists who successfully met the challenges of evolving chemical industries. Lavoisier was only the most famous chemist the among his students; the list also included Jean Darcet, Pierre Bayen, A.-L. Brongniart, Jean-Baptiste-Marie Bucquet, Jacques-François Demachy, Jean Philippe de Limbourg, Pierre-Joseph Macquer, Guillaume-François Venel, and Richard Kirwan.[3] Venel summarized Rouelle's assets as a teacher succinctly:

The courses which M. Rouelle has given at Paris for about twenty years, are, even in the opinion of strangers, among the best of this kind. The order in which particular objects are presented, the abundance and choice of examples, the care and exactitude with which operations are performed, the origin of and relation between the phenomena observed, the new luminous, broad insights suggested;

Figure 4.1
Guillaume-François Rouelle. Courtesy of Académie des Sciences Archive.

the excellent manual precepts taught, and finally, the good, sound doctrine which sums up all the particular notions; all these advantages, I say, make the laboratory of this capable chemist such a good school, that one can in two courses, with ordinary dispositions, emerge sufficiently instructed, to deserve the title of distinguished amateur, or of artist able to engage successfully in chemical researches. This judgment is confirmed by the example of all the French chemists, for whom the first taste of chemistry followed the first courses of M. Rouelle.[4]

Rouelle also preached chemistry to the philosophes and notables of the High Enlightenment. Among his students were Jean-Jacques Rousseau, Denis Diderot, and Anne-Robert-Jacques Turgot, Baron de Laune. Rousseau, who cultivated a keen interest in chemistry as a youth, attended Rouelle's lectures in 1744 and later composed his own chemical work, *Les Institutions chymiques*.[5] Diderot, also an aspiring scientist in his youth,[6] took Rouelle's course at the Jardin in the years 1754–1757 and edited the lectures for publication.[7] He opened the pages of the *Encyclopédie* to Rouelle's students, who characterized chemistry as a rational, independent, public science.[8] Outside the lecture hall, Rouelle found more opportunities to preach his chemical gospel at the salon of Paul Henri Thiry, Baron d'Holbach, who initiated translations of German chemical and mineralogical texts as well as contributing many articles on metals and minerals to the *Encyclopédie*.[9] Diderot and Grimm were regular members of the group, while other notable figures such as Buffon and Turgot frequented the scene along with Rouelle and his disciples Darcet and Venel.[10] Rouelle's propensity to contradict authorities and colleagues, portrayed vividly in Grimm's correspondence, was probably shaped in this conversational culture of the salon.[11] Morellet, a perpetual salon-goer and a regular at Baron d'Holbach's gatherings, singled out the penchant to contradict as a cultivated trait of "civilized" people and strong motive for conversation.[12]

Rouelle assumed his teaching post at a time when the demise of the five principles had left French theoretical chemistry in disarray without a clear alternative. The Enlightenment public sphere also demanded a more rational chemical discourse than a mere apothecary's manual.[13] Although the decline of the Cartesian paradigm called for a new textbook to replace Nicolas Lemery's, the last authenticated edition of which was published in 1716, French teachers did not step up to fill the gap. In the meantime, foreign authors such as Newton, Stahl, and Boerhaave came to enjoy significant reputations.[14] While Newton's physics helped modify and ultimately eclipse the Cartesian fashion in natural philosophy,[15] Stahl offered chemical treatises filled with new operations. A definitive testimony that Newton's and Stahl's doctrines were in circulation among French chemists comes from Senac's textbook *Nouveau Cours de chymie, suivant les principes de Newton & de Sthall* (1723). This rather incoherent text reveals the fractured state of French theoretical chemistry in the early 1720s, which made it susceptible to foreign influences. It also indicates a new kind of intellectual milieu that shunned excessive imagination and promoted disciplined reasoning in scientific explanations. Herman Boerhaave, who incarnated Boyle's sceptical chymist with a command of logical discourse based on experiments, became a venerated authority in the French medico-chemical scene for this reason. He wove together various strands of chemical and physical experiments to present a systematic discourse of chemistry as a part of natural philosophy. A methodical reader of classical texts and well versed in the contemporary development of experimental physics, he had surveyed a wide spectrum of literature, including Boyle's works and the *memoires* of the Academy, in an attempt to create a learned discourse that would bring various branches of knowledge into a coherent whole.[16] Whereas previous generations of French teachers had catered primarily to the practical operations of pharmacy, Stahl and Boerhaave presented chemical knowledge as an academic discourse for the university audience, initiating the transformation of chemistry from an apothecaries' trade into a scholarly knowledge.[17]

Rouelle's popular lectures at the Jardin set the stage for chemistry as a public science in Enlightenment Paris, much as Abbé Nollet's lectures at the Collège did for experimental physics.[18] While public lectures in chemistry had a long history by this time, Rouelle faced a rather delicate task of teaching a mixed audience. On the one hand, he had to maintain the traditional mission of the Jardin to educate prospective apothecaries and physicians in basic chemical operations.[19] He relied heavily on the established repertoire of chemical operations in the French didactic tradition, especially Nicolas Lemery's. On the other hand, he had to appeal to the Enlightenment public's taste for universal knowledge. To this end, he introduced a simplified version of Stahl's and Boerhaave's academic discourse without trying to correlate them in a consistent manner. As one can discern from the popularity of Boerhaave's lectures and books among the European medical students and literary public, the enlightened chemical audience demanded an exceptional literary capacity of the author who could cultivate their quest for philosophical knowledge as well as their grasp of technical procedures. Diderot, who granted Rouelle the "title of founder of chemistry in France" and recommended his lecture manuscripts as "the most complete, the most linked, the most analytic" of all in his proposal for the university in Russia,[20] also faulted him for the lack of a "rigorous method" and for losing sight of "the most general and the most profound":

I have followed his course three years in a row. It was not given to all the world to profit from his lessons; his impetuous spirit was incapable of being subjugated to a rigorous method. He began a subject, but was soon distracted from it by a group of ideas which were presented to him; the most general views and the most profound escaped him. He would apply his experiments to the general system of the world; he encompassed the phenomena of nature and the works of art; he linked them by the finest analogies; he got lost, one was lost with him, and one never returned to the particular object of the day's demonstration without being astonished by the immense space one had surveyed. He committed the grossest mistakes against the French and Latin grammar, but only the foolish noticed them. He is concerned here, he said one day to him [audience], with elegance and purity: are we at the Academy of good speech?[21]

Though Rouelle tried to meet the demands of his more articulate audience by drawing on Boerhaave's lectures, he did not share the latter's vision of physical chemistry. Instead, he sought to establish an identity for chemistry independent of physics by arguing that chemistry was a science more useful, more powerful, and more penetrating than physics.[22] To this end, he enshrined Stahl as a venerable chemical authority (comparable to Newton in physics), starting something like a fashion of German chemistry in France.[23] His students Macquer and Venel helped spread the Gospel of Stahlian chemistry through their writings, albeit with divergent visions.

Rouelle performed exceptionally well as a public and private lecturer by reconstructing the defunct theoretical discourse of French chemistry for the literary audience of the High Enlightenment. He did not fulfill, however, another function of the chemical teacher: that of a textbook writer. It was one of Rouelle's early students, Pierre-Joseph Macquer, who filled this gap in French didactic discourse with *Elémens de chymie*.[24] His *Dictionnaire de chymie*, anonymously published in 1766, enjoyed even greater popularity.[25] His chemical texts were widely used throughout Europe, riding a new wave of enlightened interest in chemistry.[26] From 1768 to 1776, Macquer also edited sections of the *Journal des Sçavans* dealing with physics, medicine, pharmacy, surgery, chemistry, astronomy, and natural history, reaching a broader audience in this literary capacity. In a culture that was increasingly preoccupied with the literary representation and linguistic control of practical knowledge, Macquer's writings became the main vehicle of promoting chemistry as a public science. Condorcet acknowledged Macquer's contribution in his official eulogy:

> M. Macquer found himself at an epoch in which chemistry began to be delivered from the dreams of alchemists which still infected the works of the restorers of this science; but the clarity, the method was a merit unknown in the books of whose who treated it, & on the whole in France. A residue of Cartesianism added to the obscurity of the science by overloading it with pretended mechani-cal explications.
>
> M. Macquer is the first who gave the elements of chemistry in which one finds the same clarity, the same method which already reigned in other branches of natural philosophy. Before him, one regarded chemistry as a science isolated, intricate, obscure, and full of secret operations and enigmatic recipes, nearly as a dangerous occupation in which one risked compromising his health, his fortune, and even his reason. It appeared in the works of M. Macquer a simple science, founded on facts, proceeding by operations to which a sound method prescribed all the details, useful for all the needs of human life & linked to the general system of our knowledge. Thus his elements contributed to expanding the taste for chemistry by showing how easy it was to learn.[27]

Historians have placed much emphasis on Rouelle's lectures as the epicenter of French Stahlism. Macquer's texts contributed as much to the same cause, as Diderot observed in a matter-of-fact fashion.[28] Macquer actually exceeded his former teacher in his capacity to join chemical theory to practice, in part thanks to his close association with the apothecary Antoine Baumé.[29] Instead of simply reenacting the authorities as Rouelle did for the most part, Macquer integrated Stahl's ideas into operational chemistry. Noticeably different in Macquer's writings was the classification of the different types of chemical operations under the rubric of affinity chemistry. By providing a systematic classification and language for a broad range of chemical experience, he cultivated chemical affinity as the frontier of theoretical chemistry. Macquer also praised Geoffroy for listing the principal effects of combinations and decompositions that formed the foundation of all chemistry in a "quite cóncise & quite short" table. Although incomplete and sometimes overgeneralizing, it presented the actual state of chemical knowledge in a most succinct summary. If the table seemed sometimes inadequate, it was only because chemistry was "still at an immense distance from its point of perfection, not because the tables of the different orders of affinity were badly imagined, useless or dangerous."[30] In a similar vein, Baumé regarded Geoffroy's table as "a masterpiece" that presented "a kind of connected series of knowledge, which has thrown more light upon Chemistry and Physics than all the criticisms which have been made upon it to the present day." The revision of affinity tables offered "one of the most important services" to chemistry.[31] Affinity tables soon became a fashion. Chemists throughout Europe began to produce more and more elaborate, up-to-date tables encompassing the current theoretical and empirical developments in chemistry. To the same degree that Macquer dominated French chemistry in the 1770s as a preeminent academician, a director of royal industries, and a prolific writer with an encyclopedic knowledge of chemistry, affinity constituted the main theory domain of French chemistry.

As a popular lecturer and a prolific writer, respectively, Rouelle and Macquer occupied prominent places in the Enlightenment public sphere of chemistry. They both believed that chemistry was not dark, hermetic labor, that it was a rational science with its own mode of representation, that it was worthy of pursuit by a new generation of youths committed to uncovering the truths of nature, and that it had geniuses like Stahl and required more. Whereas the seventeenth-century Paracelsian physicians and apothecaries had developed a single identity as a chemist through a shared struggle against the Faculté, Rouelle and Macquer in Enlightenment France occupied different niches in a more diversified social space of chemistry. Whereas the seventeenth-century academies and universities had tended to exclude apothecaries from their social sphere, Rouelle could mingle comfortably with the utilitarian philosophes of the Enlightenment. Though these ideologues sought to assimilate the obvious utilitarian potential of chemistry into their visions of ideal society, the encounter awakened the chemists to the intellectual demands for a more comprehensive yet precise representation of nature. Rouelle's and Venel's pitch for the independence of chemistry from medicine and physics betrayed their insecurity in the literary public sphere. Macquer, trained at the prestigious Faculté, cultivated a strategy for legitimating chemistry that differed substantially from Rouelle's. By the 1720s, the Faculté regarded chemistry as an integral part of medical training that gave them an edge over surgeons and other medical practitioners.[32] Macquer thus took it for granted that the positive development of chemistry was closely linked to its medical application. For him, chemistry's importance was a given, rather than a cause to fight for. Instead of emphasizing chemistry's independence from medicine and physics, therefore, he sought to define its place among the other sciences. In the commerce of words that structured the Enlightenment public sphere, the identity of chemistry and chemists became fractured, in part because of the different standpoints of apothecaries and physicians.[33]

4.1 A Hierarchy of Principles

Although Louis Lemery dealt the doctrine of five principles a fatal blow, the principalist approach to composition survived in a more sophisticated form offered by Georg Ernst Stahl (1659–1734), and was passed down to later French chemists, including Lavoisier and Berthollet. Stahl was educated in medicine at the University of Jena. After obtaining a doctoral degree in 1684, he worked as a court physician to Johann Ernst von Sachsen-Weimar (1687–1694) before taking the second chair in medicine at the newly created University of Halle in 1694. With Friedrich Hoffman, his former classmate at Jena, he built a first-rate medical school at Halle.[34] His first chemical work, *Zymotechnia fundamentalis* (1697), included the first exposition of the phlogiston theory. A better-known work was his introduction to a new edition of Becher's *Physica subterranea*, published in 1703. Friedrich Wilhelm I called him to Berlin in 1715 to serve as a personal physician, a post vacated by Hoffman.[35] Stahl's move to Berlin precipitated a gradual migration of top German chemists and a consolidation of strong analytic school in Berlin. Neumann moved there as a royal apothecary in 1716. Johann Heinrich

Pott (1692–1777) and Johann Theodor Eller (1689–1760) followed as professors at the Medical-Surgical College in 1723 and 1725, respectively. With the revival of the Berlin Academy in 1744, Pott became one of the paid academicians, along with Andreas Sigismund Marggraf (1709–1782), who replaced Neumann in 1738.[36] This constellation of chemists known throughout Europe for their analytic acumen made German chemistry as attractive as ever. Beginning in the 1740s, the *Memoirs of the Berlin Academy*, published in French, strengthened the channel of communication between German and French chemists.[37] The rising reputation of German chemistry served to strengthen Stahl's already significant fame among the French, inducing translations of Stahl's and Stahlian texts during the 1760s.[38] Rouelle and Macquer, representative mid-century French chemists, both endorsed Stahl as the reigning chemical authority.

Though Stahl's towering presence in the French chemical scene is undeniable, his influence on chemical theory is not easy to trace. Nor can it be argued that the direction of influence ran only from Germany to France. Although Stahl's chemical works in Latin had been known among the elite Parisian chemists at least since the 1720s, thanks largely to Geoffroy's initiative,[39] his academic style and his past-oriented theory system did not allow immediate assimilation into the French didactic discourse.[40] French chemists cited Stahl during the 1730s mostly for his innovative chemical operations rather than for his theory. Moreover, his most influential texts, *Traité du soufre* and *Traité des sels*, were published in German after Neumann's extended visit to Paris, which established a communication between Stahl and French elite chemists.[41] Although Stahl cited mostly Becher for theory and Kunckel for operations in these books, he made occasional references to Homberg's works on salts, indicating his familiarity with the latter's works. The interaction between the German chemists, who commanded more refined analytic skills acquired in metallurgical practices, and the French chemists, who possessed an organized didactic discourse along with the budding research tradition at the Académie, merits a much more detailed study than I can provide here.

Various aspects of Stahl's theoretical edifice seeped into the French soil only gradually. Two features deserve special mention for their relevance to the French development: the phlogiston theory and the hierarchical composition of bodies. Stahl's thoughts on composition included a strong critique of the "Cartesian" corpuscular speculations such as Lemery's and Homberg's.[42] Although phlogiston became the hallmark of Stahlian chemistry during the polemical phase of the Chemical Revolution, its utility for French chemists lay mostly in the role it played in the transformation of salts. In other words, phlogiston was easily assimilated into the French investigations of salts, thanks to the precedents set by W. Homberg and E. F. Geoffroy in their investigations of sulphur. As we have seen, Geoffroy identified Stahl's phlogiston with his sulphur or oily principle, providing it with a network of positive experiments in the transformations of salts. The strength of Stahl's reputation in France depended mostly on his contribution to organizing the rapidly evolving chemistry of salts. Nevertheless, Geoffroy's identification of the sulphur principle with phlogiston also paved the ground for its metamorphosis from a kind of earth to the element of fire by linking a set of procedural identification for the sulphur principle to Homberg's and Lemery's speculative discourse on fire. Rouelle's and Macquer's characterization of phlogiston as fixed fire drew on this French legacy.

The more philosophical part of Stahl's chemistry, which struck a chord with French chemists, was the hierarchical composition of bodies. Homberg's exhaustive efforts to stabilize the five principles had exposed the limitations of such an excessively speculative exercise. A systematic reflection on the analytic identity of principles was in order, as could be seen in Louis Lemery's works. Geoffroy adopted the three principles of fire, water, and earth.[43] At this juncture, Stahl proposed in his *Fundamenta Chymiae* (1723) a cogent system of ideas concerning chemical composition, albeit cloaked in scholastic language, to dispel "all the darkness and disputes about *Principles*."[44] Although Stahl's reverent addiction to German authorities, especially to Becher, makes it difficult to discern his debt to the French tradition, indications are that he kept abreast with the French discourse on principles. He endorsed, for example, the analytic/synthetic ideal:

1. *Universal* Chemistry is the Art of resolving *mixt, compound, or aggregate* Bodies into their *Principles*; and of composing such Bodies from those Principles.[45]

Stahl elaborated a more complex, hierarchical composition of bodies. All natural bodies were either simple or compound. Simple bodies were principles. Compound bodies had three levels of composition: mix'd, compound, and aggregate. Principles composed mixts directly. Mixts, in turn, composed compounds and aggregates. In other words, one had to make a distinction between the "original" mixts ("mixts consisting of principles") and the "secondary mixts" ("bodies compounded of mixts"). Ideally, principles had to be simple substances that existed in the mixts before chemical analysis and to which mixts were resolved after the analysis, as had long been prescribed in the French didactic tradition:

3. A Principle is defined, *à priori*, that in mix'd matter, which first existed; and *à posteriori*, that into which it is at last resolved.[46]

Stahl was familiar not only with the analytic contingency of chemical principles stipulated in French didactic discourse but also with its problems. Because "a pure, natural resolution" was not always possible, he surmised, one had to distinguish "physical principles" from "chemical principles," as Nicolas Lemery had done earlier. However, Stahl did not endorse Lemery's position that chemical principles, as the limit of chemical analysis, were all that mattered. He did not believe that chemical principles were simple bodies, as physical principles were supposed to be. Three chemical principles—salt, sulphur (or oil), and mercury (or spirit)—were those "into which all bodies are found reducible by the chemical operations hitherto known." Stahl rejected the addition of Phlegm and Earth. The three chemical principles could compose "compounds," but not "mixts." What chemists commonly called chemical principles, or the "common chemical principles" of vegetables, animals, and minerals, were in fact "artificial mixts." In contrast, compounds in bare resolution produced "natural mixts." Physical principles composed mixts, although one could not yet determine them. Among the traditional candidates for physical principles, the four peripatetic elements did not deserve this title. In Stahl's hierarchical theory of composition, then, chemists dealt only with "chemical principles"—that is, natural or artificial "mixts" which made up compounds and aggregates. The analysis of mixts into proper "principles" or "physical principles" lay beyond their analytic control. Stahl's hierarchical scheme stipulated several layers of composition:

	The limit of chemical analysis		
Principles	mixts	compounds	aggregates
Physical principles	chemical principles		
	Primary mixts	secondary mixts.	

For the composition of mixts, Stahl discussed two "opinions": Van Helmont's idea that water was the only "material principle" of all things was just "with regard to the ultimate resolubility of things." Nonetheless, Stahl preferred the opinion of Becher, "a Man who seems to have been design'd for the real Improvement of natural philosophy." Becher had proposed water and three kinds of earth: vitrifiable (or fusible), inflammable, and liquifiable (or specifically mercurial). These four "original" principles, or "the immediate material principle," combined in different proportions to produce a few "primary mixts," not many in number. In sum, principles composed primary mixts which in turn composed secondary mixts. Aggregates were composed of several mixts and/or simple matters.

Though Stahl's textbook is extremely pedantic and difficult to decipher, the advantage of his new terminology becomes clear if one compares it to Homberg's *Essais*. Homberg's "principle" traveled freely between the domain of projected analysis and that of actually accomplished analysis, as we saw in chapter 2. In Stahl's new terminology, principles properly belonged to the former domain, mixts and compounds to the latter. Mixts would then be the substances actually separated out in chemical analysis. Stahl thus consigned principles to the realm inaccessible by chemical analysis:

26. Having allow'd Earth and Water for the two *material Principles* in the generation of *Mixts*, it might be expected we should first shew where these *Principles* may be found in their purity; but as they are very rarely to be obtained in that state, and, by reason of their extreme minuteness, scarce otherwise than in the form of vapour; and as they are with difficulty, and not without the most exact analysis, obtainable from Compounds; 'tis proper that we first take a view of *Mixts* and *Compounds*, and afterwards examine in which of them these *Principles* lie the loosest.[47]

By taking the "physical principles" out of chemists' analytic control, Stahl made it possible for them to maintain the principalist approach to chemical composition and reaction. Compounds partook of the properties of their principles, a belief Fourcroy later attributed to Stahl's authority.[48] The affinities of substances were also explained by the same principle they contained, or by their "analogy." Vitriolic acid acted on metals because of the earthly part the metals contained or because of the "analogy that subsists between these substances" or "their earthly nature." Various salts did not act on different metals "in a uniform fashion" because their action depended on the "analogy" between the earthly substances contained in the metal and in the salt.[49] Despite the demise of the five principles in didactic discourse, therefore, Stahl's theory of hierarchical composition retained the principalist outlook in the chemistry of salts without adequately addressing the question of affinity. Early in his career, Lavoisier complained about this vague understanding of chemical selectivity by way of "analogy."

Stahl also made a distinction between compounds and aggregates, which Macquer utilized later. Although his thoughts on this issue are drawn out and confusing, we can discern with hindsight that an aggregate was a simple collection of small particles or atoms, while the production of a compound was accompanied by a qualitative change:

6. The second general Part of the Object of Chemistry, or Combination, is also concerned about Continuums in Quantity, or Aggregates, and Continuums in Quality, which are specific Continuums, or a new kind of Compounds. In the first, several small bodies are joined into one larger; and in the second the like is done, with the addition of a new specific quality.[50]

The "collective combination" brought together a number of smaller parts into an aggregate, while the "method of union" combined the "parts that differ from each other in *number, figure*, or even in *mixture*" to produce a new "Concrete, differing specifically from the parts that make it."[51] Consequently, the resolution of aggregates produced "integrant parts," while that of compounds yielded "constituent parts":

11. We have already observed, that *Destruction or Resolution* resolves *Mixts* and *Compounds* into their *constituent*, but *Aggregates* into their *integrant parts*. The latter depends upon and rests in a bare *dissolution of Continuity*; without any regard to the *homogeneity* or *heterogeneity* of the separated parts: but the former is effected by a necessary separation of the *heterogeneous parts* dissolved by the other.[52]

The term "integrant part" had been long in use among French chemists. Stahl coined a much more consistent usage of the term, however, by differentiating it from the "constituent parts" that made up compounds. Macquer later exploited this distinction fully with his classification of affinities—affinity of aggregation versus affinity of composition—rendering Stahl's thoughts on composition much more accessible. In the process, however, he transformed Stahl's principalist approach to composition into an operational one contingent upon the play of affinities. Stahl could not explain, for example, why there should exist only a limited number of "mixts" despite the infinite possibilities the different proportions of principles could create. Such a question became unnecessary if the composition was determined by specific affinities. Macquer inverted Stahl's principalist approach, therefore, to consolidate an affinity approach to composition.

4.2   Nouveau Cours

Stahl had been well known and well respected among the French elite chemists since the 1720s for his repertoire of inventive operations. Whereas Boerhaave became a venerable medical authority and famous chemical teacher, Stahl emerged as the premiere chemical authority on experiential grounds. The proof that his works were in active circulation comes from an anonymously published textbook, *Nouveau cours de chimie, suivant les principes de Newton & de Sthall* (1723), generally attributed to Jean-Baptiste Senac (ca. 1693–1770). Although historians have noted this text mostly for its role in spreading Newtonianism,[53] French chemists later remembered it more for its introduction of Stahl's ideas into France. The search for alternative authorities and the concomitant confusion in chemical theory is evident in the book, which promised it a relatively short shelf life. Lacking in consistency and clarity, it was by no means a popular textbook. It provides us with valuable historical information, however, on the state of French didactic chemistry in the 1720s.

The most interesting aspect of Senac's book for the historian is perhaps the title which accurately reflects its implicit organization. In the introductory part of the book, Senac discusses several subjects with no apparent overarching organization—matter, the principles of bodies, the sulphur or oil, the mixture of elements, the magnetism of bodies, the cause of the magnetism of bodies, solvents, the alkahest. His intention here seems to have been to match the fundamental concepts of natural philosophy, such as matter and motion, with the chemical notions of principles and affinity. In the section titled "Matter" he discusses the general, philosophical meanings of matter. In the next section, "The Principles Of Bodies," he deals with chemists' principles. Sulphur is one of the principles he includes, albeit under a separate heading. "The Mixture of Elements" begins with a general, philosophical discussion of movement, which is followed by the discussion of chemical movements such as fermentation. "The Magnetism of Bodies," which contains affinity relations stipulated in 105 propositions, is followed by a short section on the cause of magnetism. "Solvents" provides a main category of chemical actions exhibiting affinities. Next comes a brief section on "Alkahest," the universal solvent. This implicit organization reflects Senac's belief that a close meshing of chemical and philosophical concepts was necessary to build a new system of chemical theory.

The juxtaposition of the philosophical and the chemical authority bespoke the need for a new public language to replace the "Cartesian" code and for a new theoretical system to supplant the doctrine of five principles. The weakening of the "Cartesian" grip on chemistry was unmistakable. While Louis Lemery carried on the torch of corpuscular chemistry after the deaths of Wilhelm Homberg and Nicolas Lemery, he was overshadowed in the Academy by his more competent colleague E. F. Geoffroy, who steered French elite chemistry away from excessive speculation. Even so, they did not have an alternative chemical master or textbook to replace Nicolas Lemery's dated and presumably Cartesian masterpiece. In place of the defunct alliance, Senac sought to institute Newton and Stahl as the new philosophical and chemical authorities, respectively. His failure to produce a coherent text indicates the haphazard manner in which chemists read these foreign authorities. The fact, however, that a physician saw it fit to integrate the basic concepts of chemistry with those of natural philosophy reveals a significant change in the intellectual environment for chemistry in the 1720s. Apparently, Senac was not alone in this belief; the reviewer for the *Journal des Savants* praised his book for having divested chemistry of its mystery and obscurity "in favor of the light of physical argument."[54]

Senac's "Newtonian" stance critical of Cartesian philosophy comes through clearly in his critique of Homberg's and the Lemerys' corpuscular discourse. He castigated them for imagining the five principles and pointed acid particles to be real, which provided the "asylum of chemists."[55] Although Boyle was "a great reformer of chemistry who reduced everything to experiment," Senac contended, he focused mostly on correcting the errors rather than on erecting a new foundation of true chemistry. Keil attempted for the first time to reduce chemical operations to the laws of mechanics by relying on the principles of Newton, the great philosopher. Senac identified attraction as the positive feature of Newtonian chemistry. Chemical action was due to the attractive force of the "magnetism of bodies." All of chemistry provided "evident proof" that attraction existed and caused the composition or decomposition of bodies, although the notion was not immune from philosophical abuse.[56] Senac left the cause of affinity unanswered, as Newton had done for other attractions or "magnetism":

From everything I have said, it follows that it is nearly impossible to discover the cause of the magnetism of bodies; all we can say is to discover the laws according to which it acts. M. Newton has worked thereupon; he is the first who has explicated physics & chemistry alike by magnetism; like him, let's look for the effects of this attractive cause, after which we could come to the cause.[57]

Senac identified Newton's attraction with Geoffroy's "rapport" or disposition to unite. In four propositions, he restated Geoffroy's laws, using "affinity" and "attraction" interchangeably. He also praised Geoffroy for having "rendered more service to chemistry than an infinity of authors" by his affinity table alone. His discussion of displacement reactions indicates a rather extensive knowledge of chemistry, probably drawn from Stahl's works, since they did not follow the order prescribed by Geoffroy's table.

Senac's rhetorical devices deserve some attention from historians for their reference to the ideal of universal knowledge. Ironically, he used the well-rehearsed Cartesian criterion of clarity to espouse the Newtonian turn. Ignoring the decades-old rhetoric that chemistry had finally emerged from its dark, hermetic past, Senac declared in the first page that chemistry still lacked clear principles:

Chemistry offers a vast subject, little light, much work; philosophy has not yet extended any clarity to it; the principles one has followed are obscure or uncertain; the books only present the terms more proper to hiding the ignorance of their authors than to clarifying the mind. To this darkness, chemists often join the fabulous; finally, always in dispute among themselves, they agree neither among themselves nor with nature.[58]

The new system had to be built on sensory experiences and our reflections thereof, rather than on "imaginations." Just as the Cartesian notion of extension was prone to a variety of philosophical imaginations on matter, Senac contended, chemists' five principles proved to be imaginary. Instead, he espoused the three principles of earth, water, and fire, as Geoffroy did. In addition, he acknowledged three kinds of earth, as Becher had stipulated. Salt was a "concretion" of three elements: fire, water, and earth. Nature gave the acid salt; art gave the alkali salt. From these two, all the rest were made. The mixture of an acid and an alkali gave the "salted salt." Although Senac accepted common procedures for distinguishing acid and alkali such as effervescence and a series of color tests, he sharply criticized the "imaginations" of philosophers who ascribed various shapes and sizes to their parts or principles:

One says ordinarily that the acid salt is an assemblage of round and oblong parts pointed at two ends. But on which foundation does one advance this? Here it is. The acid salt dissolves most solid bodies. Its parts therefore have to be round, says one. The acid stings the tongue without scraping it like the acrid salt. Its parts are therefore pointed & stinging. The acid always penetrates bodies easily; it is necessary therefore that the two ends are pointed in the parts of this salt: one has reasoned likewise on the formation of this salt. Some parts of water, one has said, pasted with earth & fire, would form acid parts. One has diversely arranged the aqueous & the earthly parts in order to place in each part two pyramids which are found by its base: but all this is without proof. For me, I give to the acid parts only the figure that the microscope discovers for us, that is to say, small points.[59]

4.3   Herman Boerhaave, Savante-Chimiste

The call for a new chemical system that combined positive experience and basic axioms of natural philosophy was answered by Herman Boerhaave (1668–1738), a popular professor of medicine, botany, and chemistry at the University of Leyden from 1702 to 1729.[60] Son of a country pastor, he was well versed in classical Latin texts before he enrolled as a prospective theology student at the University of Leyden. During the first years of philosophical studies, his versatile intellect was nurtured not only in classics but also in natural philosophy, experimental physics, and mathematics.[61] Switching to medicine in 1691, Boerhaave devised an extensive course of self-study while working in the library and obtaining practical instruction in chemistry from David Stam, a student of Sylvius who had occupied the chair of practical medicine at Leyden. He attended lectures by Archibald Pitcairne, one of Newton's early disciples, who was chosen for the Leyden chair in 1692. Pitcairne lectured for a year on the iatromathematical approach to medicine.[62] After obtaining a medical degree from the Academy of Guelders at Harderwijk, Boerhaave led a quiet life as a physician before he was called as a lector of medicine at the university in 1701. He began to offer private lectures on chemistry in the next year. His fame as a teacher of medicine and chemistry attracted students from all over Europe (particularly from Scotland), who in turn set up distinct medico-chemical traditions in their own countries.[63] Boerhaave's students praised his ability to construct a rational system out of multifarious facts:

I can no more judge of the *Genius* and Temper of the *Dutch*, than if I had never lived amongst them, for I knew no Dutchman, but my Professors; but, if I am allowed to take Dr *Boerhaave* for a Sample of the whole, I do say, that he was the most extraordinary Man of his Age, perhaps in the whole World; a clear understanding, sound judgment, with Strength of Memory that nothing could exceed, and indefatigably laborious: It is true, he had not that Brightness of Invention, that some Authors may have; but with these his Talents he has done more Service to the World in the Knowledge of *Physick*, than all his Predecessors in the whole World put together; by digesting a huge Heap of Jargon and indigested Stuff into an intelligible, regular, and rational System.[64]

Boerhaave set the example for the caste of the savante-chimiste (as the French liked to call the elite breed of philosophically minded chemists) or the philosophical chemists (as the English would have it).[65] The ability to scan a wide range of literature, not limited to chemistry, and to shape the problem and theory domains of general significance set this elite brand of chemists apart from ordinary apothecaries and physicians who were concerned primarily with more effective remedies. As an influential medical teacher, Boerhaave's chemical teachings found a broad audience in France, contributing to the transformation of chemical didactics.[66] At a time when the French didactic tradition was at a low ebb, Boerhaave utilized his scholarly training and inclination as a methodical reader to survey a broad range of literature and to construct a theoretical discourse that appealed to the more learned audience of the early Enlightenment. The popularity of his lectures can be gleaned from the publication history of his textbook.[67] Whereas the previous generation of French textbooks had catered largely to pharmaceutical operations, albeit sprinkled with incoherent philosophical language, Boerhaave devised a systematic discourse, akin to natural philosophy, which introduced the uninitiated to the basic concepts and practices of chemistry.[68] He advocated a mechanistic approach to medicine and chemistry.[69] He was familiar with the works of Boyle, Homberg, Geoffroy, and the Lemerys as well as with those of older alchemical and Paracelsian authors. He drew chemistry closer to experimental physics by correlating these two genres of experimental practices. His familiarity with and partiality toward the instruments of experimental physics, such as the thermometer and the air-pump, are evident in his discussions of fire and air. Abbé Nollet picked up these aspects of Boerhaave's didactics.[70] In many senses, Boerhaave carried out Boyle's vision of philosophical chemistry based on positive experimentation and corpuscular philosophy, although he added the Newtonian conception of matter and force to the mix. He also incorporated many elements of French corpuscular program into his lectures, especially Louis Lemery's concept of fire. Boerhaave's physicalist outlook had a lasting effect on the later generations of French chemists, especially on Lavoisier.

Boerhaave's scholarly and physicalist outlook served to differentiate philosophical chemists from ordinary apothecaries. His inaugural address for the chemistry chair at Leyden in 1718, "On Chemistry Purging Itself of Its Errors," presaged the attitude of these philosophical chemists during the Enlightenment. Speaking before the notables of the Dutch Republic and the learned professors of the Leyden university, he deemed it necessary to vindicate the academic status of chemistry, "that troublesome, uncouth, laborious, far beneath the notice of scholars, unknown to, or regarded with suspicion by, the erudite—breathing fire, smoke, ashes, dirt, not distinguished by any kind of saving grace." He envisioned that chemistry could become a real science if it was delivered from its past errors. Two stood out: the alliance with theology elaborated in alchemy and the indiscriminate application of chemical explanations to medicine. He criticized Paracelsus, van Helmont, and Sylvius for these reasons while praising Bacon, Boyle, and Newton for their philosophical approach to chemistry.[71] Boerhaave's inclination in approaching chemistry as an outsider to the apothecary's vocation is clearly expressed in his "Design" of the *Elements of Chemistry*. He claimed that presenting chemistry in a clear and methodical manner for instruction was difficult because it was "an Art cultivated by men taught rather by accidental discoveries, than acting according to rules of Art" who had left behind "only a confused collection of the things they observed," neglecting at the same time many important things. Even more confusion ensued when "these Artists began to introduce their disputations into it; came to coining their general principles; and went about to explain the causes of the different appearances they met with." Boerhaave wanted to remove these difficulties "by collecting together the genuine Experiments, which have been performed in this Art; from thence deducing some general rules; and then disposing those rules the most to advantage."[72] In other words, Boerhaave's vision of philosophical chemistry rested on a symbiosis between authentic experiments and rational rules. At the fundamental level, his mission echoed Boyle's. He wished to devise a systematic discourse of chemistry based on experiments proven in accordance with the laws of physics to teach chemistry to the learned audience. As Boerhaave's voice grew stronger in the Enlightenment public sphere, chemical practitioners with close ties to apothecaries' culture felt the need to protect themselves against his physicalist and scholarly condescension, as we shall see in Venel's spirited defense.

The organization of the *Elements* reveals how Boerhaave envisioned to teach chemistry as a rational knowledge. It consisted of three parts: historical, theoretical, and experimental.[73] In the "History of the Art," he dealt with the etymology of chemistry, historical locations in which it flourished, and the principal authors and sects as well as their positive or negative contributions. A selective history served to naturalize what he regarded as positive components of chemistry.[74] The "Theory of the Art" had to be reconstructed extensively to ally chemistry with natural philosophy. Boerhaave's text speaks for his power of elocution and persuasion. He vowed to "lay down certain indisputable, chemical

Positions, collected from such evident physical truths, as have been discovered by the Chemists; and these will be chiefly of a general nature, and so contrived, that by the help of them may be performed all the operations, that truly belong to the chemical Art." He did not wish to include "any other Theory, than what is built upon such general Propositions, as have first been deduced from many common undoubted Chemical Experiments, from which, as they always succeed in the same manner, some general truth may be fairly inferred." In doing so, he would "take the liberty to make use of such truths, as are demonstrated in Physics, Mechanics, Hydrostatics, and Hydraulics; since the properties, which belong to all bodies in common, must hold good in chemical ones too."[75] Boerhaave took it for granted that chemical theory should be built on chemical and physical experiments. Ideally, chemical theory constructed in this manner should be demonstrated perfectly by experiments. Thus, Boerhaave discussed actual demonstrations of the methods of chemistry in the third part, taking care "that the most common things, if it is of any consequence to know them, shall not be omitted." In order to display "the most perfect knowledge of chemistry," the demonstrations should engage "both the head and the hand":

In the prosecution of which, the Theorems explained in the second part, will be made great use of; by the help of which all the operations in Chemistry will be easily understood, and both the head and the hand fitted for the exercise of the Art: At the same time also, every one of these Processes will be a Demonstration of the truth of the particulars, from which the general Theorem was at first constructed. By this method, Gentlemen, I propose to lead you into the most perfect knowledge of Chemistry; nor have I loaded with difficulties a Science, which in its own nature is sufficiently troublesome. Without this, all the labours of the Chemists, which they call Processes, indeed are of no real service, but waste our time, and instead of being an advantage, are really a prejudice to those that are fond of them.[76]

Chemistry was for Boerhaave "an Art, that teaches us how to perform certain physical operations, by which bodies that are discernible by the senses, or that may be rendered so, and that are capable of being contained in vessels, may by suitable instruments be so changed, that particular determin'd effects may be thence produced, and the causes of those effects understood by the effects themselves, to the manifold improvement of various Arts."[77] This long-winded definition is merely descriptive, avoiding any essential characterization of the art or discussion of the causes. Boerhaave's exposition of chemical art is something akin to that of natural history, as if he is standing at the door of the chemical laboratory observing chemists at work, but without communicating with them. The natural history of chemistry continued with the description of its objects: the three kingdoms of nature. Only after giving a complete history of the objects of chemical art did Boerhaave proceed to chemical action and composition. All chemical changes were due entirely to motion. Exceedingly simple in principle, the "vast number, and great variety of the constituent Particles of Bodies" nevertheless caused "an infinite number of new and surprising appearances." Only at this point, and only in passing, did Boerhaave give a succinct characterization of chemical art: "The whole business of Chemistry is therefore to unite, or to separate; nor is there a third thing that it is capable of performing." The "simplicity of the actions of the chemical Art" did not preclude the infinite variety of effects it could produce. Even a simple mechanical union of different bodies produced a surprising variety of compounds. As seen in the interactions of lodestones and actions of menstruums, furthermore, it was obvious that "a greater number of Bodies have this mutual tendency towards each other, which whilst they are at a distance is not perceptible, but upon their near approach, discovers itself immediately."[78] There was no need, therefore, to assume something mysterious in chemical action. Boerhaave's explanation for chemical action here moved almost imperceptibly from mechanical motion to Newtonian attraction, linking these concepts of natural philosophy to chemical action.

Boerhaave incarnated Boyle's sceptical chymist in his cautious approach to chemical art with the naturalist's gaze and with the philosopher's mind. His thoughts on elements convey his skeptical attitude toward chemists' philosophical claims. He admitted that there existed in nature certain immutable, hard corpuscles that constituted the limit of any analysis. Philosophers named these principles the "elements of bodies." One could not trust chemists, however, when they "pretend to exhibit to us the first Elements of bodies, and think they can determine the nature of Compounds, from the knowledge they have of the Elements which by chemical Operations may be extracted from them." Although chemists often claimed that they reduced bodies into these elements, what they obtained in chemical analysis were never collected and exhibited in pure form. Fire could perhaps be perfectly pure, but one never saw a "drop of pure simple Water." The products of chemical analysis (water, spirits, salts, oils, earths) were not "themselves of a simple nature, but mutable, and capable of farther division." Nor could one assume that the same parts persisted in chemical actions. Nor could chemists deduce the nature of the compounds from the knowledge of the elements separated out through chemical operations. They could scarcely put them back together to produce the original compound. Boerhaave thus sought to circumscribe chemical art "within a narrow compass" and to stipulate the limits of its analysis carefully. Only then would it become a "valuable, excellent, useful and necessary" art.[79]

The utility of chemistry lay in its relation to other branches of knowledge such as natural philosophy, medicine, and mechanical arts. Since chemistry examined all sensible bodies, Boerhaave asserted, it had to be useful to all branches of natural philosophy. The first and principal part of natural philosophy consisted in "collecting together all those Phaenomena of Bodies, which our senses are able to discover; and then reducing them into a natural History."[80] One could conduct this process either by simply taking note of the way things are in nature or by consciously applying different bodies to one another with a design to generate new phenomena that would not have been noticed in nature (as in chemistry). Strangely, it is in discussing chemistry's utility for natural philosophy that Boerhaave provides us with a much more succinct definition of chemical art:

Chemistry is almost the only Art, that seems suited to cultivate this second, and most valuable method of making physical Observations. 'Tis this that resolves compound Bodies into their simple parts, and know, what new appearances, and powers, will thence arise: 'Tis this, that separates, or compounds various Bodies, and then examines them nicely with a determinate, and well observed degree of Heat, in order to find out if possible, what it is in them that nature is chiefly engaged about: And lastly, 'Tis Chemistry that by these means discovering how it may exactly imitate the natural and common Phaenomena abovementioned, hence truly explains, and exhibits to us the instruments by which nature so efficaciously operates; and thus pries into her most secret methods of working, and very often prudently directs and improves them to its own advantage.[81]

Chemistry's role in deciphering nature lay not in its philosophical aspiration, but in providing the instruments of philosophical inquiry. The language of instruments and their centrality in Boerhaave's lectures reflect his judgment that chemistry "very justly deserves the name of an Art" rather than a philosophy.[82] The instrument of an art was something that gave to the body "a certain peculiar motion" that produced the desired effect. Chemistry had six: fire, water, air, earth, menstruums, and the furniture of the laboratory.[83] Boerhaave placed the four traditional elements in continuum with the other agents of chemistry without explicitly acknowledging their elementary status. One can sense his circumspection in this regard throughout the text: On the one hand, he did not believe that there existed only four elements since fire decomposed various bodies to their elementary particles. On the other hand, he was not certain of their elementary status in chemical analysis (i.e., that they composed other bodies). In the analysis by fire, for example, he took it for granted that vegetables produced five principles. The inherent ambiguity in matching the philosophically posed elements and the chemically analyzed principles remained unresolved in Boerhaave's scheme. He simply pushed chemical principles to the background to create a philosophically viable discourse of chemistry.

Boerhaave's discussion of fire in the *Elements* (commonly referred to as the *Traité du feu*) became an integral part of the Scottish and the French discourse of chemistry and experimental physics. Of particular interest for our story of the Chemical Revolution is the way in which he fused the concepts of fire and heat, traditionally associated with the different instrumental practices of chemical analysis and thermometry, to craft a singular ontology of material fire with a spectrum of meaning applicable to both chemistry and physics. By staking an elaborate claim embedded in a methodical discourse that fire and heat must be the same "thing," he forced the later generations of chemists and experimental physicists to scramble for a common instrumental domain to stabilize the ontology of fire. In other words, Boerhaave did not make a neat distinction between feu-chaleur and feu-combustion, as Metzger interpreted, but broke down the barrier between the previously disparate instrumental domains of heat and fire to invent a material being that catered to both domains.[84] By doing so, he made heat an integral part of chemical constitution. He articulated the constitution of bodies as an equilibrium of attractive and repulsive forces, which became a stable part of French theoretical outlook after the Chemical Revolution. It was this symmetric, discursive infiltration of heat and fire into each other's instrumental domains that Lavoisier had to sort out later in his characterization of gases.

Advising "utmost caution" and "extraordinary circumspection" so as to avoid laying a wrong foundation for such an important subject, Boerhaave began his discussion of fire with the "physical changes" that were sensible: heat, light, color, expansion or rarefaction, and the powers of burning and melting. He screened them carefully to see if one of them "will serve us as a certain token, and a proper measure of the presence, and quantity, of this most active Element." From "the most faithful, and diligent inquiries," he could discern only one effect "which is always, and every where the same, perfectly inseparable from it, and constantly invariable in every kind of Object." He fastened on the "expansion of Bodies by Fire . . . effected in a Glass hermetically sealed" as "a true, certain, individual, and proper mark of Fire." Wherever rarefaction occurred, there was "a proportionable degree of Fire as the cause of it." In other words, Boerhaave identified the expansive property of heat as the true indicator of fire. With the thermometer, one could begin experimental inquiries to ascertain the properties of fire.[85]

Despite the long circumspection, Boerhaave's concept of fire drew largely on the instrumental practice of thermometry, to judge from the extensive list of experiments he included to illustrate the concept of fire: expansion by heat, contraction by cold, thermometric properties of various fluids, the notion of absolute cold, heat radiation, and so on. He maintained a close contact with Fahrenheit, who constructed thermometers for him and conducted investigations according to his instructions.[86] Boerhaave tried to match, however, the thermometric measurements and conceptions of heat to the chemical notion of fire through a careful reflection on the composition of bodies. His efforts in this regard can be discerned in many of the experiments included in his book. The first experiment, illustrated with an iron bar, dealt with the expansion of solid bodies by fire, which would ultimately turn them into fluids. The expansion of a solid body was deemed proportional to the quantity of fire admitted and inversely proportional to the density of the body. Boerhaave brought this experiment to bear on the issue of chemical decomposition by asking whether fire divided bodies to the level of ultimate particles of elements:

Does Fire then so attenuate and divide such a Body, that those corpuscules which are thus fluid, are in reality the very Elements of the Bodies, so long as they continue in this state? And is this the reason, that the particles of Metals, when in fusion, are so intimately mingled one among another, that it is impossible to reduce them to the same degree of fineness by any other method?[87]

Fire seemed to have the "surprising power" to divide bodies into their "ultimate parts" and to reach "the most intimate nature of Bodies."[88] Boerhaave did not believe, however, that fire could destroy the ultimate particles or "Elements themselves."[89] In the second experiment, he used the effect of cold or the absence of fire to establish the existence of contractive force inherent in the particles of bodies. The constitution of bodies would then be determined by the "action and reaction" of the expansive force of heat and the contractive force of union:

. . . it seems exceeding probable, that this Fire is contained *in vacuo*, and in the vacuitites that are dispersed through the most solid Bodies, as in a kind of vessel, where it is frequently agitated, and always in action; and that hence it is of necessity continually producing certain Operations, the effect of all which principally tends to remove the Elements of Bodies from one another, that so the Fire may expand itself more equally. In the mean time, however, it is not less certain, that the Elements of corporeal matter constantly endeavor to unite themselves together more and more, to lessen the vacuities between their impenetrable particles, and consequently by the excess of their power to expel the Fire that is there continually endeavoring to dilate itself. Here, therefore, will be a constant *action and reaction*, between the Fire contained in these Pores, and the constituent Elements of the Bodies; the former tending always to separate these Elements from one another; the latter from a natural propensity perpetually attracting one another into the strongest union.[90]

Boerhaave's experiments on combustion, designed to exhibit the role of the "Aliment or pabulum of Fire" in making the flame, contained little of value or insight in comparison to his discussion of heat. His efforts to decipher the true function of alcohol as pabulum in creating flame ended in the confession that "there is hardly any thing in all natural philosophy that is more difficult to understand, than what corporeal substance that really is, which is solely inflammable."[91] He was adamant, however, that just as friction did not create fire de novo, combustion did not convert the pabulum into "the very substance of elementary Fire." He believed, partly on theological grounds,[92] that "the very same Fire does always exist, in the same quantity, and without alteration." Otherwise, the quantity of elementary fire would constantly increase at the expense of other bodies, until fire alone "remained superiour, and alone in the Universe."[93]

One tricky issue in crafting a consistent ontology of fire and heat involved the weight equation. Boyle, Du Clos, and Homberg had all argued from their burning glass experiments that fire possessed weight. Metallic calces weighed more when burned at the focus of the burning glass. Boerhaave questioned the assumption that only fire contributed to the process. He relied rather on the comparative weight measurements of hot and cold iron bar which indicated no weight change, arguing that "the degree of Heat, now, or quantity of Fire in the *Focus*, mentioned in M. *Homberg*'s Experiment . . . was by no means so great as that of this Iron."[94] Boerhaave's rather unusual struggle to contradict the authority of Boyle and a host of French academicians on this issue points to the difficulty he encountered in trying to bring together different genres of instrumental practices for a coherent picture of nature. His preference for the experiment involving hot iron bar over the burning glass experiment may indicate his partiality to thermometric experiments.

Boerhaave also brought together different genres of experiments on air to devise a consistent picture of its nature and operations. He was familiar with Boyle's and Mariotte's experiments with the air-pump, which characterized the elasticity of air, as well as with Hales's experiments on the fixation of air.[95] On the one hand, this elastic medium functioned as an instrument of many chemical operations. On the other hand, air particles exhibited a "tendency towards a union." The air fixed in bodies seemed to take up a very small space when divided into smaller particles of its proper substance, but it expanded considerably when these particles were collected together. Boerhaave's discussion of air, like the one on fire, is remarkable for the facility with which he mixed different genres of instrumental practices to create a coherent ontology of these supposedly elementary bodies. He discussed various effects produced by the external air, a gravitating fluid that made up the atmosphere of the earth, "with an eye to Chemistry."[96] In almost prophetic words, he detailed the multiple functions of air in chemical operations: All bodies undergoing chemical operations were simultaneously subject to the effects that air could perform "by the concurrence of all its powers together." Encompassing all bodies, it penetrated the interstice of bodies and united "its Elements" with the bodies it met, losing its fluidity and elasticity. It remained in the bodies, "closely confined," until effervescence, fermentation, putrefaction, or fire set it free. Through its constant movement, air "performs particularly the office of intermixing Bodies very intimately with one another, . . . it puts them in motion, and rubs, and mingles them together, producing after this manner very singular effects."[97] This thorough concurrence of chemical and physical experiments defined the goal of Boerhaave's discourse on the instruments of chemistry. The knowledge of natural philosophy was a prerequisite for understanding the chemical art:

I have principally endeavored to shew, how necessary the knowledge of Natural Philosophy is to a person who would make himself master of the chemical Art; and consequently, how necessary it is likewise to be acquainted with all the Arts by which Natural Philosophy is promoted. Without these helps the Chemist is continually falling into errors himself, as well as deceiving others, whilst he mistakes the true causes of things, and assigns false ones in their room; whereas if he is sufficiently furnished with these, he has paved a way by which he may readily arrive at the true knowledge of Nature.[98]

4.4   Guillaume-François Rouelle, Teacher

To judge from the examples of Voltaire, La Mettrie, and Rousseau, Boerhaave gained a significant following among the French medical students and literary public.[99] French chemical teachers took his discourse with a grain of salt, however, trying to preserve the native medico-pharmaceutical tradition. At the middle of the century, the most important teacher in Paris was Guillaume-François Rouelle (1703–1770), who held the position of demonstrator at the Jardin. He was born in the Norman village of Mathieu and studied medicine at the Collège Dubois of the University of Caen before moving to Paris in 1725. He was apprenticed for seven years to J.-G. Spitzley, a German apothecary who had acquired the laboratory of Nicolas Lemery. Rouelle studied chemistry with several teachers and botany with Antoine and Bernard Jussieu, professor and demonstrator of botany at the Jardin respectively. Apparently his career aspirations changed with these associations; he abandoned his apprenticeship three years shy of completion, which was required to open a business as a maître apothicaire in Paris. In 1737 he began offering private lectures at the place Maubert, not far from the Jardin. These lectures helped build his reputation and ultimately led to the appointment at the Jardin, although his previous acquaintance with the Jussieu brothers should have helped.[100] Boulduc's death created a vacancy not only at the Jardin, but also at the Academy. In the election for the position of associé in April, Rouelle lost the bid to Malouin. In the subsequent election for an adjoint chimiste in May, he was chosen over Macquer.[101]

The key question unanswered in this biographical profile is the identity of Rouelle's chemistry teachers. His familiarity with and extensive use of Nicholas Lemery's experiments in his lectures is understandable in view of his apprenticeship in Lemery's former laboratory and the lasting fame of Lemery's textbook. His knowledge of Stahl's and Boerhaave's doctrines cannot be as easily accounted for. Although Rouelle and his biographers are silent on this issue, it would be odd if he did not attend the free lectures at the Jardin, in view of his association with the Jussieu brothers and his choice of career as a chemical teacher. The administration of this century-old institution had gone through some changes. Guy-Crescent Fagon (1638–1718), initially appointed to the position of démonstrateur et opérateur pharmaceutique pour l'intérieur des plantes in 1672, acquired greater control of the facility during his years as intendant. Until 1695, his teaching duty was assigned to Simon Boulduc, who then acquired an additional chair in chemistry. From this point on, chemistry had two chairs at the Jardin. Simon Boulduc's new chair passed down to his son Gilles-François in 1729. Rouelle assumed this position in 1743 under the title démonstrateur en chimie au Jardin des Plantes, sous le titre de professeur en chimie.[102] In parallel with Simon Boulduc's lectures, Etienne-François Geoffroy, Louis Lemery, and Berger substituted for Fagon over the years until Geoffroy was appointed démonstrateur de l'intérieur des plantes et professeur en chimie et pharmacie in 1712. Geoffroy's chair passed on to Louis Lemery in 1730, and then to L. C. Bourdelin in 1743. (See the appendix.) Buffon assumed the intendancy of the Jardin in 1739 and embarked on an ambitious project of expanding its perimeters and activities.[103]

According to this genealogy of the Jardin's chemistry chairs, E. F. Geoffroy, Louis Lemery, Simon Boulduc, and G.-F. Boulduc are among the candidates for Rouelle's chemical teachers, in addition to any private courses he might have attended (such as Jean Grosse's). While their lecture notes have not come into light, we know their specialties from the research papers which indicate some continuities between their works and Rouelle's lectures. Simon Boulduc was an expert on vegetable analysis. Gilles-François Boulduc conducted sophisticated analysis of mineral waters. Geoffroy advocated a move away from the corpuscular excess, while praising Stahl's chemical works. Lemery developed a sophisticated conception of solution chemistry and of fire. Grosse advanced the theoretical understanding of middle salts. Rouelle came to be known and remembered for his analysis of vegetables, which apparently surpassed Boerhaave's, and participated in the Academy's project of mineral analysis.[104] He enshrined Stahl as the supreme chemical authority, which could be easily accounted for if he had contact with Geoffroy or Grosse. Rouelle's notion of fire incorporated some of Lemery's. The French chemical tradition played an important role in shaping the practical part of Rouelle's teaching, even if he relied heavily on imported theories.

Rouelle's few publications indicate that he kept abreast with investigations of salts, both French and German. His "general theory" of salts mixed various theories, old and new, to provide an order for the expanding territory of salts. Two features stand out. First and foremost, it introduced the classificatory language of natural history into chemistry. Second, it utilized relatively new theories of crystallization as the basis of classifying salts. In order to justify this move, Rouelle gave a careful reflection on the dissolution model of chemical action, elaborated earlier by Louis Lemery, while insisting on the chemical, rather than the mathematico-physical explanations. Rouelle's first paper, presented to the Academy as an election bid on December 18, 1743, and published in 1744, dealt with the classification of neutral salts. He defined the "neutral, middle, or salted" salt more broadly than before, to accommodate the extension and sophistication the subject of salts had acquired during the previous decade:

> I give to the family of middle salt all the extention it can have: I call middle, neutral, or salted salt all the salts formed by the union of some acid whatever this may be, either mineral or vegetable, with a fixed alkali, a volatile alkali, an absorbent earth, a metallic substance, or an oil.[105]

Rouelle's acquaintance with the Jussiueu brothers must have been more than perfunctory, to judge from his interest in classification. In order to "make some progress on the theory of the crystallization of salts," Rouelle deemed it "essential or absolutely necessary" to have a good classification that would group similar crystals together and differentiate dissimilar ones. Some classifications of neutral salts already existed in Lemery's memoirs, in Stahl's *Traité sur les Sels*,[106] and in Gulielmini's "Dissertation on Salts." Stahl, in particular, paid much attention to the circumstances of crystallization and to the effects of water as a solvent. None of them focused specifically on crystallization, however, which left relevant information in scattered places. Rouelle wanted to gather these fragments of thought on crystallization and to match them with his own observations and experiments to "form a whole." First he divided all neutral salts into two classes according to the different quantities of water they required for dissolution. Each of these classes was then subdivided into two sections depending on the quantity of water that entered into the formation of their crystals. This method lined up most of the crystals analogous in figure and in the manner of crystallization, but kept some similar crystals too far apart. Rouelle thus proposed another method, which took into consideration the figure of crystal, its manner of formation, and other phenomena. In this scheme, all neutral salts were divided into six sections according to the figure of their crystals. Each section was subdivided into genres by the acids and into species by the bases they contained. The first section, for example, consisted of middle salts whose crystals had the figure of small foils or very thin shells. These salts crystallized perfectly on the surface of their solvent at insensible evaporation. They were easy to crystallize because they had the least amount of water in their crystals and demanded the most water to be dissolved. So far, only one genre belonged to this first section: that of vitriolic acid.

Rouelle introduced a new way of characterizing and classifying salts on the basis of their properties of dissolution and crystallization. He disputed the mathematical model of dissolution that postulated an "equiponderance" or the same specific gravity between saline molecules and water. He claimed that chemists generally disliked this model, since saline molecules, much more composed than water, could not be specifically as light as the parts of water. Chemists supplemented this mathematical model by devising the "pores" of saline molecules or by recourse to the movement of water, which held the saline molecules suspended. Rouelle preferred Stahl's explanation which stipulated the "union" of saline molecules with water. The size of saline molecules and the movement of water were auxiliary causes. On the cause of crystallization, Rouelle deferred judgment, reciting various opinions. Descartes attributed it to the "impulsion of subtle matter," Newton to attraction, and Becher and Stahl to the nature of the faces of saline molecules.

Rouelle's "general theory" of salts was nothing more than a classification of salts according to their modes of crystallization.[107] Nevertheless, it extended the scope of neutral salts to include not only the combinations of acid with alkali, volatile alkali, absorbent earth, and metallic matter, which had been the working definition since Glauber, but also the combinations of acids with "whatever substance that serves as the base and gives it a concrete and solid form."[108] The classification based on crystal forms had some problems, however, since the figure of the crystal could change with different ways of evaporation: insensible, moderate, and rapid.[109] Such complication may have hampered Rouelle's ambitious project of building a general theory of neutral salts. In 1754, Rouelle extended the scope of neutral salts still further.[110] He showed that there existed neutral salts that contained much more acid than their base could absorb and thus were easily soluble. Accordingly, neutral salts had to be divided into three classes: those with excess acid (which dissolved quite easily), the perfectly neutral or salted salts (which were less soluble), and those with little acid (which were nearly insoluble). The historian of the Académie concluded that this research would serve as the "key to an infinity of embarrassing phenomena" and "throw a new light on the nature of neutral salts."[111] To judge from Lavoisier's later praise, Rouelle made a mark in French chemical scene with his investigation of salts:

There is a certain order which is observed in human knowledge that is impossible to invert, and on which depends all the success of our discoveries. It is thus that the celebrated doctrine of M. Rouelle on the different quantities of acid which can enter into the composition of the same salt must necessarily precede the history of gypsum and its varieties. This discovery so fecund, the greatest that has been made in theoretical chemistry since Stahl, will serve as basis for all the aetiologies one will find in this memoir. Furthermore, I do not doubt that it will serve in the future as the foundation of an infinity of others and that it will unveil for the posterity the most impenetrable mysteries of nature.[112]

Though Rouelle relied on Stahl as an authority on the subject of salts, his lectures drew heavily on Boerhaave. Without particular regard for coherence, he made more extensive use of Boerhaave's instrument-element theory than of Stahl's theory of composition. He did not invoke their systems intact, but selectively adopted various parts to gather a loose congeries of concepts that catered to his repertoire of experiments, thereby adapting them to the existing French didactic tradition. Of the theory domains of composition and affinity, Rouelle's hybrid system favored the former, probably because of his emphasis on vegetable analysis and his aversion to Newtonian natural philosophy. Rouelle regarded vegetable analysis as "that in which all the parts are better linked, and the most proper to make the true spirit of chemical analysis known."[113] Diderot called Rouelle's lectures on the vegetable kingdom a chef-d'oeuvre. Apparently, Rouelle claimed that chemical analysis could furnish a general botanical method, or that one could determine the true class of a plant by way of chemical decomposition, thereby reviving a paradigm long in hibernation.[114]

The mark of the French tradition is unmistakable in Rouelle's lectures, albeit modified by Boerhaave's and Stahl's theories. True to the long-standing didactic tradition at the Jardin, Rouelle began with a short theoretical section in which he dealt with principles and instruments. He reiterated, for example, the analytic-synthetic ideal by defining chemistry as "a physical art which by means of certain operations and instruments teaches us to separate from bodies several substances which enter into their composition and to recombine them anew among them or with others to reproduce the original bodies or to form new ones."[115] Chemists divided the bodies of nature into three kingdoms: vegetable, animal, and mineral. Decomposition and recomposition of the bodies of these three kingdoms thus constituted the practical part of Rouelle's lectures.

Rouelle's discussion of principles allows us to gauge the difficulty French chemists faced in reconstructing the theoretical discourse of chemistry after the demise of the five principles. He offered a brief history of principles and elements seen through Stahl's notion of hierarchical composition. He claimed that ancient philosophers distinguished "elements" from "principles": Elements were the "molecules of first composition, that is to say, indivisible and simple molecules that enter into the composition of all bodies." Principles were the "compounds of elements" and made up more complex bodies. In other words, Rouelle reviewed the history of the terms "element" and "principle" from Stahl's perspective of hierarchical composition, in which elements were more fundamental than chemical principles. Such a complex history did not deter Rouelle from asserting chemists' control over the material world, although he made some necessary concessions. Notwithstanding the historical/philosophical distinction he outlined between elements and principles, he admitted four principles or elements—phlogiston or fire, earth, water, and air—while hinting at the existence of the fifth, mercurial principle of Becher. He defined principles or elements in particulate terms:

For us, we call principles or principal bodies (principle here is synonym of element) bodies simple, indivisible, homogeneous, immutable, and insensible, more or less mobile according to their different nature, their figure and their mass. They differ among themselves by their volume, their particular figure and their nature. It is impossible for the eyes to perceive them alone and to separate them from other principles unless they are reunited in quite great numerical quantity; thus one should ignore what is their particular figure and their nature. What is certain in some ways is that they are in small number and that nevertheless their different combinations suffice to form all the bodies of nature.[116]

These principles did not form bodies immediately by themselves, as Aristotle had stipulated with his elements and Epicurus with his atoms. Rather, they followed the order of composition stipulated by Becher and

Stahl: First, they formed "mixts" by a "juxtaposition" of two, three, or four principles entering in larger or smaller quantity. This variation could produce a prodigious number of different combinations, as Stahl had mentioned, but Rouelle estimated the number of mixts at about ten or twelve. The combination of principles into mixts took place instantly without altering the principles. Their union was quite strong and difficult to rupture because they were so small and escaped the action of chemists' instruments. The only means of separation was combination. Second, several mixts of different natures formed "compounds." This union was not as strong as those of the principles in mixts. Third, Rouelle modified Becher's notion of "supercompounds"—mixts joined to other principles or mixts—to refer to a higher order of composition, that is, the union of several compounds of different natures. Fourth, "aggregates," according to Becher and Stahl, were mixts or compounds united in great number by cohesion. There were several kinds of aggregates. They could be composed of principles alone, of mixts alone, or of compounds or supercompounds alone. It was only under this form that all the bodies of nature were presented to our eyes. One ruptured the aggregation of bodies without decomposing them. Regal water separated out the molecules of gold without decomposing the gold. Each molecule was gold. Although this elaborate Stahlian hierarchy of composition did not endorse the five principles, it provided (as Fourcroy later pointed out) a stronger justification for the principalist view of composition, which granted that compounds retained the properties of their constituent principles. Roulle's modified Stahlian hierarchy looked as follows:

principles   mixts   compounds   supercompounds
        aggregates.

Rouelle's thoughts on chemical analysis reflected his expertise in the distillation analysis of vegetables. He accepted Stahl's premise that chemistry had long been occupied with the decomposition of supercompounds into compounds without penetrating more deeply into the composition of compounds or mixts. Such analysis would require well-proportioned instruments and a well-adjusted movement. Rouelle did not believe that solution methods could provide proper means for this kind of fundamental analysis. Chemists used gross menstruums because their actions were more sensible and prompt and agreed well with the levity of chemists' spirit. They did not have patience for long digestions, repeated cohobations, triturations, sublimations, precipitations, etc. These were means unknown to ordinary chemistry but more capable of attacking and rupturing the last combinations of matter when one applied them for a long time and constantly on the same mixt. In nature, two great changes operated on the mixts by the single means of continued movement—fermentation and putrefaction—the powerful destructive agents of vegetable and animal nature. Rouelle hoped that with variations in distillational techniques one could get at the principles more easily than with solution analysis. He was aware, however, that one could never demonstrate the principles separately. One had to pass them to another body and make new combinations in order to discern their properties.

Rouelle followed up his discussion of principles with that of "Instruments," without specifying the relationship between these two categories. He named six instruments, as Boerhaave had stipulated. The four "natural" instruments were fire, air, water, and earth. The two "artificial" ones were menstruums and vessels. The knowledge of these instruments and their usage constituted one of the most essential parts of chemistry. Unlike Boerhaave, Rouelle made clear that the four natural instruments also served as the elements of bodies. For this reason, Rappaport has characterized his theory as an "instrument-element" theory.[117] Although Rouelle followed Boerhaave's conception of instrument closely, right down to the examples, he introduced a being that Boerhaave had not endorsed: phlogiston. Fire, or phlogiston, acted as an instrument of chemical action when it changed the physical state of substance as in the rarefaction of air. Fire could also act as a chemical element when it combined with other principles. Similarly, air acted as an instrument of chemical action, since most chemical operations took place in the air. It was a "fluid, elastic, and mobile element, which owed all its mobility to fire without which it was never found."[118] At the absolute cold, therefore, air would become a solid substance. Elasticity was a property of air exhibited only at the aggregate level. When it entered into chemical combination, it assumed the "state of separation, and reduced to the unit" that was inelastic. The fluidity of water also derived from the movement of the fire contained in it. Losing fire, water became ice or the pure state of water. Water was an essential instrument in solutions, crystallizations, and combustions. Rouelle did not formulate as clear a theory of earth as for other three elements. Out of Becher's three earths (vitrifiable earth, sulfureous earth or phlogiston, and mercurial earth), he transferred phlogiston to fire. He identified vitrifiable earth as "the principle of stability and solidity and the basic constituent of most complex substances." The mercurial principle was responsible for the property of metals. As Rappaport has pointed out, earth did not have a clear qualification as an instrument-element in Rouelle's theory. In fact, even air and water did not have clear qualifications as instruments, since their capacity as instruments derived from the presence of fire. Although Rouelle followed Boerhaave's scheme rather closely in expounding his four-element theory, he rendered these elements chemical by filtering out the instrumental basis of Boerhaave's conceptions, such as the thermometer and the air-pump. His hybrid theory of elements revived the principalist approach to composition that was rooted in the distillational analysis of vegetables.

Rouelle gave more thought to the chemical effects of fire by mixing in Stahl's idea of composition. While Boerhaave left some ambiguity as to exactly at which level of composition fire acted in producing chemical combinations, Rouelle limited the action of fire to breaking up the molecules of aggregates. Fire in rarefying bodies separated the molecules of aggregates, but not those of mixts. Gold in fusion lost its aggregation, but not its mixtion. Each molecule of the fluid still retained the properties of gold. Fire produced fusion by gliding into the pores or the intervals of the parts of aggregates, but it could not penetrate those of mixts. In explaining combustion, Rouelle acknowledged the existence of the "inflammable principle," a mixt composed of fire and other principles. It existed in the oily part of the body, expanded into the air upon combustion, and entered new combinations afterward. Too subtle to be isolated in chemical analysis, it united with certain bodies, albeit with different strengths. This principle, or phlogiston, was the same in animals, vegetables, and fossils. In fossils, one found a quite pure form of the inflammable principle. In sulphur, phlogiston or the matter of fire was united to quite pure acid. In metals and semi-metals, it was united to other principles which did not yield oil. One could not draw oil from metals, but calcined metals regained their metallic form when replenished with phlogiston.

Chemical affinity did not play a prominent role in Rouelle's theoretical lectures, perhaps because of his aversion to Newtonian explanations. He endorsed the premise of solution chemistry that chemical action began with the action of solvents which separated the particles of bodies. He further distinguished between dissolution and resolution, referring to chemical combination and solution in our terms, respectively. Dissolution involved the parts of solvents interposing between the parts of dissolved bodies, thus dividing them and uniting with them to form new compounds. There were "two actions in each dissolution; one is the division of the parts of two bodies, and the other, their combination. This union is made in the same instant of the division."[119] This simultaneous action of division and unification was reciprocal between the parts of solvents and those of dissolved bodies, but the name "solvent" usually was reserved for the liquid body. When the two bodies did not form a new compound, as in the case of water and salts, Rouelle argued, it should be called resolution rather than dissolution. The causal explanation of dissolution as a chemical process was more problematic. Rouelle did not endorse a purely mechanical explanation or Newtonian attraction. Fire, as the "promoter" of dissolution, acted in two ways. Through its property of rarefaction, it separated the parts of bodies from one another. As a mechanical cause, it gave more movement to the bodies. Experiments in the pneumatic machine gave some idea as to how air was involved in chemical actions. Since fire and air had to "act indifferently on all bodies," their actions could not account for the selectivity of dissolutions. The cause of union in dissolution had to be a "virtue inherent in these two bodies," albeit unknown. Ancients called it "sympathy"; moderns called it "affinity or rapport." Rouelle asserted that this "attraction did not follow the law of the square of the distance as the Newtonian attraction, but that of the homogeneity of the surfaces."[120] Although chemists only had an imperfect knowledge of it, this cause explained all the phenomena of dissolution better than any other system based on the figure of points and the pores of bodies.

If Rouelle carefully circumscribed the physicalist explanation of chemical action, he used E. F. Geoffroy's affinity table without reservation. This reflected his interest in salts. The expanding list of salts demanded a systematic revision of Geoffroy's table already in the 1740s, as Claude-Joseph Geoffroy contended. The younger Geoffroy had discovered in 1744 that the behavior of vitriolic acid did not conform to his elder brother's predictions.[121] Rouelle pointed out several exceptions to the first and the third columns of Geoffroy's table. His own table, reproduced in the *Encyclopédie* and shown here as figure 4.2, did not differ greatly from Geoffroy's except for the three additional columns.[122] He split the column headed by metallic substances into two (columns 8 and 9) and added at the end of the table two columns headed by spirit of wine. Geoffroy's "oily principle" now became phlogiston, represented by the same symbol.

Rouelle reinvigorated the French didactic tradition by pulling out disparate components of competing chemical systems, matching them to the existing repertoire of experiments, and thereby making them accessible to the diverse Parisian audience. He sought to rehabilitate chemistry as a rational science with its own principles and methods, rather than an art in need of a philosophical discourse, as Boerhaave had perceived. If some of his more articulate students missed philosophical depth and methodological rigor in his lectures, they also appreciated the view of a vast chemical territory which could not be otherwise traversed. Rouelle's assets as a teacher obviously did not include beautiful speech and methodical arrangement of the subject matter. His effectiveness as a teacher derived more from his ability to keep abreast with the new developments and fashions in chemistry and to refine the relevant analytic techniques. Aside from the popularity they enjoyed among the learned audience, Rouelle's lectures also contained genuine improvements in vegetable analysis. He "added much to Boerhaave's beautiful system of vegetable analysis, first distinguished with more care the immediate components of vegetable matter, divided and characterized by their better known properties the different kinds of extracts; discovered the glutinous material in green leaves; compared gums and sugar to starch; pub-

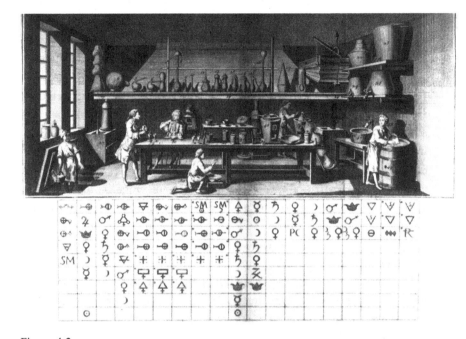

Figure 4.2
Rouelle's affinity table in Diderot's *Encyclopédie*.

lished in his processes a more complete and especially more methodical outline of vegetable analysis than had been done until that time, and revived the hope of experimenters."[123] He also adopted and "perfected" Hales's apparatus for collecting air.[124] His lecture demonstrations introduced the next generation of French chemists to the basic techniques of pneumatic chemistry well before Priestley's experiments began to cross the channel in the early 1770s. Venel, Lavoisier, and Bucquet built on Rouelle's innovation.[125] Rouelle also invented various didactic techniques to enhance his pupils' understanding. In the private lectures, for example, he prepared an exhibition of bottles labeled with "very simple, very brief, but very clear information about each of his experiments" to present "a short but adequate description of all the operations which formed the links of his demonstrations, and served as the foundation of his doctrine." These "Processes of the Course in Chemistry," which traced his lectures step by step and increased in number as his course proceeded, remained on the shelf of his laboratory throughout the duration of his course to refresh students' memory.[126] Rouelle was an inspirational teacher, if not an original thinker:

When Rouelle spoke, he inspired, he overwhelmed; he made me love an art about which I had not the least notion; Rouelle enlightened me, converted me; it is he who made me a supporter of that science which should regenerate all the arts, one after the other. . . . Without Rouelle, I would not have known how to look above the mortar of the apothecary.[127]

4.5   Pierre-Joseph Macquer, Writer

Despite his fame as a teacher, Rouelle's influence was limited more or less to the Parisian audience and to those who had access to the manuscript copies of his lectures. The French textbook tradition found a new champion instead in Pierre-Joseph Macquer (1718–1784), one of Rouelle's early students. Macquer was born in Paris of an émigré Scottish noble family.[128] He received an early education with his brother Philippe from Le Beau who laid the foundation of his clear and elegant style in writing.[129] Graduated from the Faculté in 1742, he practiced medicine for a few years before his election to the Academy in 1745 as an adjunct chimiste. Although he had no chemical work to his credit at the time of the election, he caught up quickly, soon presenting a paper on the solubility of various oils in spirit of wine.[130] Macquer must have settled on a course of studying chemistry during the early 1740s. He soon published *Elémens de chymie* (1749), which proved extremely popular.[131] Oddly for an author of a representative textbook, his teaching activity began some time after he published the *Elémens*. Aside from a year's teaching as professor of pharmacy at the Faculté in 1752, his formal teaching did not begin until 1770, when he filled in for Bourdelin at the Jardin.[132] Between 1757 and 1773, however, he taught a private course with Antoine Baumé, an apothecary and instrument maker. Macquer's *Dictionnaire de chymie*, originally published anonymously in 1766, enjoyed even greater popularity than the *Elémens*, and a second edition was published in 1778.[133] Macquer is known to posterity mostly for these books, but he meant a great deal more to his contemporaries. He rose rapidly in the Parisian scientific scene. In 1750 he began to work in the

Figure 4.3
Illustration on title page of Macquer's *Élémens de chymie—théorique* (Paris, 1756).

Figure 4.4
Illustration on title page of Macquer's *Élémens de chymie—pratique* (Paris, 1756).

royal porcelain and dye works, both directed by Hellot. Most of Macquer's research throughout his active life was devoted to improving these arts and to serving as a consultant for the Bureau de Commerce.[134] By 1766 he had a commanding position within French chemistry. In that year, he was advanced to an associé in the Academy and succeeded to both of Hellot's posts at the porcelain and the dye works. From 1768 to 1776 he edited sections of the *Journal des Sçavans* dealing with physics, medicine, pharmacy, surgery, chemistry, astronomy, and natural history. His literary activities brought him in contact with many domestic and foreign chemists.[135] He became a pensionnaire of the Académie in 1772 and took up Bourdelin's post at the Jardin in 1777. Macquer was at the summit of his career and influence during the 1770s, when Lavoisier began to work seriously in chemistry. As a master academician, a prolific writer, and a director of royal industries, Macquer represented French chemistry a generation before Lavoisier's.

The capacity for reading and writing shaped Macquer's career. Since he wrote the *Elémens* before he began any serious teaching activities, one might suspect that he drew on the existing French didactic tradition and on written sources. He took Rouelle's course in the early 1740s, but never acknowledged his debt to Rouelle. Since Rouelle and Macquer joined the Academy at about the same time and shared the intellectual environment of Parisian chemistry, it would be difficult to determine whether Rouelle provided a model for Macquer. The extant copies of Rouelle's lecture notes all date from the mid 1750s—several years *after* the publication of Macquer's *Elémens*, which differed substantially in its organization and emphasis from Rouelle's lectures. Rouelle preferred to proceed from the most complex to the simpler subjects because it conformed to the "analytic order"; Macquer chose to proceed from the simpler to the more complex, ostensibly following the method of geometry. Whereas Rouelle revived the principalist approach to composition by drawing on Boerhaave and Stahl, Macquer mixed Stahl and Newton to frame affinity chemistry. The younger generation of chemists inherited both of their approaches and used them creatively, although one can discern certain preferences in their respective students. Venel and Lavoisier carried on the principalist outlook in French chemistry; Guyton, Fourcroy, and Berthollet took affinity as the central project of theoretical chemistry. Their approaches to composition and affinity differed significantly, as we shall see.

Macquer's thoughts on principles and elements show that he studied French textbooks carefully. He endorsed the four "principles or elements," but with an emphasis on their analytic rather than their philosophical status. The "object and principal goal" of chemistry was to separate different substances that entered into the composition of a body, to study each of them, and to "reunite & rejoin them in a new ensemble to make reappear the first mixt with all its properties."[136] This "analysis, & decomposition" of bodies were limited, however; one could proceed only to a certain point beyond which all efforts failed. Substances that stopped all means of analysis—earth, water, air, and fire—were given the name of "principles or elements." Macquer did not simply revert to the Aristotelian four elements, therefore, in naming these four substances as principles. He suspected that these four might not be the "primordial elements or matter or the most simple elements." Because the senses tended to deceive, and because chemists had not yet reached the simplest elements, he thought it expedient to regard these four substances as principles or elements instead of trying in vain to get at the ultimate elements. Macquer's rationale for choosing the four principles echoed the analytic ideal defined in the French didactic tradition, in contrast to Boerhaave's and Rouelle's instrument-elements. The first elements composed secondary principles, while the union of both primary and secondary principles resulted in proper compounds or mixts. This hierarchy in the organization of chemical substances resembled Stahl's system. The next generation of French chemists tacitly accepted Macquer's analytic definition of principles and simplified hierarchy of bodies well before Lavoisier stated it explicitly in his *Traité*.

Although Macquer did not utilize Boerhaave's conception of instruments, he took the four elements not as quality-bearing, elusive principles, but as material bodies subject to experimental control. The ramifications of Boerhaave's efforts to reconcile the chemical and the physical properties of four elements become obvious in Macquer's discussion. Air was the fluid that we respired. It surrounded the globe. It was susceptible to condensation and rarefaction caused by heat or cold (that is, by the presence or absence of the parts of fire). This gave it elas-

Figure 4.5
Pierre-Joseph Macquer. Courtesy of Académie des Sciences Archive.

ticity, or the ability to occupy different volumes of space. Air penetrated all spaces and entered into the composition of several substances, particularly vegetables and animals. The large quantity of air released from these bodies made it difficult to believe that it retained its elasticity while in combination. Ordinarily a fluid, water became a solid at a certain degree of cold, which was its natural state. When heated, it began to evaporate at a certain "degree of heat." Water entered into the combination of many bodies. Like air, it did not combine with metals or with most minerals. Air and water were volatile principles; fire could dissipate them. Earth was a fixed principle when absolutely pure, and it resisted the greatest violence of fire. Usually one encountered it mixed with other substances which rendered it somewhat volatile. One thus divided it into fusible or vitrifiable earth and non-fusible or unvitrifiable earth.

Macquer's discussion of fire, which was built on Louis Lemery's, deserves a careful reading for his clear distinction between two different kinds of fire: The pure, elementary fire acted on bodies to separate their parts. Phlogiston was a secondary principle composed of elementary fire and some other principle. The fact that the element of fire had many names—the matter of sun or of light, phlogiston, fire, sulphur principle, inflammable matter—indicated that there did not exist "a sufficiently exact distinction of the different states in which it is found; that is to say of the phenomena it presents, & of the name it truly merits when it enters effectively as principle in the composition of a body, or even when it is alone & in its natural state."[137] The greatest change it could make by its presence or absence was rendering a body fluid or solid. One could thus argue that all bodies were solid in nature, and that fire, essentially fluid, was the principle of fluidity. Even air would become solid if it were sufficiently deprived of fire. The principal properties of "pure fire" were to penetrate bodies to establish a kind of equilibrium and to dilate the bodies it penetrated. As such, it was the most powerful agent of decomposition. The greatest change it could make by its presence or absence was rendering a body fluid or solid. One could thus argue that all bodies were solid in nature, and that fire, essentially fluid, was the principle of fluidity. Even air would become solid if it were sufficiently deprived of fire. Macquer's discussion of fire contained several elements later utilized by Guyton and Lavoisier. Guyton took up the notion of essential fluidity for his theory of phlogiston. Lavoisier picked up the function of fire in phase changes.

The combustion of inflammable matters indicated that they contained the matter of fire as one of their principles. By what mechanism such a penetrating and active fluid became fixed in bodies was difficult to determine, however. In order to distinguish this "fixed fire" from "pure fire," one gave it the name of "inflammable principle, or sulphur principle, or phlogiston." It differed from the elementary fire. First, when united to a body it did not communicate any heat or light. Second, it did not contribute to the solidity or fluidity of the body that contained it. Third, it could be transported from one body to another. This property in particular made it plausible to distinguish phlogiston from pure fire, or to consider it "as the element of fire combined with some other principle, which serves as the base to form a kind of secondary principle."[138] Phlogiston could not be separated in pure form. It could only be transferred from one body to another or be dissipated completely in calcination and combustion. The inflammability of a body gave "certain indication" that it contained phlogiston, but the lack of inflammability did not imply the deprivation of phlogiston. Charcoal, an intimate combination of phlogiston and the earthly substance in vegetables and animals, was a substance quite proper to the transmission of phlogiston to other substances.

Macquer gave a much more prominent place to affinity than Rouelle. In his second chapter, he listed six "fundamental truths" concerning the "convenance, rapport, affinity, or attraction" that explained the selectivity in chemical action.[139] First, as Geoffroy had stipulated, if one presented to a compound of two substances a third substance which had a greater rapport, the third substance would decompose the compound and form a new union. Second, the third substance could join the compound without decomposing it. Third, a substance that could not decompose the compound by itself could succeed nevertheless when it was combined to another via "double decomposition." Fourth, when the substances were united they lost some of their properties. Fifth, one could establish it as a general law that all similar substances had affinity for one another and were thus disposed to join (as water to water, earth to earth). Sixth, the simpler substances were, the more sensible and considerable were their affinities and the more difficult it was to separate them. This expanded set of "truths" about affinities allowed Macquer to classify diverse chemical operations at the abstract level and to map chemical experience in a more intelligible manner.

Macquer also took the affinity table much more seriously than Rouelle. He regarded the chapter on the affinity table as the most fundamental part of his textbook, emphasizing the "great utility of assembling under a single point of view the principal truths enunciated in the treatise." He reproduced Geoffroy's table, instead of giving a new one "containing all the changes and innovations," since it included all "fundamental affinities" suitable for an elementary treatise.[140] He advised the students that it would be quite advantageous to consult the table "every 'time there arose a question of some affinity" rather than wait until one had read through the book. By representing the affinities directly to the eyes, the table would fix them in memory. Macquer did explain some problems associated with Geoffroy's table, however, indicating that it had been subject to critical examination as chemical analysis became more refined. The most problematic column was the first, which prescribed the behavior of all acids in the same manner. Probably drawing on Rouelle's well-publicized table (discussed above), he also proposed to add a short column headed by spirit of wine, followed by water and oil. For the fourth column, Macquer identified Geoffroy's sulphur principle with phlogiston, but he pointed out the inconsistencies between this and the first column. The first column dictated that any acid would have more affinity with fixed alkali than with any other substance. In the fourth column, however, phlogiston was shown to have more affinity with vitriolic acid than fixed alkali. This exception to the order of affinity derived from the "famous experiment" in which one decomposed vitriolic tartar and Glauber's salt by the intermediate of phlogiston. In this experiment, phlogiston separated fixed alkalis from neutral salts and united with the vitriolic acid they contained to form sulphur. As for Rouelle, phlogiston was for Macquer only a part of a more comprehensive chemical theory based on the definition of elements, their combinations, and their reactions.[141]

Macquer's identification of affinity with attraction in his *Elémens* indicates his favorable disposition toward Newtonian language. He had experienced firsthand the debate between Clairaut and Buffon over the terms of attraction that had erupted in 1745, soon after he joined the Academy. In the *Dictionnaire*, he introduced an effective transition from the Cartesian to the Newtonian paradigm in the philosophical language of chemistry, once again adapting the formal discourse of chemistry to the prevailing intellectual fashion.[142] As Condorcet put it, Macquer delivered chemistry from the false pretense of Cartesian mechanical explanation juxtaposed on lingering alchemical obscurity.[143] Macquer's use of Newtonian language made it an important element of the public discourse of chemistry, enhancing chemistry's image as a rational science.

More important for chemical practitioners was the way in which Macquer identified affinity as the cause of all chemical combinations without engaging in causal speculation. He defined affinity as "the tendency" the "constituent, or integrant parts" of bodies had toward one another and "the force that makes them adhere together when they are united." He insisted that affinity was not an "empty word" but a force whose effects were quite sensible and palpable, as demonstrated in the details of all chemical phenomena. He considered it a property as essential to matter as extension and impenetrability. That is, affinity was what constituted the object of chemical inquiry: chemical combinations. He advised others against searching for its cause, however.[144] This was, of course, a familiar rhetorical move. Newton had used it in presenting gravity as an efficient cause for natural philosophy, so as not to be encumbered by the causal discourse of traditional philosophy and theology. One should, Macquer contended, be content with observing the principal laws that bodies followed in their various unions and combinations by virtue of this property. Just as Newton asserted the independence of mathematized natural philosophy from the traditional philosophy with his concept of gravity, Macquer sought to make chemistry independent from natural philosophy with the concept of affinity, bypassing the ontological and causal concerns.

Shifting the field of argument from philosophical discourse to empirical inquiry in this way, Macquer proceeded to offer a systematic classification of affinities. His classification, based on abstract characterization of chemical operations, mediated between the evolving practice of technical chemistry and its philosophical aspirations. He first distinguished simple affinity from complicated affinity. Simple affinity was further subdivided into the affinity of aggregation and that of composition. The affinity of aggregation worked between the "integrant parts" of the same body to bring them together in a state of aggregation, or a homogeneous mass. The affinity of composition, in contrast, worked between the integrant parts of different compounds to form a new compound. It was the force that united the "constituent parts or molecules" of a compound body. Macquer's distinction between the integrant and the constituent parts of a body conceptualized chemists' analytic experience: They could either divide them mechanically, the limit of which produced integrant parts, or decompose them chemically to produce the constituent parts. In so doing, Macquer also translated Stahl's principalist distinction between compounds and aggregates into an operational distinction of affinities:

The constituent parts are, properly speaking, the principles of bodies. These are substances of different natures which, by their union and mutual combination, really *constitute* mixed bodies which partake of the properties of their constituent parts. For example, the constituent parts of common salt are an acid and an alkali, of which this salt is composed, & which one has to regard as its principles, at least as its proximate principles. . . .

On the contrary, the *integrant parts* of bodies do not absolutely differ from each other; nor do they differ, as to the nature and principles, from the body into whose mass they enter. One has to understand by integrant parts of a body the smallest molecules into which this body can be reduced without being decomposed.[145]

The complicated affinity brought together more than two bodies and was necessarily a kind of the affinity of composition. Macquer subdivided it into five categories, but without labeling each of them clearly. It is easier to follow his fellow lecturer Baumé, who simply listed seven different species of affinity. The first two, affinity of aggregation and simple affinity of composition, referred to Macquer's two kinds of simple affinity. The next five dealt with different species of complicated affinities. In the following, I have mixed Macquer's explanation with Baumé's names and examples for better comprehension. First, compound affinity in Baumé's usage referred to the cases in which the two principles in union were approached by a third principle which had an equal or nearly equal affinity with the principles in combination. This resulted in one combination. Second, intermediate affinity in Macquer's usage occurred when the third principle had an equal affinity with only one of the principles in combination. In this case, the three resulted in one combination with one principle serving as a link or as an intermediate between two others which would not otherwise combine together. Third, the affinity of decomposition in Baumé's usage applied to the classic case of displacement; when the third principle had a far superior affinity with one of the two principles in combination, a total decomposition of the existing combination and the formation of an entirely new combination followed. Fourth, reciprocal affinity in Macquer's usage applied to the cases when the contending affinities were nearly equal. The principle that had been separated from another could reclaim it under different circumstances. Vitriolic acid, for example, displaced nitrous acid from niter to form vitriolated tartar with vitriolic acid. But nitrous acid could displace vitriolic acid from vitriolated tartar to form the original niter.[146] Fifth, two compounds could exchange their four principles to form two new compounds; this was called "double affinity."[147]

Macquer and Baumé's classification of affinities thus introduced a typology of chemical operations, allowing them to expand the discourse of affinity as a general explanatory scheme. Through their classificatory efforts, the notion of affinity became an integral part of chemical explanation. Chemists could simply invoke different kinds of affinity to explain multitudes of chemical operations without further regressing into complicated causal explanations or ontologies.

Macquer's formal discourse of affinity played a central part in forging an autonomous science of chemistry. It became a standard part of chemical, pharmaceutical, and medical textbooks, and it remained so long after the investigative program of affinity was no longer deemed viable. His texts propagated affinity chemistry beyond the French didactic tradition to form a broad frontier of theoretical chemistry. Bergman, Guyton, Kirwan, and Richter all learned Macquer's program from his texts and devoted their efforts to characterizing this elusive concept. Macquer's discourse of affinity fostered an understanding of chemical constitution that was contingent upon physical circumstances as well as upon chemical forces. This dynamic understanding of chemical constitution evolved into a much more sophisticated form in the hands of Macquer's most outstanding student: Claude-Louis Berthollet.

4.6   A Fractured Identity

French chemists entered the Enlightenment with a stable institutional niche in medicine. Whereas the struggle against the Faculté during the previous century provided chemical physicians and apothecaries with a singular identity and purpose, the Enlightenment public sphere demanded a more universal linguistic representation of their "useful" knowledge. French elite chemists tried to rehabilitate chemistry as an independent science, mobilizing a set of resources: the affinity table, sophisticated knowledge of salts, a new chemical authority in the person of Stahl, and literary devices. Their standpoints differed markedly, however.[148] Macquer, as a graduate of the Faculté and as a literary persona, wielded a triumphalist rhetoric to win over the educated audience, emphasizing the relevance of chemistry to physics, the relevance of mathematical language to chemistry, and the recent dramatic developments in chemistry. Rouelle and Venel, with closer ties to the apothecaries' culture but exposed to the intellectual atmosphere of the salon, felt insecure in their role as bearers of much-sought-after practical knowledge. They felt threatened by an intellectual milieu that ignored chemists' distinct identity and, in particular, by the scholarly and physicalist condescension permeating Boerhaave's learned discourse. They attacked the "sovereignty of physics," which obliterated the distinct realm and the identity of chemistry and chemists. The fractured identity of chemists during the High Enlightenment provides an important context for the Chemical Revolution, which accelerated the bifurcation between the philosophical chemist and the apothecary.[149]

In the preface to his *Elémens*, Macquer emphasized that the sciences were advanced not by "vain reasoning" but by experiments. Although this was a well-rehearsed rhetoric that Rouelle used often in his lectures, Macquer's interpretation of sound reasoning based on facts differed significantly from Rouelle's. In his view, chemical experiments and facts had to acquire a taste of "true physics" in order to serve as the foundation of scientific chemistry. The use of experiments had allowed physics to progress much more rapidly in the past 150 years than it had in the preceding thousand, and it could do the same for chemistry. Although chemistry had never been destitute of experiments, those who cultivated it made their experiments according to speculation [raisonnemens] and imaginary principles. The confused facts thus multiplied and collected by alchemists made chemistry "an occult & mysterious science; its expressions were only of figures, its turn of phrase was of metaphors, its axioms were of enigmas, in a word the character proper to its language was to be obscure & unintelligible." Chemists made their art useless to mankind and were justly scorned. However, the taste of true physics finally came to prevail in chemistry, as in other sciences, because great geniuses believed that their knowledge would be respected only insofar as it was useful to the society. They rendered chemistry public, useful, and communicative, deserving of the name "science, having its principles & its rules founded on solid experiments & consequent reasonings."[150] In its present state, chemistry could be compared to geometry, both offering ample material which was increasing daily, both being the foundation of useful and necessary arts, and both in possession of certain axioms and principles, one demonstrated by evidence and the other supported by experiments. Consequently, chemistry also deserved a book that would present its fundamental truths with order and precision. Macquer wished to follow the order of geometry in his elementary treatise by leading from the simplest truths (which assumed no knowledge of chemistry) to the more complex truths.

History served as another means of justifying chemistry's place among the more established sciences, as Boerhaave had shown. In his "Discours historique sur la chymie," Macquer sketched out various "stages" chemistry had traversed in its history, the "revolutions" it had experienced, and the circumstances that affected its progress. Such a "summary table" of chemistry's past would support its current status as a science. The ancients could only practice it as an art or a "métier," not as a "true science." Given that science was "the study & the knowledge of rapports that could pull together a certain number of facts," even the ancients with the most active and penetrating minds could not have accumulated enough facts to form a science. Stahl, who elaborated the most sublime chemistry in modern times, would have done no more than forge an ax at this time. Newton, who discovered how to measure the universe, would have been able to count no higher than ten. Chemistry made some progress in Egypt, a civilization that gave birth to sciences, which allowed true savants or philosophes to be distinguished from simple artisans. Despite their modest progress, Hermes and other "first chemists" in Egypt knew of only isolated facts and could not form a science.[151]

The small progress made in ancient times was followed by a "singular mania which attacked the head of all chemists," a kind of "general epidemic," a "folly of human mind," which was a great obstacle to the advancement of chemistry. Alchemy, a movement "quite opposed to philosophical dispositions," cloaked chemical knowledge in a "ridiculously enigmatic" language. As such, it defined the "middle age of chemistry, which is the darkest & the most humiliating part of its history." Only a small number of alchemical authors—Geber, Roger Bacon, Raymond Lulle, Bazile Valentin, and Isaac le Hollandois—attained a measure of intelligibility in describing certain experiments. In the sixteenth century, a "nouvelle folie" began with a "famous alchemist named Paracelsus, man of quick, extravagant, & impetuous spirit" who advocated a "universal medicine." This pretension renewed the "mania of alchemists," filling the library with numerous books on the elixir of life, panaceas, etc., all in indecipherable language.

Despite its folly, the pursuit of universal medicine began to establish "rational chemistry" on the ruins of alchemy. Paracelsus's daring critique of traditional pharmacy and his advocacy of chemical remedies opened a new route in the art of healing. Some of his critical followers in medicine denounced his folly but pursued the benefits of chemical medicine. These "true citizens," including most French textbook writers, were the "inventors of a new art of chemistry," properly speaking. They wrote clearly of the preparation of chemical remedies in accordance with the medical precept that they should always be prepared "in a uniform manner" and by fixed procedures. In the meantime, various authors of "a truly philosophical mind" also began to write up the arts related to chemistry: metallurgy, glassmaking, etc. Although they could not be completely exonerated of "illusions of alchemy," they made up for their fault with a great quantity of interesting experiments. Through their writings, they contributed to delivering chemistry from "this leprosy which disfigured it & opposed its progress."

Only during the second half of the seventeenth century did chemistry begin to renew its progress toward perfection. In this "most brilliant epoch of chemistry," men of genius began to collect, examine, and compare various parts of chemistry to put them together into a "body of symmetric & reasoned doctrine," and to lay the true foundation of chemistry considered as science. Becher claimed his place of honor as "the greatest & the most sublime of all chemists-physiciens"; Stahl was, however, an even greater genius, with a lively, brilliant, active imagination. Stahl's "luminous" writings conformed best to chemical phenomena. His theories, confirmed by numerous experiments, were the most certain guide to chemistry. Next to Stahl stood Boerhaave, a powerful genius who shed light on all the sciences he pursued. In chemistry, Boerhaave developed "the most beautiful & the most methodical analysis of vegetable kingdom, admirable treatise of air, water, earth, & above all that of fire, an astonishing chef-d'oeuvre" that seemed to leave nothing to add. Chemistry could now enjoy the most beautiful success in a truly philosophical age, supported by powerful princes and a multitude of amateurs.

The triumph of modern chemistry and chemists depicted by Macquer stands in sharp contrast with Venel's gloomy picture. In his article "Chymie" in Diderot's *Encyclopédie*, Venel portrayed chemists as rather isolated figures even among the savants who advocated the "universality of knowledge": Chemists were still distinct persons, quite few in number, with "its [chemistry's] language, its laws, its mysteries," living among the people incurious about its business and about the industry it required. This "incuriosity" led them to accept limited conceptions of chemistry, to denigrate its principles, and to avoid reading the works of chemistry. They did not distinguish a chemist from a souffleur who maintained a laboratory and prepared perfumes, phosphorus, etc. They identified chemistry with its medical use. They thought that chemists dealt only with composed beings, and that their principles were creatures of fire. The works of physiciens such as Boyle, Newton, Keil, Friend, and

Boerhaave were known, cited, and praised; the chemical works of Becher and Stahl were not.

Venel envisioned a "truly philosophical" chemistry, a chemistry "reasoned, profound, and transcendent" and equipped with an "issue" that would make it more generally applicable to other sciences. Chemistry enjoyed several philosophical fashions since it "took more particularly the form of science, that is to say, since it received the ruling systems of physics"—Cartesian, corpuscular, Newtonian, academic, or experimental. Venel reproached the chemists who had tried to express chemical ideas in the current philosophical fashions, however, for their "enthusiasm" to embrace the esprit systématique and for producing errors with their flight of fancy. They failed to assert chemistry's independence and its particular "right" or "liberty." The esprit de détail did not necessarily fare better; it produced only mediocre treatises which were no more than judiciously ordered collection of facts and which thus helped characterize chemists as mere manipulators and laborers. They lacked the "knot, the ensemble, the system, and the whole" or the "issue" that would situate chemistry among the other sciences. To elaborate a philosophical chemistry, chemists had to rebel against the "sovereignty of physics," which Fontenelle had stipulated in 1699. Fontenelle had characterized the difference between chemistry and physics in terms of their capacity to penetrate the innermost organization of matter. Chemistry dealt with gross parts, while physics further divided them through "delicate speculations" into smaller bodies that moved and figured in an infinity of ways. Venel argued that infinity or romans physiques was not chemists' prerogative. This did not mean that the spirit of chemistry was confused, enveloped, less neat, and less simple than that of physics. On the contrary, chemistry penetrated into deeper interior than physics, which only touched the surface and the exterior. Chemists would have to concede only that they had not articulated their doctrines quite as distinctly, making all its advantages clear.

The misconceptions about chemistry persisted, however, depriving it of "elevated & vigorous geniuses" who could become "zealous chemists." The "revolution which would place chemistry in the rank it merits," at least on the side of physique calculée, would require a "skillful, enthusiastic, & bold" chemist in a good position and circumstances who could awaken the attention of savants and dispel their prejudice. While waiting for this "new Paracelsus, who will make of chemistry the science that understands nature and displaces geometry from that pretension," Venel wished to render chemistry worthy of philosophers' respect by correcting the singular source of all errors in physics, or the attempt to explain through philosophy the natural phenomena that could be understood only through chemistry.

Venel walked a tightrope in trying to establish the foundation of la Chimie générale philosophique. He had to shun both the overly speculative tendencies of Cartesian or corpuscular chemistry and the merely empirical content of common chemical textbooks. While asserting that chemistry's place of honor among the sciences did not depend on its "philosophical side," he had to establish its relevance to other sciences. It was in this rhetorical context that Venel articulated a general classification of sciences based on the nature of the changes they dealt with. Natural philosophy could be divided into two groups, each dealing with the exterior properties and changes, respectively. There were three kinds of changes, each involving organic economy, composition of non-organic bodies, and the movement of aggregate bodies. Buffon had already established that the changes involving organic molecules and organized bodies could not be explained by mechanical laws and that they were subject to the laws essentially different from those of non-organic bodies. It was also true, Venel argued, that the changes involving principles differed from those of aggregate bodies—that is, the laws of chemistry were not the same as those of physics. He thus distinguished "external or physical" properties from "internal" or "chemical" properties. Physical properties (motion, elasticity, etc.) belonged to the mass or aggregate (that is, to the body as a whole) but left its integrant parts intact. Chemical properties were particular to the body and constituted it as such. Water was simple, turned volatile, and dissolved salts to become a material component of their mixtion. The physics of unorganized bodies should be divided into two independent sciences: "ordinary physics"[152] and chemistry. Venel took pains to spell out how chemical changes differed from physical ones, reiterating the definition, the goal, and the utility of chemistry. Notably, he dissociated the two concepts of fire and heat earlier fused by Boerhaave. Fire considered as heat was not a "chemical object," but a general agent of chemical operation.

If Macquer and Venel diverged sharply in their standpoints and strategies, they shared the goal of legitimating chemistry in the public sphere. Their rhetorical efforts to this end bespeak chemistry's new audience and the new intellectual demand placed on chemistry at this time, which led to the construction of a theoretical chemistry further removed from pharmaceutical practice. The march toward a "rational chemistry" based on solid facts, devoid of "imaginations," and conversant with the other genres of scientific and intellectual discourse required a new kind of chemist, one more literary and more calculating at the same time. Venel's call for a new Paracelsus could be answered only by an anti-Paracelsus.

5

A Newtonian Dream in the Province

Newtonian philosophy transformed the Enlightenment language of sciences and of the public sphere. The contour of "Newtonian" sciences in France has been well documented. Newton's works began to make an inroad into the French scientific scene during the last decades of the seventeenth century, paving the way for his election to the Academy of Sciences in 1699. Malebranche's group in the academy assimilated them in an eclectic fashion, breaking down "the initial barriers of the Cartesian fortress,"[1] well before the emergence of "militant Newtonianism" in the 1730s.[2] Along with Voltaire's popular expositions,[3] the Dutch Newtonian textbooks of 'sGravesande and Musschenbroek spread the message further.[4] Fontenelle, despite his strong reservation about the concept of attraction, wrote a respectful eulogy of Newton, acknowledging the excellence of his contribution to mathematical physics.[5] Subsequently, Maupertuis established the Newtonian paradigm of rational mechanics in the academy.[6] The debate between Clairaut and Buffon over the exact mathematical terms of Newton's law of attraction in the late 1740s signaled the onset of confident Newtonianism, which was exemplified by Benjamin Franklin's approach to experimental physics.[7]

Physicians articulated a variety of iatromechanical and iatromathematical approaches based on Newton's works, albeit with modification.[8] The relevance of Newton's philosophy to chemistry, clearly articulated in the writings of John Keill and J. Friend, did not go unnoticed.[9] Senac drew freely on their writings in his *Nouveau Cours* (1723). Boerhaave's influential textbook *Elementae chymiae* (1732) incorporated the Newtonian matter theory into chemical didactics. Buffon translated Stephen Hales's thoroughly Newtonian work *Vegetable Staticks* (1727) in 1735.[10]

He also preached the virtue of perceiving affinity as a part of the unified system of attractions in his *Histoire naturelle*[11]:

The laws of affinity . . . are the same with that general law by which the celestial bodies act upon one another. Their exertions are mutual, and proportioned to their masses and distances. . . . All matter is attracted in the inverse ratio of the square of the distance; and this law seems to admit of no variation in particular attractions, but what arises from the figure of the constituent particles of each substance; because this figure enters as an element or principle into the distance. Hence, when they discover, by reiterated experiments, the law of attraction in any particular substance, they may find, by calculation, the figure of its constituent particles.[12]

Pierre-Joseph Macquer ushered in the "Newtonian" phase of French chemistry by characterizing affinity as attraction in his popular *Dictionnaire de chymie* (1766). Guyton de Morveau articulated a Newtonian program of affinity chemistry in the early 1770s. Berthollet carried this Newtonian interpretation of chemical affinity into the next century with his *Essai statique de chimie* (1803).

This long, distinguished lineage of "Newtonian" scientists makes it all the more puzzling that Newton's theories did not play a significant role in the episode of the Chemical Revolution, particularly in Lavoisier's works. Arnold Thackray found that the research enterprise of "Newtonian" chemists yielded only "few and unrewarding" results in comparison to Lavoisier's. He concluded that Lavoisier's "brilliant experimental work and conceptual reformulations . . . were not directly aimed at either denying or replacing the Newtonian categories of thought, views of matter, or research priorities."[13] Robert Schofield also judged that Lavoisier's revolution was part of a revolution against rather than within Newtonianism.[14] It was this move away from Newtonianism or the reigning tradition of natural philosophy, according to Arthur Donovan, that established chemistry as a positive science. Donovan argues that "Newton's achievement should be seen as the culmination of the tradition of natural philosophy as it had developed up to the end of the seventeenth century, whereas Lavoisier's achievement stands at the beginning of the tradition of the positive sciences that became increasingly prominent in the latter half of the eighteenth century."[15] While Donovan's thesis sounds intuitively true and merits further exploration, especially in characterizing the second scientific revolution, it does not resolve the tension in the historiography. Many chemists throughout the period of the Chemical Revolution gave at least a nod to the Newtonian language of attraction and hard corpuscles. If it is important not to strap the Newtonian "straitjacket" on eighteenth-century chemistry, it would be helpful to address the function of Newtonian language in chemical discourse.[16]

In order to understand the shifting presence of Newton in eighteenth-century chemistry, we must pay attention to the literary public sphere and to the geography of European Enlightenment.[17] Newton emerged as a cultural icon in the Enlightenment public sphere. His philosophy functioned as a shared idiom of rational discontent with the reigning system of knowledge, authority, and socio-political order. Despite its success in infiltrating the public sphere, however, Newtonian natural philosophy did not dictate the practice of chemistry or experimental physics in Paris.[18] The juxtaposition of Newton and Stahl in Senac's text is quite instructive. Even as chemists searched for a new philosophical authority and language to supplant the Cartesian, they adapted it for chemistry drawing on Stahl's experiential authority. The two leading Parisian chemists, Rouelle and Macquer, both enshrined Stahl as the leading chemical authority, a genius who had ushered in the era of modern chemistry. If they agreed on the importance of Stahl's chemistry, they differed substantially in their attitude toward the Newtonian language.[19] Rouelle, while teaching Hales's experiments on air in his courses, made no explicit reference to the accompanying Newtonian conception of matter. His student Venel spread a strong anti-physicalist, anti-Newtonian message through his articles in the *Encyclopédie*. In contrast to the explicitly anti-Newtonian stance of the "coterie d'Holbach," Macquer effected a transition in the public language of chemistry from the Cartesian to the Newtonian through his writings. His endorsement was limited, nevertheless, to the public discourse of the textbook and the dictionary and to the obviously physical aspects of chemistry, such as weight.[20] He was much more enthusiastic about Stahl's chemistry than about Newton's philosophy, which could not offer guidance for his day-to-day industrial research.

Notwithstanding his limited influence on chemical practice, Newton's presence in the literary public sphere did exercise a selective pressure on the theory domains of chemistry. Affinity acquired a stronger legitimacy in the language of attraction, while the principles languished in the scholastic and Paracelsian past. Chemists used "principles" mostly in its analytic sense, dissociating it from past philosophical connotations. In adopting the Newtonian language, therefore, Macquer placed more emphasis on affinity than on principles as a theoretical issue for chemistry. Others less restricted by the rigorous standards of Parisian elite chemistry went further, creating a thoroughgoing Newtonian discourse for chemistry. For these outsiders, Newton's inclusion of chemical action as an instance of universal attraction provided sufficient justification to impose the Newtonian imperative on chemistry. The provincial academies played an important role in this Newtonian transformation of affinity chemistry, indicating the presence of a broader public attuned more to philosophical/literary discourse than to esoteric chemical practice.[21] George Louis Le Sage, a Genevan mathematician who won the Rouen academy's 1758 prize competition, drew mostly on Newton's theories and experiments to reconceptualize chemical affinity. Written in the style of mathematical demonstration which was of little use to most practicing chemists, his work nevertheless set an important precedent for other philosophical/literary chemists. It was in this circle of philosophical chemists who took Macquer's texts seriously that Newtonian language acquired legitimacy as the reigning chemical philosophy. Torbern Bergman, a Swedish professor of chemistry who corresponded with Macquer, also adopted the Newtonian stance at least formally. Richard Kirwan, an important British interlocutor of chemistry between 1767 and 1777, followed Guyton's lead in quantifying affinities. By the 1770s, affinity had become the frontier of theoretical chemistry among the leading philosophical chemists in Europe.

5.1 Affinity in Fashion

The affinity table caught the imagination of the leading European chemists around the middle of the eighteenth century, no doubt thanks to Rouelle's lectures and Macquer's writings.[22] The increasing complexity in the chemistry of salts also demanded expansion of the table. In England, Joseph Black proposed minor revisions of the table in his

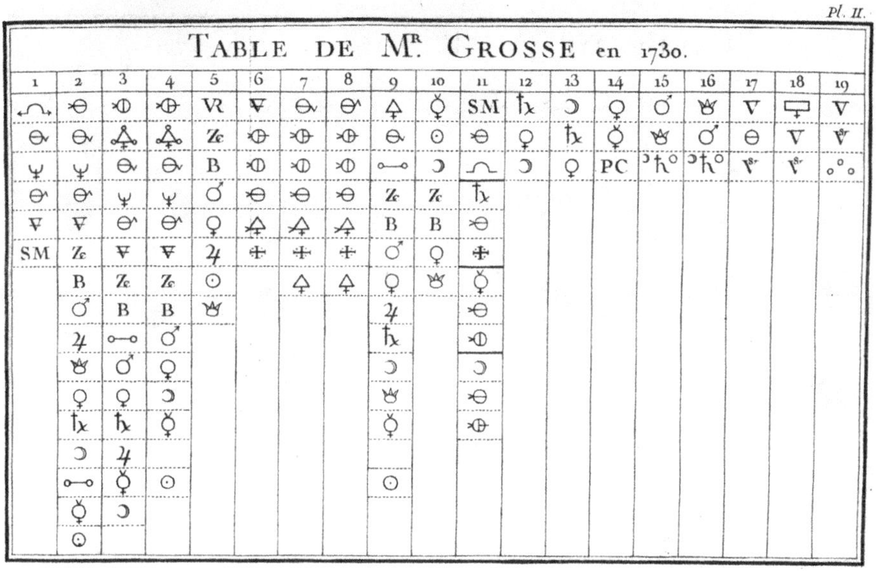

Figure 5.1
Grosse's affinity table, 1730. From Demachy, *Recueil de Dissertations physico-chimiques, présentées a diffrentes académies* (Paris: Nyon and Barrois, 1781).

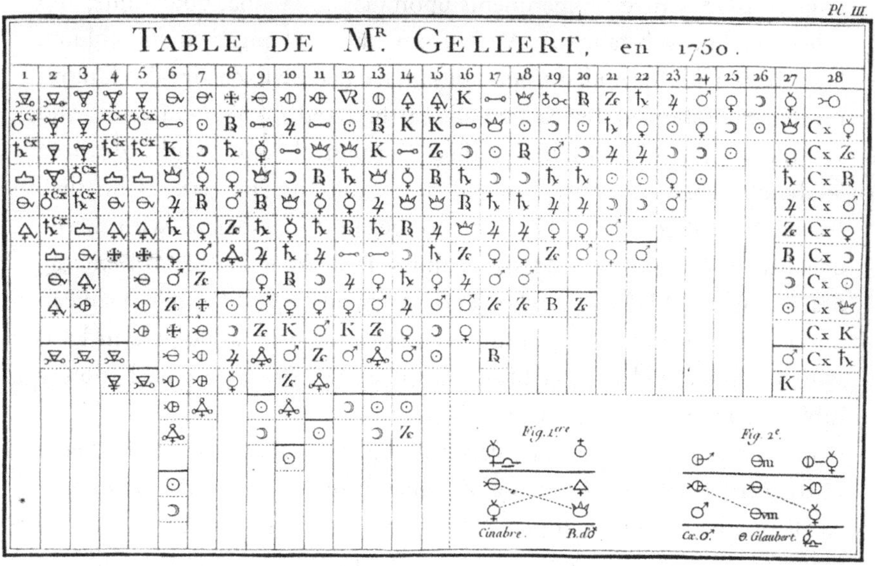

Figure 5.2
Gellert's affinity table, 1750. From Demachy, *Recueil de Dissertations physico-chimiques, présentées a diffrentes académies* (Paris: Nyon and Barrois, 1781).

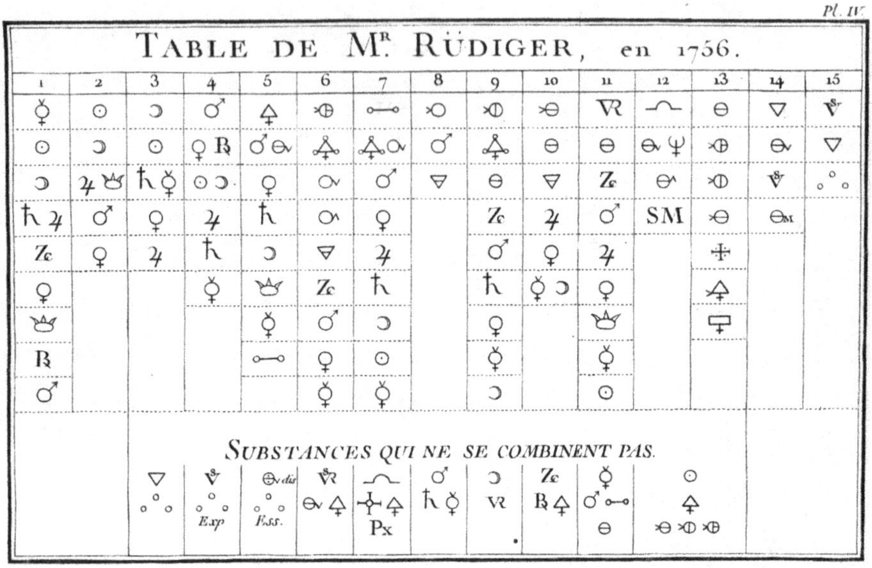

Figure 5.3
Rüdiger's affinity table, 1756. From Demachy, *Recueil de Dissertations physico-chimiques, présentées a diffrentes académies* (Paris: Nyon and Barrois, 1781).

famous 1755 article "Experiments upon magnesia alba, quick-lime, and other alcaline substances." Aside from minor changes to the column headed by acids, he added a new column headed by fixed air—his famous invention—followed by calcareous earth, fixed alkali, magnesia, and volatile alkali. Some of the leading German chemists with advanced analytic techniques also brought the table up to date. Christlieb Ehregott Gellert (1713–1795), who began to associate with Berlin chemists in 1744, produced a 28-column table in 1751. Jacob Reinbold Spielmann (1722–1783), who learned chemistry at Berlin's Medical-Surgical College from Pott and Marggraf and spent some time in Paris with J. Grosse and C. J. Geoffroy in the early 1740s, proposed another table, also with 28 columns, in 1763.[23]

In France, chemical affinity captivated a somewhat different audience in the provinces. The Academy of Rouen offered a prize competition in 1758 on the following topic: "Determine the affinities that are found between the principal mixts [principaux mixtes] as M. Geoffroy did, &

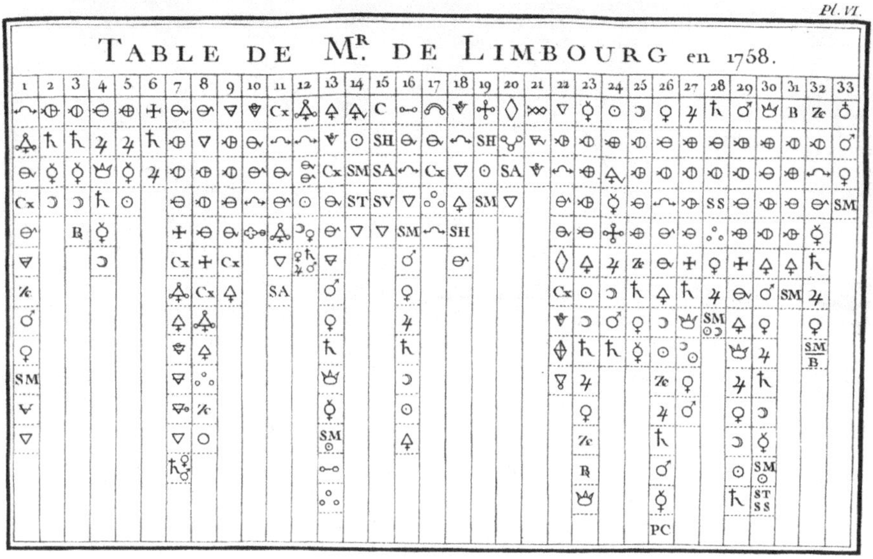

Pl. VI.

TABLE DE M.ᴿ DE LIMBOURG en 1758.

1	2	3	4	5	6	7	8	9	10	11	12	13	14	15	16	17	18	19	20	21	22	23	24	25	26	27	28	29	30	31	32	33

(Table of chemical affinity symbols; cells contain alchemical and chemical notation symbols, abbreviations such as Cx, SH, SM, SA, ST, SV, PC, B, etc.)

Figure 5.4
Limbourg's Nouvelle Table des Affinités chymique, 1758. From Demachy, *Recueil de Dissertations physico-chimiques, présentées a diffrentes académies* (Paris: Nyon and Barrois, 1781).

find a physico-mechanical system of these affinities." The question expressed the academicians' wish that the physical and the chemical side of the affinity issue should be united. Not finding an essay that addressed both aspects, they divided the prize between Jean Philippe de Limbourg (a doctor of medicine from Liége and a former student of Boerhaave, Musschenbroek, and Rouelle) and George Louis Le Sage (a master of philosophy and mathematics from Geneva). Each had answered only half of the question successfully.[24] Limbourg had produced an extended affinity table of 33 columns; Le Sage had supplied a mathematical exposition of affinity in line with the concept of gravity. Though both Limbourg and Le Sage subscribed to the Newtonian language, Limbourg's use of the new master in natural philosophy was peripheral to the practical issue of determining the order of affinities. In contrast, Le Sage's essay shows how the theory domain of affinity began to acquire a philosophical and mathematical dimension through the intervention of Newtonian mathematicians.

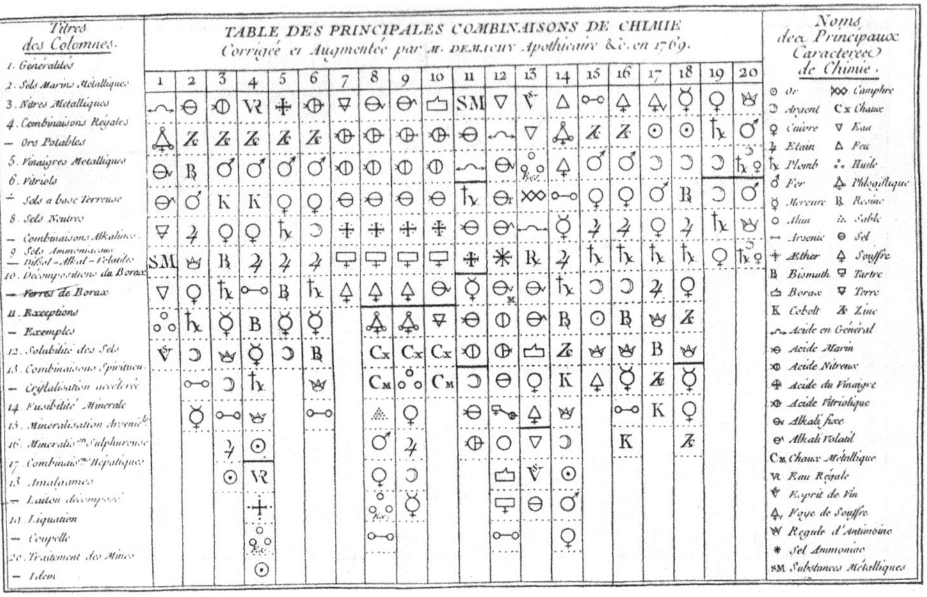

TABLE DES PRINCIPALES COMBINAISONS DE CHIMIE
Corrigée et Augmentée par M. DEMACHY Apothicaire &c. en 1769.

Titres des Colomnes.

Noms des Principaux Caracteres de Chimie.

Figure 5.5
Demachy's Table des principales combinaisons de Chimie, 1769. From Demachy, *Recueil de Dissertations physico-chimiques, présentées a diffrentes académies* (Paris: Nyon and Barrois, 1781).

Limbourg purported to establish a "physico-mechanical system" of affinity, obviously stretching his competence to satisfy the wording of the prize question. He invoked Newton's authority. Denouncing Lemery's and Homberg's corpuscular explanations as "purely imaginary suppositions," he offered a Newtonian "system" that related the rapports or affinities of bodies to general attraction. Affinity consisted in "attractions, in the facility diverse substances have to touch more exactly & at more points." From this general position, Limbourg derived "laws & general rules of affinities" that regulated the behavior of the compound.[25] In general, the affinity of a compound represented the affinities of its constituent parts, which determined the fate of a third substance approaching the compound. The third substance could displace one of the constituent substances or could unite with both of them, as Macquer had stipulated. Sometimes, however, a compound could exhibit new rapports which were not proper to its constituent parts.

The main part of Limbourg's essay consisted of a new, extended affinity table built on a slightly different set of principles than Geoffroy's. He took into consideration practical complications in determining affinities by displacement reactions alone. One could judge the "degrees of affinity" from diverse points of view—the "promptitude or the facility" with which matters united, the "constancy & firmness of their union," or the displacement reactions. For example, vitriolic acid was seen to have much affinity with iron because of the promptitude and facility of their union, with alumineuse because of the constancy and firmness of their union, and with phlogiston because it released phlogiston from all other combinations. These three rules, used most often by chemists to evaluate the degrees of affinities, frequently led to contradictory results. Bodies that united easily could also abandon one another with the same facility. In other cases, the most fixed unions were sometimes easily ruptured. Finally, displacement reactions often exhibited reciprocity. While he was not satisfied with any of the rules as the single criterion for determining the degrees of affinity, Limbourg gave a practical guide that displacement reactions should be given priority whenever possible. The other two could substitute for this rule, especially when one could not effect the displacement reaction. Limbourg's discussion offers a rare glimpse of the difficulties chemists encountered in deploying Geoffroy's table to the expanding territory of salts.

To compensate for the lack of generality in Geoffroy's table, Limbourg devised a table of 33 columns presented in the same manner as Geoffroy's. He explained each column with a list of exceptions. It might be useful for us to take a closer look at the table using the most famous substance: phlogiston. Limbourg considered phlogiston a concrete substance that possessed a set of determinate affinity relations with other substances. In the first column headed by acids in general, phlogiston appears right below the top line, predicting that it should have the strongest affinity with any acid. While Geoffroy had only considered the affinity of phlogiston for sulphuric acid, Limbourg listed its affinities with all acids:

There is such a great affinity between the acids & phlogiston that diverse metals, iron, lead, &c. are precipitated by themselves from solutions, deprived of their phlogiston which the acid releases from them & takes hold of. By reason of this same affinity, zinc and tin join saltpeter in fusion and are inflamed; thus phlogiston leaves the metallic substance to be united with nitric acid.[26]

Limbourg characterized the precipitation of metallic calces by acids—later called humid calcination—as a process in which metals lost their phlogiston to acids. This parallel behavior of metals in humid and dry calcinations became a critical area of reflection by Guyton and subsequently by Lavoisier in their phlogistic versus antiphlogistic theories of calcination. Limbourg's table also had a separate column headed by phlogiston, although it reproduced the information given in other columns headed by acids, alkalis, metals, and calces. Phlogiston had become a stable substance in chemical practice, characterized by a network of well-defined operations.

Le Sage (1724–1803), a mathematician who regarded chemical labor beneath his intellectual station, devised a radically different Newtonian approach in his *Essai de chymie mechanique* (1758). Identifying chemical affinity as a kind of attraction, he devoted a large portion of his text to general mathematical theorems on attraction. He argued that chemistry lacked less "physiciens capable of making ingenious applications of general principles, than mechaniciens who takes the trouble of looking for the first material cause of these principles." Although most of Le Sage's theorems executed in the style of mathematical proofs were of little interest to other chemists, they indicate how Newton's inclusion of chemical attraction in his system made it possible for mathematicians and physicists to stake a claim on the chemical issue of affinity. Of particular interest in Le Sage's *Essai* is the theorem that the attraction between two bodies is proportional to their respective densities rather than to their absolute masses. Le Sage argued that the proof of this theorem could explain why cohesion and chemical attraction overrode weight in specific instances such as crystallization. The particles immersed in a fluid approached one another in proportion to the difference of their density and the density of the fluid. If the density of salt particles was three times that of water, the total attraction in a salt solution (salt-water and salt-water) would be six times that of two water particles. When the salt particles approached one another, however, the total attraction (salt-salt and water-water) would be ten times that of water particles. This increase of the total attractive power in the solution caused crystallization. The density theorem would eliminate the need "to imagine an attraction distinct from universal gravitation, & subject to another law."[27] The idea of effective mass was to surface later in Guyton's discussion of affinity and in Berthollet's conception of mass action.

The difference between the mathematician's approach to affinity and the physician's consisted not only in the method but also in the selection of relevant facts. Le Sage extracted most of his chemical examples from Newton's writings rather than from the contemporary chemical literature, which set him apart even from the philosophical chemists. He also regarded chemical labor as unbefitting the mind of the mathematician. Although he deemed it necessary to multiply experiments in order to perfect affinity tables, such an immense work would demand "less genius than dexterity, exactitude, patience & external facility" and should be left to people "less proper to the abstract research of the first causes of all this mechanism: seeing that these two classes of lights & faculties, which are often found separated, are seldom encountered united." While he suggested adding some simple substances such as white gold and light or elementary fire to the affinity table, finding a numerical expression for the affinities of various substances was higher on his agenda. This endeavor would require constructing two tables, "one, of the relative degrees of forces between the affinities observed, expressed as exactly as it would be possible; the other, of densities, reported in the same unit or measure."[28] At the least, one should indicate the equality or inequality of the operating forces in an algebraic form. For example, in order to say that fixed alkali was united to acids more strongly than water was joined to sulphurs, one should write the string of symbols shown here in figure 5.6.

5.2   A Lawyer-Chemist

Guyton de Morveau (1737–1816), a provincial lawyer-chemist, answered the call for a Newtonian affinity chemistry that combined chemists' and physicists' outlook. Louis-Bernard Guyton, as he was named at birth, enjoyed a normal childhood in a bourgeois family that could boast a few physicians, lawyers, and minor government officials.[29]

Figure 5.6

Born to Antoine Guyton, an avocat with great oratorical powers and vast juridical knowledge,[30] he received private instruction in religion and literature before entering the Collège de Godrans, a Jesuit institution that numbered among its alumni Buffon and the musician Rameau as well as a number of local notables.[31] Like many other students at the Collège, Louis-Bernard picked up a serious knowledge of Latin and was exposed to the contemporary literature. During the family's frequent sojourns to Morveau, a region southeast of Dijon, he also acquired a taste for botany. Apparently, Louis-Bernard liked to fix mechanical devices such as horologues and watches. In 1753 he entered the Dijon faculty of law, another school connection he shared with Buffon. Acquiring the license in 1756, he became a barrister. In 1762 he was appointed an avocat-général du roi at the Burgundy parlement after his father purchased the office through a difficult litigation. He carried out his duties successfully until 1782, when he retired with a pension, apparently to pursue chemistry full-time.

In the eighteenth century lawyers were numerous and powerful, situated at many critical positions in the households of princes and nobles. Although they did not enjoy the prestige or the power of the traditional aristocracy, this elite class of the Third Estate had a strong presence in the political landscape of the ancien régime. Well educated and highly respected, they wielded their legal skills to forge a vision of "judicial monarchy" in which lawyers, as the technicians of administration, would play a leading role.[32] While some of the lawyers were financially secure and well connected, the prospects of social advancement in the ancien régime were limited. They deployed their vocational skills to sway public opinion, therefore, while serving the households of the powerful. Guyton provides an excellent example of this insider/outsider identity and status of lawyers in the ancien régime, which helped them assume a leadership later in the revolutionary politics. While serving the parlement with distinction, he sought to reform the judicial system by modifying the complexity of existing customs and laws to develop a "simple, uniform, universal and constant" jurisprudence. He was equally involved in educational reform. During the campaign against the Jesuits in the early 1760s, which took a violent turn in Paris but unfolded in Dijon mostly among the literary circle, he anonymously published a poem satirizing the Jesuits, "Rat iconoclaste ou le Jésuite croqué," and circulated a memoir on the reform of public education.[33] Guyton's proposal for curriculum reform was based on a meticulous study of other such proposals, including those of Montesquieu, Locke, Rousseau, and a host of classical authors. Criticizing the current curriculum geared toward preparing young men for the church, he pitched for a college open to everybody. The content of college education should include not the subjects that served a particular état, but those that belonged to all états such as philosophy which included logic, metaphysics, moral, mathematics, and physics. Guyton's interest in science stemmed from his belief in its value for public education, as can be seen in his activities and speeches at the Dijon academy. Like many other lawyers, he later became deeply involved in revolutionary politics.

Guyton's association with the Dijon Académie des Sciences, Arts, et Belles-Lettres began in 1764, when he became an honoraire. It was a relatively new institution, created in 1740 with the endowment of Hector-Bernard Pouffier.[34] Dijon had a well-established reputation among the French literary circle by this time, having been dubbed the "capital of the Republic of Letters" by Bayle and Voltaire, but lacked an institutional setting to sustain it. The petition to establish a university in 1722 had resulted in the single faculty of law, without the equivalent faculty of arts. Pouffier, who had been a delegate of the parlement for the petition, left most of his fortune for the establishment of an academy to fill this gap. Under his provision designed to exclude the noblesse de robe, the academy recruited members in the classes of physics, morals, and medicine. The membership consisted of six honoraires, twelve pensionnaires, and six associés. With the exception of Buffon, who accepted the place of honoraire without attending the meetings, the members were avocats, prêtres, and physicians. The academy was never a purely scientific body, however, as can be seen in the prize competition that brought Rousseau instant fame.[35] It turned to a more literary direction when

Richard de Ruffey joined it as an honoraire in 1759 and merged it with his literary society. Suppressing the original classes, he sought out notable literary figures such as Voltaire for membership. This literary turn of the Dijon academy opened the door for Guyton who had no established credit in science.

The Dijon academy provided the institutional context and resources for his evolution as a chemist through the debates, lectures, prize competitions, and the library and laboratory facilities. Soon after he joined the academy, Guyton was drawn to the debate over the weight gain of metals upon calcination, which led to the publication of his "Dissertation sur le phlogistique." In 1772, when he became vice-chancellor of the academy, he campaigned for a public lecture course in chemistry. Such a course was initiated in 1776. He also installed a laboratory to carry out experiments for his lecture course and demonstrations at the meetings. Throughout the 1770s and the 1780s, the academy held prize competitions having to do with various actions of acids. When it failed to award the prize in 1785, Guyton diverted the money to purchase a complete set of the *Philosophical Transactions of the Royal Society*.[36]

After the death of his father in early 1768, Guyton purchased a large house and installed in it a fine laboratory stocked with chemicals and apparatus purchased from Baumé's shop in Paris.[37] Soon he was doing serious industrial research involving analysis of the charcoal of Montcenis. The ensuing publications on metallurgy and mines attracted the attention of Buffon, who was building an iron foundry at his Montbard estate. A close regional and school connection and similar circle of friends around the Burgundy parlement must have drawn these men into a close, if respectful, relationship.[38] Guyton admired Buffon's broad vision; Buffon numbered Guyton among the best French chemists[39] and invited him to work on the mineralogy section of the *Histoire naturelle*.[40]

Guyton, who had not had an opportunity to benefit from Rouelle's lectures, proclaimed himself "the disciple of Macquer's books." He began to correspond with Macquer in 1769,[41] and he chose his research topic directly out of Macquer's *Dictionnaire*.[42] Learning chemistry at a distance, without a vocational interest, Guyton was drawn to affinity, the most theoretical subject in Macquer's textbook. In 1769 he presented to the Dijon academy an essay proposing a table halotechnique. He adopted Buffon's idea that the shapes of particles played a role in determining chemical affinity. In *Digressions academique* (1772), Guyton elaborated a dissolution model of chemical action determined by attraction. In the ensuing years he continued to push the frontier of affinity chemistry while trying to absorb Lavoisier's new discoveries into his overall scheme. Guyton's persistent yet elusive dream of quantifying affinities induced generations of research on the subject. A narrative of the Chemical Revolution built around Guyton, in contrast to the conventional one centered on Lavoisier, should make obvious the importance of affinity chemistry, into which Lavoisier's reform was assimilated.

5.3 A Field of Argument: The Negative Weight of Phlogiston?

When Guyton joined the Dijon academy, one of its pensionnaires, the physician Chardenon, was engaged in the debate over the weight gain of metals upon calcination.[43] Although the phenomena had been known for some time in the case of certain metals, such as lead, chemists had not paid serious attention to it.[44] Chardenon shaped this dormant subject into an urgent puzzle by pointing out the deficiencies of the existing explanations, which incited a polemical rebuttal from a man named Ribaptome. The debate between Chardenon and Ribaptome, carried out on the pages of various journals, indicates how the participation of more literary and non-vocational chemists served to shape new fields of argument within chemistry, leading to the reformulation of its problem domains. Their debate revolved around the facts harvested from the existing literature, supplemented by discursive logic, rather than around new discoveries. In addition to the intrinsic merit of this debate as a precursor to Guyton's "Dissertation sur la phlogistique," therefore, the argumentative and rhetorical strategies of the participants deserve a close look.

In the memoirs presented to the Dijon academy on July 15, 1763 and on December 9, 1764, Chardenon criticized two existing hypotheses concerning the weight gain and offered his own alternative.[45] The first opinion attributed the weight gain to the fire accumulated in the calces. He concluded that Boerhaave's and Voltaire's experiments showing no difference in the weights of cool and hot iron refuted this opinion. The second opinion, espoused by Duclos and Béraut, stipulated that fire caused foreign corpuscles floating in the air to be deposited on the bodies exposed to calcination. He offered four "equally decisive" objections against this "ingenious" hypothesis. First, why did one not observe the weight gain every time one exposed fixed substances to the action of fire? Second, should this augmentation of weight be sensibly less in the void than in free air? Third, should the weight gain be correspondingly larger as it required more time and heat to reduce the body into a calx? Finally, in the reduction of calces, fire chased off the corpuscles it had accumulated on the bodies during calcination. One would, therefore, have to attribute diametrically opposite effects to the same cause of fire. In characterizing the problem of weight gain, Chardenon took various experiments out of their instrumental context—thermometry and combustion with fire or the burning glass, respectively—and tried to impose a coherent discursive logic on them.

Judging that neither of the two options supplied a satisfactory answer, Chardenon offered a solution "with the circumspection of a philosopher," since it was "easier to recognize an error than to discover a truth." Focusing on the diminution of weight in the reduction of calces to metals, he attributed the weight loss to phlogiston which was "nothing other than fire itself." In other words, the presence of fire caused the diminution of weight. Physiciens failed to recognize this because they believed that the absolute weight of a body could be increased only by the addition of new parts. If it was astonishing that the addition of a new part could render a body much less heavy, Chardenon argued, "one has to pay attention to the nature of the added parts." Fire seemed to provide "wings" for the bodies:

[The parts of fire] are in fact prodigiously less heavy than air; they tend always to be elevated, even with such a great force that one can easily suppose that they play this "species of centrifugal force," even when they are imprisoned in different corporeal substances. From this it follows that fire probably renders metals lighter after their reduction by giving, so to say, wings to the terrestrial molecules which enter into their composition.[46]

Chardenon knew that his idea would sound much more credible if metals became lighter in proportion to the quantity of the phlogiston used in reduction, but he failed to provide decisive experiments. He could only state his hope that other physiciens better situated to conduct experiments would try to determine if the ratio he suspected between the quantity of phlogiston and the diminution of weight really existed. He suspected that in the detonation of metals with niter the weight gain would be proportionate to the quantity of niter necessary to deprive the metals of all phlogiston. These facts, if confirmed, would prove that the presence of fire rendered metals lighter and that its absence augmented their weight. Chardenon shied away from doing the experiments himself, however. He had accomplished the philosopher's task of reflecting systematically on the facts and offering a method for finding the solution.

Chardenon's ideas drew a sharp attack Ribaptome, who argued that Laurent Béraut had already answered these objections in his 1747 memoir.[47] Ribaptome issued a point-by-point rebuttal to Chardenon's objections.[48] His rebuttal is interesting for its speculative, rather than experimental, ingenuity. In other words, Ribaptome was no more interested in instigating a new line of empirical investigation than Chardenon. First, to the objection that one did not observe the weight gain every time one exposed fixed substances to the action of fire, Ribaptome attributed it to the different "internal composition" of bodies. Vegetables with weak internal texture lost much of their proper matter in calcination. In contrast, metals with tightly linked internal parts, such as lead and tin, acquired much more foreign matter than they lost proper substance, which caused the weight gain. Second, to the objection that the augmentation of weight also took place in the void, Ribaptome answered that aerial salts easily penetrated the glass vessel. Third, to the objection that the weight increase was not proportional to the heat or the time required for calcination, Ribaptome, drawing on Homberg's burning glass experiments, blamed the different degrees of heat used in the process. If the heat was too low, the rarefaction of air would not deter abundant precipitation of aerial molecules. If it was too high, the dissipation of proper parts would be greater than the fixation of the foreign parts. Fourth, to the objection that fire also reduced calces, Ribaptome answered that fire volatilized the aerial particles in calces via the molecules of oily matter containing phlogiston.

Chardenon's and Ribaptome's arguments lacked new experimental evidence. Both men had sifted through the existing literature to gather well-known experimental facts and disputed each other's claim by this literary evidence. Such a text-based debate had an inherent limitation, especially since the participants mixed different genres of experiments. The confusion generated by Boerhaave's identification of heat and fire became evident in their exchange, especially since they had to deal with another substance not considered by the venerable savante-chimiste: phlogiston. For example, Ribaptome objected that phlogiston was not fire but an inflammable substance that could be exhaled in calcination. The loss of phlogiston did not affect the quantity of fire natural to the body. The experiments with the thermometer indicated that all bodies, except living organisms, had the same degree of heat in the same air. The calx, therefore, had the same quantity of fire as the metal. The identification of heat and fire, respectively rooted in the instrumental practice of thermometry and in burning glass chemistry, created a problem that could not be resolved without a neutral instrumental venue. Ribaptome also objected that fire was not always elevated in the way Chardenon prescribed, since a bar of heated iron warmed its surrounding equally in all directions. This concept of fire, drawn from thermometric experiments, was not the same as chemists' flame. Insofar as physicists' conception of heat relied on thermometric measurements and chemists' conception of fire on the flame-producing operations, they did not have a shared instrumental basis for a unified ontology of heat and fire. Chardenon in his response thus distinguished between two different kinds of fire, as Macquer did. He argued that only the "isolated fire" was equally expanded in all bodies, while the "combined fire" was more or less abundant according to the particular nature of bodies. Phlogiston was the latter, the fire in bondage [feu in puissance] or the fire "enchained by a matter," according to the "most celebrated opinion of modern chemists."[49]

Chardenon and Ribaptome both endorsed the principle of conservation, but they disagreed over how to implement the principle in measuring practice. Ribaptome charged that Chardenon had invoked the "strange paradox" that "a body, after having lost some of its proper substance, can, without receiving a new matter, augment the absolute gravity." He mocked the idea that phlogiston gave a "wing" to the substance through its "specific levity." Chardenon invoked in his defense an instrumental analogy drawn from the hydrostatic balance. Appealing to the "principles of the manometer," he argued that the same body could weigh differently in different media. If one plunged the balance and two equal weights of lead attached to air-filled bladders in perfect equality, they would be in equilibrium in the water. If one suppressed one of the bladders, however, the weight attached to it would prevail. Likewise, bodies that lost phlogiston would gain weight in the air.[50] Chardenon's argument makes complete sense if one is accustomed to handling the hydrostatic balance, which was the most accurate means of measuring the specific gravities of solid bodies then known. Since bodies changed weights depending on the medium, one could expect yet another change if one placed them in a medium much more rare than the atmospheric air.

In fact, Chardenon envisaged a system of "absolute" weights measured in phlogiston, a medium supposedly much lighter than the atmospheric air. In his published memoir of 1769, he introduced a distinction between the "specific or respective" weight and the "absolute" weight. The specific weight varied according to the different degrees of condensation or rarefaction of bodies; the absolute weight of a body could "only be augmented by the addition of new parts of matter." The weight, or "the force by which bodies tend to move along the line perpendicular to the horizon," seemed essential to all bodies and to exist in each elementary particle. It did not follow from this, however, that it was distributed equally in the bodies of different natures. Different bodies were constituted by diverse combinations of "different elementary & constitutive matters whose properties, if common, may be in different degrees." From this "different density" of elementary corpuscles followed the "different degrees of the relative weight" of bodies. If a body contained elements whose gravity was less than that of air, these elements would destroy a quantity of gravitation proportional to their levity in the air. In other words, Chardenon believed in the conservation of diverse chemical elements and their particles rather than in the conservation of the universal particles that made up all bodies.[51]

Chardenon's instrumental analogy hints at the conflicting loyalties of chemists and physicists—conservation in kind vs. universal conservation.

If his explanation of the weight gain drew on a reasonable instrumental analogy, it had an obvious fallacy in the eyes of physicists. It did not take into account the necessary volume changes that should accompany changes in specific weights. For the physicists who believed in the uniform construction of nature, any change in the density of a body constructed of identical parts induced a change in volume that reflected the change in the state of aggregation. In contrast, Chardenon assumed a diversified composition of matter. He illustrated his point again with the analogy of a drag-net. One attached rings of lead at the bottom of this net to drag it down to the bottom, and pieces of cork at the top to keep it on the water's surface. The more pieces of cork were tied to the net, the less it would sink to the bottom. Phlogiston could be compared to these corks attached to the drag-net. One removed phlogiston, or this "tenuous & fugitive part," from the metals in calcination, and one restored it to them in reduction. Just as the pieces of cork increased or decreased the weight of the drag-net in the water, phlogiston would increase or decrease the weight of the body in the air. If this illustration was to work in Chardenon's favor, however, one had to assume that different bodies were composed of different kinds of particles. Physicists did not subscribe to such a heterogeneous construction of the world.

Contrary to the common historiographical assumption that phlogistic chemists accorded a negative weight to phlogiston and thus did not subscribe to the conservation of matter principle, Chardenon did not question the validity of conservation. He envisaged an instrumental practice that would accommodate chemists' experience, or a system of "absolute" weights measured in phlogiston, a medium supposedly much lighter than the atmospheric air. In other words, Chardenon did not seek to apply an established principle of natural philosophy to the chemical system. He began with a theory based on stabilized chemical systems and sought to construct a measuring scheme around it. His measuring scheme could not be implemented in practice, however. He sought, therefore, a "more decisive & more appropriate" proof by determining the quantity of phlogiston in calcinable bodies. Since niter had the property of detonating with phlogiston, one could detonate metals with niter and deduce the

respective quantities of phlogiston they contained from the corresponding weight changes. Unfortunately, lead, the metal chosen for this experiment because of its noticeable weight gain, did not detonate with niter.
Chardenon had to estimate the necessary quantity of phlogiston from
the reduction of lead calx. Despite the experimental detour, he concluded
that "phlogiston contributes by itself to diminishing the weight of metallic substances" and that the augmentation of weight in calcination and
the diminution of weight in reduction were due solely to "the absence
or presence of phlogiston." Similarly, "the simple privation of phlogiston" from phosphorus augmented its weight sensibly, as Pott had shown.
Chardenon thus placed an algebraic grid on the calcination and reduction processes of metals as well as on the combustion reaction of phosphorus. The weight changes of metals and phosphorus in these processes
were due exclusively to the addition or subtraction of phlogiston.

The debate between Chardenon and Ribaptome illustrates how the
participation of a broader public created new problem domains in chemistry. Although the weight gain was a well-known phenomenon, it was
not a pressing concern among practicing chemists. It became a focus of
debate because of a logical fallacy these provincial amateurs of science
detected in literature. They pursued an explanation that could correlate
the instrumental practices of chemistry and experimental physics. The
tenor of discursive logic they brought to bear on the problem demanded
a correlation of the concepts of heat and fire, respectively nurtured by
thermometric measurements and chemical analysis. It also invited
reflection on a universal measuring scheme for "absolute" and "specific"
weights. They exposed thus the limitations of localized instrumental
knowledge, making it necessary to invent concepts that would work
across the instrumental and disciplinary boundaries. In other words, the
participation of a broader public nurtured a transdisciplinary perspective by forcing the practitioners of chemistry and experimental physics
to negotiate the instrumental boundaries of their knowledge and to
create a coherent discursive ontology that would be applicable to all
instruments and sciences. Interpretations of instrumental practices by the
educated public contributed to sharpening the basic axioms of the
science, much as Boerhaave would have liked.

5.4   The Essential Volatility of Phlogiston

Guyton began his career in science with a strong physicalist outlook, although he applied it to chemical problems. The first paper he presented to the Dijon academy dealt with the role of air in combustion, a topic discussed by Macquer in the *Dictionnaire* (1766). In the article "air," Macquer confirmed Hales's experiment which showed that a burning candle became extinguished under a sealed bell jar, leaving a void. He also accepted Hales's explanation that air provided the food or aliment for fire. Explicitly refuting their opinion, Guyton attempted to subsume these phenomena under a general "theory of fire" that assigned to air only a physical role in combustion process. Relying on the experiments and the combined authority of Desaguliers, Boyle, Musschenbroek, Nollet, and Boerhaave, he advanced "the most simple explication" that the void was created by the rarefaction and the subsequent condensation of the air enclosed under the bell jar. Invoking this "incontestable and demonstrated principle in physics," Guyton also explained why the flame was extinguished and why charcoal did not burn in an enclosed space regardless of the intensity of fire. Continuing fire expanded the air inside, compressing it and reducing its elasticity. Combustion required, however, an elastic fluid that constantly approached and receded from the body. A dense medium would not circulate as easily and thus kill the flame. Burning charcoal plunged in spirit of wine was extinguished immediately, despite the inflammability of the spirit, because it was too dense.[52] Guyton's initial approach to the problem of combustion was distinctly a physicalist one, although he took the problem from Macquer's *Dictionnaire*. He soon began to communicate with the master chemist, who advised him on future course of research and possible problems.[53]

After Chardenon's death, Guyton continued with the problem of weight gain, which culminated in his "Dissertation sur le phlogistique." He wished to elaborate Chardenon's system in a more convincing manner, first by gathering all relevant experiments and evaluating the relative merits of various solutions to the problem through a critical history and second by establishing new experiments not only in calcination but also in all chemical operations for which the presence or absence of phlogiston was the unique cause of the weight variation of metallic earths. He also wished to compile a table, as Chardenon had, that would list the proper density of bodies measured in the most exact manner by estimating the quantity of phlogiston contained in them.[54] With a complete historical account and exact quantitative measurements, Guyton's "Dissertation" consolidated the phenomena of the weight gain in metallic calcinations as an incontrovertible experimental fact, at the same time pinning its cause on the loss of phlogiston. Submitted to the Paris academy in 1771, it brought a high acclaim for this provincial chemical novice. Macquer in his report to the academy concluded that Guyton's "Dissertation" established "the fundamental fact of the augmentation of the absolute weight of bodies by the subtraction of their inflammable principle." He praised it as a "great work . . . on one of the most interesting and most elevated objects of chemistry," particularly for its "most beautiful experiments which form the basis of his dissertation." He had some reservations about the "systematic" aspect of the work, or Guyton's explanation of the weight gain.[55] According to Condorcet, Guyton had "proven that this augmentation was real and general for all the metals."[56] In March of 1772, Guyton was elected a corresponding member of the Paris academy.

Guyton's "Dissertation" actually accomplished more than establishing the "fact" of weight gain. It brought the field of argument concerning the weight of phlogiston to the attention of Parisian elite chemists, particularly Lavoisier. By focusing sharply on the problem of weight gain and by organizing relevant experimental facts around it, Guyton provided a framework for debate and dissension concerning the identity and the function of phlogiston in a series of interconnected operations: calcination, combustion, and acidification. Although the involvement of phlogiston in these operations had been known for some time, chemists had not consciously organized them into a core set of reactions that supported the phlogiston theory. Furthermore, Guyton interpreted these operations as a set of algebraic equations involving phlogiston. Through these literary and linguistic maneuvers, he stabilized a field of argument in which the weight gain of metals became an outstanding chemical riddle in a way it had never been. Phlogiston, identified as the key to this urgent riddle, soon came under intensified scrutiny.

The value of Guyton's "Dissertation" lay, therefore, as much in the exhaustive literary exposition of the problem as in the quantitative measurements. Historians' interest in this work should also be directed to its organization or structure of argumentation. In the introduction, he presented the weight gain of metals upon calcination as a central problem in characterizing phlogiston. He devoted the rest of the volume to proving this thesis and establishing phlogiston as the sole cause of the weight gain. The first chapter scanned the available literature to harvest qualifying facts—the phenomena that properly belonged to the category of weight gain. One can sense a lawyer's mind at work in the way he sifted through the printed literature for the evidence pro and con, setting up the rules to screen out irrelevant facts and to discern authentic ones. The second chapter sketched out a "historical and critical examination" of the existing solutions to the problem. The third chapter provided the "proofs" that phlogiston was the sole cause of the augmentation of weight and outlined his system based on an a priori assumption of the volatility of phlogiston. The fourth chapter ostensibly contained more empirical support for Guyton's system. It accomplished a lot more. Arguing that the calcination of metals by fire was not the only operation in which the loss of phlogiston caused a weight gain, Guyton examined calcinations by niter, calcinations by arsenic, dissolutions by acids, and reductions and cementations to prove the same point. By doing so, he ran a common thread through various procedures previously underconnected, attributing them to the same cause. In particular, he asserted that the metallic calces resulting from the action of acids on metals or the "true humid calcination" had the exact same nature as those produced by fire or detonation. Lavoisier's theory of calcination became inextricably intertwined with that of acids through his experiments on humid calcination. In the fifth chapter, Guyton sought to extend his system to other phenomena, such as calcinations of mercury and Prussian blue, combustion of sulphur, and the effect of phlogiston on density. He perceived that these phenomena offered a significant challenge to his theory, and he wished to explain them away. By doing so, however, he unwittingly provided his future opponents with the means of undermining his theory. The calcination of mercury by acid and its subsequent reduction without charcoal, as well as the combustion of sulphur and phosphorus, soon provided ammunition for the budding antiphlogistic camp.

Guyton's explanation of the weight gain of metals turned on the "essential" volatility of phlogiston.[57] The immediate cause of volatility was the "excess of gravity of the medium over that of the volatile body," which allowed it to move away from the center of gravity. When a volatile body was united to a fixed body, the "compound density" of the two bodies would determine the movement toward or away from the center of gravity. The "essentially volatile" body was always lighter than the subtlest medium, either because its elementary form could not be changed or because its volume could not be compressed to increase the density. Phlogiston was such a body. Although it could establish an equilibrium with the medium by the "compound gravity" when it was united to a heavier body, it never ceased to be volatile. The mechanism of calcination followed from this assumption: When lead was exposed to fire, its aggregation was ruptured in two ways, either weakly and successively so that only the phlogiston at the surface could escape or strongly so that the metal itself melted. In the first case, phlogiston had to leave alone because the adhesion between the molecules of the metal was stronger than the adhesion between those molecules and phlogiston. Phlogiston thus left at the moment when its volatility was able to "vanish the adhesion" that held it to the metal. In the second case, the pores of the metal were enlarged so that its molecules were touching only at some points. Phlogiston, still tied to these molecules surrounding it, carried them along when it left. Volatility was "neither essential nor intrinsic" to these molecules of the metal, since they lost it as soon as they were detached from phlogiston. In other words, Guyton disagreed with "the most celebrated physiciens" who thought that all bodies in nature could be volatilized at a certain degree of fire, that volatility was only the effect of dilatability, and that any body could be sufficiently rarefied to be sustained in the subtlest medium. Refuting their "hypothetical principle," he regarded such rarefaction as possible only through "the destruction & the transmutation of elements." In other words, volatility was not a physical state caused by heat; it was a singular property of phlogiston.

Guyton did not assign a negative weight to phlogiston, as Lavoisier later charged against phlogistic chemists, but followed Chardenon's path of hydrostatic analogy. The excess or shortage of the specific gravity of a body over that of the medium determined its fall or rise in the medium. It followed, therefore, that the "absolute weight" of a body was not the one measured in the air, which was still "specific or relative to that of air." Consequently, he argued, "although all the addition of whatever matter augments the *strictly absolute* weight of a body, it is possible that this addition does not augment or diminish its specific gravity in the air."[58] Phlogiston diminished the weight of the body in the air in proportion to the excess of its levity over that of air. Guyton illustrated his argument with the hydrostatic balance. If one took four perfectly equal cubes of lead, weighing 563 grains each, and put two each on the two basins of a balance, they should attain an equilibrium either in the air or in the water. If one placed on one basin of the balance a thin plate of cork weighing 6 grains between the two cubes of lead and plunged the balance in water, however, one had to counterbalance the other basin with 28 grains to reestablish the equilibrium. The experiment illustrated that the addition of matter could produce a diminution of weight in water. Aside from the volume considerations, one could imagine exactly the same situation for phlogiston in the air. Phlogiston was so subtle that by being fixed in the smallest interstices it served to tighten them rather than to extend them. Guyton concluded that "the presence or absence of phlogiston is the true cause of the diminution or augmentation of the weight of bodies susceptible of being combined with it."[59]

The "essential" volatility of phlogiston had disturbing implications for the internal structure of matter. It could jeopardize the uniform construction of bodies in identical particles, as Roux had observed in his evaluation of Chardenon's work. In order to explain the weight gain solely by the absence of phlogiston, Roux noted, "it would appear in fact that M. Chardenon admits in different elements a different degree of weight; he believes necessarily in the variety of existing or possible bodies."[60] Guyton did not subscribe to the view that "each elementary being would have a different specific weight," which he dismissed as a gratuitous assumption on Chardenon's part that contradicted the law of universal gravitation. He did not wish to confer a new, hypothetical, and imaginary faculty on phlogiston. In order to avoid this unsettling inference from the "essential" volatility of phlogiston, he identified the difference of specific gravity as the unique cause of the weight gain. The diminution of specific gravity was nothing but the effect of the expansion of matter or of the augmentation of volume by fire. More precisely, phlogiston had to contain less matter in a given volume than the most rarefied air. In other words, the notion of essential volatility applied only to the phlogiston in mass, not to the elementary atom of phlogiston, on which the same cause of weight acted as with all other matter. The supposed essential volatility of phlogiston was "not a property essential to the element as element, but only to the aggregative mass of elementary atoms." Phlogiston always combined with metallic earths in aggregative mass, rather than molecule by molecule as chemists usually supposed in all elementary combinations.[61] In trying to satisfy both the general laws of natural philosophy and chemical theory, therefore, Guyton threw into sharp relief the disparity between the specific weight of body as a measured quantity and the presumed internal construction of the body in identical particles. The universal applicability of gravitation did not automatically imply the uniform construction of bodies in identical particles. His lengthy response did not resolve this acute problem.[62]

According to an anonymous reviewer, Guyton's theory of phlogiston satisfied neither chemists nor physicists.[63] While praising the "fine views, researches and experiments well made," the reviewer found Guyton's notion of the essential volatility of phlogiston "repugnant" to the ideas of gravity and even to the received ideas of phlogiston.[64] In a succinct summary, he listed all the relevant arguments: First, phlogiston was volatile without the intervention of fire, although fire increased its volatility. Second, phlogiston was the principle of dilatability responsible for evaporation, fusion, or incineration. Third, Guyton did not decide whether phlogiston was pure elementary fire or a compound. Fourth, he reified the concept of absolute weight. Fifth, he illustrated this with a hydrostatic balance and cork without considering the volume change. In short, the reviewer continued in a sequel, Guyton's hypothesis comprised two principal points: the essential volatility of phlogiston and the reification of the notion of absolute gravity.[65] The first contradicted chemists' accepted notion of phlogiston as fixed fire which added weight to the bodies; the second ignored the volume changes. Instead of the metal cubes, if one placed hollow metal spheres in the two arms of the balance to attain an equilibrium in air or water and placed a cork ball in one of the spheres, the arm carrying this cork ball would carry all the gravity of the ball, regardless of the medium. That is, the "addition of matter specifically less heavy than water" would produce an augmentation of weight in water, contradicting Guyton's argument that the cork in water was like phlogiston in air.

Another anonymous review differed somewhat in focus and style. It was not so much a review of Guyton's work as a polemical piece aimed at discrediting the phlogiston theory altogether. From the beginning, the reviewer characterized phlogiston theory as an "error" imposed by Stahl's illustrious authority and "consecrated" by habit. He discussed in detail two cases that provided the crucial proof of the phlogiston theory. The first showed that sulphur was composed of vitriolic acid and phlogiston, the second that all metals were composed of a vitrifiable earth and phlogiston. By pointing out the "errors of supposition & of definition" in the interpretation of these two experiments, the reviewer sought to dismiss the necessity of phlogiston:

One has never been able to obtain it [phlogiston] alone: no person has ever seen it; it is only through inductions one suspects that it enters into the combination of a body; & if these inductions are founded on false suppositions, as I have shown, the entire system collapses.[66]

5.5   The Dissolution Model of Chemical Action

Guyton's *Digressions academique* contained another major work, titled "Essai physico-chymique sur la dissolution et la crystallisation et les précipitation," that laid out his program of affinity chemistry.[67] He had been reading the literature on affinity, including those of Geoffroy, Gellert, Limbourg, Homberg, and Rouelle, as a part of his plan to construct a comprehensive table on the properties of salts. On June 2, 1769, he presented to the Dijon academy a "table halotechnique," which consisted of 13 perpendicular columns intersecting with 24 horizontal columns. The top horizontal column listed the category of base, four simple acids (vitriolic, nitrous, marine, and vegetable acids), and eight "compound acids" [acides composés]; the first perpendicular column listed 24 different kinds of base, including phlogiston. At each intersection, Guyton listed the compound resulting from the union of an acid and a base. Guyton's halotechnic table thus visualized the composition of middle salts, pointing to the possible compounds not yet discovered in empty slots. It formed a kind of "chemical world map" that identified the countries discovered, those yet to be discovered, and the necessary route to arrive at them[68]:

I formed the project several years ago of reuniting in a single synoptic table all the saline compounds of two or three parts: all the known acids would have to be placed in the first horizontal line, divided into as many cases; the first perpendicular column equally divided would have to offer all the known bases; the simplest first, then compounded, & the corresponding cases, that is to say, the one which would be found at the summit of the angle, formed by a perpendicular column & a horizontal line, would have to indicate the salt produced by the combination of the substances named at the extremity of each side of the same angle. I presented to the Academy of Dijon in 1769 an essay of this table which I named *halotechnic*; the occupations multiplied in more than one genre have not permitted me to follow up this work, but I believe to be able to assure that a table drawn up under this plan would perhaps also be useful, at least as convenient as the table of affinities; it would form a kind of *chemical world-map*, in which one would perceive at first sight the countries known & the space that remains to be discovered; it would announce all the results of the substances which have been until this day presented one to another; it would indicate those whose combination have not yet been tried, those which refuse absolutely to be combined; & under this last point of view, it would serve as the *table of negative affinity*.[69]

In the subsequent "Essai physico-chimique sur les dissolutions, les cristallisations, et les précipitations," Guyton sought to arrive first at "the mechanical explication of affinities," second at the "determination of the law of attraction in relation to the union and adhesion of the constituent parts of bodies," and third at "the knowledge of the sign of these constituent parts."[70] Although his attention was soon diverted to the role of phlogiston in metallic calcination, phlogiston offered only a critical subject that demanded a clear formulation of the laws of affinity. Aside from establishing phlogiston as the unique cause of the weight gain, Guyton hoped eventually to calculate the quantity of phlogiston in bodies, to uncover the principles of dissolution and fusion, and to arrive at the "mechanical explication of affinities" by elucidating the figure of the constituent parts of bodies.[71] Phlogiston occupied an important place in Guyton's affinity chemistry, that is, as a pervasive solvent of nature. Crystallization offered a potential route to the figure of the constituent parts of bodies.

Guyton's comprehensive vision for theoretical chemistry consisted of a dissolution model of chemical action. He added the Newtonian component of attraction to Louis Lemery's mechanistic model of solution process, drawing on a host of chemical authors: Macquer, Boerhaave, Hoffman, Spielmann, Cadet, Boyle, Friend, Keill, Barchusen, Lemery, Bohn, Le Sage, and Limbourg. He also accepted Buffon's premise that chemical affinity or attraction depended on the shape and the relative position of particles, and that it followed the inverse square law:

I would therefore start with these principles established by M. Buffon in his sublime views of Nature that *the laws of affinity are the same as the general law by which celestial bodies act on one another; that these particular attractions vary only by the effect of the figures of the constituent parts because this figure enters as element in the distance.*[72]

Guyton hoped that the study of crystallization would help him to unpack the shape of the ultimate particles of matter and to calculate the forces of attraction.

Since crystallization occurred when similar particles were brought into contact by attraction, one could infer the shape of these particles from the crystals they formed. Guyton stipulated three principles to attain the figure of the constituent parts through the examination of crystals. First, all matter was attracted to the center of the earth, and all bodies had to weigh absolutely. Second, as Le Sage had articulated, attraction was proportionate to density, and a body could cease to act by its weight in a more dense medium. Third, attraction was constant between all parts of the matter, and any body could "cease to obey the law of attraction to the center of the earth, if it is sufficiently attracted by another contiguous body so that this first attraction may be vanished."[73]

Guyton's theory of crystallization was embedded in a more general theory of chemical action. He assumed that all chemical action depended on prior dissolution of the reacting substances. *Dissolution* was "an operation by which substances are sufficiently attenuated in order to be found in the exact ratio of gravitation with the dissolving fluid."[74] *Division* and *equiponderance* defined the two necessary conditions of dissolution. The principle of division was attraction, Guyton insisted, rather than pores and hooks as Lemery had believed. The idea of attraction was less vague, agreed better with the diverse phenomena of alkaline dissolutions, and was subject to fewer objections than the idea of pointed acid particles. With the latter, one had to ask what was the cause of action these pointed particles of acid impressed on metals, opening the door to occult causes. Attraction sufficed to explain the phenomena of division entirely, Guyton contended, picking up a theorem Le Sage had once entertained.

The process of dissolution took place "from particle to particle." An "atom" of metallic earth united with an acid only by quitting its association with phlogiston. Even the slightest contact could, in continual succession, derange and destroy the structure of the metal. The contact between an acid and a metal produced such a strong union that one could not break it by any mechanical means except the "natural mechanics of affinity."[75] As before, Guyton relied on imagery to illustrate the "mechanism of dissolution." If one had a piece of wood whose linear fibers were weakly adhered to, one could paste another body on this wood and then, in removing this body, lift a part of the wood. The paste provided the "force necessary to vanish" the "simple adherence" of the linear fibers of the wood. One could use this model even when the force of adhesion of dissolved bodies was much stronger, as long as it stayed a little below the "full attraction" or the "strongest adhesion resulting from a most perfect contact."[76] Although the parts of a metal adhered strongly to one another, for example, they acquired their particular arrangement only through the "intermediate" of phlogiston which resisted solidity. The solidity of metal did not exclude a contact still more perfect, such as the one between a metal and an acid. Guyton regarded the "paste" not as an "intermediary and foreign" agent but as a means of contact between two bodies. It was the means of adhesion by attraction, or the force with which all the substances in changing state of gravitation looked for a new equilibrium.

The second precondition for dissolution was an "exact ratio of gravitation" or equiponderance between the parts of dissolved bodies (solute) and those of the dissolving liquid (solvent). Guyton believed that this "ratio, which is itself the principle of natural division, produces &

conserves the dissolution."[77] An equiponderance between the most concentrated acid and lighter metals seemed difficult to conceive, since the gravitation of a body was only the sum of the gravitation of all its parts. Guyton reckoned that this difficulty probably perpetuated the belief that "affinities are veritably laws particular to certain substances." The perceived difficulty was not insurmountable, Guyton advised, since chemical dissolution decomposed bodies only to the level of supercompounds, rather than to the level of elements. The parts of a metal could be extended over the surface of water, insofar as they attained equiponderance with the parts of water.[78] In proposing a purely mechanical notion of dissolution, Guyton criticized the chemical notion of affinity, particularly that of specific quality and reactivity. The language of affinity only substituted a name for a phenomenon, presenting no specific idea.

The new definition of dissolution served as the "basis for the mechanical explication" of chemical phenomena, lending a "new force, new light" to the phenomena of crystallization, an operation in which an infinity of similar parts, at equilibrium with the solvent fluid, approached one another to form a regular mass determined by their reciprocal attraction. By "similar parts," Guyton meant not the integrant parts of a specific chemical substance such as the metal but "the homogeneity resulting from the same density or from the same figure of molecules." He did not wish to add the condition that the parts should have the same nature since it would invoke affinities. Crystallization required the presence of fluid throughout the process to maintain the equiponderance. So long as they satisfied the physical requirement of form, therefore, other parts could partake in the formation of crystals whether they were simple, compounded, or super-compounded. Guyton wished to found his theory "uniquely on the general laws of gravitation & of movement" and to guard himself from "admitting qualities inherent to such or such matter, independently of the density or the figure."[79] Whatever their nature, the particles of the same density and figure would form a solid, hard, regular mass which characterized crystals. He did not think it quite probable, however, that heterogeneous parts could have the same density and figure, since these conditions allowed for the homogeneity of composition in the first place.

Guyton illustrated his "mechanical explication of dissolution & crystallization" with the example of needles floating on water. If one placed a needle on the surface of water in the vase, and another at the circumference of the vase, the second moved toward the first, accelerating as it approached, until it lay parallel to the first. The process would go on as long as the water could carry the needles. A number of needles thus grouped could represent a nitrous crystal. This action of the needles was caused by their reciprocal attraction, just as the action of "insensible parts" was. Such "simple & common" experiments should reveal the "figure of the constituent parts of the body." With the single law of attraction, one could explicate all phenomena of dissolution and crystallization:

With the help of this principle, one can hope to determine, with certitude, the figure of the primitive parts of all crystallized bodies, because all regular mass can only be composed of the parts that have a form generative of the form which result from their union, according to a known law: just as it is not possible to conceive that there can ever result from an infinity of cubes, however small they may be, a kind of spherical mass, not even one that has a gross appearance of it, as soon as one admits in the mechanism of their union the necessity of the most perfect contact, one can likewise be assured that a cubic mass results necessarily from an infinity of molecules of the same form, that a cylindrical crystal is the product of several insensible cylinders; & the same holds for the more composed figures of other solids.[80]

While Guyton insisted that the figure of the molecules alone would provide the "key to chemical affinities," he knew that they did not come under experimental scrutiny and control. Neither could any working chemist hope to incorporate his ideas into their practice. In many ways, his thoughts on dissolution and crystallization were a throwback to Homberg's and Lemery's corpuscular program. Although he explicitly criticized the "Cartesian" chemistry in which the particles were endowed with physical hooks, his "Newtonian" discourse did not differ much in its speculative quality and in the use of mechanical images. Both articulated a *public* discourse of chemistry that rehabilitated the esoteric world of chemists for a broader audience. Nevertheless, the Newtonian language was more positively disposed to the theory domain of affinity than the Cartesian. Even if the primary functions of chemical philosophy were social legitimation and pedagogical efficacy, an established philosophical discourse could exert a selective pressure on the theory domains of chemistry and the direction of future investigations.

5.6   The Quantifying Dream

Following his vision of a mechanistic chemistry based on Newtonian attraction, Guyton soon set out to quantify affinities with a physical method. In the October 1772 issue of the *Observations* he found a report by M. Cigna of the Academy of Turin that the mercury level rose unequally in barometers of different diameters.[81] Attributing the difference to some kind of action between the glass tube and the mercury, Cigna used Taylor's method to verify whether this was attraction or repulsion. Taylor, a member of the early Newtonian group, had measured the weight necessary to separate thin pieces of fir-board from the surface of water by tying them to one branch of the scale. He found them to be "exactly proportional" to the surface of the board.[82] Cigna placed a piece of glass on the surface of mercury to estimate the weight necessary to detach it and concluded that there was attraction between glass and mercury. Upon communicating this result to M. de la Grange, however, he was advised that the action was probably due to the atmospheric pressure. Although Cigna remained convinced that the atmospheric pressure could not cause such an effect, the editor of the journal, M. Rozier, chose to suspend judgment on the issue, since it touched directly on "the fundamental laws of the system of the universe," and proposed it as a question to the physicists. Guyton set out to prove that it was indeed attraction, not repulsion, determined by the mass and the figure of the parts.

Guyton had already announced in the "Essai physico-chimique" that there existed no repulsion in the proper sense of the term. All phenomena that indicated repulsion depended on attraction and on the ratio of equiponderance. Such an opinion would conform to "the simplicity & harmony of physical laws." As an unequivocal experiment, Guyton wished to use two plane surfaces interposed by a volume of liquid so that they could approximate the ratio of the liquid and the interior circumference of the capillary tube. M. de la Lande had already used the approach, however, demonstrating that this phenomenon was due entirely to attraction. For the question under examination, Guyton took two glasses 2 pouces in length and 3 pouces in height and coated one of the glasses with tallow. By plunging these two glasses in water, he could observe the water rise between the two surfaces, which he attributed to the attraction between tallow and water. Using a glass disk 2½ pouces in diameter, he created the following table of adhesions:

to mercury	756	
to water	258	334
to oil of tartar	210	294
to oil of olive	192	280
to the spirit of wine	162	226

Taylor's method of estimating attraction should work in a similar manner. The inequality of the force depending on the nature of bodies should indicate its independence from the atmospheric pressure. Guyton thus concluded that Cigna's original observation that mercury and glass attracted each other was fully verified. He further confirmed his conclusion by comparing the weight necessary to detach the same glass disk from mercury in the air and in the vacuum pump. They required the same weight (9 gros) to detach the 2½-pouce glass disk from the surface of mercury according to Taylor's method. In sum, Guyton concluded that glass had a quite sensible action on mercury, that this was an attraction rather than repulsion, that Taylor's method in estimating this force was exact, and that the pressure of air had nothing to do with this phenomenon.

5.7 The "First Chemist of France"

Guyton was entrusted with the chemical articles for the supplementary volumes of Diderot's *Encyclopédie*,[83] which indicates his growing reputation as an interlocutor of French chemistry. In the articles "Affinité," "Crystallisation," "Dissolution," and "Équiponderance" one can discern his dissolution model of chemical action essentially intact. His preference for the physical explanation of chemical action comes through clearly. The article on phlogiston also exemplifies Guyton's physico-chemical approach. He argued that the more chemistry made progress, the more its terms became confounded with those of physics. Experiments and observations brought closer these two sciences previously divided by "a false spirit of the system." Nature had to be the same "for the one who admires it in its grand works, & for the one who studies it in the insensible parts of compounds." Nature had "only one law for the grand as well as for the small effects." This unity, simplicity, and harmony of nature provided the "infallible types" according to which chemists and physicists sought their discoveries. The unity of nature dictated, for example, that phlogiston and fire must be the same. Since not all physiciens accepted such an identity yet, one had to retain the "indeterminate denomination" of fire and phlogiston:

> The fire that burns is nothing other than a matter placed in movement; but all matter is not proper to receiving, keeping and communicating this movement of ignition, *the proximate cause of heat*. One has been forced to recognize that there is in the nature a substance essentially endowed with this property, & bodies more or less provided with the inflammable principle. It is this principle, considered in the composition of bodies, abstraction made from the movement, that Stahl has named *phlogiston*.
>
> According to some, phlogiston is a secondary principle, composed of the element of fire & of a vitrifiable earth: others on the contrary regard it as the pure matter of fire, not that they pretend that it can never be considered as already combined with other substances when it enters into the formation of a compound; but since, by examining its nature & its characters in all the mixts in which it exists abundantly, in all the operations in which it plays the principal role, they have it always recovered similar to itself, they think that it is a simple being whose properties are independent of the different matters in which it is engaged; & this system would appear to us founded on reason & observation.[84]

Guyton's identification of phlogiston with fire, heat, and electric fluid put him in the company of Boerhaave and other physicalists rather than that of other chemists. Phlogiston or fire became an essential part of his dissolution model of chemical action as a kind of universal solvent. Phlogiston was to metals what water was to salts.[85] Just as salts retained the "water of crystallization" from their dissolution in water, metals retained "the fire of crystallisation" from their dissolution in fire. Since fire was the greatest solvent, it should be fixed in all bodies in their transition from the fluid to the solid state. Water itself received its fluidity and dissolving power from the fire it contained. Combustible and noncombustible bodies would differ then only in the proportion of the fire they retained. The action of fire in calcinations consisted not in simple relaxation of aggregation, but in true dissolution. The identity of the effects produced by calcination and acid dissolution proved it. Guyton intercalated in this manner a mechanical explanation of chemical action with the interplay of affinities more commonly used in chemical explanations. Phlogiston or fire worked as a kind of universal solvent, dissolving all bodies. It therefore existed in most bodies as the "fire of crystallization" and interacted with them through its affinities.[86] He accepted phlogiston within the framework of affinity chemistry developed by Rouelle and Macquer, making it a veritable chemical substance endowed with affinity. At the same time, he constrained its actions by the mechanical condition of dissolution caused by gravitational attraction prescribed in the Buffonian form of Newtonian philosophy. If Guyton deviated from Buffon as to the identity of phlogiston, he endorsed the Newtonian language. His conception of phlogiston as a universal solvent was quite similar, in fact, to Lavoisier's later conception of heat as a part of chemical constitution. His later rejection of phlogiston did not cost him the dissolution model of chemical action.

Guyton put Dijon on the chemical map of Europe through his literary activities.[87] He also brought chemistry to the Dijon public. When he became vice chancelier of the Dijon academy in 1772, effectively in charge of its functions, he actively sought to popularize sciences and to spread the taste for mechanical arts. On November 17, 1774, he read to the academy a "Mémoire sur l'utilité d'un Cours de chimie" to gather support for public lectures in chemistry. He wished to open the academy's laboratory free of charge to everybody who wanted to learn. Chemistry was no longer that occult science scorned by true physiciens, written in hieroglyphics and looking for philosopher's stone or universal medicine. Instead of these chimeras and secrets, it now produced useful truths, had clear and fecund principles, and sought proofs in experiments and observations, establishing itself as "the most curious & the most important part of natural philosophy." A public course of chemistry would offer many advantages for the fashionable audience who wished to follow the contemporary development in science, for the aspiring alchemical adepts who squandered a fortune on useless endeavor, and for those involved in its manifold applications to manufacture, pharmacy, metallurgy, agriculture, and the arts. For the learned citizens in the provinces who wished to familiarize themselves with the new order of true physics and follow the light of their century, only a public course could bring the "revolution" already quite advanced in the capital to them since they could not expend the necessary time, effort, and laboratory facility to learn the science. For those who had consumed their time, fortune, and patrimony of their children to chase the vain promises of alchemical secrets, a public instruction offered the best remedy for their folly guided by ignorance. The "intimate & necessary relationship" chemistry had with all kinds of health professions supplied the most pressing reason for opening a public school of this science. Chemistry alone could predict with any certitude the effect of remedies, combine their action, proportion the doses, and prepare the ingredients. Although physicians were taught in the university, apothecaries required instruction. Should there be an epidemic and the government dispatched physicians, the treatment of the affected would be left to surgeons who dispensed remedies without the requisite knowledge. Chemistry could also provide the foundation for the theory of agriculture, dyes, and metallurgy. It had variegated relationships with mechanical arts and manufactures. All these applied fields called for public instruction in chemistry.[88]

The courses began in 1776 with the collaboration of Maret, Durande, and Chaussier. The lectures, published as *Élémens de chymie théorique et pratique*,[89] differed noticeably from the Parisian textbooks. Although Guyton began with the customary discussion of generalities, an intelligible overview took precedence over the step-by-step instruction of chemical art. An overgeneralization coupled with careful linguistic control set Guyton's lectures apart from the more traditional chemical textbooks aimed at prospective apothecaries and physicians. He asserted, for example, that all of chemical theory consisted in the two words of attraction and equiponderance, while all of its practice was reduced to the two words of dissolution and crystallization. The difficulty lay in devising a "method" to pull together disparate ideas and facts into an orderly ensemble to help the memory and to render the principles of nomenclature intelligible. Nothing seemed more capable of facilitating this job, Guyton surmised, than a table that placed at once the most important details and theoretical series before the eyes, although it was far from perfection. He intended to exploit the pedagogical efficacy of the chemical table to the maximum. His "synoptic table," much like the "halotechnic table" he presented to the academy in 1769, displayed the composition of neutral salts at a glance.

Guyton's power as a reader is evident in his ability to pick out the most salient part of French didactic discourse. He defined chemistry as "the science of the properties of simple bodies, of the properties of compound bodies, of the properties they acquire or lose by new compositions, of the means nature employs to unite or separate them, of procedures by which the hand of man has come to act on nature instead to produce changes and infinite advantages these knowledges procure it."[90] He believed that there existed probably only one matter whose modification in density, porosity, and figure produced all other bodies. Nevertheless, it was important to distinguish the "natural elements" of fire, air, water, and earth from the "chemical elements," or simple bodies one could no longer decompose. Physicists were not agreed on the nature of the four natural elements. Chemical elements were likely to be compounds, sometimes quite complicated; however, they were simple for the art, since one could not disunite their principles to destroy their actual characters. Currently, they included acids, metallic substances, vitrifiables, calcarous earths, alkalis, oils, etc. A simple body became a compound when it was united to another simple or compound body with different properties as in the case of the mixture of strong water with water. In contrast, two drops of water did not make a composition, but a simple aggregation.

A careful reader became a concerned teacher. Guyton wished to transmit chemical knowledge to a broader audience in an easily accessible manner. The lack of rational nomenclature preoccupied him for this reason. Although chemistry did not have as detailed a nomenclature as other sciences, the multiplicity of the names given to the same thing made it quite awkward. Improper terms that had been created or adopted at a time of ignorance carried only false ideas. Guyton vowed to list only a minimum number of words necessary to follow the course, pertaining to the categories of theory, quality of bodies, and operations. With a set of carefully controlled terms, he expounded the dissolution model of chemical action premised on attraction. Soon he was engaged in a sustained effort to devise the foundation of a rational nomenclature in cooperation with Torbern Bergman. Their desire for a nomenclature reform bespoke their existence in the Republic of Science. Guyton's claim to the title "first chemist of France" and Bergman's position as the representative Swedish chemist depended on their network of correspondence and literary contributions. As philosophical chemists whose reputation and capital depended on the commerce of words, they found it intolerable to deal with irregularities in nomenclature and units of measurement, which hindered the traffic of words.

5.8   The Limits of the Affinity Table

Guyton's efforts to quantify affinities was a step ahead of his times. The dominant project of affinity chemistry during the 1770s was the classification of affinities or the revision of affinity tables, following Macquer and Baumé's cue. By providing a shared vocabulary as well as a wealth of practical operations under the rubric of affinity chemistry, they offered a research guide for the next generation of chemists. They ushered in a vogue of affinity tables. In addition to Rouelle and Limbourg, Gellert also modified the table. In 1773, De Fourcy produced an even more elaborate table consisting of 36 columns.[91] At the same time, the analytic sophistication in the chemistry of salts alerted chemists to the limits of the tabular representation. Baumé pointed out already in 1765 that a simple augmentation of the table in the conventional way did not account for "an infinite number of circumstances" that changed the outcome of chemical action. Mindful of such inherent limitations of affinity tables, he proposed a double affinity table to improve its relevance to chemical practice.[92] He "imagined" that a double table for the "moist way" and "dry way" would be "extremely useful."[93] Macquer also conceded that the current tables fell short of meeting all the demands of chemistry. The table was incomplete and unjustified when giving too general predictions. He did not believe that it was "ill-conceived, useless or dangerous," but that chemistry was still "at an immense distance from its point of perfection." He projected that as experimental chemistry progressed the table would improve.[94]

Macquer and Baumé did not conduct a systematic inquiry, however, to remedy these shortcomings.

It was the Swedish chemist Torbern Bergman (1735–1784) who put himself to this enormous task and produced in 1775 the most elaborate affinity table so far, comprising 50 columns and constructed for two different analytical methods: dry and wet ways.[95] In the revised Latin edition of 1783, his table expanded to 59 columns, adding the new acids discovered in the meantime.[96] Bergman's table represented a giant stride in chemists' capacity for analytic control. It listed 25 separate acids, 15 earths, and 16 metallic calces, in comparison to the four acids, two alkalis, and nine metals in Geoffroy's original table. Translated into French, German, and English, it won immediate recognition from the European chemical community, which indicates that the affinity table had a secure place in contemporary chemical theorizing.[97]

Bergman's monumental project drew on Swedish analytic chemistry as well as on French theoretical chemistry. He had been educated at the Uppsala university in everything except chemistry, but was appointed to the chemistry professorship there in 1767.[98] He applied himself to excel at his new task, no doubt reading up the relevant literature. He could rely on the native, rather advanced tradition of analytic chemistry for practical guidance. One can gauge the depth of this tradition through the lecture notes of Henrik Theophil Scheffer (1710–1759), a student of George Brandt at the mining college in Stockholm. Bergman studied and edited them minutely to prepare a textbook which he published in 1775 along with his affinity table.[99] He singled out for praise Scheffer's ability to work with a tiny amount of material, a skill relayed to him by Cronstedt.[100] He also initiated correspondences with the leading European scientists, including Macquer,[101] scanning French chemical literature including Senac's *Nouveau cours*, Demachy's *Instituts de chimie* (1766), and Macquer's *Dictionnaire*. Macquer recommended *l'avant coureur* for "prompt and exact" information. By April, Bergman had succeeded in electing Macquer as a foreign associé of the Stockholm Academy.[102] Through Macquer, Bergman tapped into the current development of French chemistry. He took upon himself the most theoretical task in Macquer's book—the affinity table and the classification of neutral

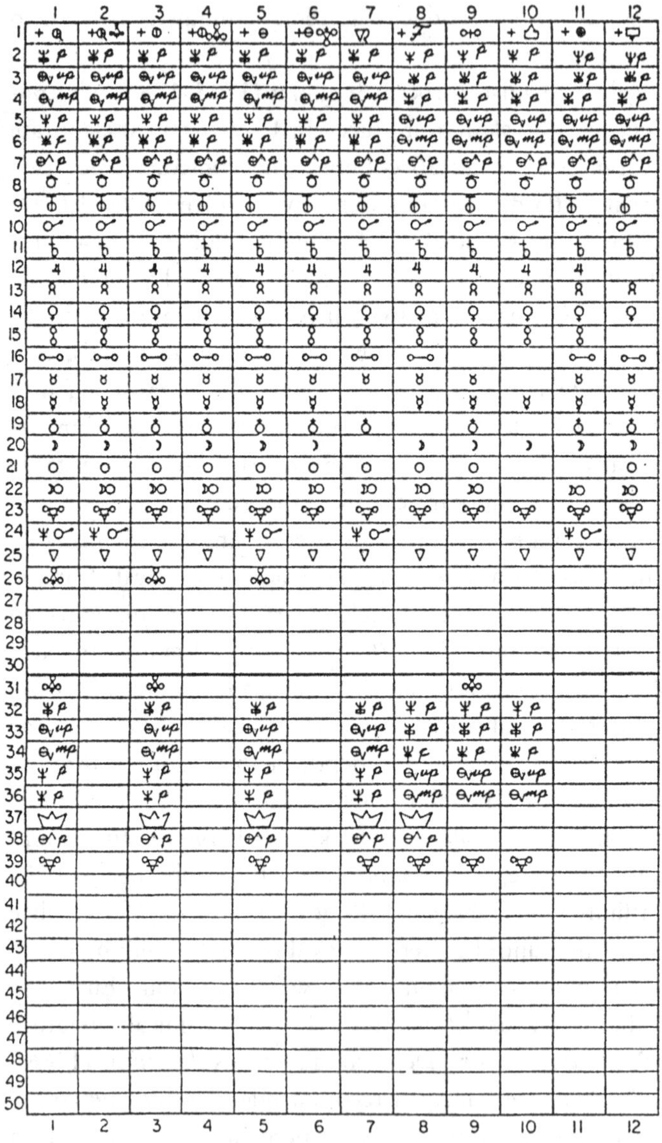

Figure 5.7
The left half of Bergman's affinity table of 1775.

Attractiones   *electivae*

13	14	15	16	17	18	19	20	21	22	23	24

(table of alchemical and astrological symbols)

| 13 | 14 | 15 | 16 | 17 | 18 | 19 | 20 | 21 | 22 | 23 | 24 |

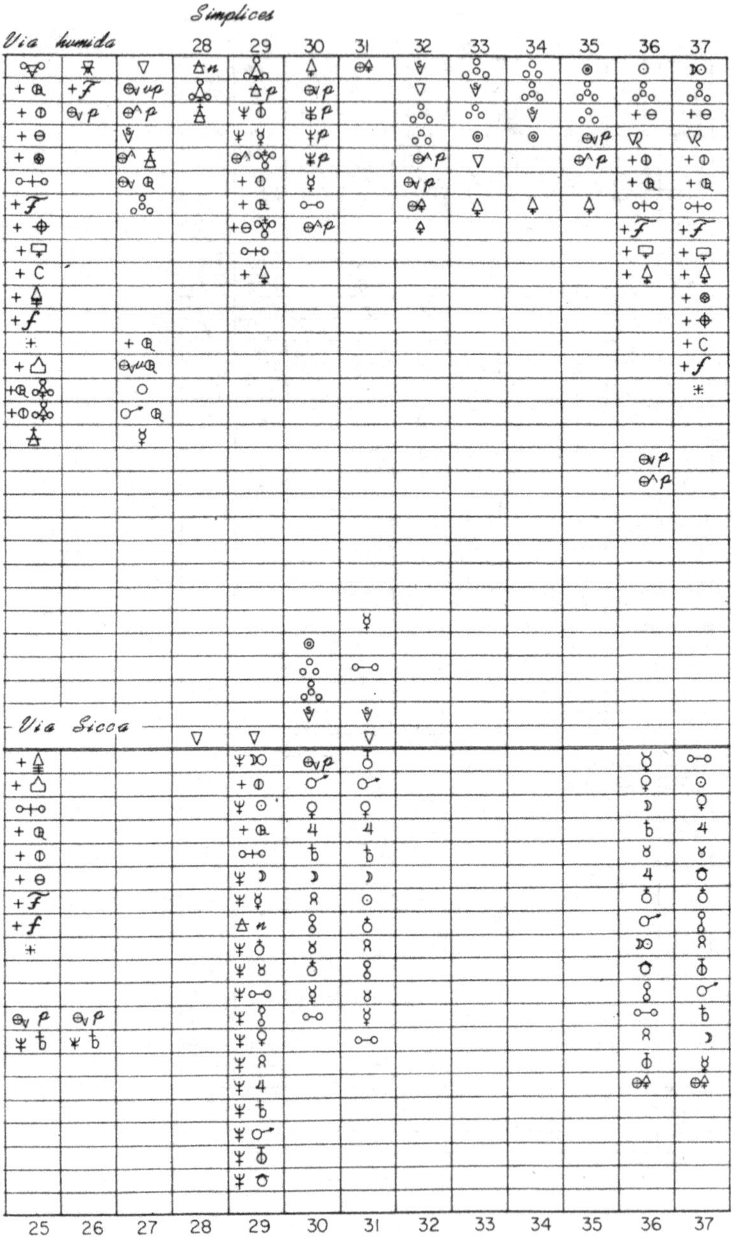

Simplices

Via humida

			28	29	30	31	32	33	34	35	36	37

Via Sicca

25	26	27	28	29	30	31	32	33	34	35	36	37

Figure 5.8
The right half of Bergman's affinity table of 1775.

38	39	40	41	42	43'	44	45	46	47	48	49	50	
☽	☿	♄	♀	♂	4	♉	♊	o–o	♌	♏	♍	♅	1
+⊖	+⊖	+♌	+⊕	+⊕	+⊖	+⊕	+⊕	+⊖	+⊕	+⊕	+⊖	+⊕	2
+⊕	+⊕	+⊕	+♍	+♍	+♌	o+o	+♃	+⊕	+♃	+♌	+⊕	+♃	3
+♌	o+o	o+o	+⊖	+♌	+⊕	+♃	+⊖	+♌	+⊖	+⊖	+♌	+𝓕	4
+☉	+♃	+♍	+♌	+⊖	+☉	+♍	+♌	+☉	+♌	+☉	+☉	+C	5
o+o	+♄	+♄	+☉	+☉	o+o	+♄	+☉	+♄	+☉	+♍	+♍	+♍	6
+𝓕	+♌	+⊕	o+o	o+o	+♄	+♌	+♄	+𝓕	+♍	+⊕	+⊕	+♄	7
+♍	+♍	+⊖	+♄	+♄	+𝓕	+☉	+𝓕	+𝑓	+𝓕	+♄	+C	+⊖	8
+⊕	+C	+☉	+⊕	+⊕	+	+⊖	+C	+♍	+C	+C	o+o	+♌	9
+♄	+☉	+𝓕	+𝓕	+𝓕		+𝓕	+𝑓	+⊕	+𝑓	+𝓕	+𝑓	+☉	10
+C	+𝓕	+C	+C	+C		+𝑓	✳	+C	+	o+o	+	o+o	11
+𝑓	✳	+𝑓	+𝑓	+𝑓		✳	o+o	+	o+o	+𝑓		+𝑓	12
✳	+♌♋	✳	✳	✳						+,		+	13
+♌♋													14
	+△	+△	+△	+△	+△		+△		+△	+△			15
⟊	⟊	⟊	⟊	⟊	⟊	⟊	⟊		⟊	⟊	⟊	⟊	16
													17
		⊖ᵥ ρ	⊖ᵥ ρ		⊖ᵥ ρ								18
⊖^ρ			⊖^ρ		⊖^ρ	⊖^ρ	⊖^ρ		⊖^ρ	⊖^ρ			19
		◉	◉					◉					20
													21
													22
													23
													24
													25
													26
													27
													28
													29
													30
♄	☉	☉	☉	♀	♀	♄	♂	♊	♂	♀	♂	♀	31
♀	☽	☽	☽	♌	☿	☽	♌	♌	♊	♅	♀	♂	32
☿	♆	♀	o–o	o–o	♀	☉	o–o	♀	o–o	4	4	☉	33
♉	♄	☿	♂	♀	♅	☿	♀	♂	♀	♀	♄	☽	34
4	4	♉	♅	♅	☉	♅	☉	☽	☉	☽	♊	4	35
☉	♅	4	♅	☉	☽	4	4	4	♆	☉	☽		36
♅	♉	♅	♅	☽	♄	♀	♅	♄	4	♌	♉		37
♂	♀	♆	♆	4	♂	♆	♆	☉	♅	o–o	♅		38
♅	♅	o–o	4	♅	♅	♊	♉	♆	♅	♆	☉		39
♅	♂	♅	♄	♆	♊	♂	♄	♅		♉	♆		40
o–o		♊	♆	♉	o–o	♅	☽	♅		♄	♊		41
♊		♂	♉	♄	♆		♅			♊	o–o		42
♆			♌	♅	♉						♌		43
			♅		♌								44
⊖♅	⊖♅	⊖♅	⊖♅	⊖♅	⊖♅	⊖♅	⊖♅	⊖♅	⊖♅		⊖♅	⊖♅	45
♀	♀	♀	♀	♀	♀	♀	♀	♀	♀		♀		46
													47
													48
													49
													50
38	39	40	41	42	43	44	45	46	47	48	49	50	

salts—as a research project. Macquer complimented Bergman on these works as early as 1770.[103] His growing regard for Bergman probably induced Guyton to initiate correspondence on January 20, 1779. As Macquer's long-distance student, Guyton had also started his excursion into chemistry by constructing an extensive table of salts, although his duties as a practicing lawyer did not allow him to pursue it further. Guyton translated into French Bergman's *Opuscules*, which he thought showed above all the "doctrine of affinities explained by the universal physical law of attraction."[104] Thanks to Macquer's and Guyton's endorsements, Bergman enjoyed a growing reputation in French chemical community. "You enjoy here with good justice," Macquer wrote to Bergman, "the most brilliant reputation in the whole scientific world as one of the most illustrious promoters of profound and transcendental chemistry and your name will make an epoch in the hall of this sublime science which will last with it."[105]

Bergman's *Dissertation on Elective Attractions* reads much like Geoffroy's memoir of 1718.[106] Out of a text that amounts to 320 pages in English translation, about 75 pages are devoted to general introduction and the rest to the explanation of the individual columns. Generally known as a staunch Newtonian, Bergman dwelled only briefly on the Newtonian doctrine in the first section (about 5 pages), largely trying to distinguish the "contiguous" attraction from gravitation. The first chapter was titled "There Seems to Be a Difference between Remote and Contiguous Attraction." Bergman admitted the existence of general attraction between "all substances in nature"—"a mutual tendency to come into contact with one another"—but argued that the contiguous attraction between small particles followed a law rather different from gravitation:

It has been shewn by Newton, that the great bodies of the universe exert this power directly as their masses, and inversely as the squares of their distances. But the tendency to union which is observed in all neighbouring bodies on the surface of the earth, and which may be called *contiguous attraction*, since it only affects small particles, and scarce reaches beyond contact, whereas remote attraction extends to the great masses of matter in the immensity of space, seems to be regulated by *very different laws*; it seems, I say, for the whole difference may perhaps depend on circumstances.[107]

Chemical attraction differed from gravitational force.[108] Unlike gravity, for which one could ignore the actual shape of bodies because of the "vast distance" between the heavenly bodies, contiguous attraction depended on circumstances such as the "figure and situation, not of the whole only, but of the parts," which produced a "great variation in the effects of attraction."[109] Since one did not have the means of evaluating the figure and the position of the particles, however, the relations of bodies had to be determined by experiments.

Shifting the field of argument in this way from philosophy to chemistry, Bergman proceeded to distinguish "several species of contiguous attraction" in line with Macquer's classification. He also offered an elaborate apologia for constructing a new affinity table, which indicates that the criticism of the table was intensifying. He was particularly concerned with the objection that affinities were not governed by fixed laws, but that they were variable according to the circumstances. Since all chemical operations depended on attraction, he argued, it was "of great importance to determine this dispute." Instead of rejecting the whole doctrine on the basis of a few irregularities, he would rather proceed with "caution and care" to examine these attractions. Even if they depended on circumstances, it would be "of extensive utility" to know these conditions. If there did exist a "fixed-order," it would serve as "a key to unlock the innermost sanctuaries of nature, and to solve the most difficult problems, whether analytical or synthetical." "I maintain, therefore," Bergman wrote, "not only that the doctrine deserves to be cultivated, but that the whole of chemistry rests upon it, as upon a solid foundation, at least if we wish to have the science in a rational form, and that each circumstance of its operations should be clearly and justly explained. Let him who doubts of this consider the following observations without prejudice, and bring them to the test of experiment."[110] In the third chapter, "Whether the Order of Attractions be constant," Bergman affirmed that "there prevails a constant order" of affinities, invoking the authority of "Experiment, the oracle of nature" as "the proper clue to guide us out of this labyrinth." Although some of Geoffroy's general assertions proved untenable, he argued, most of the difficulties disappeared upon closer examination. A few remaining deviations could be attributed to insufficient observation. As in the case of comets, he hoped that "repeated observations and proper experiments will in time dispel the darkness."[111]

Bergman acknowledged heat as "the only external condition, which either weakens or totally inverts the affinities of bodies subjected to experiments." In the fourth chapter, "A Difference in the Degree of Heat sometimes produces a Difference in Elective Attractions," he maintained that heat was an external agent which could sometimes disrupt the existing, "usual order" of "genuine attractions, which take place when bodies are left to themselves." It either forcibly weakened the "real affinities" or altered them completely. Since heat was necessary for many chemical actions, the "power therefore of this moist subtile fluid is highly worthy" of observation. Accordingly, he divided his affinity tables into two, the upper one for the "free attractions, that take place in the moist way" and the lower one for those "effected by the force of heat."[112]

The notion of double affinities served to mitigate much of the criticism that had been leveled at the affinity table. In chapter 5, dealing with "apparent irregularities from a double attraction," Bergman put forward the interpretation that many apparent contradictions to the order of affinities in fact involved four, instead of three, substances. The congregation of four substances offered "a very different and more complicated case than where only three are concerned." He illustrated this point with a well-known objection against Geoffroy's table which stipulated that fixed alkalis adhered to acids more strongly than calcareous earth. The opponents would "drop a solution of chalk in nitrous acid into a solution of vitriolated tartar, upon which a precipitation of gypsum immediately takes place," which gave them a "clear proof . . . of the superior attraction of calcareous earth." That is, tartar as a fixed alkali should have more affinity for vitriolic acid than for chalk, a calcareous earth, and therefore should not be precipitated. Bergman proposed a different understanding of the reaction with the notion of double affinity. When chalk was dissolved in some mineral acid, four substances came into action. The earth, now aided by the acid combined with it, effected what it could not accomplish alone. Thus, when vitriolated tartar, a neutral salt, was mixed with muriated lime, the reaction took place as if there were four substances in water (vitriolic acid, tartar, muriatic acid, and lime), producing a pair of combinations whose sum of attractions was greater than before. In the later edition, Bergman declared: "Chemists, in determining the single elective attractions, are often deceived by double attractions."[113]

Bergman's table indicates to us how portable and communicable the affinity table had become in the eighteenth-century European chemical community. The presence of a strong analytic tradition in Sweden no doubt helped Bergman to unpack the table and to develop it further. The expanded list of acids and the inclusion of vital air show that Bergman added on new substances as soon as he could confirm their existence in a reliable manner. His table published in 1783 added seven more acids, the matter of heat, and siderite, indicating that he was closely following the current development in chemistry. The table did not simply list newly discovered substances, however. In the case of phlogiston, Bergman had to digest various theories. In 1775, he defined phlogiston as the "most subtle material, which alone escapes all our senses, being known only from a proper study of its combinations." It entered into many substances and changed them in many ways according to the different quantity. In addition to discussing various operations which involved phlogiston, he also stated Scheele's hypothesis that the matter of heat was "nothing else but phlogiston intimately joined to pure air."[114] By 1783, he had to review carefully various views concerning phlogiston in adding the column headed by the matter of heat. In other words, not only did Bergman's table reflect accumulated empirical discoveries; it also embodied various theoretical developments which began to fragment the discipline by this time. The research on airs and the proliferating studies on phlogiston and heat phenomena all found their way into the affinity table.

If Bergman's table represented the finest achievement in the tradition of affinity tables, it also accelerated its ultimate demise. By trying to present the sum total of chemical knowledge in an all-encompassing table, he made visible the fragmentation of the paradigm that governed the affinity table. Displacement reactions could no longer determine the order of affinities. In addition to the problems caused by diverse theories, the compilation of a comprehensive table required immense labor.

In order to "perfect" his table with 59 columns, Bergman projected that he would need to perform more than 30,000 exact experiments. Such an endeavor exceeded his capacity, given his "many functions." Through his methodical perfection of the affinity table, he also showed the limit of such a representational form in the face of the explosive growth of chemistry. As chemists produced more and more substances with distinct qualities, it became increasingly difficult to contain the whole sum of chemical knowledge in a single table. Nevertheless, they did not abandon the research impetus provided by the affinity table—a systematic ordering and prediction of chemical actions. As affinity tables became more and more complicated, the efforts began to emerge to reinforce its predictive function by means of quantitative measurements. As Bergman noted in the revised *Dissertation* of 1783, Guyton initiated such research, and Richard Kirwan followed the cue. The trio of Bergman, Guyton, and Kirwan represented the frontier of theoretical chemistry in the early 1780s, indicating the success and the popularity of affinity chemistry immediately before the onset of the Chemical Revolution. They kept alive the research impetus of the affinity table or its function in instigating "exact and precise" research for the prediction of chemical combination and reaction. They defined the most refined frontier of late-eighteenth-century chemistry, embodying the hope of making chemistry a rational science.

5.9   A Gravimetric Measure of Affinity

Swedish chemists led the analytic revolution of the 1770s and the 1780s through their discoveries of new acids. After Scheele's discovery of tartaric acid (1770), J. G. Gahn brought into existence pyrotartaric acid (1774), oxalic acid (1776), lactic, mucic, pyromucic, and uric acids (1780), prussic acid (1782–1788), citric acid (1784), malic acid (1785), and gallic and pyrogallic acids (1786), forecasting a vast territory yet unexplored. Bergman's heroic efforts to find a place for these new acids in his affinity table won him the esteem of the leading European chemists. At the same time, the limits of the affinity table as a tool for prediction became increasingly obvious. It was in this context that various European chemists responded to Guyton's call to quantify affinities.

Franz-Karl Achard, Carl Friedrich Wenzel, Richard Kirwan, Laplace and Lavoisier, and John Elliot offered various schemes to this end.[115]

The most significant chemist drawn to Guyton's sphere of correspondence at this juncture was Richard Kirwan (1733–1812). He was born of an English family settled in Ireland in the fifteenth century.[116] As a second son, he was destined for the Catholic Church and received education in Poitiers, France. He loved to read chemistry books. In 1754–55 he attended Rouelle's lectures in Paris, where he apparently spent time as a Jesuit novitiate.[117] The death of his older brother saved him from the ecclesiastical career with a comfortable income. Soon he began chemical experiments at his mother-in-law's residence. After a brief excursion into the study and practice of law, he turned to chemistry in earnest in 1769, installing a laboratory at his Irish estate. The most active period of his career was 1777–1787, when he resided in London. Less noted than Priestley for specific inventions, Kirwan was the center of chemical communication in Britain during this period, corresponding with Bergman, Guyton, and Crell. After Kirwan was elected to the Royal Society (1780), he, J. H. de Magellan, and Adair Crawford founded the Chapter Coffee House Society, which discussed a selected topic every fortnight. Honorary members included Joseph Priestley, James Watt, and James Keir.[118] Kirwan also cultivated relationships with other British chemists, including Henry Cavendish, and Joseph Black. He had a broad range of interests gleaned from extensive reading. His strong interest in mineralogy explains in part his active correspondence with Bergman and Guyton.[119] He later returned to Dublin to head the newly founded Royal Irish Academy.

Kirwan's measurement of affinities, which won him the Copley Medal of the Royal Society, was directly stimulated by Guyton's work.[120] The three papers he presented to the Royal Society contained a number of innovations that lasting effects on European chemistry. Though Kirwan usually emerges in the historiography of the Chemical Revolution as a loser who supported the phlogiston theory, his focus on the saturation capacity of acids and bases as the true measure of affinities opened a new frontier of analytic chemistry which developed into nineteenth-century stoichiometry. He also tapped the true revolutionary potential of pneumatic chemistry by enlisting marine acid air as the analytic standard of acid-base neutralization reactions. The use of gas as the analytic standard enhanced the resolution of chemical analysis, paving the way for the much more refined compositional analysis of the nineteenth century. Kirwan also brought to chemists' attention the complications caused by temperature variation in the measurement of specific gravity and affinity.[121]

Initially, Kirwan tried to use the difference between the specific gravity of a compound and that of its constituents as a measure of affinity. He thus calculated the "mathematical specific gravity" of a compound as the average of the specific gravities of its constituents and compared it to the experimentally determined specific gravity of the compound, which was often greater than the mathematical mean. He reasoned that such an increase of density upon chemical combination indicated a closer union of the parts in the compound and thus could serve as a measure of affinity. He was soon "undeceived" in this endeavor:

... where the specific gravity and absolute weight of the ingredients of any compound are known, the specific gravity of such compound may easily be calculated as it ought to be intermediate betwixt that of the lighter and that of the heavier, according to their several proportions: this I call the *mathematical* specific gravity. But, in fact, the specific gravity of compounds, found by actual experiment, seldom agrees with that found by calculation, but is often greater without any diminution of the lighter ingredient. This increase of density must then arise from a closer union of the component parts to each other than either had separately with its own integrant parts; and this more intimate union must proceed from the attraction or affinity of these parts to each other: I therefore imagined this attraction might be estimated by the increase of density or specific gravity and was proportional to it, but was soon undeceived.[122]

Instead, Kirwan proceeded in a different direction, measuring the capacities or respective quantities of acids and alkalis required in making neutral salts, or the "point of saturation." In doing so, he followed closely Homberg's precedent, which reveals his intimate knowledge of French chemical literature. Fluent in French, he had a complete set of the academy's memoirs in his library.[123] In order to circumvent the necessary imprecision in such measurements deriving from the variety of acids and alkalis used by chemists, he invented a standard—spirit of salt or marine acid produced by saturating distilled water with marine acid air:

From the time I first read in Dr. Priestley's Exzperiments on Air (That inexhaustible source of future discoveries) of the exhibition of marine acid in the form of air, free from water; and that this air, reunited with water, formed an acid liquor in all respects the same as common spirit of salt; I conceived the possibility of discovering the exact quantity of acid in spirit of salt of any given specific gravity, and by means of this the exact proportion of acid in all other acid liquors; for if a given quantity of pure fixed alkali were saturated, first by a certain quantity of spirit of salt, and then by determined quantities of the other acids, I concluded, that each of these quantities of acid liquor must contain the same quantity of acid, and this being known, the remainder being the aqueous part, this also must be known; but this conclusion intirely rested on the supposition that the same quantity of all the acids was requisite for the saturation of a given quantity of fixed alkali; for if such given quantity of fixed alkali might be saturated by a smaller quantity of one acid than of another, the conclusion fell to the ground.[124]

As Kirwan was keenly aware, his measuring scheme rested on the shaky premise that a given quantity of fixed alkali combined with the same quantity of all acids, at least pure mineral acids, to produce neutral salts. This was a variation of Homberg's approach of 1699. Homberg assumed that a given quantity of alkali combined with the same quantity of "real acid" in each acid solution. Kirwan's variation was theoretically less sound, but he had a distinct analytic advantage over Homberg: He could isolate gases with much higher analytic resolution than solutions. In a rather inconspicuous way, therefore, Kirwan put to use the full analytic potential of pneumatic chemistry—a high-resolution analytic chemistry. Another analytic concern apparently supported Kirwan's assumption: While Homberg tried to discern the "real" quantities of acids by drying the neutral salts thoroughly, Kirwan reasoned that one could never fully eliminate the water of crystallization they contained. If one dried them thoroughly with heat, different acids could be volatilized differently. He thus used an expedient:

1st. I supposed the quantity of nitrous and vitirolic acids, necessary to saturate a given quantity of fixed alkali, exactly the same as that of marine acid whose quantity I determined; and to prove the truth of this supposition, I observed the specific gravity of the spirit of niter and oil of vitriol I made use of, and in which I suposed, from the trial with alkalis, a certain proportion of acid and water; I then added to these more acid and water, and calculated what their specific gravities should be upon the above supposition, and finding the result to tally with the supposition, I concluded the latter to be exact.[125]

Kirwan produced the standard solution of marine acid (spirit of salt) by saturating a known quantity of distilled water with marine acid air for 18 days at a relatively constant temperature. From the increased specific gravity of water, he also calculated the specific gravity of "pure marine acid" in "such a condensed state as it is in when united to water" by ignoring possible condensation due to the attraction of marine acid to water. From the proportion of the acid and water in his solution, Kirwan prepared a table of specific gravities of spirit of salt corresponding to different concentrations. Although the table was inaccurate as a result of his inability to find the "point of saturation as nicely as was requisite," Kirwan assumed that the error was small.[126] He then calculated the proportion of the acid, water, and fixed alkali in digestive salt by mixing a tolerably pure vegetable alkali with solutions of spirit of salt. He borrowed the method of calculating the composition of alloys to complete the calculation. For the other acids which did not assume gaseous form, Kirwan used the combining ratios of fixed vegetable alkali and marine acid to calculate their specific gravities and saturation capacities.[127]

Kirwan's work was obviously inspired by Homberg's 1699 work on the quantity of "real" acid in neutral salts, which he cited throughout the paper as a unique precedent to his own. For acetous acid, he simply extrapolated from Homberg's results without making his own experiments. He concluded that fixed vegetable alkalis took up an equal quantity of the three mineral acids, and probably of all pure acids. More generally, alkalis had a certain "determinate capacity" of uniting with acids, which was "equally satiated by that given weight of any pure acid indiscriminately." One could use these ratios to determine the quantity of "real pure acid" in other acids as well. He also speculated that the force of affinities depended on the respective quantities of substances, anticipating Berthollet's later formulation of affinity. The density of a compound ranged from a minimum to a maximum value. At the minimum density, its component parts existed in very different quantities, and their attraction was at a maximum. At the maximum density, the component parts were matched in proportionate quantities, and their attraction was at a minimum. The "point of saturation" was reached, then, at the maximum density of the compound, which held minimum attraction among its constituents. This explained why displacement reactions were seldom complete in the way Geoffroy supposed them:

. . . no decomposition operated by means of a substance that has a greater affinity with one part of a compound than with the other, and than these parts have to each other, can be complete, unless the *minimum* affinity of this third substance be greater than the *maximum* affinity of the parts already united. Hence few decompositions are complete without a double affinity intervenes; and hence the last portion of the separated substance adheres so obstinately to that to which it was first united, as all chemists have observed.[128]

In the second paper, Kirwan examined "the quantity of pure acids taken up at the point of saturation by the various substances they unite with," such as mineral alkali, volatile alkali, various earths, and phlogiston. The main focus of the paper was, however, on the identity of phlogiston with the inflammable air, which supposedly explained its elusive nature.[129] Much like fixed air, the inflammable air could exist in two states: fixed and elastic. Since it assumed aerial and elastic form as soon as it was released from bodies, it could never be produced "single and disengaged from other substances." The fixed form of the inflammable air had to be phlogiston:

[Phlogiston] can never be produced in a *concrete state*, single and uncombined with other substances; for the instant it is disengaged from them, it appears in a fluid and elastic form, and is then commonly called *inflammable air*. These different states of the same substance arise, according to the immortal discoveries of Dr. Black, from the different portions of elementary fire contained in such substance, and absorbed by it, while its sensible heat remains the same, and hence called its *specific fire*. For want of attention to these different states, the very existence of phlogiston as a distinct principle has been frequently called in question, and chemists have been required to exhibit it separate in its fixed state, without recollecting, that neither can fixed air be shewn separate in a concrete state, nor that phlogiston may also be in the same predicament; while others have totally mistaken the nature of inflammable air, and imagined it to be a combination of acid and phlogiston.[130]

Kirwan articulated the "identity and homogeneity" of the inflammable air with phlogiston in some detail. Phlogiston was the principle of combustibility, the principle that gave metal its malleability and splendor, the principle that combined with vitriolic acid to form sulphur, and the principle that diminished respirable air. Inflammable air satisfied all of these criteria. It existed in many different states, depending on the portion of the elementary fire it contained. He speculated that it might constitute the electric fluid in a state "perhaps 100 times rarer than inflammable air."[131] He proceeded to calculate the "quantity of phlogiston" in nitrous air, fixed air, vitriolic air, sulphur, and marine acid air.

In the third paper, Kirwan explored the combining weights of acids with metallic substances. He surmised that the "exact determination, as well of the quantity and proportion, as of the quality of the constituent parts of bodies" would have important applications in the practical sphere of pharmacy, dye, mineral water analysis, etc. The principal end he had in view was, however, to quantify affinities which would provide the foundation of chemistry "considered as a science."[132] Chemical affinity or attraction was "that power by which the invisible particles of different bodies intermix and unite with each other so intimately as to be inseparable by mere mechanical means." In this respect, it differed from magnetic and electric attraction as well as from the attraction of cohesion which took place between the particles of all bodies. Chemical attraction was elective. The determination of affinities by displacement reactions was problematic because many decomposition reactions involved more than a pair of affinities. In order to attain any certainty, it was necessary to "ascertain the quantity and force of each of the attractive powers, and denote it by numbers." Guyton's pioneering efforts in this regard was not generalizable, while Wenzel's method was defective. Kirwan deemed that his method of calculating the quantity of real acid and its proportion to different bases at the point of saturation would provide "the true method of investigating the quantity of attraction" between acids and bases. He proposed two rules: "First, That the quantity of real acid, necessary to saturate a given quantity of each basis is inversely as the affinity of each basis to such acid. Secondly, That the quantity of each basis, requisite to saturate a given quantity of each acid, is directly as the affinity of such acid to each basis."[133]

The saturation capacity could serve as an adequate expression of the "quantity" of the affinity between an acid and a base because saturation referred to the state of most "intimate" combination in which an acid or a base lost its "peculiar characteristic property, which it possess when free from that other." These numbers of saturation capacity could predict the outcome of more complex chemical actions. Kirwan differentiated between the quiescent affinities, "the powers which resist any decomposition, and tend to keep the bodies in their present state," and the divellent affinities, "the powers which tend to effect a decomposition and a new union." A decomposition would always take place, he stipulated, "when the sum of the divellent affinities is greater than that of the quiescent." A decomposition necessarily happened when one mixed the solutions of tartar vitriolate and nitrous selenite, producing selenite and niter, because the sum of divellent affinities was greater than that of quiescent affinities,[134] as could be seen in the following table:

Quiescent affinities

Vitriolic acid to fixed veg. alkali	215
Nitrous acid to calcareous earth	96
Sum of the quiescent affinities	311

Divellent affinities

Vitriolic acid to calcareous earth	110
Nitrous acid to vegetable alkali	215
Sum of the divellent	325

In this way, Kirwan put a number on Bergman's notion of double elective affinity. Although Guyton and Wenzel provided precedents for his quantifying effort, their physical measurements were not as generalizable or reliable. Kirwan offered extensive measurements that covered most of the substances included in the affinity table and ostensibly a more general and reliable method of predicting chemical actions. More than the numbers, his experiments are remarkable for their attention to precision, which had far-reaching consequences for the next generation of chemical investigation. Particularly noticeable is his effort to obtain the "real" quantity of acids and bases involved in the reaction. For his model system, that of the reaction between marine acid and fixed vegetable alkali, he used purified acid obtained in the form of gas instead of commonly available acid. He obtained the marine acid for his experiments by saturating water with marine air. The "exact real quantity" of the acid in the solution was determined from the increase of specific gravity. He tabulated the proportionate quantity of real acid for each increment of the specific gravity of marine acid solution made from marine air. Kirwan pushed chemists' analytic control to a new level through these exacting procedures. Such precision also made visible the significant role of temperature in conducting chemical experiments. Kirwan's experiments introduced innovative techniques in the measurement of specific gravities that shaped Berthollet's conception of affinities later on. Nevertheless, they were built on the premise that the "real quantity" of the three mineral acids that saturated fixed vegetable alkali was the same. The affinity of fixed vegetable alkali for the three mineral acids was presumed to be the same from the beginning. If this premise proved unfounded, the whole enterprise "fell to the ground," as Kirwan himself admitted.

The trio of Guyton, Bergman, and Kirwan represented the frontier of European philosophical chemistry in the early 1780s. They read widely in the available literature, and they synthesized various strands of experimental practices to craft new problem domains and to work out new solutions to outstanding problems. Their capacity as readers and their networks in the Republic of Science distinguished these chemists from ordinary apothecaries, physicians, and industrial chemists, whose interest in chemistry was largely vocational and grounded in practical sphere. As scientists whose reputation and symbolic capital depended on the commerce of words, they had little tolerance for esoteric language, secret recipes, and diverse units of measurement. Lavoisier shared these traits of philosophical chemists who advocated openness of chemical language and uniform standards of measurement. He pushed their agenda further by streamlining chemical operations with an algebraic logic.

Recruits from a different class of youths such as Guyton and Lavoisier helped French chemistry to acquire a more orderly linguistic structure as well as an enhanced public visibility. While the chemists trained in pharmacy or medicine continued to refine the analytic skills necessary for the technical development of chemistry during this period, those recruited from outside these conventional tracks helped bring chemistry to the drawing rooms of the upper bourgeoisie, weaving it into the conversational culture of the Enlightenment. The emergence of a new species of chemists such as Guyton and Lavoisier transformed French chemistry from a primarily medico-pharmaceutical trade to a public science. Guyton's career trajectory illustrates well the rise of chemistry in the Enlightenment public sphere. Chemistry was no longer the exclusive province of apothecaries and physicians; now it belonged to a broader public, which demanded an intelligible representation of the art. Guyton responded to this demand by writing articles for the *Encyclopédie*, by presenting a theoretical discourse of affinity in tune with the reigning paradigm in natural philosophy, and by initiating a nomenclature reform. Despite his role in this enduring monument of the Chemical Revolution, historians have not paid sufficient attention to Guyton.[135] The stories of the Chemical Revolution still revolve mostly around Lavoisier, recognizing Guyton only as one of the "converts" to Lavoisier's chemistry. Lavoisier's exclusive claim to the Chemical Revolution owes much to the perception that he introduced quantitative methods into chemistry, thereby transforming it from an art to a science. In addition to the unified position on the role of oxygen in chemical operations, however, the collaboration of the four French chemists also entailed the acceptance of many of Guyton's ideas on chemical affinity. In fact, a more conscious and problematic frontier of quantitative chemistry in the 1780s involved the measurement of chemical affinities. Guyton, not Lavoisier, guided these efforts, continuing the Macquerian tradition of affinity chemistry with a quantitative bent. The rapprochement of Guyton and Lavoisier produced a beautiful vision of theoretical chemistry centered on the problem of "constitution," as Berthollet later articulated explicitly, which became an enduring legacy for French elite chemistry.

6

An Instrumental Turn

Antoine-Laurent Lavoisier, a well-worn hero of the Chemical Revolution, dominates the historiographical landscape of eighteenth-century chemistry. Seen as the "paradigmatic example of a revolution in science,"[1] the Chemical Revolution continues to fascinate historians for the sweeping, thoroughgoing, and self-conscious reform Lavoisier and his fellow academicians effected. As a result, we now possess a detailed knowledge of Lavoisier's life[2] and work.[3] Increasingly sophisticated studies of Lavoisier have led, however, to diverse interpretations of the Chemical Revolution.[4] The distinct image of a "classic" revolution in scientific theory, dubbed the "overthrow of the phlogiston theory," has disintegrated into a mosaic of inchoate interpretations.[5] New interpretations focus variously on the problem of composition, the Stahlian Revolution, gravimetric analysis, experimental physics, the chemistry of salts, the problems of language and communication, and so on, without an overarching scheme of "greater range, cogency, and coherence." If there is a point of consensus among the Lavoisier scholars, it is that the "revolutionary" nature of the episode needs a careful reevaluation and that any synthesis of current scholarship is likely to yield a multi-faceted and multi-layered account of the episode.[6]

Lavoisier cuts a unique figure in his training, career path, and intellectual outlook among the contemporary Parisian chemists.[7] He was not trained in the conventional tracks of medicine or pharmacy, but he received a broad range of education from a number of elite Parisian scientists: mathematics from de La Caille, chemistry from Rouelle, experimental physics from Nollet, botany from Jussieu, and geology from Guettard. Growing up during the High Enlightenment and groomed early on by these leading academicians, Lavoisier developed a strong commitment to public service and an ambition to become a scientist worthy of posterity's notice. Chemistry happened to offer a suitable venue through the membership at the Academy. A broad-angled approach allowed Lavoisier to construct a different theoretical structure for chemistry—one more compatible with other sciences. He was interested less in inventing new procedures for chemical substances than in forging a more comprehensive, trans-disciplinary approach to the truth of nature. He cultivated habits of searching the literature exhaustively and taking exact quantitative measurements, a broad theoretical agenda, and a grammatical understanding of language and nature.[8]

Although Lavoisier shared these traits of ambitious Enlightenment bourgeois youths and education in law with Guyton, he had to meet much more exacting standards of the Paris Academy in his research. While he conducted much of his innovative investigations with relatively simple modifications of the existing apparatus, he spared no expense in staging the experiments designed to persuade his academic colleagues. Much has been written about his precision instruments and their role in the Chemical Revolution.[9] What characterized Lavoisier's innovative path to chemistry was as much the kind as the precision of the instruments in which he placed his trust. From the beginning of his career, he dreamed of constructing general theories of the earth, chemical combination, heat, etc. with a series of metric measurements. Lavoisier's science cannot be understood apart from his sincere attachment to a variety of "meters": the barometer, the hydrometer, the thermometer, the gasometer, the calorimeter.[10] Metric sensibility underlined the technology of late Enlightenment rationalism in scientific and political spheres.[11] More immediately, metric measurements allowed Lavoisier to streamline complex social and natural phenomena according to an algebraic logic, as was embodied in his famous balance-sheet method. He shared this algebraic gaze at chemical phenomena with other philosophical chemists such as Guyton and Kirwan. The language of algebra facilitated their conversation despite their antagonistic theoretical stances in chemistry. Lavoisier grounded this algebraic mentality on concrete metric measurements, thereby establishing new standards of instrumental practice in chemistry and a new subdiscipline of physical chemistry.

One of the most divisive and critical issues in the interpretation of the Chemical Revolution concerns the relationship between physics and chemistry that Lavoisier supposedly sought to forge. Some historians argue that it was his use of the theories and techniques of physics that transformed chemistry from an art to a science;[12] others place him squarely within the Stahlian chemical tradition.[13] Though Lavoisier scholars have successfully portrayed Lavoisier's binary identity as a chemist/experimental physicist,[14] there remains much to be done to configure pre-Lavoisian chemistry as an organized field of inquiry.[15] The difficulty here lies in the deeply held historiographical assumption that pre-Lavoisian chemistry was "not yet constituted as a science." It does not help much to identify Lavoisier's contribution to the Stahlian revolution if the structure of Stahl's chemistry remains obscure.[16] Lavoisier's unique training and his trans-disciplinary perspective on science should not deter us from tracing his evolution as a chemist, however. If he entered the venerated Academy without an established reputation in chemistry, his knowledge of the field deepened through his many duties as an academician, as the director of Gunpowder Administration, and as the gracious host of the research group at the Arsenal. His naive metric vision of chemistry gradually gave way to a more sophisticated conception of chemical constitution as an interplay of chemical and physical forces. The phlogiston theory became the polemical focus of his battle for new chemistry because it interfered with this dynamic conception of chemical constitution. In my portrayal of Lavoisier, therefore, I will emphasize his evolution from an amateur *physicien* with a trans-disciplinary perspective into a chemist concerned with the issues of affinity. This was by no means a linear process, since he took serious interest in the current affairs of the Academy as well as in his duties as an academician and as the director of the Gunpowder Administration. Lavoisier's was a chemistry à la mode. In this chapter I will selectively discuss his theoretical investigations during the 1770s, which consolidated the oxygen theory of combustion and acidity.[17]

6.1   The Anti-Paracelsus

Lavoisier became the answer to Venel's call for a new Paracelsus who would possess the technical and social skills to place chemistry among the ranks of other sciences. He could do so, however, only because he assumed an identity antithetical to Paracelsus. He was an insider rather than an outsider, a physicalist rather than a chemist, and a man of algebraic sensibility rather than a mystic. Like Guyton, he was born into a lawyer's family that had steadily climbed the social ladder from a minor position in the king's service to a respectable place in the Order of Barristers at the parlement of Paris. Although his family was financially secure and well connected, Lavoisier faced limited prospects of social advancement as a bourgeois youth in the ancien régime.[18] Having lost his mother at the age of 3, he grew up in the relatively quiet, comfortable household of his maternal grandmother and his aunt. At age 11, he was sent to the Collège Mazarin as a day student for a well-rounded education. The curriculum devoted the first six years to traditional subjects in humanities, a year to mathematics and exact sciences, and two more years to philosophy. Lavoisier took interest in all subjects, attempting to write a play based on Rousseau's La nouvelle Héloise and competing for a variety of prizes offered by the provincial academies. Overall, he seems to have been a serious student who imbibed Enlightenment philosophies and ideals and cultivated an ambition to rise above his social station through the application of his intellect. The question he examined for the competition offered by the Besançon Academy was "whether the desire to perpetuate one's name and actions in men's memory conforms to nature and reason."[19]

At the time, the Collège Mazarin offered a unique curriculum in the sciences and in mathematics. Completing the humanities curriculum in the summer of 1760, Lavoisier entered the classe des mathématiques and received instructions in mathematics and exact sciences from Abbé Nicolas Louis de La Caille (1713–1762), a renowned astronomer and academician.[20] Working in de La Caille's well-equipped observatory at the Collège, Lavoisier had close contact with a notable academician who personified science in the service of the state. He must have appreciated the precision instruments of astronomy,[21] and he is said to have acquired a "taste for exact calculation and logical reasoning, freed from all pedantry."[22] He later recalled de La Caille's lessons fondly:

I had taken a good course of philosophy, I had followed the experiments of Abbé Nollet, I had studied with some success elementary mathematics in the works of

Abbé de la Caille and I had followed a year of his lessons. I was accustomed to this rigor of reasoning mathematicians place in their works. They never take up a proposition until the one preceding it has been discovered. Everything is linked, everything is connected, from the definition of the point and the line to the most sublime truths of transcendental geometry.[23]

Lavoisier left the Collège in the summer of 1761 and transferred to the Faculty of Law to prepare for the family profession,[24] but he continued to explore a wide range of scientific subjects, including experimental physics, chemistry, geology, and botany. According to Jean Etienne Guettard (1715–1786), a family friend and an academician, Lavoisier had a "natural taste" for the sciences and wanted to be acquainted with all of them before choosing one.[25] After the death of Abbé de La Caille, Lavoisier came under the sustained influence of Guettard, who had an ambitious project of making a mineralogical and geological atlas of France. He began daily observations of barometer and thermometer readings, trying to standardize these instruments, and he started to collect rock and mineral specimens.[26] Although he had been introduced to chemistry by Laurent-Charles de La Planche while at the Collège Mazarin,[27] Lavoisier also took Rouelle's chemistry courses for three consecutive years, probably beginning in 1762.[28] He seems to have attended Abbè Nollet's public lectures on experimental physics at about the same time. The summers of 1763 and 1764 found Lavoisier in the company of the famous botanist Bernard de Jussieu (1699–1777) on the promenades philosophiques of the Paris region.[29] By 1766 he had formulated a grand research plan to work out a "theory of the earth" through systematic measurements with the barometer. In 1767 he went on a four-month geological expedition with Guettard during which he took barometric measurements, collected samples of mineral and spring waters for analysis, and sketched geological strata. Geology remained one of Lavoisier's favorite subjects for years to come.[30] More important, he developed an instrumentalist vision that the metric measurements would yield an orderly representation of nature and society, a vision he shared with other late Enlightenment savants.[31]

Lavoisier explored a variety of options during the early 1760s. While sampling many scientific subjects, he completed two years of study for a baccalauréat in law. After another year of study, he obtained the licence in 1764 and was admitted to the Order of Barristers at the parlement of

Paris.[32] By that time, however, he had settled on science as a serious career option, not one that would earn him a living, but one that would preserve his name for posterity. He wished to take up an "honorable profession and, in a way, a public function."[33] For a bourgeois youth, the need "to become a public person and to please and influence in a larger circle" manifested in an arena different from those of traditional aristocrats, as Goethe observed in *Wilhelm Meister*.[34] Lavoisier initiated an active campaign to gain admission to the Academy, a logical choice for a youth determined to make a name for himself in science, apparently with his father's blessing.[35] To this end, he embarked on an analysis of gypsum and entered the Academy's prize competition having to do with street lighting.[36] In these works, we can already detect what was to become Lavoisier's lifelong research habit of exhausting a subject through a methodical survey of the literature and through careful experimentation.[37] He was awarded a special gold medal for "a paper full of curious research & best physics" on street lighting.[38] Afterward, he put himself up for the position of adjoint chimiste made available by Macquer's promotion to associé, but he failed to secure the position.[39] However, maneuvering by his father's friends continued within the Academy.[40] Returning from a four-month trip with Guettard in early 1768, Lavoisier secured his financial future by joining the Ferme générale and proceeded to present a series of papers on the analysis of various waters to the Academy in preparation for another election in the chemistry section.[41] Théodore Baron, an adjoint chimiste, had died in early March. After a favorable report by Nollet and Macquer, Lavoisier won the election, but was appointed as a surnumeraire adjoint chimiste when the king took Jars's age and long service into consideration. Lavoisier's unyielding campaign to gain admission to the Academy shows that he viewed it as a career choice worthy of his talents and his ambition. He remained devoted to this institution until its last days during the French Revolution.[42]

6.2 A Trans-Disciplinary Approach to Chemistry

At the time of his election, Lavoisier was less a chemist than a well-rounded amateur physicien with a strong instrumentalist bent. He had developed a habit of exhaustive literature search and precise measurements in his research, but did not yet possess an intricate knowledge of chemistry.[43] Chemistry secured him an entry to the Academy, however. The "obscure" and "vague" nature of the science also seemed to offer exceptional opportunities for a logical mind. Looking back on his early chemical education with La Planche, Lavoisier complained about the lack of logic and exactitude in chemistry in comparison to mathematics or experimental physics. His expressive indictment of chemical teaching, though tainted with the wisdom of older age, conveys the bewilderment of an Enlightenment youth who wished to learn this vast material practice as a logical discourse:

I was surprised to see how much obscurity surrounded the approaches to the science. In the first steps, one began by assuming rather than proving. I was presented with words that could not be defined for me, or at least could be defined only by borrowing from subject matter that was completely unknown to me, and that I could acquire only by studying all of chemistry. Therefore, in beginning to teach me science, he supposed that I already knew it.[44]

Even after taking three years of Rouelle's courses, Lavoisier still felt that he had "spent four years of studying a science founded on only a small number of facts, that it had been formed from absolutely incoherent ideas and unproven suppositions, that the method of teaching chemistry was inexistent, and that no one had the slightest idea of its logic." He apparently recognized "the necessity of starting my chemical education afresh, of getting rid of everything I had learned except facts, and of arranging them methodically in my mind so as to conform to the inherent course of nature."[45]

Exactly how Lavoisier might have wished to shape a logical discourse of chemistry can be gleaned from his 1764 manuscript.[46] As was customary in French didactic discourse, he began with a definition of chemistry and a short discussion of principles. He endorsed the analytic-synthetic ideal, defining chemistry as the art of decomposing and recomposing bodies. Echoing Venel's discussion in the *Encyclopédie*, he identified chemistry's difference from physics as the search for the "nature of principles" which constituted the essence of a body. A physicist would define cinnabar as a red, brittle body almost as heavy as gold, but a chemist would see it as a combination of sulphur and mercury. To provide a complete proof, the chemist would combine sulphur and mercury to form an artificial cinnabar entirely similar to that of nature. Chemists' discussion of principles fell far short of Lavoisier's expectations, however. He found in their systems only "ideas more ingenious than true, the hypotheses nearly always deprived of foundation and often destroyed by experiment." He conceded that Stahl offered the best option available in the hierarchical composition of bodies starting from the "simple, indivisible, immutable bodies" to mixts, compounds, super-compounds, and aggregates.

Lavoisier followed the outline of Rouelle's lectures in juxtaposing Stahl's principles with Boerhaave's elements. He would consider the four elements only as "instruments" chemistry utilized to arrive at its goal, rather than as "integrant parts" of bodies. His extensive treatment of fire resembled Boerhaave's rather than Rouelle's. He did not endorse phlogiston, but tried to characterize fire in terms of the thermometric properties of heat. He had been unhappy with the way La Planche handled the issue:

The professor was not much better in the course which he subsequently gave on the elements, the notion of fire, and phlogiston. He spoke to men instead of children. He confounded perpetually what we knew with what we did not yet know.[47]

La Planche began his discussion of principles with Becher's second earth, or the "sulphur principle, phlogiston, this inflammable principle." [48] Fire as a chemical substance or phlogiston also occupied a prominent place in Rouelle's and Macquer's courses.[49] Instead of identifying fire with phlogiston, however, Lavoisier grappled with the general nature of fire that would be applicable to all sciences. In particular, he was concerned with heat, or the fire measured by pyrometers and thermometers, much as Boerhaave had been. He followed Boerhaave's cue in identifying fire as heat whose main effect was the dilatation of bodies, but sought to update the instrumental basis of the master's treatise. He thus proceeded to discuss in detail pyrometers and thermometers, relying on various physics texts including Musschenbroek's and Abbé Nollet's.[50] Lavoisier's preference for physical instruments over chemical operations in detecting fire owed to his perception that they "rendered the truth sensible to the eyes" and had the potential to produce "exact results," even if they did not currently yield anything "absolutely exact" on the dilatation of bodies. At least, the pyrometer registered the unequal dilatability and the "different degrees of elongation" of metals. Lavoisier offered an even more detailed account of the thermometer, indicating his strong interest in this instrument. After a brief history of thermometers reaching back to Drebel, Amontons, and Réaumur, he presented Réaumur's method as the best, albeit not without inconveniences. To judge from his attention to the outstanding problems in thermometry, he must have studied the literature on thermometry with care; different metals did not heat or cool with the same speed, bodies at the same temperature did not take equally the same degree of heat, and so on.

Lavoisier wished to construct an ontology of fire that could explain both physical and chemical phenomena. He acknowledged that there existed in nature "a being [être] which according to different modifications it takes produces fire, heat, light and some other phenomena." Unable to separate this "being" from the usual designation of "fire," his discussion became rather incoherent, intimating his struggle to construct an ontology that transcended the instrumental boundaries of heat and fire. On the one hand, fire was the "being" that incited various phenomena; therefore, heat and light were good "signs" for the presence of fire in a body, although they did not always manifest together. On the other hand, fire had to be the same as heat, since the thermometer detected "the presence of fire or to say it better of heat." Since "one of the principal properties of fire or rather of heat" was to dilate bodies, fire had to "occasion in bodies a movement of continual oscillation" to cause an infinity of phenomena in nature. Although it was not easy to give "a neat idea of the manner in which fire produces these phenomena," the most clear-minded physicists viewed fire as "a species of extremely subtile fluid expanded in the entire nature, which surrounds the bodies in all parts and which penetrates them as it would happen to a sponge immersed in water or any other fluid."[51] The "molecules of the element of fire" were infinitely smaller than those that composed other bodies, allowing them to penetrate the bodies with great facility. Nevertheless, following Rouelle, Lavoisier thought it more than probable that the "matter of fire" did not penetrate the parts that constituted the mixtion of bodies, nor the parts that formed composition, but only the aggregative parts, except in the state of ignition. By "heating" a body, one separated its aggregative parts by impregnating them with fire in the way a dry sponge placed in water would augment its volume to the degree that water penetrated its interior. By "burning" a body, one not only disunited its aggregative parts but also separated "the principles that constitute mixts or compounds." The element of fire produced combustion or ignition only by impressing a certain movement, mostly through friction.

Like many other chemists before him, Lavoisier acknowledged a "quite great analogy" between fire and light. Unlike other chemists who based their identification on the evidence of the burning glass, however, he used Euler's analogy between the transmission of light and sound in their respective mediums of fire and air.[52] Light was a modification of the element of fire that surrounded us. It was nothing other than a movement impressed by the luminous body in the middle of an igneous fluid expanded in nature, like a sonic body in the air. When one flapped a cloth or some other sonic body in the air, it made a sound because the molecules that composed it entered into an oscillation which was transmitted through the air, conveying to us the idea of sound. Likewise, a luminous body impressed on igneous fluid a movement of oscillation which was communicated to the eye, transmitting the idea of light to us. Although this simple idea of light contradicted the celebrated theory of Newton, which was essential to physics, it conformed well to chemical experiments. Lavoisier also regarded electric matter or fluid as "the element of fire combined with some other quite subtile fluid." After a brief discussion of the electric properties of bodies, he focused on the instrument for collecting electricity, the glass globe, and on the theory of lightning.

Lavoisier transformed in this manner the discussion of phlogiston common in the Parisian chemical lectures into an inquiry on heat. If he valued the difference of chemistry from physics in probing the interior of bodies, he also wished for a rational procedure and a universal discourse that encompassed the instrumental practices of both chemistry and experimental physics. If this 1764 manuscript is indeed his, it brings us closer to young Lavoisier's trans-disciplinary perspective stemming from his exposure to a wide range of subjects. The manuscript differs noticeably from the contemporary chemical lectures in its omission of affinity as well as phlogiston. Although Rouelle did not emphasize affinity, Macquer did.[53] According to Lavoisier's own admission, La Planche began the first day of his course with affinities, dealing with the problem of metallic dissolution in acids:

> I experienced in the sequence of course what M. de la Planche had announced to me. He supposed from his first course many things he promised to demonstrate in the subsequent courses and the courses took place without demonstrating the suppositions made. From the first day, he spoke to us of affinities, which is the most difficult to understand in chemistry. In order to demonstrate the game to us, he made experiments of metallic dissolutions although we did not yet know what was a dissolution, what was an acid, what was a precipitation, although none of us could conceive how a molecule of matter had more *convenance*, more affinity, more tendency to unite with one substance than with another.[54]

Lavoisier had enough education in chemistry to know the central problems and theory domains of the discipline. He knew that the "usual order of course of lectures and of treatises upon chemistry" began with the principles of bodies and the tables of affinities.[55] Of these two general theory domains of chemistry, he chose to pursue principles and elements. A keen interest in the nature of elements, coupled with an instrumental expertise, was to be the hallmark of Lavoisier's chemistry. In a way, his obsessive pursuit of elements betrays him as an outsider to the contemporary practice of chemistry. Although Rouelle and Macquer presented the theory of four instruments-elements in their lectures and writings, they seldom mixed these doctrines with their understanding of concrete chemical substances and operations. Lavoisier took the four elements as the primary theoretical foundation of chemistry and sought to construct a uniform composition of nature which included the earth's atmosphere. This naiveté of a novice chemist yielded an extraordinary fruit thanks to his instrumental expertise and to the contemporary developments in pneumatic chemistry. It is not surprising that Lavoisier approached chemistry from a theoretical rather than a practical point of view. Like Guyton, he was not a vocational chemist bound to a practical sphere. He did not have to invest his entire intellectual energy in memorizing numerous plants deemed useful for medicine or mastering tricky chemical operations to produce a better-tasting medicament. He approached chemistry with a theoretical and systematic perspective to ascertain the transcendental truth of nature.

6.3 A Novice Chemist

In July of 1764, Lavoisier undertook an extensive study of gypsum, which he presented to the Academy on February 27, 1765 and March 19, 1766 in an election bid.[56] The choice of mineral analysis bespoke his entry into chemistry via mineralogy which had become a major field for the Enlightenment legitimation of chemistry.[57] Johann Heinrich Pott and Jean-Etienne Guettard had called for the study of common domestic minerals in the 1750s. Following on their trail, Jean Darcet presented several works to the Academy in 1766 on the classification of earths.[58] Lavoisier hoped to cultivate an extensive field of inquiry that would translate his interest and experience in natural history into chemical expertise. Despite chemistry's many discoveries, he declared, there still existed "an infinity of bodies in the mineral kingdom whose nature is entirely unknown to us." The examination of earths, stones, and crystals should, therefore, provide "a source of inexhaustible experiments and discoveries." He would start with a particular gypsum found near Paris in the first memoir and extend the analysis in the second memoir to all other kinds of gypsum to discern their commonalities and differences. In the third memoir, he would describe plasters, the arrangement of gypsum in nature, the phenomena it occasioned, and the mystery of its formation. Lavoisier regarded the analysis of minerals as a boundary area between mineralogy and chemistry that could shed a considerable light on both subjects:

Nearly everybody who has worked (on this subject) until now seems to have omitted this part so essential and common to natural history and chemistry, the most proper to shed light in one and the other of these two sciences.[59]

Gypsum, in particular, was a substance found in a large quantity near Paris. The chemistry of gypsum was also something amenable to Lavoisier's skill level. Although he was not yet an accomplished chemist, he knew at least the basic tenets of the chemistry of salts. La Planche had included a detailed investigation of gypsum in his lectures.[60] "When the chemistry course ended," Lavoisier wrote, "I wanted to take stock of what I had learned. I felt fairly confident about everything concerning the composition of mineral salts, the preparation of mineral acids—the only things about which there then existed *exact and positive knowledge*. But it seemed that I had retained only the vaguest ideas about all the rest of the science."[61]

Lavoisier's method was also one long advocated in French chemistry: solution rather than distillation methods. While he acknowledged Pott's outstanding contribution to the field of mineral analysis, even repeating some experiments, he argued that Pott's distillation method supplied little information on the nature and composition of the substances before their analysis. Instead, he would "copy nature" and use water, the "nearly universal solvent." He also deemed it necessary to prove the composition of gypsum through analysis and synthesis, again upholding the long-standing ideal in French didactic tradition:

It would not suffice to have decomposed gypsum, to have demonstrated separately the mixts that compose it, to have demonstrated that it was formed by the union of vitriolic acid with a calcareous earth, in a word that gypsum was nothing other than selenite; it would still be necessary to take the materials nature employed, recompose a new gypsum which would produce the same effects, which would give the same phenomena.[62]

Lavoisier's analysis of gypsum reveals his familiarity with the chemistry of salts and the role phlogiston played in it. First, he understood the role of water in the crystallization process. Using Homberg's areometer, he measured the specific weight of water before and after it dissolved gypsum and that of calcinated gypsum or plaster. Since plaster required a larger quantity of water for dissolution, Lavoisier concluded that it lacked the water of crystallization. In other words, gypsum turned into plaster of Paris through calcination by losing the "water of crystallization." Second, Lavoisier knew that not all salts were neutral, or contained a balanced amount of acids and alkalis. Neither gypsum nor plaster effervesced with acid or alkali. He thus concluded that it had to be a neutral salt, which reduced its analysis "to the operation most ordinary and common, the simplest in all of chemistry."[63] Third, Lavoisier knew that the signature operation for phlogiston was its combination with vitriolic acid to produce sulphur. The plaster, when inflamed on a crucible with some charcoal powder gave out an odor of volatile sulphurous acid. Pouring an acid on the calcined matter released the odor of rotten eggs, indicating that it was liver of sulphur. The acid of gypsum had to be vitriolic acid, he reasoned, because it alone could produce volatile sulphuric acid and sulphur by uniting with the "matter of fire." Lavoisier thus understood phlogiston as most chemists did: Vitriolic acid gave rise to sulphur by combining with the matter of fire or phlogiston. "Some chemists," he wrote, "have announced that in the calcination of plaster one sometimes perceives a sulphurous material which inflames. This observation could not be genuine except for the plaster calcined by an enclosed fire, like that of our plasters; in that case a portion of the vitriolic acid uniting with the phlogiston of charcoal or empyreumatic oil of wood, forms a true sulphur."[64]

In the second memoir, which began with a tribute to Rouelle's quantitative approach to the classification of salts, Lavoisier distinguished three different kinds of gypsum depending on their composition and solubility and compared them to the three kinds of selenite made in the laboratory. Just as these products of the laboratory had a determinate composition, he asserted, different kinds of gypsum were produced in the "physical system of the earth" not by chance, but by "constant and invariable laws." Macquer, who had tried his hand at gypsum a few years earlier, promptly included Lavoisier's work in his *Dictionnaire* (1766). Lavoisier's papers on gypsum illustrate his familiarity with the current chemistry of salt as well as his research habits, which combined an exhaustive search of the existing literature with meticulous instrumental investigation. The authors he consulted for his first paper included a constellation of German authors (J. H. Pott, A. S. Marggraf, J. T. Eller) as well as the leading French authorities (Macquer, Baumé). His knowledge of chemical literature was expanding rapidly, reflecting perhaps his growing interest in chemistry.

6.4 Elements

Lavoisier expressed his dissatisfaction with chemists' systems of principles as early as 1764. Nevertheless, he regarded the four elements as the most important subjects in the entire course of chemistry, if only as the instruments of chemical operations. In the ensuing years, he pursued the subject intermittently yet persistently.[65] An early trail of his thoughts can be found in the two notes he wrote in 1766. In the process of conducting the gypsum experiments, he came upon J. T. Eller's papers on the transmutation of elements. Eller had studied with Boerhaave before Frederick the Great enlisted him in 1746 to head the class of experimental physics at the Berlin Academy. He shared Boerhaave's inclination to fuse chemical and physical doctrines, especially in his discussion of fire. Defining elements as the "material principles of all bodies which compose this vast Universe," therefore, Eller supposed that all bodies were resolved by decomposition into these four elements, which were made of simple, inalterable, and homogeneous parts. In addition, he reviewed the entire history of the doctrine of elements going back to the ancients, van Helmont, Bruno, Descartes, and Leibniz.[66] Eller took the four elements much more seriously than most chemical practitioners, as the true analytic principles of bodies and as the building blocks of the universe.

In his discussion of fire, Eller skillfully bridged the physical manifestations of heat and the chemical properties of fire as phlogiston. As the only active element, fire resided in all bodies, but only in an "imperceptible manner" until its parts were placed in movement, which caused heat, electricity, flame, and body heat. The "different degrees of force" of its movement allowed us to sense the "different degrees of heat." An additional degree of movement inflamed the body, leading to its total destruction. This was why fire acquired the name "phlogiston." The "phlogistic matter" was never found "in its simplicity & in its natural purity," but was "enveloped" in mineral, vegetable, and animal bodies, which gave it "different modifications & forms according to the diversity of the matters to which it is united." Chemists posited the existence of phlogiston because the combustion and restitution of metals seemed to prove "that this inflammable substance, or igneous matter is always the same in nature." Eller's discussion of fire moved effortlessly from the physical properties of heat to the chemical characteristics of phlogiston, fusing the two. The igneous or inflammable matter was responsible for the state of fusion. It penetrated a body to join its parts and to augment its volume. By hindering or diminishing the "points of contact" between the parts of water, for example, it placed water "in a state of fusion." It was necessary "to regard this element as a body melt & mixed with igneous parts, whose most subtle molecules stay in perpetual agitation as long as it remains in this state of fluidity."[67] The true cause of evaporation was also fire or the igneous parts that divided and separated "the last & the smallest molecules of water." In other words, "by means of heat, water can be converted into air."[68] Even more interesting is the way Eller proved his transmutation argument. He heated a small quantity of water in vacuum and obtained an elastic fluid that could sustain the mercury in a barometer. He identified this elastic fluid as air because he thought that vapors could not exist in vacuum. According to the contemporary understanding, air was an inherently elastic body, while vapor consisted of foreign particles dispersed and dissolved in air, just as the particles of salts were dissolved in water. For Eller, air was something that was detected by the barometer.

Historians have noted Eller's memoirs mostly for the transformation of water into air, which supposedly influenced Lavoisier's thoughts on elements. However, the idea that water and many other bodies could become elastic through heating was easily available to Lavoisier in the French chemical literature. Louis Lemery, E. F. Geoffroy, Macquer, Nollet, and Turgot all expounded similar ideas.[69] Neither did Lavoisier take all aspects of Eller's memoirs seriously. After supposedly demonstrating the convertibility of water into air, Eller also recited Boyle's willow tree experiment to illustrate the convertibility of water into earth. According to Eller's scheme, water was a passive element which could be converted into air or earth, depending on its fire content. He argued that only fire and water were true elements that entered into the composition of all bodies. Air and earth were produced from these two. Lavoisier did not accept Eller's two-element theory. He soon undertook to disprove the transmutation of water into earth. It is unlikely, therefore, that Lavoisier bought the idea of the transmutation of water into air. What he thought were "très bien faits" in Eller's memoirs is, therefore, a matter of speculation. Three points are worthy of consideration. First, Eller blurred the boundary between the thermometric concept of heat and the chemical concept of phlogiston in a skillful manner to craft a coherent ontology of fire. Despite his later attack on phlogiston, Lavoisier maintained this ontology of fire with his concept of caloric by dislocating it from metals to gases. Second, Eller proved his contention with barometric measurement. This instrumental proof of chemical transformation probably appealed to Lavoisier's inclination to streamline chemical processes with metric measurements. All three elements in which Lavoisier took interest were amenable to "metric" quantification: fire through the thermometer, air through the barometer, and water through the hydrometer. Third, probably following Boerhaave's example, Eller wrote an extensive historical section on elements. Lavoisier soon adopted this practice in his discussion of water and air.

The first note Lavoisier composed after reading Eller was titled, "Physical chemistry/ On the elements/ On fire, water, and air."[70] On a rapid reading, he apparently noted only the argument that air was not an element, but a compound of water with the matter of fire. This composite understanding of air differed fundamentally, however, from Eller's transmutation argument. Lavoisier deemed it necessary to examine if the matter of fire derived from the atmosphere of the sun, or if it was the same matter that entered into the state of acidum pingue. His own thoughts on elements are contained in the longer second note, titled "Chemistry/ On the Matter of Fire and Elements in General." He stipulated the existence of a solar atmosphere that contained the entire planetary system. Continuing the thought in his 1764 manuscript, Lavoisier argued that just as air was the medium that transmitted sound, solar atmosphere was the medium that transmitted light. The solar atmosphere was made by the mutual dissolution of three elements: air, water, and the igneous fluid. While air and water occupied space in exclusion of each other, the igneous fluid was made of molecules so "prodigiously fine" that they could insinuate themselves among the molecules of water, air, and all other bodies. Admitting that water, air, and fire dissolved one another "to the point of saturation" to form a vast chemical combination (since dissolution was a chemical process), or the atmosphere, Lavoisier was struggling to formulate a more precise relationship between the three elements. He wondered, in particular, if water vapor was made by the dissolution of water in the air or in the igneous fluid, and if air itself was not a fluid in expansion (that is, a vapor). The "aerial fluid" exhibited much analogy [rapport] with the "igneous fluid" in fixation. Both lost a part of their property in entering into the composition of bodies. Air ceased to be elastic occupying an infinitely smaller space.

Lavoisier's thoughts on elements differed significantly from Eller's idea of transmutation. The three elements of fire, air, and water functioned as solvents for one another, rather than being converted into one another. His struggle in trying to discern the relationship between the three elements is not surprising. In his lectures at the Jardin, Rouelle did not work out clearly the relationship between the four elements as chemical substances and their functions as solvents. He simply juxtaposed Boerhaave's conception of elements as instruments with the conventional notion of elements as the constituents of chemical compounds. The scheme worked well for water and fire, superb solvents that became fixed in chemical combinations. Rouelle also stipulated, however, that the fluidity of water was due to the matter of fire it contained; water, when deprived of the matter of fire, turned into ice. Following the same train of thought, Macquer speculated that "air itself might even become solid if it were possible to deprive it sufficiently of the fire it contains."[71] The instrument-element theory, as Rappaport has dubbed it, became somewhat entangled when one included air. One could regard fire as something of a universal solvent that provided fluidity by penetrating all other bodies, including water and air. In turn, water and air were excellent solvents of other bodies. But air could also dissolve water. These complex relations of solubility were not easily accommodated by contemporary theory. Turgot, who had attended Rouelle's public lectures at the Jardin, distinguished between eváporation and vaporization in his article "Expansibilité" in Diderot's *Encyclopédie*. Vaporization for him was a process of dissolution in fire, whereas evaporation was a dissolution in air. Vaporization thus happened in all parts of the body, while evaporation occurred only at the surface exposed to the air.[72] Similarly, Lavoisier regarded evaporation as a two-stage process. First, the liquid vaporized by combining with the matter of fire. Since fire passed from the free to the combined state, the process produced a cooling effect. In the second stage, the vaporized fluid dissolved in the air to the saturation point. In sum, vaporization was a chemical process in which fire combined with water. Evaporation, or the dissolution of water vapor in air, was not. If evaporation was a chemical process, one should be able to fix air in the body of water, since dissolution was a prelude to chemical combination. Later, we find Lavoisier pondering exactly this question—the connection between dissolution and combination.[73]

Historians have bestowed much attention on Lavoisier's early discussion of air as a prelude to his later resolution that all bodies could assume the three physical states of solid, liquid, and gas. However, the two notes of 1766 reveal his concern with the composition of the atmosphere more than his preoccupation with the general composition of matter. That Lavoisier left on an extended geological expedition with Guettard on June 14, 1767 indicates that he had a strong interest in geology during this period. He sought to apply Boerhaave's dissolution theory of chemical action to the composition of the atmosphere. It took years of painstaking work for him to transform the complex relations of solubility into a viable theory of physical states. Nevertheless, Lavoisier's careless move from "matter of fire" in the title of the note to "igneous fluid" in the text deserves attention. He thus implicitly identified the chemical element phlogiston with the cause of heat, which was bound to cause severe conceptual difficulties in understanding air. If Lavoisier believed in the Rouellian instrument-element scheme, as Guerlac has pointed out, he was aware of the complex demand made on air in this scheme.

6.5   A Metric Vision of Chemistry

When Lavoisier began to prepare for his second bid for election to the Academy, he chose to work on the hydrometric analysis of water.[74] He had been collecting numerous samples of water during his trip with Guettard. Antoine de Parcieux's (1703–1768) second memoir on the Yvette aqueduct proposal had also aroused a stronger interest in hydrometry as well as in the analysis of water.[75] Lavoisier was familiar with the method of hydrometry since he had used Homberg's areometer to measure the solubility of gypsum. His duties at the Ferme générale also called for a hydrometer that could measure the density of alcohol in an easy manner. In sum, he had the opportunity, the motive, and the capability to engage in the new fashion of hydrometric measurement.

Lavoisier's memoires on water projected his instrumentalist vision.[76] Just as he had conducted barometric measurements to formulate a theory of the earth, he wished to pursue hydrometric measurements to unpack the "science of combination in general." In his first presentation on a hydrometer of his own design,[77] Lavoisier emphasized the importance of measuring specific weights for the arts as well as for physics. It was "the most certain" method for determining not only the quality of alcohol but also the concentration of acids and saline dissolutions. The existing methods (including Homberg's) allowed only a "mediocre exactitude," not the "degree of precision" Lavoisier wished to accomplish. The analysis of water demanded "the most scrupulous exactitude." While physics thus prescribed exactitude, commerce required simplicity of method. Lavoisier sought to design an instrument that combined sensitivity and convenience. After discussing the construction of hydrometers in detail, he turned his attention to their application, particularly to understanding chemical combinations. In contrast to his methodical work on hydrometry, his thoughts on this issue were tentative and somewhat speculative. In the following passage, we find him thinking about the way in which acids and alkalis united at the molecular level and how air became fixed in them:

It is principally in the art of combinations that the knowledge of the specific weights of fluids can shed most light. This part of chemistry is much less advanced than one thinks; we scarcely have the first elements. We combine every day an acid with an alkali; but in what manner is the union of these two beings made? Are the constituent molecules of the acid lodged in the pores of those of the alkali, as M. Lemery thought, or rather do the acid and the alkali compose different facets which can be engaged with each other or be united by simple contact, in the way the hemispheres of Magdebourg do? How do the acid and the alkali sustain themselves separately in water? How do they remain there after their combination? Does the newly formed salt occupy only the pores of water? Is it a simple division of parts, or rather is there a real combination, either part to part, or one part to several? Finally, from whence comes this air which escapes with such vivacity at the moment of combination and which, playing its natural elasticity, occupies all of a sudden a space enormously larger than that of the two fluids which it left? Would this air exist primitively in the two mixts? Was it in some fashion fixed, as M. Hales thought and as most physicists still think, or rather is this an air, so to say factitious and which may be the product of combination, as M. Eller thought?[78]

Chemistry did not offer any useful idea on these questions, Lavoisier asserted; it offered only the "vain names of rapports, analogies, frictions, etc." In other words, he did not think that the classificatory language of affinity chemistry answered the question. In contrast, the hydrometer could "penetrate these mysteries" by revealing "the quantity of real saline matter contained in the fluids one wants to combine together, their specific weights before and after the combination, the comparison of their average specific weight with the one resulting from its mixture." In short, Lavoisier hoped that precise quantitative measurements of specific weights, not of affinities, would lead to an accurate understanding of chemical combination. His reference to the "real" quantity of saline matter indicates that he knew of Homberg's earlier research with the areometer on the neutralization of acids and alkalis.

Lavoisier opened his second, longer memoir on the nature of water with some grand rhetoric concerning the importance of his work to the public good.[79] While the analysis of mineral waters or "salutary waters" had commanded much medical interest in the past, the waters citizens used in everyday life deserved as much interest, because "the force and health of citizens" and the "order and equilibrium in animal economy" depended on them. The knowledge of these waters could also provide more information on the interior of the earth which could not be directly reached. As the "favorite agent of nature," water infiltrated every interstice of the earth, forming crystals, metallic dissolutions, and precious stones. Despite the importance of the subject, the existing means of analysis were insufficient and "infidel." The method of combination, often used to detect the salts contained in water, gave only "a quite imperfect idea of the quantity of these same salts and of the proportion they observe between them." The method of evaporation, so far regarded as the most certain method and the one used most frequently, often resulted in very distorted results, as could be seen in the analysis of artificial mineral waters. Salts evaporated with water in different quantities, depending on their affinities with water. Furthermore, the fire used in evaporation could decompose the salts dissolved in water, altering the composition. Unspecified procedures, such as how long to dry the salts, also invited experimenter's error, as Macquer had acknowledged.

Lavoisier's "remedy" called for the measurement of specific weights. One could determine the salt content of water from the increase in its specific weight, he argued, although the method worked only for waters containing one kind of salt. Hydrometric measurement could lead to the "science of combinations in general," which formed the foundation of all chemistry. In the decomposition of two salts involving double affinities, for example, many questions concerning their relative quantities remained unanswered. All these difficulties would be resolved if one could construct tables of the quantities of the acid, alkaline or metallic oily base, and phlegm contained in each species of salt. Only then could one "predict with certitude what will be the precise result of our combinations." Although an immense task, it would bring "justice and precision" to the arts. Lavoisier's own work on compiling the "successive augmentation of the weights of water, relative to the quantity of salt" already formed a "nearly complete branch of this work."[80] His vision of quantitative chemistry expanded Homberg's earlier efforts to determine the "real" quantities of acids.

The hydrometer carried Lavoisier over the threshold of the Academy. He was admitted as adjoint chimiste surnuméraire on June 1, 1768, losing the place of adjoint chimiste to Jars. As a young, ambitious junior member new on the scene, Lavoisier followed the Academy's activities with absorbing interest.[81] He did not yet possess a substantial reputation as a chemist. His strength was in his systematic, meticulous hydrometric measurements, which he immediately put to use, working with Mathurin-Jacques Brisson on a number of proposals and inventions related to hydrometry.[82] In August, Lavoisier picked up an issue earlier raised by Grandjean de Fouchy while summarizing Le Roy's work: the transmutation of water into earth.[83] He designed an experiment to disprove the transmutation, while preparing to write a historical section of the memoir.[84] Using an accurate balance made by Chemin, adjusteur de la Monnaie, he weighed a pelican before and after putting in distilled water and heated it for 100 days (October 24, 1768–February 1, 1769). From the fact that the entire apparatus changed little in weight (0.25 grain), he concluded that fire particles did not enter the system. Nor did "any other exterior body" penetrate the glass to combine with water and to form earth. Unlike Boerhaave, he took it for granted that fire particles should be subject to the rule of the balance. Therefore, the earth in the pelican came not from water but from the glass vessel. Lavoisier concluded that glass, like salt, dissolved in water up to a limit, but that water itself was unchangeable.[85]

Historians have praised "Lavoisier's experimental ingenuity and those gravimetric procedures that were to characterize his later work."[86] If the transmutation experiment fully exhibited his mastery of gravimetric methods, the chemical part of the work was that of a "beginner,"[87] "incredibly shoddy and amateurish."[88] Historians' harsh judgments of the chemical aspect of Lavoisier's investigation are in part justified, since he ignored completely the qualitative differences between a variety of salts and other substances dissolved in water. He compressed chemical experience into the single scale of specific weight, envisioning naively that hydrometric measurement would provide a kind of universal, uniform measure of the bodies dissolved in water. He reduced the qualitative differences of chemical substances to the precise degrees of a linear scale. If this does not necessarily imply that Lavoisier's chemical skills were amateurish, it does indicate his insensitivity to the usual protocols of chemical practice. Just as he wished to measure the matter of fire with the thermometer rather than to detect its presence through a web of chemical operations, he wished to infer the presence of various chemical substances in water by means of the single scale of the hydrometer rather than through a series of chemical operations.

Lavoisier was also planning to write a historical memoir covering "what have been since the origin of philosophy the various opinions on the nature of water, what different philosophers have thought concerning the transmutability of the elements one into another, finally, on the possibility of changing water into earth."[89] Considerably delayed, he seems to have completed the historical section between May 1769 and November 1770 by limiting the scope of the discussion to the issues at hand.[90] He outlined a history of the two genres of experiments that supported the transmutation of water into earth: the willow tree experiments of van Helment and Boyle, recently repeated by Eller, and chemical distillation experiments. In refuting the first kind of proof, Lavoisier suggested various substances dissolved in water and air as the two alternative sources of plant growth. He regarded the air in the lower part of the atmosphere as an "extremely composed being" that dissolved water and all other volatile substances of nature. Aside from the substances dissolved in air, air itself entered into the texture of vegetables "in a quite considerable proportion" and contributed to their solid parts. He was familiar with Hales's experiments through Rouelle's lecture demonstrations.[91] Air existed in two fashions:

There results from the experiments of M. Hales, and from a great number of others which have been made in this genre, that air exists in two fashions in nature: sometimes it is presented in the form of a quite rare, quite dilatable, quite elastic fluid, such as the one we respire; sometimes it is fixed in bodies, combined intimately there; it loses then all the properties it previously had; the air in this state is no longer a fluid, it makes the office of a solid, and it is only by the destruction of the body whose composition it entered that it recovers its original state of fluidity. One can see, in this regard, quite ingenious experiments reported in the Vegetable Staticks; ... These experiments are too constant to be called into doubt, they have been above all repeated a great number of times in the eyes of all public in the lectures of M. Rouelle.[92]

In his early chemical investigations, Lavoisier applied his instrumental expertise to the topics of current interest. Whereas his experimental enterprise was more or less governed by the contingent factors of fashion and opportunity, his theoretical thoughts maintained a continuity, revolving mostly around the behavior of elements—heat or fire, water, and air. In addition to a complete discussion of water, he was planning to write another memoir on the nature of air, arguing that "evaporation or dissolution in air are two practically synonymous things." As he stipulated in the 1764 manuscript, he investigated the nature of elements as instruments or solvents. The metric approach carried Lavoisier only so far into the terrain of chemistry, however. He hoped to gauge the actions of fire and water by means of the uniform scale of the thermometer and the hydrometer. Such a metric gauge, intended to compress chemical action into a one-dimensional scale, did not yield a new insight or a new investigative path. Meldrum has observed a noticeable slowdown in Lavoisier's productivity after May 1769, although his travel duties as a tax farmer compounded the situation.[93] As late as May 11, 1771, when he was able to turn his attention more to science, his thoughts ran in all directions. The subjects he intended to pursue comprised the fractional precipitation of salts from water by alcohol, the action of water on mercury, glass, niter, and indigo; a mineralogical atlas; the cause of variations in the barometer; a new table of corrections for his areometer; and the completion of his memoir on street lighting.[94]

6.6 Burning Glass Chemistry

In 1772, the convergence of Guyton's "Dissertation sur le phlogistique," the evaporation of diamonds, and Priestley's *Directions for Impregnating Water with Fixed Air* focused Lavoisier's attention on the chemical role of air.[95] Guyton's "Dissertation" established the weight gain of metals as general phenomena caused by the addition or subtraction of phlogiston. Through this algebraic reconceptualization of the phenomena, Guyton coordinated various genres of experiments—combustion, calcination, and humid calcination—around the problem of weight gain, providing a framework for debate and dissension. Although his work established the weight gain as a general fact, an immediate dissension arose as to the cause of the weight gain. Even Macquer, a strong advocate of the phlogiston theory, did not subscribe to Guyton's hypothesis of phlogiston's essential volatility. Lavoisier, who regarded the measurement of specific weight as the key to unpacking the nature of chemical combination, was not persuaded by Guyton's explanation. He believed that even fire was subject to the rule of the balance. It is more than likely that he regarded the weight gain of metals as an established experimental fact that required a better explanation than Guyton's.[96]

At about the same time, Lavoisier was preoccupied with the disappearance of diamonds, which had caused a sensation since Jean Darcet's (1725–1801) report to the Academy in May of 1768. Public demonstrations continued intermittently until April of 1772. Early in 1772, to settle the dispute among several contending interpretations for this phenomenon, the Academy set up a committee consisting of Macquer, Cadet, and Lavoisier. They examined three hypotheses—volatilization, combustion, and decrepitation—respectively favored by Darcet, Macquer, and Cadet. The answer hinged on the experimental control of air in the process. Volatilization or vaporization did not require the presence of air; combustion and decrepitation would require the chemical or the mechanical role of air in the process, respectively. Darcet had initially proposed volatilization, since a diamond heated in a crucible with a closely fitting cover evaporated completely, that is, despite the apparent absence of air. Daubenton and Tillet's report on Darcet's work raised the possibility of decrepitation, however. The diamond could have turned into a white, impalpable powder adhering to the crucible.[97] In a follow-up experiment on August 19, 1770, Darcet heated a diamond in a covered crucible and another one enclosed in a ball of porcelain paste. Both disappeared without a trace, apparently confirming his hypothesis. The committee designed an experiment to test Darcet's hypothesis. If diamond was truly volatile, they reasoned, it should disappear upon distillation in closed vessels. The diamond lost a part of its weight, but was recovered largely intact after the process without yielding any vapor. The committee thus ruled out volatilization. Attributing the small loss of weight to the air contained in the distillation vessel before the process began, they designed an experiment that excluded air. A jeweler named Maillard packed several diamonds in a clay pipe stem with powdered charcoal, sealed it, placed it in a crucible surrounded by powdered chalk, and put it inside two larger crucibles connected mouth to mouth. Although intense heat deformed the crucibles and transformed the pipe stem to porcelain, the charcoal and the diamonds inside it were found unaltered. Lavoisier's report, read on April 29, 1772 at the public meeting of the Academy, left the question undecided between combustion and decrepitation.[98]

Dissatisfied, Cadet decided to resort to the Academy's burning glass—the one Homberg and Geoffroy had used earlier in the century. He invited Macquer and Lavoisier to join in the experiment, and he entrusted Brisson to set up the equipment. Cadet believed that the destruction of diamond required free air, since it could evaporate even at a mild heat in free air, while it resisted even the strongest heat without air. Macquer suspected combustion since he had observed a diamond glowing brightly "in a manner like phosphorus." Packing the diamond in charcoal contributed to its conservation, as Mitouard proceeded to prove, because the charcoal continued to restore phlogiston to the diamond throughout the process, much as metallic substances placed in charcoal powder conserved their properties.[99] In the continuing controversy over the destruction of diamonds, then, the control of air and the nature of the intermediates surrounding the diamond became critical issues.[100] Working in the company of the leading chemical authorities, therefore, Lavoisier learned that combustion was a process that required air and phlogiston. Phosphorus and sulphur offered the best examples.

While Lavoisier was engaged in the diamond affair, the news of Priestley's work on fixed air began to reach Paris. Magellan sent a copy of Priestley's pamphlet *Directions for Impregnating Water with Fixed Air* to Trudaine de Montigny at the Bureau de Commerce. Montigny wrote in turn to Lavoisier on July 14, 1772, suggesting its translation as well as a repetition of the experiments. Lavoisier read Magellan's letter to the Academy and may have suggested Rozier, the editor of the *Observations*, as a suitable translator. The presumed medical value of the mineral water impregnated with fixed air awakened French chemists to the importance of pneumatic chemistry. Beginning with the translation of Priestley's pamphlet in August, the Magellan-Trudaine-Rozier channel soon began to pour out Blackiana, repairing the French ignorance of British pneumatic chemistry. Bucquet, Bayen, and Lavoisier soon began to work on fixed air. The race was on to become the French pneumatic expert.

By the summer of 1772, then, Lavoisier was introduced to various genres of chemical phenomena that focused his attention on the role of air in *chemical* processes, especially combustion and calcination, as well as on the means of controlling air. He probably saw in the subject of air an opportunity to establish his own expertise. He was a junior, supernumerary member of the Academy without a secure reputation in the chemical community. His quantitative analysis of water had not led him to a productive field of inquiry. Air lay outside the traditional domain of French chemistry—an unclaimed territory, as it was. Lavoisier was already familiar with Hales's experiments through Rouelle's lectures; he also may have read the text *Vegetable Staticks*, which Buffon had translated into French in 1735.[101] He had been thinking about the way in which air was fixed in bodies, planning to write a memoir on the nature of air and its role in the evaporation of other bodies. The evolution of Lavoisier's thoughts on air in the ensuing years shows how creatively he mixed the concepts that belonged to different instrumental domains and disciplines to bring air into the chemical system. In the process, "air" as an instrument-element gradually gave way to "airs" as specific chemical species.

Anticipating the use of the most powerful furnace chemistry could command, Lavoisier drew up a plan for his own course of experiments in August (August memorandum).[102] In the meantime, he reflected on the nature of air (July memoir).[103] These notes indicate that his work habit was well established by this time. He planned to use the burning glass systematically to investigate a wide range of subjects, including fixed air, while reflecting on possible theoretical implications in light of the existing literature. He had researched the relevant literature in the memoirs of the Academy, reaching back to Homberg and Geoffroy's burning glass experiments as well as to Geoffroy's work on cold effervescence. He suspected the presence of air in many minerals, and he argued explicitly for the necessity of working out the place of air in chemical theory.[104]

The July memoir consists of four sections, the fragmented contents of which have caused historians to debate the history of its composition.[105] Lavoisier's main concern was to work out the algebra of the air and fire fixed in bodies. In the first section, "Essay sur la nature de lair: Reflexions sur lair et sur sa combinaison dans les mineraux," he stipulated that the quantities of air and phlogiston released or absorbed in chemical operations simply reflected the difference between the air and phlogiston contained in the reacting bodies and the resulting compound. In the second section, "Reflexions sur la combinaison de la matiere du feu dans les corps," Lavoisier dealt with the fixation of fire in phase changes. From the fact that the salt mixed with ice produced much more cold than the salt mixed with water, he reasoned that a certain quantity of "igneous fluid" was necessary in turning ice into water as well as in making water to boil. The phase changes of water illustrated the fixation of fire in bodies. Water was only a mixture of aqueous fluid with the matter of fire in different proportions, which determined its "hotness." When the matter of fire reached a certain proportion, water turned into vapor, assuming a form "quite analogous to air." Cooling accompanied all evaporation, which was "nothing other than the combination of some matter with the matter of fire." Water, in combining with the matter of fire, became "equiponderable" with air and consequently "volatile." From this, one could deduce that the matter of fire existed in two states: as combined with other elements or as "a stagnant fluid which penetrates the pores of all bodies . . . whose more or less great intensity produces the different degrees of heat." In the third section, "Sur les effervescences froides memoires acad 1700," Lavoisier explained Geoffroy's experiment of cold effervescence according to the algebra of fire. If a compound had more "analogy" with the matter of fire than its constituents, the process of combination produced cold as a result of the absorption of fire from the environment. In the fourth section, "Reflexions sur lair," Lavoisier finally addressed the critical question of the manner in which air was fixed in the bodies. He wondered how a fluid susceptible of such a terrible expansion could be fixed in a solid body and occupy a space 600 times less than its volume in the atmosphere, or how the same body could exist in such different states. He proposed a "singular theory": that the atmospheric air was not a "simple being," but a compound of the matter of fire and a particular fluid.

Lavoisier's July memoir indicates that he was struggling to account for the fixation of air in solid bodies and to treat air as a chemically active body. He obviously suspected that the phase change of water could be generalized to account for the fixation of air. If air was not a permanently elastic body but a compound of fire and some specific base, he could easily explain the fixation of air. His understanding of fire stood at a crossroads, however, between the chemical concept of phlogiston and the thermometric concept of heat. By identifying the "matter of fire" with the "igneous fluid," he was able to envision a more or less coherent pattern of behavior among the three elements—water, fire, and air—and to bring air into Rouelle's scheme of dissolution and combination. The tension remained, however, in pulling together the concepts of fire and heat drawn from different instrumental practices.

The August memorandum contained a broad range of experiments Lavoisier intended to perform with the burning glass to devise a chemical system that included air. Historically, this instrument had given life to Geoffroy's oily principle, later identified with Stahl's phlogiston. Now the burning glass promised interesting results for two reasons. First, it provided a "fire superior to the one we employ in our laboratories."

Second, it could act in the absence of air. The advantages the burning glass offered for diamond experiments seem obvious. Had Lavoisier been able to conduct the experiment in a vacuum, volatilization would have produced an elastic fluid, whereas decrepitation and combustion could not have proceeded without air. Although Lavoisier listed all three interpretations for the "evaporation" of diamonds, he intended to test only volatilization and decrepitation.[106] He further proposed to use the burning glass on metals, stones, and fluids to study the role of air in these processes. He wished to "measure the quantity of air produced or absorbed in each operation," but the difficulties of experimentation loomed large. Lavoisier's August memorandum exhibits more of his naive hope that the burning glass would provide the key to a pressing problem. He sketched a vast field of inquiry. His acuity in defining the experimental perimeters of phlogiston theory speaks for his concern with this popular yet protean concept at this early stage.

During the summer of 1772, Lavoisier cast a wide net, trying to figure out the air content of mineral bodies while reflecting on the nature of air and the manner in which it became fixed in bodies. He also read Guyton's *Digressions* with care, acknowledging that "all metals augment weight by calcination. M. de Morvaux demonstrates it completely in his digressions academiques pages 72–88."[107] The idea of the volatility of phlogiston probably posed a question in Lavoisier's mind. Had he accepted the fact of weight gain but not Guyton's interpretation, he could have entertained the idea that something other than phlogiston was responsible for the phenomena. In fact, this was Béraut's opinion, which Guyton took pains to refute in the *Digressions*. Lavoisier did not have to invoke a foreign substance to explain the weight gain, however. He knew from Boerhaave's treatise and Hales's experiments that air assumed a solid form through fixation.

The sequence of events from this point on is well known from Guerlac's account of the "crucial year." The collaborative experiment on diamonds commenced on August 14 and continued until October 13. On September 10, Lavoisier purchased some phosphorus from the apothecary Mitouard and began to experiment with it to "verify if phosphorus absorbed air in its combustion."[108] By October 20 he could write down more details.[109] Phosphorus exposed to daylight or heated in a small vessel produced the acid of phosphorus, while absorbing "an extremely considerable quantity of air." He noted a "singular phenomenon" that the quantity of phosphoric acid thus produced far outweighed that of the original phosphorus. When the burning glass finally became available for his own use, Lavoisier proceeded with the now-historic experiments on minium or the red calx of lead. He found that the red calx gave off a large quantity of air upon heating. A few days later, he realized the connection between the weight gain in the combustion of phosphorus and the calcination of metals, which made up the substance of the famous pli cacheté (sealed note) of November 1, 1772.[110] He concluded that the weight gain in combustion and calcination resulted from the air fixed in the process.

Historians have debated over exactly which phenomenon led Lavoisier to his crucial experiment with minium.[111] Meldrum proposed the combustion reactions of phosphorus and sulphur; Guerlac reconstructed a more elaborate pathway involving effervescence and the calcination of metals. Crosland and Kohler have argued that Lavoisier was looking for a theory of acid. Though all these points are pertinent, I am not certain that we can reconstruct Lavoisier's thoughts in a clear sequence. He was not a typical experimenter who would indulge in a particular genre of experiments until he saw the light. Despite his enthusiasm for instrumental control of chemical operations, he had a keenly theoretical mind that always sought to correlate different genres of experiments for a bigger picture and different disciplines for a uniform understanding of nature.[112] Furthermore, Lavoisier scholars have missed the link between these different genres of experiments provided by Guyton's "Dissertation." First, the combustion of phosphorus and sulphur was one of the problem areas Guyton foresaw with his hypothesis and tried to explain away. If Lavoisier did not accept Guyton's hypothesis, it was a natural place to start, especially since he knew that these substances exemplified combustion phenomena. Another problem Guyton mentioned, the calcination of mercury, was soon taken up by Bayen. Second, lead was an obvious choice in testing the weight gain of metals upon calcination because it gained the most weight. Chardenon had chosen lead for the same reason, but owing to experimental difficulties in detonating lead he worked on the reduction of lead calx instead. Lavoisier followed the same route. Third, Lavoisier's concern with the theory of acidity derived from his experiments in "humid calcination," which Guyton enlisted to establish the weight gain. In other words, Guyton's "Dissertation" accomplished considerably more than simply identifying the problem of the weight gain. It also laid out the venues of further investigation to confirm or refute his hypothesis as well as the available range of alternative hypotheses. For example, Guyton dealt at length with Hales's explanation, which included the fixation of a "good quantity of air."

A longer memoir Lavoisier drafted, possibly in preparation for the fall public meeting of the Academy, reveals more clearly his preoccupation with Guyton's problem.[113] The title, "Sur la cause de l'augmentation de pesanteur qu'acquierent les metaux et quelques autres substances par la calcination," identifies the weight gain of metals as his primary concern. He perceived clearly that the facts he presented—the weight gain in the combustion of phosphorus and sulphur and the release of air in the reduction of minium—were destined to upset Guyton's phlogiston theory. He wished to convey the "chain of ideas" that led him to such a conclusion. The combustion of phosphorus confirmed that it gained weight, although he had to take care to exclude the effect of humidity. A similar experiment with sulphur made him recognize that "the property of augmenting weights by calcination, which is nothing other than a slow combustion was not particular to metals" and that it formed a law of nature. He conducted a series of experiments to discern the cause of the weight gain. Repeating Hales's experiments, he settled on "a portion of air absorbed and fixed" as the answer. He suspected that the weight gain of metals might be due to the same cause and proceeded with the reduction of litharge to confirm it. On the basis of these two genres of experiments, Lavoisier concluded that "the theory of Stahl on the calcination and reduction of metals is extremely imperfect and demands modifications. He regarded all calcination only as a loss of phlogiston, while it has been proved that there is every time loss of phlogiston and absorption of air." Lavoisier stated it as "a law nearly constant in nature" that the release of phlogiston was always accompanied by the absorption of air. This explained why combustion and calcination required air. Air entered "materially into the combination of metallic calces" and augmented the weight.

Lavoisier obviously considered his major discovery to be the common cause of the weight gain in combustion and calcination. He could provide a more rational, algebraically sound answer for the outstanding Stahlian (in fact Guyton's) problem: the weight equation. The authors he cited in a revised version prepared for the fall public meeting of the Academy in November—Boyle, Kunckel, Boerhaave, Hales, Pott, Beraut, Chardenon, Geller, Meyer, and above all Guyton de Morveau—indicate his focus on the weight problem. In other words, Lavoisier's pli cacheté and November memoirs contained a direct answer to the problem raised by Guyton's "Dissertation." The weight gain of metals was due to the absorption of air. Despite the importance he imputed to his discovery, Lavoisier did not present his memoir in the November meeting of the Academy. While Perrin attributed this to the lack of time in that meeting, I rather suspect a different reason. The idea that air caused the weight gain was not, by itself, a radically new theory. Others had proposed various substances dissolved in air and even air itself as a candidate. In order to attain a level of credibility, Lavoisier needed to present convincing evidence. This meant, for him, that he needed to quantify the air fixed in combustion and calcination. Priestley's work had also raised some uncertainty as to the identity of fixed air. Lavoisier was reading widely in the literature of pneumatic chemistry. The more he read, the less certain he became of the identity of fixed air. We find him engaged in a series of experiments during the next year precisely to address these two problems: a quantitative proof for the fixation of air and the identity of fixed air.

6.7 Between Air and Airs

Air became Lavoisier's new passion, and he began to read Hales, Black, MacBride, Jacquin, Cranz, and Priestley extensively.[114] By February 20, 1773, according to his laboratory notebook, he was ready to begin a "long series of experiments" on the elastic fluid released or absorbed in fermentation, distillation, various chemical changes, and combustion.[115] He had a vast plan. He felt that the existing works formed a portion of "a great chain," but were insufficient "to form a body of complete theory." He wished to repeat and to confirm all the important experiments in pneumatic chemistry before proceeding with new ones so that he could place the existing works and his own in a continuous chain and "bring about a revolution in physics and chemistry." To this end, he would start with the operations that fixed air: vegetation, respiration, combustion, calcination, and chemical combination.

Lavoisier was not certain at this point as to the identity of fixed air. While he took for granted that an elastic fluid was fixed in many bodies, he sampled various opinions on the nature of this fluid. Since it came "originally from the atmosphere," he inferred "either that this substance is air itself, combined with a volatile part that emanates from substances, or at least that it is a substance extracted from ordinary air." Although he was aware of the qualitative differences between fixed air and atmospheric air, he did not yet consider them as chemically distinct species. The experiments also continued to humble Lavoisier's ambition. At every turn, he encountered a staunch resistance of the material world to his beautiful scheme according to which fixed air was responsible for the weight gain of metals. He had to devise new apparatus and methods to circumvent the experimental roadblocks, but to no avail.[116] Many theoretical and empirical uncertainties hampered Lavoisier's investigation of fixed air during the "next crucial year" until he realized that air consisted of many chemically distinct airs.

What Lavoisier set out to accomplish through his investigation of fixed air can be discerned from the ideal experiment he outlined at the Easter meeting of the Academy on April 21, 1773.[117] Although his investigation during the previous months had not yielded any decisive evidence, he projected his theory of metallic calcination as an important discovery that would add to the glory of French chemistry. Reserving the details of experiment for the regular session, he only outlined an ideal experiment in an enclosed space which would allow him to discover the relationship between the weight gain and the volume of the air involved in the process. He concluded that metals were calcined proportionate to the volume of the air used and that they would gain the weight accordingly. This proved that the weight gain was due to the fixation of air or that a metallic calx was nothing but a metal combined with fixed air.

Lavoisier's "Easter memoir" reminds us of Guyton's "Dissertation" in which the latter focused sharply on the loss of phlogiston as the sole cause of the weight gain. In a rather long discussion of Stahl's phlogiston theory which was eliminated in the final version, we find Lavoisier thinking exactly in anti-Guytonian terms: "... all the experiments that I have made on this subject lead me to believe that almost all of the phenomena that one attributes to phlogiston derive only from the absence of fixed air, and reciprocally, as I have shown for metals."[118] Lavoisier also discussed other means by which metals could absorb air: detonation with niter, precipitation from acid solutions by an alkali, etc. These were precisely the genres of experiments Guyton listed to bolster his theory.[119] Lavoisier simply inverted the conclusion from the loss of phlogiston to the addition of fixed air: "... all metallic substances without exception are susceptible of being combined with fixed air." He went further than Guyton in formulating a tentative theory of acids, perhaps spurred on by the contemporary debates. He wished to conclude the memoir with some experiments which could "shed the greatest light on the nature of acids and on the phenomena of fermentation." He suspected that fixed air was involved "in the formation of all acids." He then paused briefly to reflect on the status of Stahl's theory. Although he did not reject phlogiston outright, he intimated that chemistry was entering an epoch of a "nearly complete revolution."[120]

Lavoisier's confident performance at the Easter meeting reveals his prior conviction as to the validity of his hypothesis. He could rely on Guyton's experimental work on the weight gain of metals. Only, the positive weight of fixed air offered a more rational choice than the negative weight of phlogiston or fire, or the weight that was not registered by the balance. The subsequent course of Lavoisier's experiments provides further evidence of his conviction. He stayed mostly within the planned course, improvising new methods and devices without a spectacular success. He "sought reliability, but not great precision" in preparing a detailed account of experiments he promised in the Easter memoir. He even devised various ad hoc hypotheses to account for experimental inaccuracies.[121] From Holmes's account, we can discern that Lavoisier started the year with a rather simplistic theory of fixed air which he intuited from Guyton's and Priestley's work, that he struggled with the experimental results which did not conform to his expectations, but that he pursued them regardless because of his belief in his new theory. Lavoisier's theoretical inclinations could account for the sudden turnabouts in his experiments noticed by Holmes. He was not the kind of experimenter who would listen to his experiments to devise the next step, but a theoretician who was looking about to find evidence for his theory, which was becoming fainter at each step. Nevertheless, he gained in experimental expertise, constructing his "balance sheet method" piece by piece and making it more reliable through his experimental mastery of chemical operations that involved air.

Lavoisier's quest differed from Guyton's in one important aspect. While Guyton's algebra took into consideration only the weight equation, Lavoisier always had an algebraic equation of air and fire in sight. In April, he projected a full-blown theory of the three states of bodies in a manuscript memoir, "Essay sur la nature de l'air."[122] He stipulated that all bodies of nature could exist in three different states, as could be seen in the case of water, depending on the quantity of the "matter of fire" contained in them. Air itself was nothing other than a fluid in expansion. Being combined with a great quantity of the matter of fire or phlogiston, it was too volatile to exist as a fluid on the earth. One could separate air from the fire it contained only by supplying the bodies for which air had more "analogy." Consequently, the fixation of air released the matter of fire or phlogiston, as was the case in the calcination of metals. Like the water fixed in salts, the air fixed in calces withstood a higher degree of heat, but was revivified with the addition of the matter of fire. Lavoisier even used the phrase "the fixed matter of air" to strengthen the analogy between fire and air in the fixed and free states. In order to revivify air from a calx, one had to put it in contact with phlogiston. Such was "in a few words, the entire theory of the reduction of metals." It was not the metal (as Stahl had supposed) but the air that was revivified in the process. This transfer could satisfy Lavoisier's algebra of fire and weight.

Two features of this memoir deserve our attention. First, Lavoisier carefully differentiated fire and heat. The fluid that combined with bodies to change their state was always identified as the matter of fire or phlogiston, although this happened at a specific "degree of heat" for each

body. If Lavoisier's thoughts on the composition of air seem to anticipate his later articulation of the three states of matter, he still pointed to
the "inflammable principle, or . . . the matter of pure fire," rather than
to heat, as the substance responsible for phase changes. He wished to
dislocate phlogiston from the metal to the air according to the algebra
of fire, rather than according to the heat content of bodies. This algebra
of fire or phlogiston responsible for phase changes was bound to falter
his quest for instrumental proof. Second, Lavoisier thought it possible,
or even probable, that air was "a compound of several fluids in vapors
mixed together." This would explain in a quite simple manner "the essential differences" that seemed to exist between the fixed or mephitic
air and the atmospheric air. He took a cautious step, that is, toward
acknowledging the existence of "airs" as distinct chemical species.

Lavoisier pressed on with his experiments and presentations despite
the difficulties. He read to the Academy a historical essay and several
experimental reports, opening on May 5 the sealed note of November.
The *Opuscules* was ready by August 7, and the Academy appointed a
commission to examine the work. After witnessing many of the major
experiments, the rapporteurs praised Lavoisier's "rigorous method" in
quantitative experiments. The volume was published and deposited with
the Academy on January 8, 1774. Lavoisier had finished first in the race
to become the French pneumatic expert. He had thoughtfully prepared
the *Opuscules* to ensure his place in history. The long historical section
was necessary, he argued, to inform the French public of the development in pneumatic chemistry abroad. For the purpose, he would take
on the "simple character of an historian" with "utmost impartiality."
Despite the rhetoric, Lavoisier's historical exercise reflected his "need
to make his experiments intelligible."[123] The organizing theme was the
debate between the British "partisans of fixed air" and their German
opponents rallying around Frédéric Meyer's acidum pingue. Lavoisier
knew that the importance of fixed air for chemists lay in its relevance to
the theory of acidity.[124] The debate over the existence and the identity
of the universal acid had continued since Homberg articulated the principle of acidity. The discovery of fixed air added new fuel to this on-
and-off debate. Bergman regarded fixed air as a true acid, or the "aerial
acid." Marsilio Landriani, William Bewley, Felice Fontana, and others

entertained the possibility that fixed air might be the universal acid.[125] The debate between the British partisans of fixed air and the German proponents of Meyer's acidum pingue became a focal point of theoretical chemistry in the mid 1770s, creating a critical contact zone between the new chemistry of airs and the old chemistry of salts. Lavoisier's sympathy lay in general with the "partisans of fixed air," but not entirely.[126] He had a set of pressing questions and scanned the field across the camps for answers. These questions served as the criteria of inclusion in the historical section. A few stand out.

First of all, Lavoisier was concerned about the identity of fixed air. The most pressing questions were "1st, Whether fixed air be the same as that of the atmosphere; 2dly, Whether it be the same from whatever bodies it may have been extracted."[127] Surveying the literature, Lavoisier weighed carefully the experiments and opinions pro and con. In each chapter, he summarized the answers on this issue given by each past investigator. On the one hand, Hales, Boerhaave, Venel, Saluces, and others admitted only one species of air, which was either free or fixed. On the other hand, van Helmont, Boyle, Black and MacBride admitted the qualitative difference of fixed air from the atmospheric air. Boyle and Cavendish also hinted at diverse species of air, which de Smeth and Priestley worked out more explicitly. The notion of "fixed air" had gone through changes accordingly. For Hales, it was a term opposed to "elastic air." There existed only one species of air that was free or fixed. At present, however, Black's fixed air meant a particular species of air existing in free state. There existed even a tendency to "establish as many species of air, as there are bodies which can supply it, which can answer no purpose, but to obscure the theory of chemistry."[128] Priestley's work, "the most elaborate, and most interesting, of any which has appeared since that of Dr. Hales, on the fixation and separation of air," formed nonetheless "a train of experiments, not much interrupted by any reasoning, an assemblage of facts, most new."[129] The term "fixed air" was a misnomer in its current usage.

Lavoisier obviously suspected that the air produced in various chemical operations was not the same as common air. He did not wish to acknowledge as many species of air, however, as there were ways of making them. He preferred an explanation which maintained the status of air as a chemical principle, or an instrument-element, like water and fire. The different properties of air produced from various bodies could be attributed to various substances dissolved in air. Fixed air would then be "a common air impregnated with foreign substances which it holds in solution," just as water held foreign substances in solution[130]:

The case is the same with air as with water; both elements have the quality of dissolving, and being saturated with several substances; each of these elements acquires new properties which belong neither to water nor to air, but merely to the substances with which they are impregnated. As there are certain bodies which water is capable of dissolving, and which cannot be separated from it; the same should be the case with air; this last element may be impregnated with substances as volatile, and as dilatable as itself, and which can never be separated from it, either by distillation, filtration, or any other method; but it does not the less follow, but that these new properties which are to be found in the air, should always be attributed to the foreign substances, and not to the air itself.[131]

Second, Lavoisier regarded quantification as the key to the evolution of the field. He identified Hales as the founder of pneumatic chemistry for this reason. Hales had noticed that under a bell jar the combustion of phosphorus and sulphur stopped at a certain point determined by the size of the container. Lavoisier suspected that this type of experiment conducted in a closed space would be crucial in settling the identity and the quantity of fixed air and scanned the literature on this issue. He also regarded the specific weight of air as the key to quantification. Priestley had noticed a "very singular, and almost incredible, phenomenon": "Nitrous air, whether by itself, or combined with common air, always retained a specific gravity, sensibly equal to that of atmospheric air."[132] If there existed many different kinds of air, the measurement of their specific gravities should differentiate them. As we have seen, this is what he endeavored to accomplish with the hydrometric measurement of various waters.

Third, Lavoisier selected some authors for their contribution to the more speculative part of his thoughts on "the theory of the combination of air with bodies."[133] He quoted Boerhaave at length for the idea on the fixation and restoration of air, which was quite similar to his own: Elastic air entered into the composition of bodies "as a constituent" and resided in them "in such a manner as not to produce the effects of air, as long as it is combined and united with them." Lavoisier thought that the "aerial particles of the same nature" became fixed in bodies and liberated with fire, resuming the state of "true air, as before its union." Chemistry effected this "resolution and composition" of air.[134] Jacquin held a view quite similar to Lavoisier's. In addition to tracking Black's and MacBride's experiments quantitatively, he reflected on the manner in which air might exist in bodies. He distinguished the air that entered into the pores of bodies and was rendered sensible by the air-pump from the air that entered into the composition of bodies and was "in a state of division or dissolution which does not permit it to enjoy its elasticity." Furthermore, he observed that "whenever the air is dissolved, and combined with certain substances, it has, as in all the chemical combinations, 1st. a point of saturation; 2dly. a certain degree of adhesion, which is greater or smaller in proportion to the difference of affinity which it has with these different substances."[135]

In sum, the historical section of Lavoisier's *Opuscules* indicates where he stood at this point in his understanding of fixed air. On the theoretical side, he was concerned about the nature of fixed air and its mode of combination with different bodies. Although he suspected that fixed air was different from the atmospheric air, he was not certain whether they were chemically distinct species or simply different modifications of the same species. He wished to retain, if possible, his original conception of air as an instrument that dissolved diverse substances. Experimentally, Lavoisier sought effective instrumental practices for the quantitative treatment of air and the methods of controlling experimental errors, such as insulating air from water with an air-pump.

The second, experimental part of the *Opuscules* consists of three genres of experiments, accurately reflecting Lavoisier's work during the previous year—a review of the principal experiments involved in the Black-Meyer controversy, the calcination of metals, and the combustion of phosphorus and sulphur. In the first three chapters, Lavoisier repeated the experiments related to the main controversy to prove that "the same elastic fluid" was involved in them. He maintained a quantitative thread by tracking the specific gravity of the air produced in each experiment. In the plan and execution of these experiments, he assumed that a chemical substance was uniquely defined by its composition (constituents and their proportions), which could be proven by analysis and synthesis. The measurement of specific gravity served to differentiate chemical substances except that phlogiston defied this rule of the balance. In discussing the experiment designed to prove the existence of an elastic fluid in fixed and volatile alkalis, Lavoisier weighed carefully the relative merits of Black's theory of fixed air and Meyer's hypothesis of acidum pingue. Regarding the diminution of specific weight "observed in the alkaline solution, in proportion to added lime," Lavoisier was favorably disposed to Black's opinion that the lime attracted something from the alkali. He reflected, however, that Meyer's hypothesis could also work if the lime supplied the alkali with "some substance of a lighter nature than this solution," which might be "nothing but phlogiston." It was a "fact known in chemistry, of which no doubt can remain" that phlogiston lessened the specific gravity of the liquors with which it was combined.[136] Lavoisier understood, then, Guyton's instrumental reasoning behind the "essential volatility" of phlogiston. Although he pointed out the consequence of involving phlogiston in chemical explanation, he did not indulge in a detailed critique of phlogiston.

The second segment, chapters 4–8, dealt with metallic calcinations. Lavoisier sought to prove with quantitative measurements that metallic precipitates and calces contained the same elastic fluid. He offered a "conjecture" that the weight gain in metallic calcination derived from the fixation of "the air of the atmosphere, or an elastic fluid contained in the air." Three facts rendered this conjecture "a very great degree of probability": that the calcination of metals required the presence of air, that the increased surface area of the metal facilitated the process, and that the reduction of metals was accompanied by effervescence. To prove his case, Lavoisier presented a well-orchestrated series of experiments on lead. In particular, he took care to prove that the elastic fluid obtained in the reduction of minium did not come from the added charcoal. He spelled out (though with caution) his earlier thought on the composition of air: that air was not a single substance, but a state that all bodies could assume depending on the quantity of the matter of fire:

If it were permitted me to indulge in conjectures, I should say, that some experiments, which are not sufficiently complete to submit to public inspection, induce me to believe, that every elastic fluid results from the combination of some solid or fluid body with the inflammable principle, or perhaps even with the matter of pure fire, and that on this combination the state of elasticity depends. I should add that the substance fixed in metallic calces, and which augments their weight, would not be, properly speaking, on this hypothesis, an elastic fluid, but *the fixed part of an elastic fluid*, which has been deprived of its inflammable principle. The principal action of charcoal, and all other substances of that nature employed in reductions, would then be, to restore the phlogiston or matter of fire, to fixed elastic fluid, and with it the elasticity which depends on it.

However different this opinion may seem to be from that of M. Stahl, it yet perhaps is *not incompatible with it*. It is possible that the addition of charcoal in the reduction of metals may answer two purposes at once; 1st, That of restoring to the metal the inflammable principle which it has lost; 2dly, That of restoring to the fixable elastic fluid in the metallic calx, the principle which constitutes its elasticity. But I repeat again that it is with great caution, that an opinion on so delicate, so difficult a subject should be hazarded; a subject which is very nearly connected with one still more obscure, I mean the nature of elements themselves, or at least of what we regard as elements. Time and experiment alone can settle our opinions on these points.[137]

It seems that the experiments on humid calcination forced Lavoisier to deal with the issues of affinity. In order to effect various displacement reactions of metals, he had to rely on Geoffroy's table. He tried, for example, to "form a three-fold union of fixed air, metals and acids, with a view of acquiring some ideas of the degree of affinity between these different substances." He thus poured a solution of elastic fluid (water impregnated with fixed air) on various metallic dissolutions in acids. Regardless of the proportion, he was "never able to produce a precipitation." They seemed to "prove in general, that metallic substances have more affinity with mineral acids, than with elastic fixable fluid." Though they are tentative, these remarks indicate that Lavoisier began to worry about the problem of affinity around this time. He had studied Geoffroy's table of affinities as early as 1766.[138]

The third segment, chapters 9–11, dealt with the combustion mostly of phosphorus. Lavoisier examined the weight gain and scrutinized the air left afterward, taking note of its volume. The combustion of phosphorus in an enclosed space took the center stage because it allowed him a quantitative control of the air involved in the process. The fact that the combustion could not proceed beyond a certain limit led to "a suspicion that atmospheric air, or some other elastic fluid contained in the air, is combined, during the combustion, with the vapors of the phosphorus." Since there was "a great difference between conjecture and proof," however, he sought to determine the weight gain "with as much precision as the nature of the experiment will admit of."[139] He also attempted to burn phosphorus in vacuum with the burning glass since the failure of such an attempt would provide a direct proof that combustion fixed air.

Despite much uncertainty in theoretical outlook, the *Opuscules* established Lavoisier as the leading French pneumatic expert, in part thanks to his own effort in publicizing the work.[140] Macquer reviewed it in a competent manner, praising its experimental content.[141] Although Lavoisier made an urgent appeal that air should become an integral part of chemical classification and theory, he had to struggle incessantly in the ensuing years to bring together his previous thoughts on elements and his vision of a quantitatively transparent, metric chemistry in a coherent chemical system that included air. He had as yet much chemistry to learn. His plans for further volumes of *Opuscules* indicate that his attention was not focused on chemistry or on fixed air. He hoped that the first volume would be followed by several others dealing with the following:

1st, on the existence of the same elastic fluid in a great number of bodies in nature in which it has not been hitherto suspected; 2dly, on the total decomposition of the three mineral acids; 3dly, on the ebullition of fluids in the vacuum of an air pump; 4thly, on a method of determining the quantity of saline matter contained in mineral waters, from the knowledge of their specific gravity; 5thly, on the application of the use, either of pure spirit of wine, or of spirit of wine mixed with water, in certain proportions, to the analysis of the very complicated mineral waters; 6thly, on the cause of the cold which is observed in the evaporation of fluids; 7thly, on different points of optics, on which I have had occasion to be employed in a Memoir relative to the lighting of Paris . . . 8thly, on the height of the principal mountains in the environs of Paris . . . Lastly, I shall add a numerous train of observations on the barometer, made in different provinces of France. . . .[142]

6.8 The Oxygen Theory

The strange calx of mercury played a crucial role to change this state of affairs.[143] It had long been known to chemists for the peculiar property of being reduced to mercury without charcoal, thus by implication without phlogiston. Guyton brought this anomalous behavior of mercury to a sharper focus by listing it as a problem area in his "Dissertation." Pierre Bayen soon undertook a systematic investigation of the formation and reduction of mercury precipitates. Arguing that to explain "the augmentation of the weight of metallic calces" had been "for several years the goal of almost all of the chemists in Europe," Bayen defined his task as the identification of Lavoisier's elastic fluid.[144] He concluded in April of 1774 that charcoal was "useless" in the reduction of mercury. The Stahlian doctrine that phlogiston was essential to the revivification of metals proved false at least in the case of mercury. He warned that it was "dangerous to commit oneself to systems, no matter how authoritative they may be." Bayen ignored Guyton's effort to explain this apparent anomaly, voicing an emerging skepticism toward the phlogiston theory among the Parisian scientists. Shortly before his paper, there appeared in *Observations* an anonymous critique of phlogiston, as we saw in the previous chapter.[145]

Guyton's work on phlogiston and Bayen's criticism stimulated much speculation on the nature of phlogiston, ranging from elementary fire, pure matter of light to electric fluid.[146] Comte de Milly presented a claim to the Academy, for example, that he had reduced a metallic calx by means of "simple electric fluid." He concluded that an electric fluid could act "exactly like" phlogiston. To evaluate Milly's claim, the Academy appointed a committee, which included Lavoisier. Cadet, who carried out the experiments for the committee, reported that the red precipitate of mercury was reduced to mercury without the addition of charcoal, but Baumé dissented. This prompted the appointment of a second committee, consisting of Brisson, Lavoisier, and Sage. They reported on November 19 that the red precipitates could be reduced without the addition of charcoal.[147] Incidentally, Priestley visited Paris in October and mentioned to a group of scientists his discovery of the air that was produced by heating the "red precipitate of mercury." Although Lavoisier was in the audience, he apparently did not realize the significance of this discovery at that time. He had reached a kind of impasse. He could not identify the air fixed in the processes he had been examining for two years, nor could he decide definitively whether the matter of fire came from the elastic fluid or the combustible body.[148]

Lavoisier undertook the experiment with the mercury precipitate in February of 1775, in the midst of running other experiments related to the fixation of air.[149] By this time, he was more confident that the matter of fire came from air. He did not yet regard different kinds of air as chemically distinct species, but as common air with various proportions of phlogiston or the matter of fire.[150] In the course of experiments, however, he found that the air produced from mercury calx without the addition of charcoal rendered limewater only "slightly opaline." In sharp contrast to fixed air, a candle burned more brightly in this air, convincing Lavoisier that it was "in the state of common air." By heating charcoal with this air, he obtained "pure fixed air." At the end of these investigations, Lavoisier presented a memoir in the public meeting of the Academy on April 26. The memoir opened with a succinct statement of a question Lavoisier had been struggling with for some time:

Do there exist different species of air? Is it sufficient that a body be in a state of permanent expansibility to constitute a species of air? And lastly, are the different kinds of air which nature affords us, or which we are able to form, substances distinct of themselves, or modifications of atmospheric air? Such are the principal questions which appertain to the plan which I proposed to myself to lay before the Academy.[151]

Lavoisier answered unequivocally in the affirmative, drawing on the comparative reductions carried out with and without charcoal. The air obtained with charcoal displayed all the properties of fixed air. The air obtained without charcoal reacted negatively to each of the standard tests for fixed air, an outcome he supposedly "recognized with great surprise." Furthermore, candles burned more brightly in this new air, which also supported respiration better. All this convinced him that "this air was not only common air, but that it was even more respirable, more combustible, and consequently more pure even than the air in which we live." In other words, the air contained in calces was "an exceedingly pure portion" of the atmospheric air. The combination of this air with charcoal produced "fixed air," normally encountered in the reduction of calces over charcoal. By now, Lavoisier had all the elements of his new theory of combustion and calcination, except the precise identity of this "exceedingly pure portion" of common air.

Not surprisingly, Lavoisier was extremely receptive to Priestley's discovery of "dephlogisticated air," a new air "between four and five times as good as common air." Priestley's *Experiments and Observations on Different Kinds of Air* reached Paris by December in advance sheets. Lavoisier immediately set out to confirm the experiment. On February 13, 1776, he resumed the work on mercury calx, repeating much of Priestley's work.[152] On April 7 he began to test the idea that the air absorbed in producing mercury calx was the same as the one released in its reduction.[153] He was also investigating the composition of nitric acid as a part of his job at the Gunpowder Administration. He studied the solution of mercury in nitric acid, specifically the air contained in that acid.[154] Soon, he announced a new theory of acid which stipulated that all acids contained "the purest portion" of air, or Priestley's dephlogisticated air, which rendered them acidic.[155] He had finally realized his initial scheme of fixed air—that the fixation of air was responsible for the weight gain in acidification and metallic calcination—albeit with a different kind of air. Combustion, calcination, and acidification involved the fixation of dephlogisticated air. He had to make a fundamental transition, however, in his understanding of the elements. Air became airs. The instrument air became multiple chemical species. It took years for Lavoisier to work out the consequences of this transition, which forced him to accept the analytic ideal of chemical elements.

6.9   A New System of Chemistry

With the new theory of combustion, calcination and acidity in place, Lavoisier entered the most productive period of his life. In order to accommodate this new creation in the existing chemical system, he needed to rework many areas of chemistry and physics. In particular, he had to clarify the identity and the role of phlogiston in combustion and calcination. He could easily account for the origin of the heat and light produced in these processes by the fire or phlogiston contained in air:

Metal +   air   =   calx   + heat or light.
          (fixed part + phlogiston)   (metal + fixed part of air)   (phlogiston)

A precise algebra of weight and heat he wished to implement on chemical processes still eluded him, however, if metals contained an unspecified amount of fire or phlogiston. He could not rationally maintain the theory that bodies could assume different forms depending on their fire content if both solid metals and fluid airs could retain an excess quantity of fire or phlogiston. That is, he could no longer endorse the phlogiston theory and the fixation of air simultaneously, as he previously had.

An algebraic equation of fire required quantification of the heat involved in chemical reactions. Lavoisier began to tackle the problem on chemical and physical fronts by enlisting two talented young scientists: Jean-Baptiste-Marie Bucquet (1746–1780) and Pierre-Simon Laplace (1749–1827). When Turgot appointed him to the Gunpowder Administration, Lavoisier took up residence at the Paris Arsenal, where with Bucquet's assistance he set up a find laboratory. Bucquet, who had attended many scientific courses in Paris, began to teach in his private laboratory in 1768. He joined the Faculté as professor of pharmacy in 1775 and was made professor of chemistry in 1776.[156] Between January and September, 1777, Lavoisier and Bucquet worked on a range of subjects, mainly on heat phenomena, presumably to prepare for a lecture course.[157] They intended to repeat all the fundamental experiments of chemistry, to examine them with great care and rigor in order to form a fresh point of view, and to condense them into an "exact table" that would be approved by everybody. The collaboration produced a list of 26 memoirs, which they presented to the Academy on September 5.[158] On January 10, 1778, Bucquet presented an outline of ten memoirs to the Academy, after which he was elected as an adjoint chimiste. These included their studies on conduction of heat, passing reference to latent heat in vaporization, the temperature effect in mixing ice and oil of vitriol, the action of heat on minerals, and production of gases. Lavoisier's collaboration with Laplace also began in early 1777, focusing on the conditions under which various substances turned into vapors.[159]

At the heel of these intensive investigations on heat came Lavoisier's first, guarded attack on phlogiston. On November 12, 1777, at the autumnal rentrée publique of the Academy, Lavoisier read the "Mémoire sur la combustion en général."[160] It deserves a careful consideration here for its programmatic outlook. Although Lavoisier worded the presentation cautiously, calling his theory a hypothesis, probably as a courtesy to Macquer's leadership, the message was clear. He wished to propose "a new theory of combustion" that coordinated the known facts better. In Stahl's hypothesis, one had to suppose that the matter of fire or phlogiston existed in all combustible bodies, including metals and sulphur. Such a supposition led to a "vicious circle": "The combustible bodies contain the matter of fire because they burn and they burn because they contain the matter of fire." In other words, it was only a hypothesis that could be easily replaced by another of the same explanatory power:

The existence of the matter of fire or phlogiston in metals, sulphur, etc. is therefore really only a hypothesis, a supposition, which, once admitted, explains, it is true, some of the phenomena of calcination and combustion; but, if I show that these same phenomena can be explained in a manner as naturally with an opposing hypothesis, that is to say, without supposing that there exists either the matter of fire or phlogiston in the matters called combustible, the system of Stahl will be found shaken to its foundation.[161]

Lavoisier's alternative "hypothesis" is familiar to us. Combustion was a phenomenon in which bodies combined with the base of pure air, which caused the weight gain. Pure air, in turn, was "an igneous combination in which the matter of fire or light enters as dissolvant and in which another substance enters as base." The release of this matter of fire from pure air caused the heat and light produced in combustion reactions:

phosphorus +   pure air
            (base + matter of fire)

  =   phosphoric acid   +   heat or light.
      (phosphorus + base of pure air)   (matter of fire)

In his new theory of combustion, Lavoisier distanced himself somewhat from his earlier identification of fire with phlogiston. Yet his ontology of fire resembled Guyton's phlogiston. He defined the "matter of fire" as "a very subtle, very elastic fluid which surrounds all the parts of the planet we inhabit, which penetrates with more or less facility the bodies that compose it and which when it is free tends to be placed in equilibrium in everything." Just like Guyton's phlogiston, it was a "solvent of a great number of bodies," and it combined with them "in the same manner as water is combined with salts, as acids are combined with metals." Just like any other chemical combination, the "matter of fire" caused the bodies in combination with it to lose part of their properties in closer approximation of its own properties. All bodies could, therefore, exist in solid, liquid, or aeriform states, depending on their content of the matter of fire. In other words, Lavoisier's "matter of fire" resembled Guyton's "phlogiston" in the way it interacted with other substances. Both acted as a universal solvent and interacted with other substances via its affinity. This conception of fire, shared by Lavoisier and Guyton, later became an encompassing vision of the Chemical Revolution:

... to summarize, the air is composed, according to me, of the matter of fire as solvent, combined with a substance which serves as base and in some fashion neutralizes it; every time when one presents to this base a substance with which it has more affinity, it quits its solvent; from this, the matter of fire reclaims its rights, properties, and reappears before our eyes with heat, flame and light.[162]

The "Mémoire sur la combustion" contained all the essential elements of Lavoisier's synthesis, later seen in his *Traité*. Nevertheless, he cautioned that he was not trying to replace Stahl's doctrine by "a rigorously demonstrated theory, but only a hypothesis that seems to me more probable, more conforming to the laws of nature, which would appear to contain explications less forced and with less contradiction," a hypothesis that would explain all the principal phenomena of physics and chemistry in a quite simple manner. This posture of modesty was probably expected of a junior academician attacking the position maintained by a senior member. Throughout the paper, Lavoisier emphasized the ephemeral nature of systems in physical sciences, perhaps to soften the expected opposition from Macquer. Indeed, his assault on phlogiston met with a cool reception. Macquer acknowledged the role of dephlogisticated air in combustion reactions, but retained phlogiston as the matter of light. He thought that Lavoisier's uncharacterized "base" of pure air was just as elusive as his phlogiston.[163] The criteria of legitimate evidence divided Lavoisier and Macquer. Lavoisier had no problem with postulating the existence of a substance that he never laid his hands on, the base of pure air, because his algebra of weight and fire allocated a quantity to this unknown entity. At the same time, he thought it absurd that something could exist and not register its weight on the balance. This was precisely the argument from the laws of physics directed against Guyton's phlogiston earlier. Furthermore, Lavoisier expected that the matter of fire now dislocated to air would be fully subject to thermometric control. For Macquer, however, phlogiston was characterized by a series of chemical operations that provided it with a set of chemical properties even if it still eluded the balance. For another decade at least, chemists tried to locate this elusive substance with the balance. They had a brief success with the inflammable air or hydrogen, which further slowed the reception of Lavoisier's theory by other chemists.[164] Lavoisier fought the phlogiston theory not only with the balance but also with the logic of algebra.

Despite his circumspect rhetoric, Lavoisier was confident by 1777 that he had built a solid ground for his "hypothesis" against the Stahlian system. For this confidence, he had equipped a fine laboratory and enlisted two capable scientists to cover a vast terrain of chemistry and experimental physics. Now he could proceed to build a new system of chemistry. In 1777 he worked out many different aspects of his emerging system at a furious pace. The nature of acids was high on the list. In April he read a memoir in which he showed that phosphorus and sulphur absorbed "eminently respirable air" to become acids.[165] The list of memoirs he presented to the Academy in September included one on the action of vitriolic acid on mercury. While Lavoisier was working out the implications of his theory, the realm of acids began to expand rapidly. After the publication of Bergman's affinity table (December 1778),[166] Lavoisier turned to the composition of acids, studying the acid of sugar minutely.[167] He also repeated earlier research on vitriolic acid to confirm the existence of "eminently respirable air" in its composition.

In 1779, Lavoisier put all his research on acids together in a paper titled "Considérations générales sur la nature des acides." Noticeable in this memoir is his compassion for chemical art. No longer castigating the ambiguity of chemists' principles, he defined them as "the limits of chemical analysis." To the degree in which chemistry made progress, chemists would be able to decompose the substances now regarded as principles. The theory of neutral salts on which chemists had fixed their attention for more than a century was now "so perfected that one can regard it as the most certain and the most complete part of chemistry." The next step was further analysis of the constituent principles of neutral salts, or of acids and bases. Drawing on his earlier experiments on phosphoric, vitriolic, and nitric acids, Lavoisier stated confidently that Priestley's dephlogisticated air, or the "purest air, the eminently respirable air" was the "constitutive principle of acidity" common to all acids. To inject a degree of rigor in expression, he named dephlogisticated air in the state of combination and fixity as the "acidifying principle" or "oxygen principle." Air had become a "generic word' applied to all substances in the state of elasticity. This acidifying or oxygen principle combined with "the matter of fire, of heat, and of light" to form the purest air, or Priestley's dephlogisticated air. It produced fixed air [acide crayeux] with charcoal, vitriolic acid with sulphur, acid of niter with nitrous air, phosphoric acid with phosphorus, and metallic calces with metallic substances. As usual, Lavoisier envisioned a vast chemical territory to be conquered with his new theory—the degree of affinity the oxygen principle displayed toward other bodies and all the new compounds resulting from its combinations:

Here is nearly what the general knowledge acquired on the combination of the oxygen principle with different substances of nature has reached at this moment, and it is not difficult to see that there remains on this regard a vast field to conquer; that there exists a part of chemistry all new and entirely unknown until this day, and which will be completed only when one arrives at determining *the degree of affinity* of this principle with all the substances with which it is susceptible of being combined, and at knowing the different species of compound which result from them.[168]

Lavoisier made a remarkable transformation from an amateur physicien who despised notions such as rapport or analogy to a chemist who appreciated the importance of chemical affinity and the diversity and difficulty of chemical procedures. The simpler the substances under investigation were, he reckoned, the more difficult the means of decomposition would be. The operations involving acids were, therefore, considerably more difficult than those with neutral salts, and they demanded "a great variety in the means and the procedures" as well as an exact control of "the quality and the quantity" of acids.

6.10   Interactions

As the most ambitious junior member of the Academy, Lavoisier attracted much attention from other chemists. While historians have mostly discussed the reception of his ideas and systems by the contemporary chemists, it is important to recognize the other side of the equation—namely, how the responses from other chemists affected Lavoisier. We need a more symmetric account of the Chemical Revolution than has been offered so far, especially with regard to Lavoisier's interaction with other chemists.[169] While such a study is beyond the scope of this book, it would be useful to chronicle briefly Guyton's and Fourcroy's responses to Lavoisier's works as a prelude to their collaboration later on. Although Macquer cast a long shadow over the Parisian chemical scene, Guyton had cultivated an extensive network of correspondence that plugged him into various debates. Fourcroy emerged in the early 1780s as the leading chemical teacher in Paris. As such, they can help us identify major issues among the contemporary chemists.

Guyton received a copy of Lavoisier's *Opuscules* soon after its publication.[170] His concern is evident in the articles for the supplementary volumes of Diderot's *Encyclopédie*.[171] To write the article "air,"[172] he studied Boerhaave's *Traité du feu* and Hales's *Vegetable Staticks* as well as various articles in the *Observations*. He could thus provide a brief history of the opinions on the fixation of air by Boerhaave, Lavoisier, Newton, Boyle, Hales, Black, MacBride, Venel, and Priestley, perhaps relying in part on Lavoisier's *Opuscules*. Guyton argued that in order to be certain that air or "pure air" became fixed in bodies, one had to screen out the effects of other matters dissolved in the air. One would know the nature of this principle only when a series of further experiments could determine the system of "its proper & exclusive affinities." This was exactly what Lavoisier set out to do, as we shall see, once he identified oxygen gas as the air involved in a variety of chemical operations. In his article on calcination, Guyton retained his earlier position in *Digressions*,[173] reluctant to grant a chemical role to air. He insisted that air was necessary to combustion not as food but to sustain the oscillatory movement around the flame. He gave more thought to the contemporary debate in his article on phlogiston,[174] particularly in view of Bayen's work with mercury calx. While listing a number of objections to the fix-ation of air in calcination, he summarized succinctly the heart of the question:

One sees by what we have said that the science of chemistry presents nothing so difficult or important as this theory: all these difficulties are reduced neverthe-less to a single question which suspends at this moment the progress of our knowledge: *Is it addition, or is it subtraction of some matter which constitutes the state of calx after calcination?* M. Black attributes it to the absence of fixed air; M. Meyer, to the presence of a substance which he calls *acidum pingue or causticum.* Most physicists are occupied with the solution of this interesting problem. M. Lavoisier has published a good series of experiments on the exis-tence & the properties of the elastic fluid which is fixed, according to him, in metallic earths during their calcination; & we know that M. Macquer, to whom chemistry is already indebted for many discoveries, works to clarify this matter by developing the theory of causticity. It is necessary to hope that as such efforts excited by general interest, & directed toward the same goal, will give birth finally to a light sufficiently bright to dazzle all eyes, & bring back to the same route all those who apply themselves to the study of this part of natural sciences.[175]

Despite his oppositional stance, Guyton shared the algebraic concep-tion of chemical operations with Lavoisier. In 1775 he worked in Lavoisier's laboratory, where he gained enough skill to stage demon-strations on the manipulation of seven different kinds of air at the Dijon Academy. He did not accept Lavoisier's theory, however. He hoped that Macquer, "the father of good chemistry," would come up with a defin-itive resolution of the "great problem of fixed air and of acidum pingue" in his new edition of the *Dictionnaire.*[176]

In 1776, Guyton tried to reconcile the experiments on fixed air with the phlogiston theory.[177] He felt "alive to this inquietude" that the "modern experiments on fixed air" stirred up, which threatened to demolish rather than to extend Stahl's doctrine. He was less concerned with the necessity of phlogiston in calcination reactions, however, than with the integrity of his dissolution model of chemical action. He took it as an integral part of Stahl's theory that phlogiston acted on metallic earths as a "true dissolvant" via its affinity. If metals in calcination absorbed and appropriated a portion of air, "the natural consequences are that air is not only necessary to this operation as a mechanical agent, that the state of metallic calces is not due to the absence of phlogiston."

One could further conceive calcination as a union of fixed air and metallic earth. This meant, however, that a combination had to take place between an elastic fluid and a substance that remained "constantly under the powdery, grosser form, incapable of becoming homogeneous with a fluid," or "without equiponderance, without attraction of affinity, without dissolution." Such a violation of the "univocal march of nature" did not agree with him.[178] In other words, Lavoisier's theory violated the basic premise of Guyton's theory that chemical action required dissolution. Guyton suggested an ingenious interpretation to save his dissolution model of chemical action: Dry calcination was only an effect of true precipitation which offered a "marked analogy" with humid calcination, or the precipitation of metals from acids. Marine acid attacked lead and copper, but not silver. By dissolving silver first in nitrous acid, however, one could make it react with marine acid. The meant that "the silver reduced by some solvent into its elementary molecules can act differently & more efficaciously on the parts of marine acid than when it is in solid mass & combined with the matter of fire." The variety of "figures" changed the "sum of the reciprocal attraction which produces different degrees of affinity" by affecting the distance. Likewise, dry calcination involved "two essentially different fluids" with their proper affinities—phlogiston or the matter of fire and fixed air. The metallic earth dissolved in phlogiston could release phlogiston, seize fixed air, and produce a combination "by reason of a superior affinity." The new compound precipitated because it was less soluble. The affinity of fixed air with different metallic earths varied as that of all other solvents did. According to this "hypothesis," acids had to contain fixed air within themselves since they furnished it to metallic earths to produce metallic calces. One had to admit, then, only "the reciprocal & simultaneous affinities" to account for the involvement of fixed air in dissolutions, reductions, cementations, etc.

Guyton's two-fluid theory of calcination came remarkably close to Lavoisier's oxygen theory. According to Guyton's hypothesis, acids had to contain fixed air, since they furnished it to metallic earths to produce metallic calces. Acids did not touch metallic calces because both were neutralized by the same principle: fixed air. Metallic earths were, therefore, always united to phlogiston, fixed air, or acids. All these facts showed that pure metallic earth had not yet been obtained. This should not deter chemists from affirming "as before, with Stahl, that metals receive in fact their constituent form from the union of their earth with phlogiston." In chemistry, everything was made "by dissolution, according to the law of attraction, & these varieties of distances produced by the varieties of figures in which M. le comte de Buffon has found such a clear explication of affinities." Finally, "far from striking a blow to these truths," the experiments on fixed air made known "a new solvent whose action is subject to the same rules, whose verified existence has already rectified our conjectures on several important points, whose affinities & products can only enrich further chemistry & the arts which depend on it."[179]

Guyton's initial response to Lavoisier's pneumatic studies indicates that he was less concerned with the phlogiston per se than with the overall scheme of affinity chemistry in which the rival theories had to perform, namely, the dissolution model of chemical action determined by the interplay of affinities. He paid attention to Lavoisier's works in dealing with the topics of air, calcination, and phlogiston, but he maintained the dissolution model of chemical action intact in the articles on affinity, crystallization, and dissolution. He also interpreted these operations in an algebraic manner, pointing out the consequences of such thought process. Although he defended the phlogiston theory, his hypothesis did not differ much from Lavoisier's oxygen theory in its algebraic skeleton: Acids contained fixed air, which they imparted to metallic earths to produce metallic calces. Acids and metallic earths were neutralized by the same principle: fixed air. Guyton and Lavoisier came to the same algebraic explanation of calcination and acidification, that is, except for the identity of the air involved in these operations.

As Lavoisier's voice grew within the Academy, Guyton worried with Macquer about the unsettling consequences of Lavoisier's works for the phlogiston theory. Lavoisier apparently revealed his intention to attack the phlogiston theory in a letter to Guyton. Macquer feared that Lavoisier would soon reveal his "great discovery" which would overturn the phlogiston theory, although Lavoisier's "Mémoire sur la combustion en général" proved to be gently worded to mollify the opposition.[180] Macquer modified his phlogiston theory in the second edition of the *Dictionnaire* (1778) to address the objections earlier raised by Buffon and drew closer to Lavoisier's position shortly before his death in 1784.[181]

Fourcroy also took Lavoisier's works seriously. In the *Leçons*, he gave a concise history of chemistry divided into six epochs. After the first four epochs (Egypt, the Arabs, alchemy, and the pharmaceutical chemistry begun by Paracelsus), the modern period began with the philosophical chemistry practiced in the seventeenth century by Barner, Bohnius, and Beccher and continued in the eighteenth century by Stahl, Boerhaave, the Rouelle brothers, and Macquer. Pneumatic chemistry brought about the sixth and most recent epoch. The discovery of gas had become a fecund source of new discoveries well beyond the "sublime theory of Stahl," setting up a "great epoch in the history of chemistry."[182] While citing Boyle, Hales, and Priestley as precursors of this last epoch, Fourcroy praised Lavoisier's experimental proof of the fixation of air in calcination and combustion. This discovery gave rise to a class of chemists who began to doubt the presence of phlogiston and who attributed all the phenomena previously explained by phlogiston to the fixation or release of air. Lavoisier's doctrine had the "advantage of a more rigorous demonstration" over Stahl's. While the existence of phlogiston had "never been rigorously demonstrated," a great number of facts had already proved that all phenomena of chemistry could be explicated by the doctrine of gas without admitting phlogiston. In addition to Bucquet's favorable leanings toward the pneumatic doctrine, Fourcroy asserted, even Macquer was "quite convinced of the great revolution the new discoveries have to occasion in chemistry." Macquer thus replaced phlogiston with light, whose action and influence on the phenomena of chemistry would never be placed in doubt. Despite these endorsements, Fourcroy wisely limited himself to the "simple quality" of a historian who observed the relative force and probability of each doctrine in each case without judgment.[183]

7

A Community of Opinions

The landscape of French elite chemistry changed markedly in the mid 1780s, when Macquer's death left Guyton and Lavoisier in leadership positions. Guyton represented French chemistry abroad through his writings and correspondence; Lavoisier emerged as the master chemist of the Paris Academy. The Academy played an indispensable role in shaping the subsequent revolution in chemistry by enforcing a set of standards, by conferring authority on its elite chemists to forge a dominant representation of their science, and by providing an institutional context for intellectual exchange. Lavoisier supplemented its interdisciplinary setting of weekly meetings by hosting a kind of scientific salon at the Arsenal. Berthollet and Fourcroy, the junior authors of the Chemical Revolution, attended the Arsenal gatherings regularly along with Laplace, Monge, and Meusnier. Lavoisier freely acknowledged the role of this conversational culture in shaping his ideas.[1] The Arsenal Group, as "a community of opinions," provided an immediate audience and miniature tribunal for his new projects and theories[2]:

If at any time I have adopted, without acknowledgement, the experiments or the opinions of M. Berthollet, M. Fourcroy, M. de la Place, M. Monge, or, in general, any of those whose principles are the same with my own, it is owing to this circumstance, that frequent intercourse, and the habit of communicating our ideas, our observations, and our way of thinking to each other, has established between us *a sort of community of opinions*, in which it is often difficult for every one to know his own.[3]

The Arsenal Group came to control the most important chemical establishments around Paris after Macquer's death in 1784. In addition to Lavoisier's stewardship of the Gunpowder Administration, Berthollet was appointed to the royal dye factory and Fourcroy to the

professorship at the Jardin. Lavoisier also took visible measures to consolidate his symbolic leadership. He staged a public demonstration of the
large-scale decomposition of water in March of 1785, which occasioned
the "conversion" of Berthollet in April. Berthollet immediately began to
spread the gospel, paying particular attention to Guyton. A long-delayed
reading of Lavoisier's "Réflexions sur le phlogistique" during the
summer announced publicly the eclipse of the Macquerian paradigm.
With the reorganization of the Academy, Fourcroy joined its ranks. He
crossed over to the antiphlogistic camp the next year, completing a
unified front for the Arsenal Group. They also acquired a powerful ally
in the person of Guyton who brought in his extensive knowledge of
contemporary European chemistry as well as his vision of nomenclature reform.[4] The collaborative work of these four chemists—Guyton,
Lavoisier, Berthollet, and Fourcroy—was soon made public in the
Méthode de nomenclature chimique (1787), in their critique of Kirwan's
Essay on Phlogiston (1788), and in Lavoisier's *Traité élémentaire de
chimie* (1789). The "triumph of the antiphlogistians" is a familiar story.[5]
This public and collaborative phase consolidated the most enduring
monument of the Chemical Revolution: nomenclature reform. The coordinated efforts and the combined authority of these four chemists
authenticated the novelty of Lavoisier's achievements and the efficacy of
the new nomenclature.

In documenting this collaborative phase, historians have focused
mostly on the battle over the fate of phlogiston and the nomenclature
reform.[6] In doing so, they have missed the most comprehensive vision of
the Arsenal Group regarding chemical affinity and constitution. The conventional story revolves around the triumph of Lavoisier's antiphlogistic
campaign and the successive "conversion" of the other three authors of
the Revolution. While Lavoisier undoubtedly became the central figure
of the Paris Academy intellectually and administratively, the other three
chemists were by no means mere puppets who simply recited his opinion.
Berthollet possessed an intricate knowledge of the chemistry of salts,
while Fourcroy was an outstanding teacher who amassed an encyclopedic knowledge of chemistry. Guyton was situated at the center of
the European network of philosophical chemists, which enabled him to
discern main trends in chemical theory and practice. If the Arsenal Group

chemists deployed their communal resources to develop a new nomenclature and their combined authority to defend it against the phlogistic camp, a significant part of their communal opinion revolved around the notion of chemical affinity in relation to the problem of composition. Their divergent approaches gradually became focused on the interplay of heat and affinities in determining chemical constitution. Although the term "constitution" was introduced much later by Berthollet to designate the equilibrium of chemical and physical forces that constituted a chemical body, it captures best the theoretical vision of the Arsenal Group in the 1780s.

Instead of talking about "conversions," we should characterize the mid 1780s as a period of intensified interaction among the four authors of the nomenclature reform. On the one hand, they supported Lavoisier's antiphlogistic cause, which entailed not only accepting the oxygen theory but also transferring the role phlogiston used to play in phase changes to caloric. In other words, Lavoisier's system required an integration of heat as a part of chemical constitution. On the other hand, they sought to accommodate the new analytic frontier of European chemistry in the realm of acids and the corresponding development in affinity chemistry. Although Bergman's extended affinity table reinforced the decades-long interest in affinity, it also exposed the limitations of such a labor-intensive endeavor. A number of chemists took up Guyton's challenge of quantifying affinities, including Kirwan, Achard, Wenzel, Lavoisier, and Laplace. Guyton had retired from the Burgundy parlement to write the chemical volumes of the *Encyclopédie méthodique*. First on his agenda was the article on affinity, for which he surveyed all the relevant research. As Guyton put it, the moment had come to be occupied with theories. The rapprochement between Lavoisier and Guyton shaped the Arsenal Group's dynamic conception of chemical affinity and constitution, which guided French elite chemistry in the ensuing decades. Though Lavoisier's public campaign used phlogiston as a polemical focus, the Arsenal Group conceptualized the constitution of chemical bodies as an interplay of heat and affinities. Phlogiston merely became a necessary expenditure in stabilizing this new vision.[7] The concern to elaborate a comprehensive program of chemical affinity and constitution brought about the second theoretical moment in French chemistry.

7.1   The Affinities of Oxygen

In the 1780s, once the oxygen theory was securely in place conceptually and experimentally, Lavoisier turned his attention to two different genres of experiments with theoretical importance[8]: the precipitation reaction of metals and the measurement of chemical reaction heat.[9] These two genres of experiments, one chemical and the other physical, were probably linked in Lavoisier's outlook through the notion of affinity. Bergman's affinity table and the expanding territory of acids had gathered momentum for affinity chemistry. The presence of Guyton, Berthollet, Fourcroy, and Laplace in Lavoisier's close circle also helped steer his thoughts in this direction. Affinity was on Lavoisier's mind, once he had a stable systematics of chemistry. His works on affinity did not come to fruition, unlike his earlier projects, in part owing to their complicated nature. The onset of the French Revolution and his premature death at the guillotine during the Reign of Terror did not help the matter.

Metallic dissolution in acids was the prototypical chemical operation that helped shape the notion of affinity at the beginning of the century. Lavoisier dealt with these reactions in working against Guyton's phlogiston theory via humid calcination, but he had to leave them out of the *Opuscules* because of the "necessity of first investigating the nature of the acids themselves."[10] In the early 1780s, with the oxygen theory of acids in order, he returned to this problem to work out the affinity relations of the oxygen principle. As Guyton had argued in response to Lavoisier's *Opuscules*, one could grasp the nature of the principle fixed in bodies only by determining the system of "its proper & exclusive affinities."[11] In the "Considération générales sur la dissolution des métaux dans les acides," Lavoisier reconceptualized the humid calcination of metals as an action involving oxygen.[12] Except for the identity of the air, his theory was remarkably similar to Guyton's two-fluid theory of calcination, which held that dry and humid calcinations both involved fixation of air. He supposed that metals had a "great affinity" for the oxygen principle, as was exhibited in the facility of their calcination at mild heat. Metallic calcination "truly decomposed" air, since its oxygen principle left "the principle of heat, the igneous fluid" behind to join the metallic substance, for which it had a greater affinity. The igneous fluid became free, sometimes producing flame and light, while the metal gained a considerable amount of weight. A similar process took place in the humid calcination or the dissolution of metals in acids: Metals united with the oxygen principle by decomposing the acid or water and gained nearly as much weight as in dry calcination. Although Baumé, Macquer, Bergman, and Fourcroy had stated this fact, Lavoisier claimed, they had not identified it as calcination, nor had they understood that the metal became "saturated with the oxygen principle" in this operation. He wished to establish this truth of humid calcination and to offer some general considerations on the affinity relations of oxygen with metals. Lavoisier now made an explicit link, that is, between humid calcination and acidification.

Lavoisier deployed the oxygen theory in this manner to reinterpret one of the most difficult and most intricate categories of chemical action then known: the selective dissolution of metals in acids. Despite his confidence in the oxygen theory, he knew that he would have to account for the affinity relations of oxygen in order to explain these operations. At the same time, he held the "matter of heat or the igneous fluid," rather than "the matter of fire," to be responsible for the aeriform state, thereby eliminating the possible involvement of phlogiston. The proof of his theory required, of course, quantification. Lavoisier knew from his earlier research on the decomposition of nitrous acid into oxygen and nitrogen that, if he removed a part of oxygen from nitrous acid, a corresponding quantity of nitrogen had to be liberated. The dissolution of metals in nitrous acid should thus produce nitrogen gas while effecting the humid calcination of metals. He would prove this fact by providing a grid of quantitative measurements:

Supposing that this proof was susceptible of being attacked or weakened, it is easy to add others to support it. In fact, if I prove that in the metallic dissolutions I have cited nitrous acid loses a portion of its oxygen principle, and that there is a decomposition of acid proportional to this quantity; if I show subsequently that what is found less in the acid is found more in the metal; that it augments weight in a quantity equal to what nitrous acid loses, it will prove that the metal is calcined at the expense of the acid; finally, if I arrive at proving that this principle removed from the acid and united to the metal is the oxygen principle, I would have proven that the calcination in the humid way, which is operated at the time of the dissolution of metals in acids, is absolutely analogous to the one which is operated in dry way.[13]

The algebra of weight and heat for the humid calcination of metals was much more complex, however, than the algebra for the dry calcination. In order to form a "just idea" of the operation, Lavoisier argued, one had to know the involvement of the oxygen principle as well as the compound nature of water and acids. There operated in metallic dissolutions "a great number of forces" with different energies. In order to unpack this complicated interplay of forces, he proposed a quasi-algebraic formula utilizing the symbols used in the affinity tables. Since chemistry was far from attaining the necessary "mathematical precision," he cautioned, these formulas should be regarded as "simple annotations whose object is to aid the operations of the mind."[14] Lavoisier's method was similar to the self-consistent method. Using established symbols for each component, he assigned a proportionality constant to each combining pair to express their respective quantities, calculated numerical values of these constants from the results of his experiments, and then tested the general applicability of these numbers by applying the equation to other sets of experiments. He claimed that with many repetitions he always found "a nearly perfect agreement between the result of experiment and that of calculation," and that the formula he constructed represented "with exactitude" what took place in the dissolution of iron in nitrous acid. Although he would need an expression of heat in order to perfect this formula, he thought that such a move would produce a formula "too complicated" and introduce "a geometry too advanced, to which it [chemistry] is not yet susceptible."[15] Lavoisier's model of dissolution process involved all pairwise interactions between its constituents:

1) The action of heat, which tends to separate the molecules of water and reduce it to vapor
2) The action of this same heat, which tends to disunite the principles of nitrous acid, and convert it into gaseous substance.
3) The action of this same heat on the constitutive principles of water
4) The action of heat, which diminishes the affinity of aggregation of metal, and tends to separate it into parts
5) The reciprocal action of nitrous gas and the oxygen principle
6) Their combined action on water
7) The action of metal on the oxygen principle of acid and on that of water
8) The action of acid on the metal, or rather on the metallic calx.[16]

Calculating the energies of these forces constituted a lofty goal of chemistry. Although it would progress slowly, Lavoisier did not regard it as an impossible task. The "lesson" of his memoir was "to foresee the possibility of applying the exactitude of calculus to chemistry." In order to achieve this feat, he would have to begin with certain data, such as the exact quantities of the substances involved and their proportions. He also tried to establish an algebraic equation for the precipitation of a metal by another, drawing on Bergman's experiments.[17]

Lavoisier's subsequent article, "Mémoire sur l'affinité du principe oxygine," offers a rare glimpse of his thoughts on affinity tables. The affinities of oxygen lay at the center of his quest. He began the memoir by enumerating all the reactions he brought under the rubric of oxygen theory: The oxygen principle produced vital air in combination with the matter of heat, vitriolic acid with sulphur, nitrous acid with nitrous air, saccharin acid with sugar, phosphoric acid with phosphorus, carbonic acid with charcoal, and water with aqueous inflammable air. He then asked:

But in which order are found all these combinations? What are *the degrees of affinity* that the oxygen principle has with these different substances? According to which laws are they excluded or precipitated? This is the object that I propose to examine in this memoir; *the one that I have had in view from the beginning of the work that I have explained to the Academy; the one for which I have never ceased to assemble the material for several years.* [18]

Before answering the question, Lavoisier first offered a detailed critique of the affinity table. First, the tables of affinity presented only the results of simple affinities, although there existed in nature double, triple, and more complicated affinities. Lavoisier's critique derived directly from his conception that heat functioned as a universal solvent in chemical reactions. In order to form "precise ideas" of chemical action, he deemed it necessary to present all bodies in nature as being immersed in "an elastic fluid quite rare, quite light, known under the name of igneous fluid, of principle of heat." This fluid tended continually to separate the parts of bodies, but was opposed by the attraction between them, or the "affinity of aggregation." The chemical action between two bodies thus depended on the degree of heat contained in each of them. Bergman had introduced the affinity table in two parts, one for the humid way and the other for the dry way, but Lavoisier perceived that "in order to obtain the tables rigorously according to the experiment, it would be necessary, so to say, to form one for each degree of the thermometer." Second, the affinity table did not account for the effect of the attraction of water in humid analysis. Chemists regarded water "as a simply passive agent, while it acts with a real and perturbing force"—a force that affected the results of experiments. Third, the table could not express the variations in the attractive force of the molecules of bodies with the different degrees of saturation. For example, the combination of oxygen and sulfur gave rise to two distinct acids, sulfuric and sulfurous, without any intermediary acids.

Lavoisier's critique of the affinity table reveals an important connection to Guyton's dissolution model of chemical action. In Guyton's system, phlogiston or fire as a universal solvent became fixed in many bodies as the "fire of crystallization," just as water became fixed in salts as the "water of crystallization." The interaction of bodies necessarily involved the affinities of solvents—phlogiston and water. Lavoisier maintained this ontology and this dissolutive function of fire, as he had done in his "Mémoire sur la combustion en général," although he now transferred it to the igneous fluid or the principle of heat. If Guyton's phlogiston and Lavoisier's "principle of heat" shared the same characteristics as a universal solvent for chemical operations and as a material component of chemical combination, they differed in one important aspect: Guyton's phlogiston was not subject to a quantitative control in the way Lavoisier's principle of heat was.

What is remarkable in Lavoisier's discussion of the affinity table is not the usual acuity of his criticisms but the positive light he placed on the pursuit of affinity. Lavoisier did not condemn affinity as he had phlogiston. He found the existing tables inadequate but not useless. In fact, he prepared one column of affinity table with the oxygen principle at the top, constructed in the same manner as Geoffroy's, although he thought it "superfluous" to add it to the existing tables of affinities. His long-term vision consisted not in extending the conventional table, but in inventing a calculus of affinities:

What I have said against the tables of affinities in general naturally applies to what I am going to present; but in doing so I do not think less that they could be of some utility, at least until the time when the experiments are more multiplied and when the application of calculation to chemistry places us in a state of carrying our views much further. Perhaps, one day, the precision of data will be brought to the point where a geometrician can calculate in his cabinet the phenomena of any chemical combination in the same manner that he calculates the movement of celestial bodies. The views that M. de Laplace has on this subject and the experiments that we have projected according to his ideas to express in numbers the force of affinities of different bodies, already permit us to regard this hope as not absolutely a chimera.[19]

7.2   The "Memoir on Heat"

In parallel with his experiments and reflections on affinities, Lavoisier pursued the quantification of reaction heats. Although he had been acquainted with Black's work on latent heat in the early 1770s, he became aware of the works on specific heats only after 1780. Just as British pneumatic chemistry had made a splash in the Parisian scene a decade earlier, the studies of heat by various British chemists caused a stir in Parisian science during the early 1780s.[20] Magellan's highly laudatory *Essai sur la nouvelle théorie du feu élémentaire, et de la chaleur des corps* (1780) introduced Adair Crawford's *Experiments and Observations on Animal Heat* (1779) to the French audience.[21] Magellan's essay also included the first table of specific heats supplied by Richard Kirwan.[22] By calculating the variation in the specific gravity of chemical substances according to temperature, Kirwan brought to chemists' attention the need to control temperature in the exact quantitative control of chemical operations. Phlogiston complicated the relationship between specific gravities and specific heats because it was not subject to the quantitative control of the balance or the thermometer. Lavoisier soon unraveled this tangled web of phlogiston, specific gravity, and specific heat by radically delimiting the domain of the balance in chemical operations.

In the summer of 1781, Lavoisier resumed his collaboration with Laplace. They examined in succession the dilatability and the latent and specific heats of various substances, producing their "Memoir on Heat" in 1783.[23] They must have surveyed the relevant literature, since the memoir offered a succinct overview of the field. Their immediate objective was to control quantitatively the distribution and the flow of heat in a system of bodies by introducing a new "method" of quantifying heat: the ice calorimeter. In particular, they wanted to relate the consideration of reaction heats to the existing fields of heat studies: latent heat, specific heat, and absolute zero.[24] In fact, the inclusion of reaction heats in the total distribution of heat marked their unique contribution to the field. Their effort in this regard probably stemmed from their desire to quantify affinities. In the memoir on the affinities of oxygen, Lavoisier spoke hopefully of Laplace's work in this direction. Although his collaboration with Laplace did not yield concrete data, Lavoisier offered some thoughts on the molecular level.[25] Since the state of a body was determined by the equilibrium between the affinity of its particles and the heat which tended to separate them, the study of thermal equilibrium would lead to a "very precise means of comparing these affinities."[26]

In the "Memoir on Heat," Lavoisier and Laplace had two objectives: to establish chemical reaction heat as a part of the total distribution of heat in a system of bodies and to integrate heat as a part of chemical constitution. In order to accomplish this mutual integration, they needed an instrument to quantify the reaction heat. Lavoisier wanted to disprove the existence of phlogiston, which had become a superfluous and cumbersome entity in his algebra of weight and heat. He could account for the production of heat and light in dry calcinations by the principle of heat contained in oxygen gas. Unable to isolate the "base" of oxygen gas, however, he could not quantify the latent heat of oxygen gas to prove that all the heat produced in dry calcination came from oxygen. If metals also contained "the matter of fire or phlogiston," it would further compromise his ability to control the operation quantitatively. In humid calcinations, Lavoisier needed a means of quantifying the affinities of oxygen. Chemical reaction heats seemed to offer a way, but they were not a routine part of thermometric measurements. For this reason, the most important mission of the "Memoir on Heat" was to offer a new method of quantifying heat to supplant the existing "method of mixtures." First and foremost, the memoir promoted the ice calorimeter—the "machine," as Lavoisier and Laplace called it at the time.[27] Though imperfect, the new method could establish chemical reaction heat as an integral part of heat distribution.

The concern with the heat changes accompanying chemical reactions comes through clearly in all four sections of the "Memoir on Heat."[28] The first section, "Description of a new means of measuring heat," started with a concise statement of the problem—how to treat heat quantitatively. Thermometers had improved so much that "there remains nothing more to wish for" in the measurement of temperatures or changes in volume. The distribution and the flow of heat in a system of bodies were not yet within quantitative control, however, especially when the system experienced chemical reactions. Lavoisier and Laplace sought to mobilize the existing field of thermometric investigation to this end, examining basic concepts and methods of measurement. Their knowledge of the field is borne out by their succinct summary, in a single paragraph, of the three major lines of works related to heat: latent heat, specific heat, and absolute zero. For the theory of latent heat, they did not decide in favor of either the chemical or the physical theory, espoused respectively by Black and Irvine: "In the change from the solid to the fluid state, and from the fluid to the vapor state, a large quantity of heat is absorbed, either because it is combined in this transformation, or because the capacity of the substance to contain it increases." Regarding the measurements of specific heats and their relation to absolute heat (which would require an estimate of absolute zero), they felt that "all these estimates, although very ingenious, are based on hypotheses which must be confirmed by a large number of experiments."[29]

The measurement, rather than the ontology, of heat occupied Lavoisier and Laplace's attention. Although the notion of free heat had originated from the conception of heat as "a fluid diffused throughout nature, and by which bodies are more or less permeated according to their temperature and to their special faculty of retaining it," they deemed it equally possible to conceive heat as "the result of the imperceptible motions of the constituent particles of matter" or as the vis viva of the particles of a body.[30] Whichever hypothesis they followed, the free heat had to remain always the same in a simple mixture of bodies.[31] However, nothing in either of the hypotheses guaranteed a priori the truth of this conservation principle when there was a chemical combination. Nevertheless, Lavoisier and Laplace argued that the quantity of free heat should be the same once the system was restored to its original state:

"All changes in heat, whether real or apparent, suffered by a system of bodies during a change of state recur in the opposite sense when the system returns to its original state."[32] The detonation of saltpeter supplied them with a visible proof.

Lavoisier and Laplace defined heat capacities or specific heats as "the relative quantities of heat necessary to raise the temperature of equal masses the same number of degrees." Such ratios should vary with the temperature, but could be taken as constant between the freezing and boiling points of water. The method of mixtures assumed that the heat lost by one body in mixture had to be equal to the heat gained by the other body. If one took equal masses of two bodies at different temperatures and mixed them, therefore, the ratio of their specific heats would be inversely proportional to their respective temperature changes. The neat mathematical form given to the method of mixtures and the abstract language of mass rather than weight show Laplace's mathematical mind at work on the problem. Kirwan, in comparison, had borrowed a simple method from metallurgists for calculating specific gravities. Lavoisier and Laplace criticized the method of mixtures for several reasons. It was difficult to attain thorough mixing, corrections were necessary for the heat lost to the vessels and to the atmosphere, and substances that combined chemically could not be compared directly. Above all, they argued, the method of mixtures could not be used to measure the heats involved in chemical combinations such as combustion and respiration, which constituted "the most interesting part of the theory of heat." This defect had prompted them to devise "a new method capable of measuring them with precision."[33]

The principle and the function of the "machine" Laplace invented are well known. It is easy to detect his mathematical mind at work in the description of the principle and the function of the instrument.[34] A solid or liquid body at a certain temperature was brought into a hollow sphere of ice and brought to zero (the temperature of the surrounding ice, which served as an insulator). A number proportional to its specific heat was obtained by measuring the quantity of water thus obtained from melted ice and dividing it by the mass of the body and the number of degrees above zero at which the body originally was brought into the machine. The quantities of heat evolved in chemical combination, combustion, and respiration were measured in the same manner. Measuring the specific heats of gases required a variation in the procedure because of their low density. Lavoisier and Laplace let a stream of gas pass through the ice chamber equipped with a thermometer at the entrance and another at the exit. They then measured the number of degrees by which the gas was cooled during its transit. By using a considerable mass of the gas, they could determine its specific heat "with precision."

In the second section, "Experiments on heat carried out by the preceding method," Lavoisier and Laplace offered a table of specific heats obtained by their new machine, taking water as the standard (a common procedure). Before preparing the table, they estimated that "the heat necessary to melt ice is equal to three quarters of the heat that can raise the same weight of water from the temperature of melting ice to that of boiling water." Their estimate was presumed to be "independent of the arbitrary divisions of the weights and of the thermometer."[35] They planned to extend the table to include a greater number of substances. They also wanted to explore the relationship between the specific heats and the specific gravities of bodies, a project somewhat reminiscent of Kirwan's earlier works.[36]

The third section of the memoir, "Reflections on the theory of heat," summarized lucidly what Lavoisier and Laplace sought to accomplish with their machine: exact quantitative control of the distribution and the flow of heat in a system of bodies. In order "to frame a complete theory of heat," four different kinds of measurement were necessary: a linear thermometer,[37] the specific heats of bodies as a function of temperature, the absolute quantities of heat contained in bodies at a given temperature, and the quantities of heat evolved or absorbed in chemical combinations or decompositions. This is in fact an excellent summary of the directions in which the thermometric investigation of heat had proceeded until then, except for the last item, which Lavoisier and Laplace added. They could not measure all these quantities directly, however, as they readily admitted. Particularly problematic was the relationship between the thermometer readings and the absolute quantities of heat. The assumption that the ratio of absolute heats was proportional to the ratio of specific heats was "very uncertain" and would require many experiments for confirmation. Specific heats only indicated the difference of heat between two bodies. The portion of the heat common to both bodies remained unknown. In other words, the absolute heat of bodies was beyond experimental control without knowledge of absolute zero.

Lavoisier and Laplace surmised that chemical reaction heats might give a better indication of the absolute heat contained in a body, because "the heat given off in chemical reactions is not the result of an unequal temperature of the substances that react together." To test this assumption, they derived a mathematical equation, "a very simple expression for the number of degrees corresponding to the heat of water at zero."[38] In order to convert this equation to concrete numbers, however, two further assumptions were necessary: that the quantity of free heat was always the same before and after the reactions and that "the specific heats of bodies express the ratio of their absolute quantities of heat, or, which amounts to the same thing, that the increments of their heat, corresponding to equal increments of temperature, are proportional to the absolute quantities of heat." Using these two hypotheses, Lavoisier and Laplace calculated concrete values, which gave only poor agreement. They had to conclude that "the knowledge of the specific heats of substances and of their reaction products cannot consequently predict the heat they should develop when they combine; only experiment can enlighten us in this matter."[39] Aside from chemical combinations, the changes of state also involved the heat that was not detected by the thermometer. Since the heat involved in this process could be estimated only through the measurement of specific heats, it could not be determined independently. Despite their ambitious agenda and clear vision, therefore, Lavoisier and Laplace could not formulate a precise relationship between the absolute heats and the specific heats of bodies. Even so, they succeeded in the more immediate task of inserting chemical reaction heats as a part of the total distribution of heat. The last section of the memoir, "On combustion and respiration," starts with a brief synopsis of Lavoisier's oxygen theory as it was presented in the 1777 memoir in which he argued that the heat and light given off in combustion were due in great part to the changes in the oxygen gas. He contrasted his theory from Crawford's and wished to prove it via measurements of combustion heats and experiments on respiration.

Lavoisier and Laplace's work with the calorimeter marked a significant turning point in the Chemical Revolution. Although their instrument was cumbersome and found only limited usage,[40] their work laid out clearly the directions of future research to formulate a theory of heat for chemical combination. Guerlac has characterized their work as an effort to transform chemistry into a branch of physics, but it defies such a reductionistic explanation. The calorimeter allowed Lavoisier and Laplace to conceptualize heat in divergent ways, mediating between the two subcultures of science.[41] Their characterization of chemical reaction heat as a part of the total distribution of heat in a system of bodies served to establish the importance of chemical reactions and combinations in the studies of heat and to affirm the involvement of heat in the constitution of chemical bodies. Lavoisier actually moved toward chemistry with his work on calorimeter. Though his trans-disciplinary perspective undoubtedly favored a representation of nature that transcended the instrumental and disciplinary boundaries, he made significant concession to chemists' theories. Although he sought to quantify fire through the thermometric measurements of heat, he retained the chemical ontology of fire as a universal solvent with his "principle of heat," later termed caloric. Having mocked chemists' affinities or analogies in his youth, he came to appreciate the importance of chemical affinity by the 1780s. He sought to detail the affinity relations of oxygen as the basis of his new chemical system, supplemented by a quantitative measurement of affinities. Through his collaboration with Laplace, he sought to conceptualize the constitution of bodies as an interplay of heat and affinities. This new conception of chemical "constitution" (as Berthollet later generalized it) became the main legacy of the Chemical Revolution for French elite chemistry.

7.3   The Affinity of Chemical Operations

The rise of the affinity paradigm in French chemistry can also be discerned from a textbook written by Antoine François de Fourcroy (1755–1809).[42] Born to an apothecary once in the service of the Duc d'Orléans, he was discovered by Félix Vicq d'Azyr (1748–1794), then at the height of his career as an anatomist, who helped him to enroll in the

Faculté de Médecine. In the Faculté, he learned chemistry from J. B. M. Bucquet, who was involved in Lavoisier's systematic inquiry to overhaul chemical theories.[43] He also attended the private courses of A. L. Brongniart (1742–1804), a demonstrator of chemistry at the Jardin. During the winter of 1778–89, Fourcroy opened a private lecture course in Bucquet's laboratory at the Rue Jacob. After Bucquet's death in 1780, he purchased Bucquet's library and apparatus and set up a laboratory at the Parvis Notre Dame, where he lectured in the next three years. Fourcroy's close relationship with Bucquet partially explains his early participation in the Arsenal gatherings well before he was elected to the Academy as well as his friendly posture toward Lavoisier's pneumatic chemistry. Graduating a few months after Bucquet's death, Fourcroy may have aspired to win the professorship at the Faculté, but his failure to obtain the title of docteur régent blocked him from any future career within this medical establishment.[44] Ostensibly building on Bucquet's plan, Fourcroy soon published a course of 70 lectures in chemistry and natural history, *Leçons élémentaires d'histoire naturelle et de chimie*.[45] After the appointment to Macquer's post at the Jardin, teaching positions soon found their way to Fourcroy, who excelled in the job:

> [He had a] logical method, the ability to extemporize, and such precision and elegance in his use of words that it was as if they had been chosen only after long consideration. And yet they were spoken with such liveliness, brilliance and novelty that they seemed to be spontaneous. He had a flexible, sonorous, silvery voice that lent itself to his movements and penetrated into the innermost recesses of the largest auditorium. Nature had given him everything.[46]

Fourcroy introduced a significant innovation in the French didactic tradition by fleshing out the operational foundation of theoretical chemistry. His definition of chemistry as "a science which tries to discern the intimate & reciprocal action of all the bodies of nature on one another" was completely new, indicating his emphasis on operations and affinities rather than on principles.[47] He also discussed analysis and synthesis only as the "means" of chemistry. "True or simple analysis" produced unaltered principles; "false or complicated analysis" altered them. Synthesis or combination united different bodies by utilizing "a force & a tendency" that existed between them. Although analysis and synthesis were often found united in the operations of chemistry, true analysis was quite rare. Fourcroy regarded synthesis as the more useful chemical tool, designating chemistry as the "science of synthesis or of combination, rather than that of analysis." In other words, he favored the operations (or affinity) approach rather than the philosophical (or principalist) approach to composition. He explicitly rejected what he characterized as the Stahlian tenet that a compound always partook of the properties of its constituent principles. Chemistry for Fourcroy was no longer an effort to distill substances to get at their constituent principles; it was a science of synthesis that utilized the actions of affinities to make new compounds. He deemed it "absolutely necessary to know well the laws and phenomena of this great property [affinity] before examining the reciprocal action of all natural bodies toward each other."[48]

Fourcroy's textbook testifies to the utility of the affinity paradigm in organizing the expanding repertoire of chemical operations. Although he paid lip service to the necessary unity of force in the universe, Fourcroy did not think that "true affinity" was the same as Newtonian attraction. What chemists called affinity, rapport, or attraction could take place between bodies of similar nature or between bodies of dissimilar nature. The first affinity was the affinity of aggregation, which united the molecules of the same nature or the "integrant parts" to augment the volume and the mass and to change the form. By measuring the effort necessary to separate the parts of an aggregate, one could measure the degree of adherence or the affinity of aggregation between them. The second affinity, the affinity of composition or combination, constituted "the greatest and the most sublime" knowledge in chemistry. In negotiating this complex field of affinities, Fourcroy vowed to pay attention to all facts, separating them from the hypotheses regarding the cause of affinity, and to formulate empirical laws.

Fourcroy further elaborated the difference of affinity from Newtonian attraction by stipulating ten "constant and invariable" laws of the affinity of composition. First, it took place only between the bodies of different nature, since chemical combination entailed a change in the essential properties of bodies. Second, it took place only at the point of contact, which distinguished the affinity of composition from Newtonian attraction. The more these points of contact were multiplied, the stronger was the affinity and the more exact was the combination.

Fourcroy advised physiciens not to confound chemical affinity and Newtonian attraction:

... this law distinguishes singularly the affinity of combination from Newtonian attraction, because this latter takes place only at quite great distances, & because it is changed even into a force quite opposed, repulsion, when large bodies on which it acts are placed at a certain distance from one another. There is therefore a quite great difference between these two forces, & one does not have to confound them, as several *physiciens* of our day seem to want to do.[49]

Third, the affinity of composition took place only between small bodies. Chemical subjects differed from physical subjects or bodies whose exterior properties such as mass, volume, surface, extension, and figure could be submitted to calculation and observation. Physical subjects were aggregates of which physiciens could make an observation and compare the qualities. Chemical subjects lost their aggregation and consequently no longer offered physical properties of aggregates to the senses. These were molecules so delicate and so tenuous that one could no longer measure their extension, figure, or volume. Only when the bodies were reduced to this degree of fineness by different ancillary operations did they obey the affinity of combination. This force seemed to reside in the infinitely small parts and belonged to the last elements of bodies. Again, this showed how the affinity of combination differed from attraction between large masses. Though seductive to imaginative minds, Fourcroy asserted, the attraction modified by the figure and the surface could not explain the affinity of combination.

Newtonian attraction could not provide a comprehensive systematics of chemical operations. Fourcroy's fourth law was concerned with affinity among several bodies, or "complicated" affinity, which could not assume the mathematical form of attraction. Affinity, as an active metaphor, possessed greater heuristic and organizing power than attraction, in part because of its less-than-precise meaning. Chemical combination usually yielded a compound with a completely new set of properties, as his eighth law stipulated, not simply the properties of its principles. Seemingly innocuous, this law reveals mostly clearly Fourcroy's deep-seated conviction that affinity, not the principles, determined the composition of bodies.

Fluidity was essential to the operation of affinity. Fourcroy's fifth law stated that at least one of the reacting bodies had to be fluid in order for the affinity of composition to take place. The more fluid the bodies were, the weaker were their forces of aggregation so that they could unite more easily and intimately. Two gases in contact thus gave the strongest compound. The affinity of composition was, therefore, in an inverse ratio of the affinity of aggregation, as Fourcroy's sixth law stated. That is, the weaker the aggregation was, the stronger was the affinity of combination. Heat as the cause of fluidity was necessarily implicated in the play of affinities, as Fourcroy's seventh law laid out. When two or more bodies were united by the affinity of composition, he noted, their temperature changed at the instant of their union. According to Baumé's observations, this constant phenomenon seemed to depend on the change of the aggregation of the substances involved (i.e., their passage from the state of solidity to that of liquidity, or vice versa). Since the change of aggregation itself depended on the action of the affinity of combination, Fourcroy claimed, it was evident that affinity changed the temperature.

Affinity tables were an integral part of Fourcroy's discourse on affinity. His tenth law acknowledged that bodies had different degrees of affinity, which effected separations and decompositions of seemingly miraculous appearances. He provided a detailed classification and description of precipitates as a prelude to introducing Geoffroy's table, "the ingenious means of offering at a glance the phenomena of the most constant precipitations."[50] Fourcroy offered a rather different method, however, concerning the determination of the orders of affinity. His ninth law stated that the affinity of composition should be measured by the difficulty one experienced in destroying the combination:

... the more perfect the compounds are, the more difficult it is to separate their principles and to destroy their composition. The degrees of difficulty one will experience in this separation could indicate therefore the degree of adherence or affinity that exists between such and such bodies.[51]

As we have seen, this was one of the options Limbourg had considered. From his experience, Fourcroy believed that the vivacity of combination, far from indicating a perfect composition, often gave birth only to a quite imperfect compound. "In order to fix in an exact manner the degree of affinity with which bodies are united and remain united," he contended, "it is necessary therefore to have recourse to the measure of the difficulty one experiences to separate them or to decompose their union."[52]

Fourcroy's discussion of affinity at the level of phenomena and laws, rather than causes, exhibits his understanding of chemical practice. Unlike Guyton, who preached that affinity had to be the same as Newtonian attraction because nature is simple and uniform, Fourcroy insisted that the laws of affinity were quite different from those of attraction. Attraction worked between large bodies and at long distances; affinity worked between small bodies at near-contact distances. Since the present state of knowledge did not allow chemists to determine the cause of affinity, Fourcroy deemed it much more useful to examine the phenomena and to multiply the laws. He regarded philosophical speculation as a hazardous and misleading exercise. While distancing the theoretical discourse of affinity from the philosophical discourse of attraction, Fourcroy mapped chemical operations with his laws.

Fourcroy's laws reflected various modifications that affinity discourse had undergone to accommodate the expanding repertoire of chemical operations. The "affinity of intermediate" referred to the case in which an intermediate body was required to effect an action between two bodies. Oil, for example, did not unite with water until it was combined with a salt to produce a soap soluble in water. Fourcroy regarded the solubility of soap as a newly acquired property of the compound, which he stipulated in the eighth law. He also had to take stock of anomalies and the effects of physical conditions. Following Bergman's example, he attributed the cases that seemed to contradict the "constant and invariable" laws of affinity to "some circumstances capable of modifying them such as the quantity of matter, the temperature of the atmosphere, movement or rest, the dissolution by water or by fire, that is to say, the humid or the dry way, the particular state of aggregation of each body, &c."[53] The "reciprocal affinity" referred to the case in which a displacement reaction could be reversed, seemingly violating the determinate laws of affinity. For example, vitriolic acid had more affinity with fixed alkali than nitrous acid. It decomposed the union of this alkali with nitrous acid. But nitrous acid, in turn, could separate vitriolic acid from the alkali. Fourcroy attributed this reciprocity of effect to the heat involved in reactions, which altered the ordinary laws of affinity. Berthollet later took up this concern with the physical circumstances that affected the laws of affinity.

In the introductory part of his *Leçons*, Fourcroy discussed affinity, principles, heat, air, water, and earth. Aside from the generalities (which included the definition, the object, the means, the utility, and the history of chemistry), chemical affinity and principles and elements were the main theoretical subjects. Much as he had done for affinity, Fourcroy discussed the principles of chemistry in a more practical vein. He defined principles as "the beings more or less simple which are proper to forming all bodies and which are separated in their analysis."[54] The first analysis of compounds gave "proximate, or secondary principles" [principes prochains, principiés, ou secondaires], which were themselves compounded. Diverse matters obtained in distillation, such as fluid or concrete salts, oils, and gases, were proximate principles formed of other principles. Upon further analysis, oil produced water, air, earth, etc. The last analysis separated out the "removed or primitive principles" [principes éloignés, principians ou primitifs], which formed proximate principles by their union. These primitive principles received the name "elements" because they could no longer be decomposed and because they made up nearly all bodies in nature. In practice, however, chemistry did not live up to this analytic definition of elements:

If one has to call element everything that cannot be decomposed, we observe 1. that the number of these beings is quite large because metals, sulphur, &c. appear to be no more susceptible of decomposition than the four elements properly said, ... 2. that the four bodies named elements are not all such, because the atmospheric air can be decomposed into several gaseous substances; 3. that earth offers diverse species quite different from one another, & nevertheless equally simple.[55]

Nor could chemists trace in their practice the principles, mixts, compounds, supercompounds, decompounds, and superdecompounds the Stahlian hierarchy of composition stipulated. Fourcroy charged that this hierarchy, based on chemists' belief that primitive principles could unite in different numbers and states, existed "more in their imagination than in nature."[56]

Despite his pragmatism, Fourcroy discussed the four "elementary bodies" in some detail, especially fire. He accepted the ontology of fire as a combustible matter that radiated light and heat. Physiciens defined it as a quite mobile, active, penetrating fluid formed of tenuous and durable particles endowed with continual movement, but no one had been able to concentrate this fluid and examine its properties. Fourcroy discussed the triple effects of fire—light, heat, and rarefaction—at the phenomenal level. Heat diminished the state of aggregation, helping the force of combination. Fourcroy also provided a thoughtful reflection on phlogiston, or the pure fire fixed in bodies, noting that Macquer had recently identified it with light. First, the properties Stahl had attributed to phlogiston—odor, opacity, color, volatility, fusibility, combustibility—were not found in many bodies that supposedly contained it. Charcoal, nearly pure phlogiston, did not have an odor and was not volatile, fusible, or even easily combustible. Second, bodies that lost phlogiston often acquired the properties attributed to its presence. Many metals acquired a more pronounced color after calcination. Third, if phlogiston was the same as light, it had to be visible during its transfers. Most metals calcined without reddening, however. Fourth, no one had demonstrated the existence of light in any phlogisticated body. These and other objections gave rise to a new, "pneumatic" theory, which was "absolutely the inverse" of Stahl's.

Fourcroy was keenly aware, then, of the limitations of the principalist approach to composition embedded in the phlogiston theory. Not only did he point out the mismatch between the supposed presence of phlogiston and the manifested properties of the bodies containing it; he also explicitly criticized the Stahlian hierarchy of composition as a mere imagination. Instead, he wished to introduce a classification of chemical experience based on affinities. In other words, he sought to work out the affinity approach to composition in chemical systematics. This is why he later fought to preserve the doctrine of elective affinities in opposition to Berthollet. Although historians have interpreted Fourcroy's work as a contribution to the study of composition, his works should be placed within the framework of affinity chemistry which called for a new order of chemical composition and systematics.

7.4 The *Encyclopédie méthodique*

Fourcroy's text can serve as a sensitive indicator of the contemporary theoretical debates, owing in part to his involvement in the Arsenal

Group.[57] Although the contemporary chemists took Lavoisier's pneumatic chemistry seriously, they did not regard it as a broad theoretical foundation for practical chemistry. Salts occupied a central place in chemists' daily practice. Concomitantly, affinity constituted the most important theoretical subject. The discovery of new acids and their salts kept expanding the horizon in this rapidly evolving field. Lavoisier had a serious claim over this new territory of acids, however, if his oxygen theory proved to be true. It is not surprising, therefore, that Guyton chose acid, adhesion, affinity, and air as the subjects of as the main articles in the first chemistry volume of the *Encyclopédie méthodique*.[58] He was fully aware of the interlocking developments in these subjects. He wished to unite all ancient and modern knowledge of chemistry in a single work and to provide an order independent of the fortunes of particular hypotheses. The moment had come, he proclaimed, to be occupied with theories.

Guyton considered the definition of acid as the "key to chemistry." As the most habitual instruments of chemical operations, acids produced the most beautiful phenomena in great numbers. He thus stated that "it is the chemistry of acids which has drawn the outline of the science & which serves as the base to the general system of our knowledge in this part of the study of nature." The "true definition" of acid followed from his dissolution model: Acid was the most powerful solvent that worked on the greatest number of other substances, or according to Newton, "that which attracts strongly, and is strongly attracted." Guyton outlined briefly the history of the concept of acid, including the corpuscular theory of acid and alkali by Homberg and Lemery and the theory of universal acid by Becher, Stahl, Meyer, Lavoisier, and himself. In fact, he claimed partial credit for Lavoisier's oxygen theory. Inspired by the latter's experiments on nitrous acid, Guyton had begun to suspect that vital air entered into the composition of all acids as a constituent part, or that it was the "true universal acid, the acid element." Lavoisier developed this idea further in his general considerations on the nature of acids, concluding that vital air was the "true oxygine or acidifying principle."[59] Although Lavoisier did not acknowledge Guyton's contribution, Guyton had devised a two-fluid theory of dry and humid calcination that was identical to Lavoisier's oxygen theory in its algebraic skeleton. After these introductory remarks, Guyton proceeded to enumerate the properties of all the known acids, 28 in number. His detailed exposition of acids reflects the importance he accorded to the subject.

According to Guyton, adhesion was the same kind of power as affinity. He defined it as "the property certain bodies have to be attached to other bodies, or the force which retains them attached there."[60] Water adhered to finger, mercury to gold, and so on. One had to distinguish adhesion (which united two dissimilar bodies) from cohesion (which united the parts of a homogeneous body). Nevertheless, Guyton asserted, adhesion and affinity depended on the same cause and followed the same laws. Taylor's method of measuring the adherence of a metallic disk to mercury could quantify affinities, although he agreed with Kirwan that it was of limited use. Again reciting his experiment done in the early 1770s, Guyton concluded that adhesion was the first effect or instant of affinity. Affinity was only "an adhesion capable of producing dissolution to a certain degree." It was possible, therefore, "to estimate the ratios of affinities by the ratios of adhesions." He described in detail some experiments by Achard that extended Taylor's method further. Although Achard executed numerous experiments with courage, he did not include simple substances with the most powerful and better-known affinities, making it difficult to draw a decisive parallel between adhesion and affinities. Guyton designed some experiments to measure the adhesion of marble to various acids. He believed that the method could determine the "degrees of affinity" at least in cases that were not amenable to other methods. He cited other promising experiments in the same vein, including Dutour's experiments on capillary tubes.

Guyton's article on affinity gives every impression that he regarded it as the central theoretical subject of chemistry.[61] Since this article offers the most comprehensive survey of affinity chemistry as a theoretical venture at this time, it is worth examining in detail. Buried deep in Guyton's narrative is the assertion that the determination of saturation capacity constituted one of the most interesting subjects of research. Coming from Guyton, a central figure in the European chemical network during the 1780s, this judgment should be taken seriously. Kirwan had already focused on the point of saturation as the means of determining affinities and devised a higher analytic standard for it. Lavoisier also mentioned that the affinity table could not predict different points of saturation. In other words, the leading European chemists knew by this time that the saturation capacity of acids and alkalis was an important step in determining the composition of bodies. It was this line of inquiry on affinities that laid the foundation of nineteenth-century stoichiometry, which constituted the practical core of chemical atomism. Without acknowledging the role of affinity chemistry in determining composition, that is, we cannot understand the origin of chemical atomism. Historians have mostly taken Lavoisier's words in the *Traité* at face value and disregarded affinity as an important theoretical subject during and after the Chemical Revolution. This explains in part why they have had so much difficulty in linking Lavoisier's principalist approach to the development of modern chemistry.

If Guyton's article could serve as a useful guide for historians in studying the contemporary thoughts on affinity, it did not provide the contemporary chemists with a coherent theoretical system. His insistence on the truth of Newtonian system in depicting the uniformity of nature sat uneasily with the more empirical approaches. Despite his Newtonian tendencies, Guyton preferred "affinity" (defined as "the force with which the bodies of different natures tend to unite") over "attraction." Although Bergman deemed "elective attraction" a more suitable term for the language of an exact science, Guyton felt that affinity had been in the possession of chemists longer. It also signified a sense of selectivity that "attraction" did not. Attraction was everywhere, affinity was not. Affinity connoted the "intensity of attractive power," on which chemical phenomena depended. Notwithstanding his belief in the uniformity of nature, therefore, Guyton bowed to chemists' need to express their experience in choosing "affinity." Though undoubtedly he valued the efforts to quantify affinities most, he also included modified versions of Macquer's classification and Fourcroy's laws of affinity in his comprehensive review of the subject.

Guyton's article on affinity was divided into six parts; the history of the subject, the physical principles of affinities, different manners of considering them, apparent anomalies, their utility and application to the practice of chemistry, and finally the means of rendering the system more complete. In the first part, which concerned the history of affinity, he discussed mostly the invention and subsequent development of affinity tables as a prelude to the Newtonian formulation of the subject in the 1770s and the 1780s. He claimed that the affinity table opened the "first age of science" for chemistry, differentiating it from "the tradition of some recipes whose success would depend on servile imitation, or the art of some routine procedures to try the hazard of a new product."[62] It became a science that could explain its procedures and predict their results. A table of affinity compiled "all the knowledge gathered at the moment of its construction" and thus gave at a glance the progress of chemistry at the designated epoch. As for the cause of affinity, Guyton endorsed Buffon's version of Newtonian attraction, magnifying his own contribution to the field. In the second part, "The physical principles of affinities," he argued that there existed but one attraction whose modifications led to a variety of forces such as adhesion, affinity, and cohesion. Likewise, there existed but one law of attraction, the inverse square law, whose modifications would eventually explain all other forces. The only difference of affinity lay in the fact that the figure of interacting bodies, insignificant in the case of gravity because of the immense distance between celestial bodies, played a significant role, inducing variations in the law. Guyton suggested, invoking Macquer's and Bergman's authority, that "all chemical research should be directed to reconciling universal gravitation with the intensity of the power of our affinities."[63] Notwithstanding his conviction, it was not easy to prove that there existed but one attraction and one law. The "intensity of power" separated the three molecular forces—affinity, cohesion, and adhesion—from gravitation. Guyton belabored the point that all three classes of phenomena must follow the same inverse square law as did gravitation, but without much success. The phenomena refused to cooperate. Despite the conceptual and experimental difficulties, he did not waver from his conviction that a single law of attraction accounted for all natural phenomena. The simplicity and the uniformity of nature demanded such a conclusion.[64]

The third part, "of different manners of considering affinities," had three sections: the classification of affinities, the six laws of affinity, and the inquiries related to calculating the ratios of affinities. In the first section, "Division of affinities according to their effects," Guyton modified Macquer's classification of affinities into five categories: affinity of aggregation, affinity of composition, disposed affinity, affinity by concourse, and reciprocal affinity. The first two testify to the success of Macquer's classification in stabilizing the notion of chemical combination. The affinity of aggregation worked between "the molecules of the same nature." It augmented mass without producing new combinations. The affinity of composition united substances of different nature, simple or compounded, to form a new compound or super-compound. The new compound formed a homogeneous whole, an assemblage on which mechanical forces could effect nothing. It was destroyed only by affinity, and it possessed properties often strange and sometimes contrary to those of composing parts. The affinity of composition served as a great instrument of analysis and synthesis.

Instead of multiplying the categories of affinity according to the different types of operations, Guyton sought to establish general categories that would "clarify the theory and guide the practice of operations." He regarded denominations such as affinities of decomposition, precipitation, and dissolution as useless distinctions that overloaded the science needlessly. The distinction between dissolution and fusion also catered to "a language appropriated to diverse operations," a language that had no place in a methodical classification. Composed or complicated affinity [affinité composée or affinité compliquée] referred to the cases in which a compound united as a whole with a third substance without abandoning one of its principles. Guyton viewed this as a misnomer that only distracted the mind from the true phenomena by invoking a complicated force. He conceived these three-metal combinations in two steps. In the first step, the molecules of the two metals came into contact. In the second, "this composed molecule attracts a molecule of the third metal, with the force which is proper to its state of composition."[65] A little reflection revealed then the "error or uselessness of most ordinary divisions of affinities," always leaving the single affinity or the affinity of composition in the last analysis.

Another useful distinction Guyton adopted was the one between disposed affinity and affinity by concourse. The disposed affinity referred to the case in which one of the reacting substances had to change its state in order to effect a combination which otherwise would not have been produced. For example, mercury combined with acetous acid only after calcination, that is, after its combination with vital air. Fire and water as solvents also facilitated certain combinations, exhibiting the affinity of intermediates. Guyton generalized Bergman's notion of double affinity with the notion of the affinity by concourse, which he regarded as "one of the most important points of modern chemistry." It always supposed the actual and simultaneous concourse of at least four substances or more "conspiring forces" which gave rise to new combinations due to the circumstances of the concourse. Nothing happened, for example, if one poured nitrous acid or salt of mercury in dissolution separately on perfectly neutral vitriol of potash. Nevertheless, if one mixed the dissolution of mercury in nitrous acid with vitriol of potash, the two acids exchanged their bases, yielding two new compounds: vitriol of mercury and niter of potash. This double exchange, a "product of the concourse of several affinities," did not upset "the order of ordinary affinities of composition."[66]

In order to predict these complicated reactions, for which Geoffroy's table was useless, Guyton deemed it necessary to "determine in numbers the rapports of these conspiring forces in a manner that reconciles the results of calculation with observed phenomena." Macquer had already stipulated that "there is a mutual exchange *every time when the sum of the affinities that each of the principles of the two compounds has with the principles of the other surpasses that of the affinities that enter between the principles that form the two first compounds.*"[67] Kirwan's theory of divellent and quiescent affinities, denominations at once "clear, exact, and convenient," deserved greater attention in determining the outcome of double exchange reactions. Guyton illustrated Kirwan's theory with examples. If the affinity of muriatic acid with barytic earth was 36, one could suppose that the affinity of the same acid with pure or caustic potash was 32, that of mephitic acid with baryte was 14, and that of mephitic acid with potash was 9. Neither potash nor mephitic acid could decompose barytic muriate by itself, the quiescent force being 36, the divellent force with potash 32, and the divellent force with mephitic acid 14. If one presented mephite of potash o barytic muriate, however, the divellent force became 46, while the quiescent force was 45. The decomposition thus proceeded to produce two new compounds of muriate of potash and barytic mephite:

Muriate of potash

Barytic muriate {	Muriatic acid	32		Potash		} = 45 Mephite of potash
	36	+		9		
	Baryte	14 = 46		Mephitic acid		

Barytic mephite

Guyton formulated the principle that "every time when there are more than three bodies, it is no longer necessary to consider the simple affinity of one body to another, but the sum of all affinities that concur for the same goal."[68] He assigned the numbers for divellent and quiescent affinities arbitrarily on the basis of a rough estimate of their relative strengths. Because of the difficulty of determining these numbers absolutely, he had to rely on previous experiments to conjure up a reasonable set of numbers that did not contradict the known affinity relations. He reckoned that one could use them without inconvenience until they contradicted other results. Guyton's method was then something akin to the self-consistent method. It allowed him to circumvent theoretical uncertainty and empirical difficulties. He provided detailed rules for constructing symbolic representations of double affinity relations with numbers. He also offered a table of numerical affinities for five acids and seven bases, which differed from those proposed by Kirwan and Fourcroy.[69]

The fifth affinity Guyton endorsed was reciprocal affinity [affinité réciproque], which referred to reversible chemical actions. Marggraf had shown that nitrous acid displaced muriatic acid from common salt. This was normal since the stronger acid should displace the weaker acid. But muriatic acid in turn displaced nitrous acid from niters of potash and soda. Instead of assigning to this "reciprocity of effects" a reciprocal cause (the same cause that produced opposite effects), Guyton attributed the "apparent reciprocity" to particular circumstances. As Berthollet had shown in 1785, nitrous acid upon decomposition gave a part of its vital air to muriatic acid, changing it into dephlogisticated muriatic acid. In another case, Baumé had observed that, in dissolving vitriol of potash in equal quantity of nitrous acid, he obtained a large quantity of niter. In other words, nitrous acid, which ordinarily yielded the alkaline base to the more powerful vitriolic acid, took up the base in this operation. Kirwan and Bergman regarded this as an anomaly to be explained away. Kirwan thus interpreted the reaction as a case of double affinity involving the matter of heat. Bergman attributed the exchange to the excess portion of acid that was not attracted and retained by the same force. Since the alkaline vitriols were decomposed only partially by nitrous and muriatic acids, they were satisfied only partially. Berthollet had adopted and confirmed this explanation, showing that the decomposition of alkaline vitriols was never complete.

Guyton noted that the decomposition of alkaline vitriols by nitrous and muriatic acids brought out a completely different and rather serious issue of general theory: Were there truly "diverse degrees of saturation" of the same salt? This was a question "that merits all our attention, not only because it involves the general theory of chemical attraction" but also because a satisfactory knowledge of this kind of affinity was necessary before subjecting it to calculation and explicating other dependent phenomena.[70] Reciprocal affinity raised an important issue regarding the possibility of multiple saturation points, which Guyton found "repugnant to all notions" of chemical combination. The saturation point of a salt in water could change according to the temperature, since the attractive force could be modified by heat as well as by the respective positions, density, and figure of the molecules. When the circumstances remained the same, however, the point of saturation should not change, since it was the effect of a constant cause. The natural and necessary meaning of "saturation" referred to the point beyond which the two composing parts of a compound could neither receive nor retain in combination a greater quantity of each other. In ordinary saturation, the composing parts lost as much as possible of their particular properties in order to manifest the properties of a new compound, which were foreign to both composing parts. The point of saturation had to be a fixed point at which perfectly neutral salts were formed. Nevertheless, nothing prevented the resulting neutral salt from attracting an excess of one of its components, although such attraction was necessarily weaker than the attraction between the components up till the saturation point. The weaker attraction should not change, therefore, the existing composition or the saturation point. This distinction between the affinity of two substances between themselves and the affinity of the compound resulting from it with one of these substances could explain a great number of phenomena that remained quite obscure, including the crystallization of certain salts with excess acid or base and the incertitude of the "point of saturation." Despite these complications, Guyton believed that there was "a unique and invariable saturation point for each combination" and that the so-called second term of saturation was only the saturation of a supercomposition or of a different composition. Nevertheless, he cautioned against joining the concepts of saturation and neutralization too tightly.

The saturation point became a focus of contention because it marked the point at which the participating affinities determined the composition of bodies. The empirical or affinity approach to composition would be seriously compromised if one could not fix the point of saturation. Guyton thus insisted on fixed points of saturation despite many examples that seemed to disprove them. He criticized Macquer's distinction of "absolute saturation" (the diminution of the tendency to union of a substance by its combination to another substance) from "relative saturation" (this diminution compared to that of other combinations). Although this distinction did not override Guyton's belief in the "constant identity of the point of saturation," it could introduce a notion of the different degrees of saturation between different acids and alkalis. Saturation had to be always complete, however, because it expressed the actual equilibrium of forces. Guyton also discounted Kirwan's idea that the quantity affected the degree of affinity. For example, if 100 grains of vitriolic acid could dissolve 110 grains of calx, its affinity for 55 grains of calx was half of the total degree of affinity. Guyton asked whether the quantity of base necessary for the saturation of a given quantity of an acid was in direct ratio of its affinity with this base. If so, only the quantity of acid necessary to neutralize the base should be separated since the acid once separated could not "divide the force of union." Such an idea was contradicted by the observation that showed the separation of all available acid.[71] Guyton argued that Kirwan's system also contradicted the "physical principles of affinities." Since affinity was attraction, which was always proportional to the sum of the points of contact, it was impossible to imagine that half a quantity of matter could furnish as many points of contact or occupy the same space as the full quantity of the same matter. Above all, Kirwan's theory would mean that affinities were infinitely variable and that they would change perpetually according to the respective quantities of the solvent and the base. Guyton believed that the affinity between two bodies was as constant as the figure and the density of their molecules. Guyton's discomfort over the implications of variable saturation points and affinities proved rather prophetic; Berthollet took up these issues later. Though historians have often noted that Berthollet's experience with large-scale processes led him to his notion of mass action, in fact that notion had been extensively discussed and fussed over by Macquer, Kirwan, and Guyton decades earlier. Berthollet's exposure to industrial processes probably reinforced his memory of these problems and his attention to them.

In the second section of the third part, "The laws of affinity," Guyton gave six laws similar to Fourcroy's. First, chemical union could occur only when one of the bodies was sufficiently fluid to carry its molecules to the point of contact with another so as to obey the affinity relations. Second, affinity took place only between the smallest integrant molecules of bodies. Third, one should not deduce the affinity of one substance with another from the affinity of the compound of these substances with the excess of one or the other. Fourth, the affinity of composition was effective only when it involved the affinity of aggregation. Fifth, two or more bodies united by the affinity of composition formed a being that had new properties distinct from those before the combination. Sixth, temperature could slow, quicken, nullify, or augment affinities.[72]

In the third section, "On the manner of considering affinities for determining the power thereof," Guyton evaluated various methods for the quantitative determination of affinity values. Wenzel considered the time for dissolution to measure dissolving power; Fourcroy focused less on the facility of union than on the resistance to separation which announced the intensity of this power; Macquer took into consideration both the facility of union and the force of adherence; Kirwan estimated the affinity of acids with bases by the quantities required for their saturation. Guyton's criticism of each of these methods reveals his own preference for a quantitative scheme consistent with his modeling of affinity at the molecular level.

Wenzel, in his *Lehre von der Verwandtschaft*, assumed that the affinity of a body with a common solvent was in an inverse ratio of the time required for its dissolution. He set up an experiment in which small metallic cylinders coated in such a way that the solvent could work only on one end were placed in similar vessels with a common solvent, which in turn were placed in larger vessels filled with water to maintain a constant temperature. After an hour, he weighed each cylinder to see whether the quantities dissolved corresponded to the degrees of affinity of these metals with the common solvent. By measuring the dissolution at equal degrees of heat during equal lapses of time on equal surfaces, Wenzel made it easier to calculate the time a complete dissolution of each cylinder would take. The differences of duration would express in determinate numbers the different degrees of affinity. Guyton perceived much difficulty with Wenzel's approach, however. Affinity was an effect of attraction modified by diverse circumstances and in particular by the figure of the constituent parts of the body. What determined the intensity of the attractive power at a given distance was not, however, the figure of the parts of the body or of the solvent alone, but the relationship [rapport] between these different parts or the latitude of their disposition to contact. This "evident truth" resisted Wenzel's hypothesis, which regarded the solvent as a given force that did not change and acted more or less rapidly on the body by its mass. The varying speed of dissolution by nitrous acid, which Wenzel used as an example, did not match the order of affinities determined by precipitation, as Kirwan had objected. Furthermore, Wenzel's calculations did not yield absolute numbers for predicting the affinity of concourse, although this was where they were most needed.[73]

As for Fourcroy's idea that "the affinity is measured more by the difficulty one has in separating a compound into its principles than by the vivacity of their union," Guyton judged that there were no means of estimating this resistance other than by the usual way of displacement. Only chemical means could rupture a chemical combination. Although Fourcroy's method promised to be certain, it was not applicable to all cases. In some cases, heat and cold effected the separation of two bodies directly without decomposition, as was shown by Lavoisier and Laplace in their "Memoir on Heat." The "equilibrium between the heat which tends to separate the molecules of a body & their reciprocal affinities which tend to unite them" could furnish a "quite precise means" of determining affinities. Guyton saw a great promise in the "united works of these two great physiciens."[74] He conjectured that a compound subjected to a strong degree of heat reacted in various ways, depending on the affinity of its two constituents for the matter of heat. They could remain fixed, be volatilized together, or be separated. The separation occurred when the two constituents had varying degrees of affinity for the matter of heat. All the instances of separation effected by heat confirmed that the matter of heat combined with other bodies with different degrees of affinity. If so, one could not estimate the affinity of composition between two substances from the facts of decomposition at a given temperature. The affinity of calorific principle with the substances in question, variable and unknown every time, made it impossible to fix the affinity of composition. Therefore, there was no means of determining with any precision the resistance two constituent parts put up against their separation.

Guyton also gave a detailed description of Kirwan's method for estimating the affinities between acids and bases by their saturation capacities. By calculating the quantity of real acid and determining the precise point of saturation, Kirwan had concluded that the quantity of real acid necessary to saturate a given weight of each base was in an inverse ratio of its affinity for the acid, and that the quantity of each base necessary to saturate a given quantity of each acid was in direct ratio of the affinity of the same acid with the base. Respective quantities of bases 100 grains of mineral acids would require for their saturation were as follows:

	potash	soda	lime	ammo.	magn.	alumine
vitriolic acid	215	165	110	90	80	75
nitrous acid	215	165	96	87	75	65
muriatic acid	215	158	89	79	71	55

Although Guyton did not share Kirwan's enthusiasm for this method, he acknowledged that "the determination of the proportions of the ingredients or constituent parts of salts & all the compounds in general, is regarded today in common agreement as the most important point to advance the theory & to perfect the practice of all operations."[75]

The fourth part dealt with the "Apparent anomalies of affinities," which hindered Bergman's quest for the constancy of the order of affinities. Since Guyton believed in the invariable laws of nature, he attributed apparent anomalies to the particular causes that modified the effects of general principles, including heat, double affinity, successive changes of substances, solubility, supercomposition, and excess quantities. Sorting out constant chemical factors from variable physical factors became a main task in the later formulations of affinity programs, particularly Berthollet's.

After a brief discussion of the uses of affinity in chemical practice in the fifth part, Guyton moved on to the sixth part, "The means of completing the system of affinities." Though he believed that "the elective attraction between three bodies" was "the most certain foundation of the science of affinities,"[76] he acknowledged the defects of existing tables. In order to complete and perfect them, one would have to include several substances: phosphorus, matter of heat, vital air, inflammable gas, phlogisticated air, all acids and their radicals, neutral salts, etc. This would be an immense task, as Bergman had predicted. To approach this problem, one had to set some ground rules. One needed "a perfect knowledge" of all the substances used. They had to be pure. Water and the matter of heat had to be included in the discussions of affinity, as Lavoisier had suggested. The exact knowledge of the quantity of substances was also important since it influenced the results significantly— not the absolute weight, however, but the quantity contingent upon concentration and temperature. Only when these condition were met could one investigate the affinity relations of these substances and determine their values in numbers. In doing so, one had to keep an open mind on the order of affinities and take into consideration all contingent circumstances.

With his article, Guyton characterized chemical affinity as a broad frontier of theoretical inquiry that would ultimately absorb pneumatic studies. His discussion identified three important problems in the current investigations of affinity: the determination of saturation capacities, the effect of contingent factors on chemical reactions, and reciprocal affinity. These problems took on urgency in the efforts to classify and control the operations of salts. Guyton's systematic reflection on affinity, particularly his efforts to discern disturbing physical forces and his notion of effective quantities, provided a starting point for Berthollet's chemical statics.

7.5   A Rhetorical Moment

If Lavoisier planned to participate in shaping the frontier of theoretical chemistry with a systematic measurement of affinities, other matters attracted his attention: the water controversy, Macquer's death and the subsequent consolidation of the French antiphlogistic camp, the collaborative project on nomenclature, a concerted attack on Kirwan, and finally the writing of the *Traité*. The composition of water offered a critical testing ground for the oxygen theory, and Macquer's death allowed Lavoisier to assume the symbolic leadership of the Academy. He had been preparing the "Réflexions sur le phlogistique" for a long time. To judge from the timing, his full-scale assault on French phlogistians was not driven by purely scientific reasons; rather, it constituted a rhetorical moment when he "decided to speak out . . . and to assert an independent authority."[77] Skillfully mobilizing the decomposition of water as a crucial evidence, he openly criticized the ad hoc nature of Macquer's and Baumé's shifting theories of phlogiston. Berthollet and Fourcroy soon announced their "conversion" to Lavoisier's camp, followed by Guyton. These "conversions" were no less rhetorical than Lavoisier's. Berthollet announced his conversion, citing experiments with marine acid, the only acid then known that did not contain oxygen. Fourcroy had been favorably disposed toward Lavoisier's system long before his public "conversion." After losing Bergman and Macquer in the same year,

Guyton needed a new network to authenticate his proposal for the nomenclature reform. The Arsenal Group was fast emerging as the representative faction of French chemistry. A series of rhetorical moments thus ushered in the public phase of the Chemical Revolution.

The composition of water became the final testing ground for the respective efficacy of the phlogiston and oxygen theories, thanks in part to Kirwan's identification of phlogiston with inflammable air. The combustion product of inflammable air had been a topic included in Lavoisier's 1777 collaboration with Bucquet.[78] Bucquet suspected fixed air; Lavoisier expected an acid, in accordance with his oxygen theory of acids. Neither was proven right. Lavoisier continued the experiment in the winter of 1781–82 without being able to obtain an acid. As usual, critical pneumatic experiments had to be imported from England. At the time, Kirwan's identification of phlogiston with inflammable air was gaining ground in Britain.[79] During the spring of 1783, Cavendish obtained water by detonating inflammable air and dephlogisticated air together in a closed vessel. Priestley, who had been investigating a curious conversion of water into ordinary air in communication with James Watt, confirmed Cavendish's results.[80] Charles Blagden delivered this news in his visit to Paris. Lavoisier had been preparing to resume his experiments on the combustion of inflammable air with Laplace, using a more sophisticated combustion apparatus. He immediately repeated the experiment on June 24 in the presence of Blagden and a number of academicians. Monge, who was in Paris until he returned to Mézièrs at the end of May, also confirmed the results with quantitative precision.[81]

The composite nature of water held a significant place in Lavoisier's oxygen theory because it accounted for the inflammable air and the vital air produced in the humid calcination of metals. In the calcination of metals using vitriolic acid, Lavoisier knew that the vital air did not come from the vitriolic acid, since the operation did not produce sulphur or sulphurous acid. Water was the only possible source. Lavoisier worked hard, therefore, to prove the compound nature of water, which he presented to the Academy on November 12, 1783.[82] The importance he attached to this project becomes evident, as Daumas and Duveen pointed out, in his careful preparation for a large-scale demonstration. The public enthusiasm for ballooning had induced the Academy to appoint a committee to improve balloons by determining the best shape for them and the best ways of maneuvering them, and above all by finding a cheap, light gas that was easy to obtain and always available. Although Lavoisier was a member of this committee, he was not simply interested in the mass production of inflammable air. He had Mégnié construct two gasometers according to Meusnier's design for an exact quantitative control of the gases involved. Berthollet thus observed that "M. Lavoisier has tried to bring to this matter all the accuracy of which it is capable."[83] Lavoisier clearly wished to stage the decomposition of water as the crucial experiment to win over converts to his system.[84] The time was ripe for him to assume the leadership of French chemistry. Soon, Berthollet formally announced his "conversion" to Lavoisier's oxygen theory, opening a new, collaborative phase of the Chemical Revolution. He actively sought to convert Guyton as well, no doubt with Lavoisier's blessing:

I hold, Monsieur, the chemical language you speak quite heterodox. It is what I have given up to the single conviction: while M. Lavoisier could not give any explication of the formation of inflammable air and while he had false ideas on the nature of nitrous acid, I defended the theory of phlogiston. It rendered much better account of the phenomena; but the decomposition of water, a more exact knowledge of the nature of nitrous acid, more just ideas on the principle of each that gives elasticity to gases and fluidity to liquids, the affinities of vital air better determined have substituted the data of experiment for a useless hypothesis. . . .[85]

With a strong ally on his side, Lavoisier finally read his "Réflexions sur le phlogistique"[86] to the Academy. It was a sequel to the theory of combustion and calcination given in 1777 and contained not much new on the subject except for two aspects. First, the rhetorical stance of the paper differed drastically from his earlier papers. Lavoisier no longer deferred to the Stahlian camp, but he criticized explicitly Macquer and Baumé for their stubborn adherence to the defunct Stahlian system. He blasted phlogiston as a "hypothetical being, a gratuitous supposition." It was time for him to explain "in a manner more precise and more formal an opinion I regard as an error to chemistry and which seems to me to have retarded the progress considerably." Lavoisier's target was not Stahl, but the doctrine of phlogiston as it had been "conceived and presented by M. Macquer." He thus focused his criticism on Macquer's and Baume's ad hoc modifications of Stahl's phlogiston theory. Macquer had originally characterized phlogiston in the *Dictionnaire* as "fixed fire." In order to reconcile the fixity of phlogiston in the body and the apparent mobility of the element of fire, however, he supposed that an earthly principle served as an intermediate to unite fire with combustible bodies. The combination of the element of fire with an earthly principle thus constituted the inflammable principle or phlogiston. Faced with a striking contradiction that bodies gained weight by losing phlogiston through combustion, Baumé supposed that the element of fire and the earthly principle that made up phlogiston could combine in "an infinity of proportions," thus producing an "infinity of intermediary beings between free fire and the phlogiston properly understood." Although this extension to the Stahlian system allowed more flexibility in chemical explanation, it only aggravated the discrepancy in weight relations. When Lavoisier demonstrated that the augmentation of weight was due to the fixation of air, Macquer devised yet another theory that phlogiston was the pure matter of light. It was "astonishing to see M. Macquer, all appearing to defend the doctrine of Stahl by conserving the denomination of phlogiston, present a totally new theory, which is not that of Stahl."[87] In order to explain the weight gain, Macquer now accepted the fixation of air in calcination as well as the simultaneous release of phlogiston or the matter of light. Macquer's new version of the phlogiston theory explained in a natural and simple manner many objections to Stahl's hypothesis. Nevertheless, it still shared the basic default of Stahl's theory that phlogiston must have weight. No experiment had shown, however, that phlogiston had a "weight susceptible of being appreciated, or even perceived, in chemical experiments."[88] Lavoisier marshaled various examples to poke a hole in "the system of M. Macquer" and to break down the "partisans of the doctrine of Stahl." He targeted phlogiston's protean nature:

All these reflections confirm what I have advanced, what I had as the object to prove, what I am still going to repeat, that chemists have made phlogiston a vague principle which is not rigorously defined, and which, as a consequence, is adapted to all the explications in which one wants to make it enter; sometimes this principle has weight, sometimes, it does not; sometimes, it is free fire, sometimes it is fire combined with earthly element; sometimes it passes through the pores of the vessels, sometimes it cannot penetrate them; it explains at once causticity and non-causticity, translucency and opaqueness, colors and the absence of colors. It is a true Proteus which changes its form at each instant.[89]

Second, Lavoisier elaborated fully the molecular constitution of bodies contingent upon the interplay of heat and affinities. A clearly visible effect of heat was the increase in the volume of the body thus affected, which meant that "the molecules of bodies do not touch, that there exists between them a distance that heat augments and cold diminishes."[90] Only a fluid could cause this:

One can conceive these phenomena only by admitting the existence of a particular fluid, whose accumulation is the cause of heat and whose absence is the cause of cold; it is without doubt this fluid which is lodged between the particles of bodies, separates them, and occupies the place they allow between them. With most physicists, I name this fluid, whatever it may be, *igneous fluid, matter of heat and of fire*.[91]

The force of attraction counteracted the distintegrative force of heat. This was "a general law of nature, to which all matter would appear to be subjected." All bodies of nature thus obeyed two forces: "the igneous fluid, the matter of fire, which tends continually to separate the molecules, and the attraction, which counter-balances this force."[92] Depending on their relative strengths, bodies assumed solid, liquid, and aeriform states. A third force, the atmospheric pressure, regulated the transition from one state to another. A fine tuning of these forces at equilibrium involved the play of affinities. Although the matter of heat always distributed itself evenly to establish an equilibrium, how much of it each body could admit depended on the size of the pores, the intervals between the molecules, the attraction between the molecules, and the affinity between the molecules and the matter of heat. Lavoisier thus defined "combined heat" as the portion of heat united to a body so that one could not remove it without decomposing it. The matter of heat in this state lacked elasticity and exercised no heating effect. It was no longer in a state of aggregation, but was a constituent part of the body. "Free heat" designated all the heat that was not engaged in combination. It was not completely free, however, because of the affinity matter had for the heat. The "constant and determined" heat lost in phase changes reflected the heat that passed from the free state to the combined state.[93] Crawford's specific heat was the quantity of free heat necessary to elevate the temperature of a body by a certain number of degrees. In sum, Lavoisier brought to completion his earlier, tentative picture of chemical constitution in which heat played an integral role.

"Réflexions sur le phlogistique" was more than a frontal assault on phlogiston. It was a deliberate performance designed to consolidate Lavoisier's symbolic leadership of Parisian chemistry. Through the gatherings at the Arsenal laboratory in which Berthollet, Laplace, and Fourcroy participated regularly, Lavoisier had laid groundwork for his leadership. The decomposition of water served to quell much opposition to his new theory. More important, he had put together a theoretical structure more comprehensive than the oxygen theory—a view of chemical composition that took into account the role of heat, establishing a vision of the constitution of bodies as an equilibrium of heat and affinities. He took the initiative, therefore, to secure the fortunes of his theory as well as his symbolic position. The constitution of bodies became a shared vision of the Arsenal chemists.

7.6 The Nomenclature Reform

Although Lavoisier valued linguistic precision, the impetus for the nomenclature reform came from a different sector: Bergman and Guyton, the two leading European chemists, whose symbolic capital depended more on their writings and their correspondence than on the judgment of the Academy.[94] In addition to their interest in facilitating chemical communication, they shared a deep interest in mineralogy, a subject that was in dire need of rational classification.[95] In his *Opuscula Physica et Chemica* (1779), Bergman called for denominations more "conformable to the nature of substances" and named new salts accordingly. Macquer was warmly receptive to this endeavor, as we have seen. In translating Bergman's book, Guyton embraced the proposal for a nomenclature reform enthusiastically, albeit with modifications. In his early attempt, Guyton took it as a principle that the new nomenclature should utilize a clear conception of "true chemical elements, in the rigor of this expression." He thus criticized Demeste's contention that the earth of gypsum and selenite was an absorbent earth different from calcareous earth. Aside from the lack of decisive proof, Guyton characterized the term "absorbent earth" as "vague & uncertain." If this was indeed an elementary earth, he argued, it should be called "earth of gypsum" [terre gypseuse] so that "the entire world would understand, without pain, under this denomination the earth which constitutes gypsum by its combination with vitriolic acid." Guyton's vision of a rational nomenclature encompassed the analytic purity of each substance and the specificity of the name assigned to it.[96]

While preparing the articles for the *Encyclopédie méthodique*, Guyton took the opportunity to devise and to put into effect a more rational nomenclature. He wished to implement a scheme that would facilitate communication among European chemists. The perfection of a science depended on the perfection of its language, as Lavoisier reiterated later:

The denominations of beings which form the object of science or of an art, which are its materials, its instruments, its products, constitute what one calls its proper language. *The state of the perfection of the language announces the state of the perfection of the same science*; its progress can be certain and rapid only in so far as its ideas are represented by precise & determined signs, just in their meaning, simple in their expression, convenient in usage, easy to retain, which conserve as much as possible, without error, the analogy which brings them together, the system which defines them, & the etymology which can serve to reveal their meaning.[97]

The "revolting impropriety" of ancient words made reforming the language of chemisty imperative. False denominations hindered the communication of chemistry's discoveries and thus retarded its progress. Many famous chemists had deplored the confusion, obscurity, and discomfort a group of improper names imposed. In fact, no other science required more clarity and precision as chemistry, yet no other science had a language so barbaric, vague, and incoherent. The same substance had to be considered in many different states: united, separated, in a certain order, in a certain kind of composition, in a certain degree of combination, etc. Instead of relying on the authority of his own opinion, Guyton proposed, he would lay down the principles and the plan of his reform and solicit "the advance approbation, judgment, even objections of savants" to profit from their criticism. In particular, he called on Bergman and Macquer to offer their opinions on the matter. His five principles for a new nomenclature included the clause that the names should match the natures of the things themselves. Simple substances should be given simple names and the name of a compound was "clear and exact" only insofar as it recalled its constituent parts to designate its proper nature. These principles later served as the cornerstone of the 1787 reform.

Bergman and Macquer reacted favorably to Guyton's proposal, but their deaths in 1784 left him alone in this noble endeavor. When he began to prepare the article "Air" for the *Encyclopédie méthodique*, Guyton turned to Lavoisier, who was quite receptive to the idea.[98] In their subsequent collaboration, Lavoisier retained Guyton's basic tenets, including the importance of the analytic definition of elements and of assigning names that matched the nature of bodies. Guyton had been using the analytic definition of elements at least since his lectures at the Dijon Academy began in 1776. More than likely, it was through the months-long conversations at the Arsenal in 1787 that Lavoisier worked out his famous definition of elements as the limit of chemical analysis. Guyton, Berthollet, and Fourcroy all had been using this definition on Macquer's authority. In contrast, Lavoisier started his career with Boerhaave's theory of four instrument/elements and invested a substantial amount of time and energy trying to make the scheme to work. He probably had to abandon the idea when air became airs and water proved to be a compound, but he did not work out an alternative. The analytic definition of elements had a serious drawback for a person with a philosophical mind, as Fourcroy pointed out in the *Leçons*: One had to admit too many substances as elements.

Lavoisier played primarily the role of a metaphysician in the nomenclature reform. He dressed up Guyton's proposal in the language of the most influential French philosopher then in circulation, Condillac.[99] In a memoir presented at a public meeting of the Academy, Lavoisier acknowledged that much of the communal discussion among the four chemists involved the "metaphysics" of language and gave a detailed summary of Condillac's views on the subject. Languages were "true analytic methods, with the aid of which we proceed from the known to the unknown." Algebra, the "analytic method par excellence," had been invented to facilitate the operations of the mind, to provide a short cut for the reasoning process, to condense long discussions to a few lines, and to help one arrive at the most convenient, quick, and certain solutions to complicated questions. To the degree that languages were the "true instruments" of human reasoning, these instruments had to be perfected for the science to advance.[100] While acknowledging the works of his predecessors, particularly Guyton's role in conceiving a vast plan in 1782, Lavoisier assumed the authority of a representative academician in presenting the proposal to the academy and to the world.

The technical content of the nomenclature reform was presented by Guyton and Fourcroy in the regular sessions of the Academy.[101] Their system consisted of a single table divided into six columns, which Guyton hoped would contribute to "the uniformity of language so essential to the communication of works & to the progress of the science."[102] He explained in detail the first column of "undecomposed substances," which was further divided into five sections, each comprising simple substances, acidifiable bases or radical principles of acids, metallic substances, earths, and alkalis. The total number of undecomposed substances came to 55, which set up the horizontal divisions of the table. According to Fourcroy, the table comprised not only the known substances but also the substances yet to be discovered. The second column, devoted to the compounds of undecomposed substances with caloric thus listed only three: oxygen gas, hydrogen gas, and azotic gas. The compounds of other 52 simple substances with caloric were yet to be discovered. The remaining columns were intended for the compounds of simple substances with oxygen, oxygenated gases, simple substances oxygenated with bases or neutral salts, and the compounds of simple substances in their natural state (without oxygen). The table visualized the importance of caloric and oxygen in the new classification, but it also left room for other kinds of compounds that did not include oxygen. It provided a kind of "world map" of chemical territory, as Guyton earlier tried with his halotechnic table of neutral salts. If the new table of nomenclature shared this mapping agenda with the affinity table, its focus on systematics rendered the order of reactivities invisible. Nevertheless, the new systematics represented an important compromise between the principalist and the affinity approach to composition. On the one hand, the geometric order of composition from the simple to the complex substances retained the principalist premise that one had to start with the original building blocks of the universe. On the other hand, the combination of simple substances into their compounds was determined by the relations of affinity manifested in concrete chemical operations. One could not create all variations of composition depending on the proportion of principles. Instead one had to assume determinate composition contingent upon the affinity relations. This new compromise allowed French chemists to realize the analytic-synthetic ideal of chemical elements and compounds. In this manner, the nomenclature reform of 1787 capitalized on the accumulated practice and theory in the chemistry of salts. Although Lavoisier provided for the new classification crucial links with his oxygen theory and philosophical legitimation with Condillac's language, the overall framework had been nurtured in the didactic practice of chemistry. Fourcroy, who was quite familiar with this teaching tradition, further elaborated the scheme in his monumental *Système de connaissances chimique* (1800).

7.7 An Essay on Phlogiston

The newly constituted Arsenal team found an optimal outlet for their communal views in Kirwan's *Essay on Phlogiston* (1787). Often vilified as a remnant of the old system, this document occupies a strategic position in the history of the Chemical Revolution. Its publication provided the French academicians with a ready-made opportunity to strengthen their alliance.[103] As a polemical piece, it throws into sharp relief all the main facets of the debate, magnifying certain aspects over the others to historians' benefit. The document served a similar function for the contestants' contemporary audience. Lavoisier wished to make the message of his revolution heard beyond the Arsenal Group and to gather "public opinion" in its support.[104] We must listen attentively to their arguments, pro and con, if we wish to understand the main features of the revolution led by Lavoisier. My focus in analyzing this text is less on the question of who was right, or even on who had more persuasive tack, than on what were the rules of contestation.

First, the protagonists in this debate were not fighting over priority and personal reputation. The respect both sides held for their opponents comes through clearly, indicating that at this time Kirwan was perceived as the most influential advocate of phlogiston in Europe. The fact that

Lavoisier's group issued a detailed response to Kirwan's *Essay* bespoke the "esteem" in which they held Kirwan, whose concurrence they "earnestly desire"[105]:

> Among the philosophers who have not yet adopted the new doctrine, [Kirwan] is certainly one of those who is the most capable of producing uncertainty in the minds of such persons as decide by authority. His acquaintance with every part of natural philosophy; the discoveries with which he has enriched the sciences, and even the ingenious modifications he has introduced into the theory of phlogiston; all contribute to give weight to his opinions. If the French chemists, whom he has opposed, should destroy his objections, will they not perhaps have a right to conclude that there are not any other solid objections to be made? It is principally from this last consideration that a translation of his Essay on Phlogiston was determined upon.[106]

Kirwan exhibited no less courtesy toward his opponents. Lavoisier, the main author of the "antiphlogistic hypothesis" and "a philosopher of great eminence," was "the first that introduced an almost mathematical precision into experimental philosophy."[107] Kirwan also acknowledged many of his opponents' positive contributions, such as proving that the weight gain of metals was due to the air fixed in the process and that the atmosphere consisted of two distinct airs.

Second, the protagonists shared a great deal. They all agreed that chemical principles were material substances that abided by the rule of the balance. Kirwan could advocate the phlogiston theory because phlogiston as inflammable air was no longer a "mere hypothetical substance, since it could be exhibited in an aerial form in as great degree of purity as any other air."[108] His entire critique of the antiphlogistic camp rested on precise measurements of specific weights. He was in fact one step ahead of his French opponents in advocating the importance of these measurements for chemical theory. Lavoisier also claimed that his entire theory was proved "by weight and measure." Lavoisier differed from Kirwan not in his deeper commitment to the rule of the balance but in his algebraic vision of chemistry and in his grammatical understanding of nature. That is, the superior explanatory power of his system lay in the interlocking algebra of all the components, rather than in its application to particular cases at hand. While Kirwan argued against the antiphlogistic hypothesis on the basis of its failure to explain the phenomena at hand, Lavoisier responded with a succinct summary of his entire theory, which showed its interlocking algebra: the three states of bodies depending on their caloric content, the consequent composition of gases as combinations of caloric and bases, the calcination of metals as oxidation, the decomposition of water, and the theory of acidification. All the facts relating to these operations had led him to conclude that combustion was equally reducible to an algebraic equation. Inconsistency in the algebra of chemistry irritated Lavoisier. If one did not believe in this overarching structure of algebra, both hypotheses seemed equally valid and could not be falsified in individual instances. That is, Lavoisier's hypothesis could explain all the categories of phenomena consistently without contradiction. Kirwan had not offered an alternative formula:

No supposition enters into these explanations; the whole is proved by weight and measure. Why, therefore, need we have recourse to an hypothetical principle, the existence of which is never supposed, and has never been proved; which in one case must be considered as heavy, and in another as void of weight, and to which, in some cases, it is necessary even to suppose a *negative weight*; a substance which in some instances passes through the vessels, in some others is retained by them; a being which its maintainers dare not rigorously define, because its merit and its convenience consist even in the uncertainty of the definitions which are given of it?[109]

Third, affinity was an important, if understated, issue in the debate. Kirwan found Lavoisier's table of the affinities of oxygen "liable to numerous objections" because it ran counter the experimental facts. The critical factor was the affinity of the principle of heat with the oxygenous principle, which occupied 21st place in Lavoisier's table. Of the 20 other substances that occupied higher places, Kirwan protested, none of the first 19 united with the oxygenous principle at the common temperature of the atmosphere. Nitrous air, the only substance that united with the oxygenous principle at every temperature, occupied the 20th place.[110] In response, Lavoisier could only reiterate his earlier thoughts on the affinity table. Most important were the temperature dependence of the order of affinities, the role of water, and the variations introduced by the different degrees of saturation. In addressing the third point, Lavoisier acknowledged the facts of partition, rather than complete decomposition:

It would be a false idea of the affinities, if any one should persuade himself, that in all cases a body will take from another the whole of a principle for which it has a greater affinity. Let us clear up this enunciation by examples. If the sulphuric acid be boiled upon mercury, silver, or copper, these metals do not completely decompose the acid; they do not take from the sulphur the whole of the oxigene to which it was united; for they act upon the oxigene only by virtue of the attractive force they exercise upon it diminished by the attractive force which the sulphur exerts upon the same principle. The oxigene, in these kinds of decomposition, must be considered as obedient to two unequal forces; on the one part, it is attracted by the metal which tends to calcination, that is to say, to become an oxide; and on the other, it is retained by the sulphur: and *it is divided into two parts, until an equilibrium is obtained.* When therefore the table of affinities announces that silver, mercury, and copper, deprive sulphur of the oxigene, the event is not accurately expressed; for it ought to be said, that when these metals are presented to oxigene and sulphur, the oxigene is divided between the sulphur and the metals in a certain proportion, which constitutes an oxide and sulphureous acid.[111]

Lavoisier was aware, then, of the fact of partition in the play of affinities, an invention usually attributed to Berthollet. This does not necessarily mean that Lavoisier had been working on the problem, but it suggests that it was one of the current issues among the contemporary chemists. The complex interaction of various affinities in metallic dissolutions baffled them. Most of the substances Lavoisier listed as having strong affinities with oxygenous principle were metals. He theorized that metals dissolved in different acids and precipitated each other according to their affinity for the oxygenous principle and degree of oxydization. Kirwan objected that such a unitary cause could not explain the complexities of these reactions. In response, Fourcroy insisted on the evidentiary strength of the antiphlogistians' "simple theory" over Kirwan's "complicated explanations." Lavoisier laid down three principles. One of them stipulated that "the force with which the oxigene adheres to a metal is not the same at all the degree of calcination." In other words, "in this kind of combinations, the affinity is a variable force, which decreases according to certain laws not yet determined. The same thing does not happen apparently in combinations which have a fixed degree of saturation; such, for example, as in the neutral salts." If Lavoisier's oxygen theory of composition offered an algebraically ordered classification, it fell short of answering the larger, more complex issues of affinity that Kirwan wished to address.

Finally, the contestants and commentators made it clear that their ultimate goal lay beyond the immediate questions of chemical theory. The limited experimental genre of combustion suddenly became central to chemical theorizing because it became a hurdle in constructing the true representation of nature. In the introduction, Kirwan narrowed the focus of controversy to the question of whether the inflammable principle existed in combustible bodies. Limited in scope, the question had a disproportionate consequence for chemical theory because these bodies were the instruments of many chemical operations. It was crucial to work out the implications of the inquiry:

> The controversy is therefore at present confined to a few points, namely, whether the *inflammable principle* be found in what are called phlogisticated acids, vegetable acids, fixed air, sulphur, phosphorus, sugar, charcoal, and metals.
> Limited as this controversy appears to be to a small number of bodies, it is nevertheless of great importance, if an exact arrangement of our ideas, and a distinct and true view of the operations of nature, be of any importance. The bodies above-mentioned are the subject of many, and the instruments of almost all chymical operations: without a knowledge of their composition, and a clear perception of their mode of action, it will be impossible to form even an approximation to a solid theory of this science; the daily accumulation of facts will only increase perplexity and confusion, and if any useful discovery be made, it will be the mere result of chance.[112]

According to the English translator, William Nicholson, the controversy touched on a subject of "the most extensive importance," namely the interplay of affinities and heat in the constitution of bodies. For this reason, it was "a leading consideration, and, as it were, *the very soul of chemistry*, to determine what happens when bodies undergo combustion."[113] The highest stake of the game was less the fate of phlogiston than a general conception of bodies, which promised to deliver the true representation of nature.

7.8   Elements of Chemistry

In his programmatic *Traité élémentaire de chimie*, Lavoisier sought to establish a new order of chemical discourse.[114] While historians have regarded this text mostly as an effort to join chemistry to physics, Lavoisier's vision was less physicalist than trans-disciplinary. It is significant in this regard that he chose Condillac, rather than Newton, as the philosophical authority for his new chemistry.[115] Historians have also overlooked how Lavoisier's "revolutionary" vision was shaped by the existing didactic tradition of chemistry. He wished to transform the French didactic discourse by transgressing the "usual order," which began with discussions of the principles of bodies and the tables of affinities. Despite his earlier endorsements, he rejected the doctrine of four elements completely. In a way, Lavoisier's emphatic denial of the four elements reveals the strength of his earlier commitment: He castigated it as a prejudice "descended to us from the Greek philosophers" and carried on by their authority. It was "a mere hypothesis, assumed long before the first principles of experimental philosophy or of chemistry had any existence." Chemists such as Becher and Stahl introduced variations based on their experiments, but they did not meet the standards of "strictly rigorous analysis required by modern philosophy."[116] All the discussions concerning the elements or principles of bodies amounted to nothing more than useless metaphysical speculation. Instead, Lavoisier defined elements or principles of bodies as the substances that constituted the limit of chemical analysis. That is, he explicitly endorsed the analytic ideal of chemical elements:

All that can be said upon the number and nature of elements is, in my opinion, confined to discussions merely of a metaphysical nature. The subject only furnishes us with indefinite problems, which may be solved in a thousand different ways, not one of which, in all probability, is consistent with nature. I shall therefore only add upon this subject, that if, by the term *elements*, we mean to express those simple and indivisible atoms of which matter is composed, it is extremely probable we know nothing at all about them; but, if we apply the term *elements, or principles of bodies*, to express our idea of the last point which analysis is capable of reaching, we must admit, as elements, all the substances into which we are capable, by any means, to reduce bodies by decomposition.[117]

Although this analytic ideal of chemical elements or principles had a long history dating back to Nicolas Lemery and was currently in use,[118] Lavoisier offered a philosophical justification it had lacked in the previous philosophical paradigms, Aristotelian, Cartesian, or Newtonian. Such an adaptation of current philosophy was a long-established practice in chemical didactics. Lavoisier thus endorsed Condillac's theory of language as the foundational maxim of chemical science: Languages were true analytic methods, for which algebra provided the most simple and exact model. It was impossible, therefore, to separate the "nomenclature of a science from the science itself":

The impossibility of separating the nomenclature of a science from the science itself, is owing to this, that every branch of physical science must consist of three things; the series of facts which are the objects of the science, the ideas which represent these facts, and the words by which these ideas are expressed. Like three impressions of the same seal, the word ought to produce the idea, and the idea to be a picture of the fact. And, as ideas are preserved and communicated by means of words, it necessarily follows that we cannot improve the language of any science without at the same time improving the science itself; neither can we, on the other hand, improve a science, without improving the language or nomenclature which belongs to it.[119]

This correspondence between facts, ideas, and words would allow Lavoisier to formulate a chemical algebra or "embryonic chemical equations."[120] Insofar as each chemical substance was designated by a unique idea and a unique word, chemists could trace their operations as an algebra of these words. The definition of elements as the limit of chemical analysis provided the first step in this chemical algebra. Each substance that was stabilized in analytic practice would be given a unique identity and a unique name. Their combination would be reflected in the name of the compound in the same way that the name of a neutral salt was given in terms of its acid and alkali. Ideally, chemical nomenclature should be designed so that the addition or subtraction of chemical substances would be reflected exactly in the name.

Concerning affinity, Lavoisier's evaluation contrasted sharply with his condemnation of the doctrine of chemical principles.[121] Far from regarding it as a dead subject, he appreciated the intensifying interest in affinity as the theoretical frontier of chemistry. It was precisely for this reason that he decided to exclude it from the *Traité*. The subject of affinity lacked a solid foundation on facts and did not belong to an elementary treatise:

The rigorous law from which I have never deviated, of forming no conclusions which are not fully warranted by experiments, and of never supplying the absence of facts, has prevented me from comprehending in this work the branch of chemistry which treats of affinity, although it is perhaps the best calculated of any part of chemistry for being reduced into a completely systematic body.[122]

Lavoisier's decision to exclude affinity from the *Traité* was due, therefore, to the unsettled state of affairs. The inadequacy of the existing affinity tables and the difficulty that attended the quantification efforts probably loomed large. He acknowledged that many eminent chemists such as Geoffroy, Gellert, Bergman, Scheele, Guyton de Morveau, and Kirwan had collected a number of facts which only needed a "proper arrangement." But "the principal data are still wanting, or, at least, those we have are either not sufficiently defined, or not sufficiently proved, to become the foundation upon which to build so very important a branch of chemistry."[123] In order to complete the work on the classification of neutral salts, for example, Lavoisier deemed it necessary "to add particular observations upon each species of salt, its solubility in water and alcohol, the proportions of acid and of salifiable base in its composition, the quantity of its water of crystallization, the different degrees of saturation it is susceptible of, and, finally, the degree of force or affinity with which the acid adheres to the base." Despite their efforts, Bergman, Guyton, Kirwan, and others had made only a moderate advance. He suspected that "even the principles upon which it is founded are not perhaps sufficiently accurate." The details of these investigations would "have swelled this elementary treatise to much too great a size." The experiments requisite to complete them would have "retarded the publication of this book for many years."[124] Lavoisier was painfully aware, perhaps, of the number of years it took for him to solve the puzzle of the weight gain. Affinity promised to be a subject that would require many more years of intensive work.

In contrast to the doctrine of principles, which belonged to the past, Lavoisier saw affinity as a subject of future chemistry. It was a "higher, or transcendental" part of chemistry. More likely than not, he agreed with his fellow chemists that affinity was a worthy cause, although he expected a long, rough ride ahead, as his travel down the road of composition had been. He did not wish to pursue it at the time, since he "had more reasons than one to decline entering upon a work in which [Guyton] is employed."[125] It was an open field for the next generation of chemists. Lavoisier offered sound advice:

This is a vast field for employing the zeal and abilities of young chemists, whom I would advise to endeavor rather to do well than to do much, and to ascertain, in the first place, the composition of acids, before entering upon that of the neutral salts. Every edifice which intended to resist the ravages of time should be built upon a sure foundation; and, in the present state of chemistry, to attempt discoveries by experiments, either not perfectly exact, or not sufficiently rigorous, will serve only to interrupt its progress, instead of contributing to its advancement.[126]

Although Lavoisier declined to treat the subject of affinity systematically, he used the concept freely as an integral part of his new system of chemistry. He accepted, as other chemists did, that chemical substances (including caloric) interacted via their affinities. Oxygen, for example, "possesses a stronger elective attraction, or affinity, for phosphorus than for caloric."[127] Scattered throughout the text, one finds the usual classification of affinities—double and triple affinities, complicated affinities, compound affinity, and so on.[128] There is also a table of the combinations of most acids with salifiable bases "in the order of affinity."[129] Though Lavoisier used "affinity" mostly at the descriptive level, he also used the language of Newtonian attraction. In combustion, heat was used "to separate the particles of the metal from each other, and to diminish their attraction of cohesion or aggregation, or, what is the same thing, their mutual attraction for each other." In the state of calx, metals were not "entirely saturated with oxygen, because their action upon this element is counterbalanced by the power of affinity between it and caloric."[130] Ideally, one should be able to quantify these ratios of affinity:

Although we are far from being able to appreciate all these powers of affinity, or to express their proportional energy by numbers, we are certain, that, however variable they may be when considered in relation to the quantity of caloric with which they are combined, they are all nearly in equilibrium in the usual temperature of the atmosphere.[131]

Caloric occupied a special place in Lavoisier's system of chemistry because of its role in determining chemical constitution. All bodies in nature could assume the states of solid, liquid, and gas, depending on "the proportion which takes place between the attractive force inherent in their particles, and the repulsive power of the heat acting upon these." The cause of heat or "caloric" was likely to be "a real and material substance." It was "the repulsive cause, whatever that may be, which separates the particles of matter from each other."[132] The balance of the attractive force between the particles of a body and the repulsive force of caloric under the given atmospheric pressure determined its constitution. Thus, a systematic investigation of any body required knowledge of the nature of its elements and of their affinities with one another and with caloric:

Before we can thoroughly comprehend what takes place during the decomposition of vegetable substances by fire, we must take into consideration the nature of the elements which enter into their composition, the different affinities which the particles of these elements exert upon each other, and the affinity which caloric possesses with them.[133]

This dynamic vision of chemical constitution as an interplay of affinities and caloric became the unified vision of theoretical chemistry for the Arsenal Group chemists. Guyton did not have much difficulty embracing this vision, since Lavoisier's caloric played exactly the same function as Guyton's phlogiston as a universal solvent, counterposing the force of affinities. He did not lose much by his "conversion" (except perhaps Kirwan's friendship), and he had much to gain—including the endorsement of the Arsenal Group. As a philosophical chemist, Guyton was no more attached than Lavoisier to the phlogiston as a concrete chemical substance manifested in multifold chemical operations. He gave up phlogiston only in name. In the second preface to the first chemistry volume of the *Encyclopédie méthodique*, preceding the article "Air," Guyton endorsed Lavoisier's theory explicitly. Truth, the unique goal of sciences, demanded that he disregard the "pretension of a writer" seeking glory and side with the system that offered "the beauty of order, the unity of plan, and the symmetric order of parts."[134] This criterion of truth should be applied especially to the science that had acquired immense riches in detail every day and found itself in a "state of crisis which quite ordinarily precedes great periods of growth." The days when first-rank chemists were divided on the most important points of theory were nearly over. Despite some remnant supports for the "ancien système," everyone "who examines the facts with impartiality, who submits the consequences to the rule of exact logic" recognized that the doctrine of phlogiston was no longer sustainable. In this regard, Guyton declared, "the revolution is consummated."

The events of the French Revolution swept up Lavoisier soon after he finished the *Traité*. If he intended to pursue the subject of affinity further, he never had the opportunity, since he was guillotined at the height of the Terror. We may never know what might have been. According to Daumas, Lavoisier would have dealt with affinity in his planned *Cours de chimie expérimentale rangée suivant l'ordre naturel des idées*, a project designed to cover all aspects of chemistry from the philosophical speculations on the constitution of matter to industrial chemistry, geology, and the construction of apparatus.[135] Indications are that he approved the general direction Guyton took in trying to bring the law of affinity toward that of universal gravitation. It is to his collaborators that we have to turn for further illumination of their communal "opinion."[136]

8

The Next Frontier

The legacy of the Chemical Revolution has been ambiguous for historians. Despite Lavoisier's heroic stature in the historiography, the most enduring elements of the Revolution, such as the analytic definition of elements and the nomenclature reform, were not his. The oxygen theory of acids and the caloric theory of heat, which could be regarded as his unique contributions, quickly fell into disrepute.[1] Furthermore, historians have not been able to link the Chemical Revolution to the nineteenth-century development of chemical atomism, despite the logical connection they pointed out between the analytic definition of elements and the chemical atom. The investigative tracks of European chemists after the Revolution did not proceed from one to the other. Without a historical pathway between the two, we are left with two distinct foundational moments of modern chemistry.

If we wish to solve this riddle of two fathers—Lavoisier and Dalton—we must grant multiple voices to the Chemical Revolution. The analytic definition of elements was a deeply entrenched part of the French didactic tradition that Guyton and other chemical teachers utilized in their lectures. The nomenclature reform, built on this definition, was an integral part of Guyton's vision as a literary chemist who accumulated his symbolic capital in the Republic of Science. His most comprehensive vision, however, consisted in the systematic and quantitative pursuit of affinities. In contrast, Lavoisier focused his attention narrowly on the studies of air and heat to resolve the algebraic contradiction the phlogiston theory fostered. The oxygen and caloric theories resulted from his efforts to assimilate British pneumatic chemistry and studies of heat into French theoretical chemistry. The identification of phlogiston as the matter of fire in the French didactic tradition made it necessary for him to eliminate it. Lavoisier's revolution thus hinged on the antiphlogistic crusade.

In contrast to the two senior lawyers-turned-chemists who valued the trans-disciplinary, public appeal of chemistry, Berthollet and Fourcroy were trained in the conventional track of medicine and more in tune with Macquer's affinity chemistry. Except for Lavoisier, then, affinity remained the leading theoretical concern of the Arsenal Group chemists. Berthollet had a broad investigative agenda of establishing the theories of acidity and alkalinity so as to understand the chemistry of salts and consequently their affinities, albeit with the principalist approach. Fourcroy, a leading Parisian teacher, articulated a chemistry of operations governed by affinity relations. He used determinate affinity relations to work out a new, empirical order of chemical composition in line with Lavoisier's oxygen theory. In other words, the two junior members of the Revolution refurbished the traditional theory domains of composition and affinity according to the new developments in pneumatic chemistry. These two domains became enmeshed in their investigations, laying out divergent paths for French elite chemistry. Although Berthollet started with the principalist approach, his investigations dealt mostly with the affinities of acids and alkalis in industrial processes. Grappling with these fickle, large-scale processes that often toppled the prediction of affinity tables, Berthollet came to envision a sophisticated program of affinity chemistry that enumerated all the physical forces associated with the chemical processes.[2] Fourcroy began with the affinity approach to composition, but he merged it with Lavoisier's principalist conception embedded in the oxygen theory to forge a comprehensive chemical classification.[3] Fourcroy's new order of composition assumed that the affinities of chemical elements produced chemical compounds with definite proportions.[4]

The combined authority of these four academicians helped defeat the phlogiston theory and institute the nomenclature reform throughout the European chemical community.[5] If they came together to accomplish these immediate goals of the Revolution, their long-term vision consisted of building a more sophisticated theoretical structure for chemistry. In addition to a comprehensive classificatory scheme designed to accommodate further developments in chemistry, they crafted a notion of chemical constitution as determined by the interplay of heat and affinities. This became an enduring legacy of the Chemical Revolution for French elite chemistry. The blueprint for this new conception of chemical "constitution," as Berthollet later named it, was contained in Lavoisier's *Traité*. Fourcroy's *Philosophie chimique* and Berthollet's lectures at the École normale in 1795 indicate that they shared this vision of chemical constitution.[6] They formed no consensus, however, as to how they should quantify and mathematize affinity and heat values to foster a viable line of chemical inquiry. If the Chemical Revolution was successful in cleaning the attic and thereby putting the household in order, it was less so in opening up new lines of inquiry to push the analytic and theoretical frontiers of chemistry. Berthollet's and Fourcroy's divergent approaches to chemical affinity and composition established a physicalist and a natural-history tradition in French elite chemistry, and those traditions remained in productive tension through the first half of the nineteenth century.

Among the four authors of the Chemical Revolution, Claude-Louis Berthollet offers the best guide to trace its successes and its failures.[7] According to the conventional story, he started out as a Macquerian phlogiston chemist, was converted to Lavoisier's theory in 1785 after the decisive experiment on the decomposition of water, became a junior partner in the nomenclature reform, and produced a grand theoretical system on affinities by the turn of the century. In this story, it is difficult to assign a continuous thread or a distinct identity to Berthollet's chemistry. His earlier, principalist moves—phlogistic and oxygen theories— do not lead to his later concern with affinity. If we wish to understand Berthollet's evolution as a chemist, therefore, we must place him in the broader context of eighteenth-century French chemistry. To a certain degree, we must obliterate the towering presence of Lavoisier in the historiography. In the 1770s, when Berthollet learned the basic tropes of chemistry, the leading figure in French chemistry was Pierre-Joseph Macquer, who personified the chemist in the service of the state and who become the role model for the younger generation of chemists. Berthollet's debt to Macquer was not limited to the phlogiston theory; he learned to appreciate the chemistry of salts as a theoretical subject and, along with it, the chemistry of affinities. Trained in medicine and entering the Parisian chemical scene just as Lavoisier's pneumatic studies began to make a splash, Berthollet combined his extensive hands-on knowledge of chemistry with the new theoretical structure to conceive a sophisticated program of affinity chemistry by the turn of the century. In many ways, he was the true heir of Macquer, a chemist who understood the importance of both theoretical and practical chemistry. He absorbed from Macquer an understanding that salts served as the model system of chemistry and that their affinity relations were of the utmost theoretical importance. Though he deferred to Macquer's authority and opinions, however, Berthollet did not follow the master's footsteps blindly. His defense of phlogiston was couched in principalist terms, a perspective he shared with Lavoisier.

Berthollet's participation in the Arsenal meetings produced important hybrid thoughts, although he always grounded them in chemical experience. He developed a deeper commitment to an understanding of chemical "constitution" as determined by heat and affinities. He taught this vision to the students at the École normale, articulated it formally in his *Essai de statique chimique* (1803), and he instituted it as the research paradigm for the Arcueil Group.[8] Through these efforts Berthollet effected a fundamental transition in French theoretical chemistry. The theory domains of composition and affinity merged into that of constitution. His sophisticated conception of affinity and constitution in the *Essai* failed to induce a viable research program outside of the elite Parisian chemistry, however, indicating the distance between the Arsenal Group's vision and the stoichiometric frontier of European chemistry. Only when chemists had stabilized chemical species through stoichiometric investigations did they go back to "Berthollet's problem" and begin to rework the theory domain of chemical affinity with a new set of experimental techniques.[9]

8.1 The Principalist Approach to Affinities

Claude-Louis Berthollet (1748–1822) was born in Savoy, a town at the French border of Italy where French culture and language dominated, with a measure of Genevan influence.[10] He studied medicine in Turin, obtaining the title of doctor in 1770. A few months later, he left for Paris. Through Théodor Tronchin, a Genevan who served as primary physician to Philippe Égalité, duc d'Orléans, Berthollet became an ordinary physician in the service of Madame de Montesson. The duke's family had long been an influential patron of Parisian science, as we have seen in Homberg's case, providing living and research facilities for some of the leading scientists. Louis-Philippe, father of Philippe Égalité, lodged several scientists—mechanics, physicians, chemists, physicists and naturalists—in his mansion and provided them with laboratories. Scientific courses and demonstrations were organized at the duke's residence.[11] While engaged as a private physician, Berthollet pursued a career in public medicine, going through a long and expensive process at the Faculté to obtain the title of docteur régent in 1781. He also took chemistry classes taught by Macquer and H. M. Rouelle at the Jardin. The chronological overlap and Berthollet's knowledge of pneumatic chemistry suggest that he attended Bucquet's classes at the Faculté.

From the beginning of his career, Berthollet took a serious interest in affinity chemistry. Salts constituted the main subject of his investigation—acids, alkalis, and caustics. His early investigations revolved around the composition and affinity of acids and alkalis. While exploring the affinities of salts, he applied the principalist approach to the classification of chemical substances and actions. According to that approach, the properties of a compound were determined by the principles it contained and their proportions. The first subject of Berthollet's research was air, which had been a "burning" question since the publication of Lavoisier's *Opuscules*. He took less interest in different kinds of air per se than in their relationship to the theory of acidity.[12] His first publication, on the decomposition of tartaric acid,[13] was followed by *Observations sur l'air* (1776), which contained theoretical reflections. Although Sadoun-Goupil has found this work notable mostly for its trendy subject of air and use of phlogiston theory,[14] Berthollet aimed at proving that fixed air was the universal acid, which testifies to the centrality of the acid question at this time.

The *Observations* consisted of two main parts, one on fixed air and the other on dephlogisticated air. In the first part, Berthollet set out to prove that fixed air was the "universal acid" contained in all vegetable, animal, and mineral acids, and that it was the principle of acidity. He showed that tartaric acid was "only fixed air joined to a small portion of oil." Vinegar was a combination of fixed air and inflammable air. Utilizing the principalist approach, he explained the relative strength of the two acids in terms of the proportion of the fixed air each contained. Tartaric acid was the most powerful vegetable acid because it contained more pure fixed air than others.[15] Berthollet did not characterize fixed air as a simple substance or "first principle." (According to Macquer's definition, a first principle constituted the last point of chemical analysis.) He thought it more likely that different kinds of air were modifications of simple air. Inflammable air would then be simple air saturated with excess phlogiston. Without a clear idea of its composition, however, he deemed it necessary to regard it as a simple being. The same argument applied to fixed air, although it seemed even more composed. Berthollet's passing reference to Macquer's "first principle" in the passage quoted below indicates that the analytic definition of elements or principles was widely in use on Macquer's authority, well before Lavoisier codified it more explicitly in the *Traité*:

> This inflammable air is not a species of particular air, a first principle: it is what the celebrated author of the dictionary of chemistry calls a principle of second order: it is probably simple air saturated with excess phlogiston; but until when one has a clear idea of its composition, it is necessary to regard it as a simple being; . . . I say as much of fixed air which appears nevertheless more composed & much more removed from simple air than inflammable air.[16]

In the part on dephlogisticated air, Berthollet adopted Macquer's modified phlogiston theory, which accommodated Lavoisier's recent works. A metal subjected to the action of nitrous acid exchanged some of its phlogiston for the air in nitrous acid, which Berthollet regarded as pure atmospheric air. While accounting for the weight gain of metals by the air fixed in the process, he also defended phlogiston emphatically:

> This pure air, this dephlogisticated air, whose discovery brings much honor to M. Priestley, & which makes part of nitrous acid & part of most metallic calces from which one can draw it, would confirm, if it were necessary, as this author remarks, the absence of phlogiston in metallic calces which several moderns have placed in doubt.
> What! Have the calces & precipitates of zinc & iron not lost the principle which renders these metallic substances inflammable? Could you make a

metallic calx or precipitate reappear in its natural form without the union of this principle which forms combustion, heat, light; without the union of the principle of fire which is found accumulated in charcoal?[17]

While conceding the protean attributes of phlogiston, Berthollet sought to answer the objections raised against it. As for the objections based on the reduction of mercury calces without charcoal, he reckoned that mercury could unite easily with fixed air and atmospheric air losing little or none of its phlogiston. Trying to vindicate Macquer, Berthollet took a stronger stance against Buffon's opinion in *Introduction à l'histoire des minéraux*. Advocating the uniformity of matter, Buffon characterized chemists' phlogiston not as a "simple and identical principle," but as a "compound, product of mixture, the result of the combination of two elements, air and fire, fixed in bodies." Macquer took Buffon's critique seriously because of his stature in the public sphere.[18] Berthollet dismissed Buffon's idea as obscure and incomplete, defending Macquer's position that phlogiston was fire that became a principle of bodies. It passed from one compound to another, adding and removing different properties. As befitted a physician, Berthollet concluded the second part by discussing the salubrity of air. The more deprived the air was of phlogiston, the less fit it was for respiration.[19]

Despite its somewhat meandering organization (perhaps a sign of hurried publication), the *Observations* won praise for its "happy precision" in communicating ideas and experiments—no doubt bringing a measure of recognition from Macquer, who acted as a gracious patron.[20] Berthollet's debt to Macquer ran deeper than the phlogiston theory. Scattered throughout the second part of the *Observations* were his thoughts on affinity. He assumed that the actions of salts were governed by their affinities or rapports, and he took a particular interest in the affinities of acids. In discussing gunpowder, for example, Berthollet wished not only to confirm the theory of detonation but also to discover the "rapports of niter with other salts."[21] Priestley's "vitriolic air" was inferior in its affinities to nitric and marine acids. Although these remarks do not imply a systematic understanding of or an exclusive interest in the subject of affinity, they are evidence of Berthollet's firm grasp of the affinities that operated in concrete experimental systems.[22] Despite the infusion of pneumatic chemistry, the working foundation of French chemistry at this time consisted of the affinities of salts. Berthollet sought to establish connections between pneumatic studies and the affinity chemistry of salts in the best way he could.

After the *Observations*, Berthollet presented a rapid succession of papers to the Academy, which led to his election in 1780. Though historians have noted these papers mostly for their solid defense of the phlogiston theory,[23] Berthollet did not preclude the involvement of various airs in the weight equation of chemical processes. Besides, the chemistry of salts provided a broader framework for his early works. Phlogiston was much attended to as an important substance, but not exclusively, in his pursuit of the theories of acidity and alkalinity. First came, therefore, a series of papers on acids. In a paper on sulphurous acid, according to Sage and Cadet's report to the Academy, Berthollet prepared a variety of sulphurous salts by mixing concentrated sulphurous acid and diverse alkalis. He then compared these sulphurous salts to Stahl's (which had been prepared by exposing fixed vegetable alkali to sulphur vapor in combustion), trying to establish their identity. In distillation, they all produced nearly the same quantity of air, which he identified as atmospheric air, along with various residues. In trying to obtain sulphurous acid separate from vitriolic acid, he also came to the conclusion that sulphurous acid was a dissolution of sulphur in vitriolic acid.[24] In studying "fixed sulphurous air," he decomposed liver of sulphur in the air to observe the release of fixed air. This loss of fixed air, rather than phlogiston, accounted for the weight loss.[25] Though it is difficult to assess Berthollet's theoretical thoughts from this matter-of-fact review, it seems entirely possible that he was trying to broaden the scope of his theory of universal acid by proving the existence of fixed air in sulphurous acids and salts.

Berthollet's paper on nitrous acid was reviewed by Cadet and Lavoisier.[26] Since Lavoisier had a strong agenda by this time of establishing vital air as the acid principle, it is not possible to discern Berthollet's voice. Berthollet had apparently obtained fixed alkali and dephlogisticated air by heating pure niter. Since niter was composed of nitrous acid and fixed alkali, this meant that nitrous acid was converted to dephlogisticated air. This conclusion contradicted the experiment in which Priestley had obtained much nitrous acid as well as dephlogisticated air. The discrepancy raised the possibility that nitrous acid had escaped during Berthollet's experiment and combined with water. To settle the matter, the rapporteurs repeated Berthollet's experiment with a more complicated apparatus. Lavoisier, of course, had his own theory to test on this occasion. He presumed that dephlogisticated air was the "common principle" of all acids and that nitrous acid was composed of nitrous air and dephlogisticated air. However, he found only dephlogisticated air. Unable to theorize in a satisfactory manner, the rapporteurs recommended that Berthollet continue with his experiments to discern what had happened to the base part of nitrous acid, the "principle which, combined with dephlogisticated air, constitutes nitrous acid." They nevertheless acknowledged the importance of the subject matter:

The acids being the instruments we employ habitually for analyzing bodies, it is of a great importance to know their nature well, and it would appear that in this regard we touch on the epoch of a happy *revolution* which will not fail to have a great influence on chemistry.[27]

Contrary to historians' characterization that Berthollet started out as a staunch believer in phlogiston, his early papers reveal deference to Macquer's authority more than dogmatic allegiance to the phlogiston theory. Berthollet was a young investigator interested in the theories of acidity and alkalinity, treading cautiously in uncharted waters. In his studies of acids, he made room for the involvement of airs as the cause of weight gain or loss, thereby accommodating Lavoisier's discoveries without overstepping the boundaries of the phlogiston theory. Berthollet's early interest in the theory of alkalinity can be detected in his presentation on volatile alkali.[28] He claimed that caustic fixed alkali was converted, with repeated cohobation, to volatile alkali and an earth.[29] Lavoisier and Macquer, as rapporteurs, were skeptical: If this was indeed the case, one would have to conclude that fixed alkali was a compound of these two principles, which would be an important discovery for chemistry. They did not believe, however, that Berthollet's experiments were solidly established, and they recommended a delay in publication until the existence of volatile alkali could be demonstrated with a "degree of perfection."[30] Berthollet returned to the subject a few years later. First, he announced a sophisticated method of preparing pure caustic alkali.[31] After his "conversion" to Lavoisier's camp, he ventured a theory that volatile alkalis contained inflammable air as a constituent part.[32] Regard-less of the merit, Berthollet's sustained interest in the theory of alkalis reveals the framework within which he shaped his investigative interests: the chemistry of salts. He tackled the theory of acid, then the theory of alkali, then the theory of causticity.

In choosing to work on the problem of causticity, Berthollet entered an arena dominated in France by Macquer's authority. The choice also reflected his interest in organic realm, particularly in the relationship between vegetable and animal substances.[33] As a category of phenom-ena, causticity referred to the property that made acids, alkalis, and certain middle salts attack and destroy organic substances on contact. Joseph Black attributed the loss of the causticity of alkalis to their union with fixed air. Frédéric Meyer attributed causticity to acidum pingue or causticum. Macquer acknowledged in the second edition of the *Dictionnaire* that the element of fire "possessed causticity in the most eminent degree," but he did not single out a principle or an element as the cause. Instead, he defined causticity as "the force with which the parts of caustics tend to unite with the parts of other bodies."[34] Along with dissolution and combination, causticity defined for Macquer one of the major categories of chemical phenomena for which the operating forces of affinity had to be examined in a more general way. In view of the importance Macquer accorded to the phenomena of causticity, it is not surprising that Berthollet tackled this problem in preparation for his elec-tion bid to the Academy. In 1779 and 1780 he read four papers devoted to the causticity of metallic salts, acknowledging his debt to Macquer deferentially:

No person appears to have developed the principle of the action of caustics as well as M. Macquer has; this is an effort they make to combine, a tendency, a consequence of their affinities and I pay homage to him for my ideas if they shed some light on the subject.[35]

Berthollet's theory of causticity differed substantially from Macquer's, however. The importance of causticity for chemistry lay in the power bodies exhibited in decomposing one another by simple contact. As Bayen's investigations had shown, mercury was absolutely inactive in its natural state, but it became the most poisonous caustic in the form of corrosive sublimate. Its action was much more moderate in mild mercury [mercure doux], and even more so in calomel [calomélas] and mercury panacea [la panacée].[36] While most chemists attributed this difference of causticity to the different proportion of mercury in these combinations, Macquer attributed it to their different states of aggregation. The aggregation of the parts of mercury in corrosive sublimate was ruptured so that they could exercise their "tendency to combination" on the bodies that came into contact with them, exhibiting an extreme degree of causticity. Whereas Macquer adroitly avoided giving a principalist explanation of causticity, Berthollet embraced it wholeheartedly by attributing the causticity of corrosive sublimate to the deprivation of phlogiston. In Berthollet's theory, the absence of fire or phlogiston induced metallic salts to seek out the phlogiston contained in other substances. This accounted for their strong causticity.

Berthollet further elaborated this principalist theory in his presentations on the causticity of metallic salts.[37] He again praised Macquer's sagacity in defining causticity not as an attribute of the matter of fire but as "the dissolving action of substances . . . the force with which their integrant parts tend to combine & unite with the parts of other bodies." Vowing to apply Macquer's "luminous principle to the action of different caustics," however, Berthollet reiterated his principalist position that the deprivation of fire or phlogiston induced causticity. Ostensibly endorsing Macquer's view, he set out to prove that the phlogiston content determined causticity. Not only was the mildness of mild mercury due to its higher phlogiston content; additional phlogistication could further reduce its causticity, producing calomel and mercurial panacea. The causticity of other mercurial salts (such as corrosive precipitate and white precipitate) in Bayen's experiments also depended on their phlogiston content. In short, Berthollet endeavored to prove that the decreasing order of causticity in corrosive precipitate, white precipitate, mild mercury, calomel, and the panacea revealed their increasing phlogiston content. He obviously took phlogiston more seriously than Macquer, who hesitated to single out fire as the unique cause of causticity. He pinned the loss of causticity on phlogiston, while Macquer attributed it to the arrangement of integrant parts. His thoughts on causticity exemplified the principalist approach to composition, which was diametrically opposed to Macquer's affinity approach.

Aside from his zealous defense of the phlogiston theory, Berthollet's concern with affinities is visible throughout his works. Explicitly citing Bergman's work on elective attraction, which had given the subject a stronger appeal and urgency, Berthollet thought through his experiments with the known orders of affinity or elective attraction.[38] In order to discern his prevailing interest in affinity, we must pay close attention to his experiments and deductions. In his experiments on mercury salts with nitrous acid, the mixture of mild mercury and pure nitrous acid produced upon heating an effervescence and a large quantity of red vapor. He knew from Bergman's work that nitrous acid had a strong affinity for phlogiston, enough to detach it even from corrosive sublimate. Thus, he speculated that nitrous acid combined with the phlogiston contained in mild mercury and turned it into a red precipitate and corrosive sublimate, which were deprived of phlogiston. This was why corrosive sublimate needed running mercury to form mild mercury. In this reverse process, the "tendency to combination" of corrosive sublimate would work on the phlogiston in running mercury.

The rapporteurs, Macquer and Lavoisier, prepared a comprehensive report on all three papers. Macquer reminded the audience that he had defined causticity "in general as a tendency to combination," only hinting at his disagreement with Berthollet's principalist interpretation. When one removed from a substance one of its principles to which it had much affinity, it tended to repossess it with an active force at the expense of all the bodies with which it came into contact. Since Berthollet attributed this general principle of causticity to the deprivation of phlogiston, it had to follow that more caustic substances contained less phlogiston. Causticity, therefore, had different degrees. The facts seemed to agree strikingly well with the theory. Macquer and Lavoisier praised the memoir for "a great ensemble of new facts, very fine observations, a theory if not demonstrated, at least rendered probable and established on the facts well linked and well presented." Despite their endorsement of the factual content, the rapporteurs were skeptical of the theoretical part, perhaps reflecting Lavoisier's conviction on the futility of phlogistic explanations:

In general, although the theory of M. Berthollet may be quite ingenious, although it appears to satisfy the explication of the greatest number of phenomena, finally although it is linked perfectly with the theory of Stahl, we do not regard it as demonstrated. Phlogiston is a principle one has abused in recent times to explicate everything. While M. Berthollet attributes with some probability the causticity of corrosive sublimate to the absence of phlogiston, M. Baumé, in his Chemistry, attributes it to its presence and superabundance. When a subject is obscure and new, it is easy to link together, by plausible hypotheses, the small number of known phenomena it presents; but, it is no longer the same when the facts are multiplied to a certain point: one more experiment often suffices to reverse a theory which would appear solidly established, and it is no longer as easy to imagine new ones.[39]

Berthollet's early papers reveal that he regarded Macquer as the reigning authority in chemical opinion. He wished to usurp the master's authority in his experimental endeavors by defending the ailing phlogiston theory, explaining away its deficiencies and extending its scope, perhaps to a degree Macquer himself was not willing to endorse. Berthollet's principalist approach to chemical composition and reaction differed fundamentally from Macquer's affinity approach, however, and was closer to Lavoisier's. If he was too deferential to the authority figure to see the radical tide of theoretical change Lavoisier had initiated, he gained rapidly in experimental skills. Working through the thicket of the chemistry of salts, he had to sort out their complex affinity relations, although he approached the affinity question with the principalist view. Composition, or the principles contained in each body, was to determine its affinities.

8.2   Between Phlogiston and Oxygen

By 1780, Berthollet could look forward to a promising career in science. After his election to the Academy, he became a part of Lavoisier's circle at the Arsenal laboratory. By this time, Lavoisier had all the elements of his oxygen theory in place and was conducting a series of extensive investigations to consolidate a new system of chemical theory. Although Bucquet had perished in the meantime, Lavoisier's collaboration with Laplace continued to refine the concepts of heat and affinity. Berthollet watched and debated these efforts at a close range. One can only imagine the kind of exchanges that took place in this rather intimate setting among the reigning scientists of the time, which consolidated the distinctly physicalist outlook that French chemistry came to possess in the ensuing years. The evolution of Berthollet's thought in the 1780s offers important clues to the interaction among the Arsenal Group scientists. Berthollet had a much broader understanding of chemistry than Lavoisier. Building a theoretical system was not his primary goal. He was a versatile chemist involved in many facets of applied chemistry and practical operations. He tended toward concrete experimental systems, taking detours and revising theories when necessary. He also maintained an active interest in the study of organic nature in relation to concrete industrial applications. It is not possible, therefore, to offer a comprehensive picture of Berthollet's work until we are better equipped to deal with the relationship between chemical practice and representation.[40] In the meantime, however, we can at least attempt to modify the traditional assessment of his "conversion" to Lavoisier's theory.

In the early 1780s, Berthollet began to piece together a phlogistic theory of acids. His works made explicit the relevance of phlogiston to the theory of acids in the way Guyton's "Dissertation" brought out the centrality of phlogiston in understanding calcination and combustion reactions. By doing so, Berthollet offered the strongest and the best-coordinated defense of the phlogiston theory in France, just as Lavoisier began to step up his antiphlogistic campaign. Nitrous acid assumed a central position in Berthollet's investigation because of its involvement in a variety of phenomena linking the vegetable and animal kingdoms.[41] In 1781, he read three memoirs on the decomposition of niter, following on his earlier paper on the same subject.[42] Apparently, he took up Lavoisier and Cadet's challenge to resolve the contradiction between his and Priestley's results.[43] He also wished to test the validity of Lavoisier's hypothesis. Drawing on distillations of niter with a variety of substances, from which he obtained varying quantities of "dephlogisticated air," Berthollet sought to defend the phlogiston theory while taking Lavoisier's "ingenious system" into account. Although he acknowledged that Lavoisier had proven by "incontestable experiments" that the atmospheric air was involved in the calcination of metals, he took issue with the corollary that phlogiston was an unnecessary hypothesis, or that metals and charcoal did not share a common principle. In Lavoisier's system, the weight gain of metals depended on the portion of the atmospheric air absorbed by metals. Phlogiston, a principle supposedly shared by metals and charcoal, did not play a role in this interpretation, since charcoal only combined with the base of pure air to form carbonic acid. Berthollet endeavored to show that metals and charcoal produced the same phenomena with niter or that they combined with the dephlogisticated air contained in niter to produce carbonic acid or phlogisticated air. It was necessary, therefore, "to admit in metals & in charcoal an identical principle."[44] Since metals formed carbonic acid with dephlogisticated air, carbonic acid could not be a combination of charcoal and a part of dephlogisticated air, as Lavoisier interpreted.

Berthollet raised a strenuous objection to Lavoisier's "matter of heat," asserting the necessity of phlogiston in accounting for the phenomena of heat, light, and flame in combustion processes.[45] Had Lavoisier simply pretended that elastic fluids contained a greater quantity of the principle of heat than their liquid counterparts, he would have been in line with other physicists. However, he should not have attributed to this single principle all the phenomena one ascribed to phlogiston, "a principle that would enter into the combination of bodies, would not vary according to its temperature, but would be essential to constituting what they are."[46] The flame and other effects of combustion could not be attributed to the loss in the volume and the elasticity of air, as Lavoisier had argued. In the detonation of niter with charcoal, there was released a certain quantity of elastic fluid which, according to Lavoisier's scheme, should have produced cold rather than heat. The matter of heat and phlogiston had to be two distinct entities, at least at the operational level of chemistry:

It is necessary to distinguish the matter of heat from phlogiston, although these two substances would appear to be only a modification of the same principle, & although often one would appear to be changed into the other. Here are examples: perfect metals are revivified by heat & light, as they are by the phlogiston of another metal which precipitates them from an acid in metallic form. Nitrous acid without color being exposed to heat in hermetically sealed tubes, becomes fuming, as M. Priestley has observed, & M. Schéele says that the rays of sun suffice to color nitrous acid. It is thus that phlogiston would appear to be changed into the matter of heat, when one dissolves zinc in nitrous acid; for this metal abandoning phlogiston in its dissolution by nitrous acid gives only a very small quantity of nitrous gas, & a mediocre quantity of phlogisticated air; but there is released from this dissolution a great heat: in parallel if one dissolves iron filings in concentrated nitrous acid in a fashion that effervescence may be quite live, nearly all the gas which is released is phlogisticated air, & not nitrous gas; by its spontaneous decomposition nitrous gas is converted in great part into phlogisticated air.[47]

Berthollet understood that Lavoisier's efforts to remove phlogiston derived from an algebraic understanding of chemical operations. He thus worked out a detailed algebra of heat to criticize Lavoisier's theory of acid, which attributed acidity to the base of vital air. If nitrous acid contained only this "acidifying base," deprived of the principle of fire, the flame and the heat released in the detonation of niter with sulphur did not come from nitrous acid, but from sulphur. Sulphur should contain, then, much principle of fire that was released in detonation. The effects of detonation should be proportionate to the quantity of the sulphur one used. Vitriolic acid would then be composed of the base of sulphur, deprived of the principle of fire by combustion, and the pure part of the atmospheric air, or the part fixed in nitrous acid. Therefore, neither the detonation of niter with charcoal, sulphur, and metals nor the combustion of these substances would demand explications adapted to each circumstance, as Lavoisier's system required. If one attributed inflammation to the principle of fire contained in the aeriform fluids, the combustion of two parts of inflammable gas, which was lighter than dephlogisticated air by a factor of 10, and one part of dephlogisticated air should produce heat of which only 1/21 came from the dephlogisticated air and all the rest from the inflammable gas. In the combustion of charcoal, however, all the heat produced supposedly derived from the dephlogisticated air. Such a disparity in the quantity of heat produced by the same substance called for an explanation. Despite their opposite theoretical stances, Berthollet and Lavoisier shared the principalist approach to composition expressed in the algebra of weight and heat. Berthollet's response offers an important clue to why Lavoisier strove to quantify his caloric, particularly the heat involved in chemical reactions. As long as caloric remained a speculative entity, it did not have a decisive advantage over the rival theory of phlogiston. Condorcet was baffled as to why these two chemists debated over essentially similar theories of combustion:

[Berthollet] terminates his memoir with reflections on the existence of this principle admitted nearly generally for some years, & today regarded as useless by several chemists among whom M. Lavoisier is an example. Perhaps, these two opinions are not as opposed to each other as they would appear at first glance; one agrees on the one hand that the doctrine of Stahl has necessarily to be modified; one admits on the other, combinations or release of light & heat of which the secret is still unknown; finally, could one not observe that these opinions are already a little closer than they were in the beginning? & does not one have the right to hope that after some new experiments which remain to be tried, they will end up being reunited?[48]

Condorcet's hopes were fully justified in that Berthollet accepted a significant tenet of Lavoisier's theory: that the fixation of air caused the weight gain in acidification. Sadoun-Goupil thus interpreted Berthollet as having progressively distanced himself from the phlogiston theory and moved toward the oxygen theory between 1782 and 1783. The Arsenal gatherings probably disposed Berthollet somewhat favorably toward Lavoisier's theory. In 1782 he reexamined sulphurous and vitriolic acids, offering a "conjecture" that sulphur, sulphurous acid, and vitriolic acid were compounds of the same base with varying proportions of phlogiston and vital air. While he retained phlogiston, he also accepted the presence of vital air in vitriolic acid to account for the weight gain in the combustion of sulphur:

I regard sulphur as a combination of phlogiston with a base which is common to it & vitriolic acid, & I regard vitriolic acid as a combination of this same base with vital air deprived of its elasticity. It appears that sulphurous acid contains proportionately less of the aerial principle than vitriolic acid, & less of phlogiston than sulphur; the balance which is found between these two principles & the base common to vitriolic acid & sulphur, is ruptured by heat; phlogiston is united more intimately with a part of this base, & the aerial principle, which was a third of this part, is combined in parallel with the other part which is abandoned by phlogiston. There results from this two simpler & thus more perfect combinations: sulphur or the combination of the base of vitriolic acid & phlogiston; & vitriolic acid or the combination of the aerial principle & the base . common to sulphur & vitriolic acid.[49]

In another paper on the acidification of sulphur, phosphorus and arsenic, Berthollet again attributed the weight gain to the "aerial principle," explicitly crediting Lavoisier for this discovery.[50] In burning sulphur and phosphorus, he sought not only to quantify the weight gain accurately but also to measure the ratio of the acid to water in each product. He claimed that his results on the combustion of phosphorus agreed perfectly with Lavoisier's. The purpose of his work was less the confirmation of Lavoisier's results than a more accurate determination of the ratios of the aerial principle in acids. Nevertheless, Berthollet's increasing sophistication in quantitative measurements speaks for the interactive environment of the Arsenal Group.

Berthollet utilized the phlogiston theory for the last time in explaining the differential strengths of acetic acid [vinaigre radical] and acetous acid [vinaigre distillé].[51] De Lassone had observed that acetic acid had more action than even the most concentrated acetous acid. While confirming this observation, Berthollet concluded from his own experiments that acetic acid exhibited essential differences from acetous acid: It had a "superior affinity" with alkalis, it formed with them a "more perfect combination," and it resisted the action of heat better. The difference was due, as in the case of sulphurous and vitriolic acids, to the different proportions of phlogiston and aerial principle. The exchange of phlogiston for the aerial principle heightened its acidic properties. If Berthollet still retained phlogiston in his explanations, his principalist outlook coupled with quantitative argument was in perfect accord with Lavoisier's.

8.3   The "Conversion"

Berthollet's posture as a young, confident, energetic defender of the phlogiston theory was seriously threatened by Lavoisier's intensifying campaign and Macquer's death. He had already accepted that pure air was involved in a variety of operations, causing the weight gain. Lavoisier's system could easily accommodate his experiments on sulphuric and nitrous acids. Above all, he shared with Lavoisier the principalist outlook fortified by algebraic thinking. The interlocking algebra of chemical operations involving air favored Lavoisier's system. In addition, Lavoisier and Laplace presented in 1783 a method of quantifying caloric. Unlike phlogiston, which defied the rule of the balance, caloric now possessed a more concrete status as an object of quantitative experimental investigation, although it took some time for Berthollet to accept it fully.[52] The scene was also changing on the social side after Macquer's death in 1784. The symbolic leadership of French chemistry now lay with Lavoisier, at least within the Academy. Unless he was willing to fight Lavoisier on all fronts of theory, Berthollet had little choice but to concede. As the last straw, the decomposition of water made it impossible to identify phlogiston with inflammable air, the most promising version of the phlogiston theory then in circulation. Phlogiston still eluded chemists' analytic control.

Berthollet became the first major French chemist other than the deceased Bucquet to join Lavoisier's camp. His 1785 "conversion," followed by those of Fourcroy in 1786 and Guyton in 1787, has been a much noted event in the historiography of the Chemical Revolution. Historians agree that Berthollet abandoned phlogiston in 1785 and that the water experiment provided the impetus.[53] This was an account provided by Berthollet himself when he declared that phlogiston had finally become a "useless hypothesis" despite the long and productive service it had rendered in guiding chemical research.[54] His persistent efforts (begun in May of 1785) to win over Guyton to Lavoisier's camp leave no doubt that he was convinced of the superior explanatory power of the oxygen theory. He counted "the decomposition of water, a more exact knowledge of the nature of nitrous acid, just ideas of the principle which gave elasticity to gases and fluidity to liquids, a better determination of the affinities of vital air" among the reasons for his conversion.[55] The first paper that signaled Berthollet's "conversion" was presented in a public session of the Academy on April 6, 1785.[56] Scheele and Bergman had shown that the removal of phlogiston from muriatic acid by manganese resulted in dephlogisticated marine acid, a "substance quite avid for phlogiston." Berthollet reinterpreted their experiments, arguing that the dephlogisticated air of manganese combined with marine acid to form dephlogisticated marine acid, a "truth" Lavoisier had announced some time earlier. In December, he read another memoir on the same subject, providing a more detailed discussion. Invoking the decomposition of water and Laplace's production of inflammable gas by the dissolution of metals, he declared that phlogiston had finally become a "useless hypothesis."[57]

In 1785, Berthollet published a series of papers in which he openly supported Lavoisier's system. He deployed the new system, however, to explain the traditional domain of salts and their affinities. These papers reveal his dexterity as an experimenter with the capability to identify and to follow a common thread in navigating uncharted waters, weaving together various genres of experiments. Unlike Lavoisier, who tended to tackle well-established experimental systems with more refined techniques of analysis, Berthollet continued to expand the material realm of chemistry and to discover hidden relationships. For the experiments he conducted in 1785, dephlogisticated marine air (chlorine) provided the common thread. Bergman had explained the long-standing mystery of the dissolving power of regal water by stipulating the intermediate production of dephlogisticated marine acid. He supposed that nitrous acid extracted the phlogiston contained in marine acid and changed it into dephlogisticated marine acid. Berthollet explained it instead as the combination of vital air with marine acid. In addition to various experiments related to this claim, he also investigated the affinity relations of marine acid in comparison to other acids.[58] Dephlogisticated marine acid was also implicated in the composition of spirit of wine and ether, which assumed added significance with the controversy over the identity of inflammable air. Berthollet endeavored to clarify a series of experiments by Priestley, Landriani, Monge, Lavoisier, Gallisle, and Scheele involving spirit of wine and inflammable air.[59] His experiments were not exclusively devoted to verifying Lavoisier's system. He continued a broad line of inquiry designed to illuminate the composition of acids and alkalis, which would eventually unpack the relations of affinity and causticity. He thus turned to his earlier experiments on alkalis, hoping to gain a "more exact knowledge of the nature of volatile alkali" and thereby to shed light on the operations that involved this alkali. Refuting earlier theories that volatile alkalis contained phlogiston, he concluded that they contained the inflammable air.[60]

Despite the open display of his loyalty, Berthollet's choice of marine acid as the critical evidence in support of Lavoisier's theory makes it difficult for us to believe that his "conversion" was occasioned by scientific reasons alone. Just as he took up the cause of phlogiston vigorously for Macquer, he now fought against it on Lavoisier's side for various practical and intellectual reasons.[61] His earlier experiments on sulphuric and nitrous acids were more conducive to Lavoisier's theory since their strengths increased with higher degree of oxidation. The case of dephlogisticated marine acid did not fit the pattern. According to the oxygen theory, dephlogisticated marine acid had to contain more oxygen than marine acid because it was obtained from marine acid by the action of an oxidizing agent, manganese (manganese dioxide). It had to be a stronger acid as a consequence. Despite his own expectations to this effect, however, Berthollet had to conclude that it was "nearly entirely deprived of acidity."[62] It had a harsh taste rather than the sour taste of acids, it destroyed vegetable colors instead of reddening them, and it produced no effervescence with potash or soda.

Not surprisingly, Berthollet soon found himself on a divergent path from Lavoisier with regard to the composition of acids.[63] In a memoir presented in December of 1787, he showed that prussic acid was a compound of hydrogen, azote, and carbon. It did not contain oxygen.[64] In 1788, he tackled Lavoisier's assumption that metallic oxides, not usually acids, acted as such with respect to alkalis because of their oxygen content. He showed instead that, although they acted as acids when combining with alkalis, they also acted as alkalis when combining with acids.[65] By 1789, Berthollet could state more generally that oxygen was not the acidifying principle. Resuming his earlier work on sulphurous and sulphuric acids, he pointed out several facts that he thought should "restrain the opinion of M. Lavoisier on the nature of acids," if not deliver a blow to the theory. First, the abundant presence of oxygen in a body did not necessarily make it an acid, as was shown in the case of water. Second, several acids (muriatic, fluoric, boracic, lithic, prussic) did not contain oxygen.[66] In 1796, he added sulfuretted hydrogen gas to the list.[67] Berthollet's objections to the oxygen theory of acidity went unheeded, however, despite the long list of reliable experimental results. His alternative conception of acidity and affinity was too complex to be implemented as a laboratory program.[68]

8.4 Heat and Affinities

In contrast to his fleeting commitment to the oxygen theory of acidity, Berthollet maintained the dynamic understanding of chemical constitution laid out in Lavoisier's *Traité*, which became the conscious theoretical frontier of the Arsenal Group. In addition to Guyton, who had been nurturing a concept of phlogiston quite similar to Lavoisier's caloric in its dissolutive function, Fourcroy also incorporated heat into his chemical system. In the *Philosophie chimique* (1792), he sought to axiomatize chemical knowledge. Light, heat, and air occupied prominent places in this theoretical structure, constituting three of the twelve general phenomena that supposedly encompassed all facts and experiments of chemistry. Fourcroy admitted that caloric was the cause of heat which penetrated all bodies and dilated them by separating the molecules and diminishing their attractions. It possessed "different attractions or diverse degrees of affinity for different bodies." Many bodies existed in this state of "invisible dissolution in caloric" which prevented them from combining with other bodies. When bodies united to produce a new combination, they often lost caloric. All this proved that caloric was a "particular body & not a modification of all bodies."[69] Berthollet shared this conception of heat as a part of chemical constitution with Guyton, Lavoisier, and Fourcroy.

The French Revolution interrupted Berthollet's chemical work during the 1790s, when he worked on a variety of commissions dealing with industrial, educational, and administrative reforms. In the revolutionary turmoil, Guyton and Berthollet emerged as the leading chemical experts in the service of the government. They became the first members of the new Institute created in 1795, and professors at École centrale des travaux (which later became the École polytechnique). Despite his lack of teaching experience, Berthollet was also entrusted with the chemistry course at the École normale, which was created to train teachers in a short period of time.[70] His lectures at this elite teaching institution in 1795 deserve special mention because they reveal much more clearly what he had come to accept as the basic tenets of the new chemistry. They also provided an opportunity for him to discuss various inconsistencies in the accepted ideas on affinity. These lectures were designed to encourage the students to conduct an active "conversation" with the teacher. That provision probably made them more appealing to Berthollet, who had experienced an active intellectual exchange at the Arsenal.[71] Instead of starting with the simplest chemical properties and following a course of experiments, as he deemed was necessary for an elementary course of chemistry, Berthollet expounded the most theoretical and the most conceptual part of chemistry. The constitution of bodies as the balance of attractions and repulsions was at the core of the new chemistry.

Berthollet's lectures at the École normale reflect his changing perspective on the science of chemistry, now organized around the concept of affinity. He defined chemistry's place among the sciences in a way similar to Venel's classification in the *Encyclopédie*, except for his use of the Newtonian language of attraction. Physics comprehended all phenomena of nature. "General physics" dealt with the phenomena common to all bodies, such as movement, weight, and the action of heat, which followed general laws. "Particular physics" comprised studies of electricity, magnetism, light, and sound. There was yet another class of phenomena which resulted from the "intimate action of the molecules of bodies." Of these, the phenomena that did not belong to organized bodies composed the domain of chemistry. Affinity or chemical attraction was the principle of chemistry. It should not be confounded with general attraction because it worked at insensible distances. Berthollet's history of chemistry reflected the central importance he accorded to the concept of affinity. The origin of philosophical chemistry lay in the recognition that there existed a force that caused the molecules of bodies to tend toward one another and was variable according to the nature of different bodies. Geoffroy's table opened the second period; Bergman's ushered in the third. Affinity tables presented not only the results of a great number of experiments but also a comparison of the forces that produced the phenomena. They offered a great advantage in the elementary course: "They clear the mind, they speak to imagination, they commit facts and their causes to memory; thus, it is convenient, in each lesson in which one has to explain phenomena produced by a complex affinity to expose the figurative table for it."[72]

Affinity and heat constituted the most theoretical part of chemistry. In line with the pedagogical purpose of the new school, which was to train the students in basic methods or the "art of teaching," Berthollet assumed an elementary knowledge of chemistry on the part of the students and sought to provide "exact notions of theories which serve as the basis of all the development of teaching and lead their spirit to general ideas." These theories for chemistry comprised affinities and the principle of heat or caloric. Chemical attraction was the principle of all chemical phenomena or "the immediate cause" of dissolution, combination, and composition. If the molecules of bodies obeyed only this

force which tended to bring them together in a union, they would form
an inert mass. But these molecules tended to be moved away from one
another by another force, that of heat. The principle of heat, or caloric,
could be considered as an elastic fluid that penetrated the molecules of
all bodies and tended to separate them. From this point of view, it
belonged to general physics. The caloric also entered into the combinations of bodies. From this perspective, it belonged to chemistry. The distinction between sensible heat and latent heat derived from this dual
function of caloric. Therefore, the interplay of chemical attraction and
caloric determined the composition of bodies.

Berthollet's lectures included a classification of bodies and their affinities. Principles were simple substances that exercised most action on
other bodies and entered into the greatest number of combinations.
Rather than philosophically defined elements, they were the last point of
analysis. They comprised oxygen, hydrogen, carbon, azote, sulphur,
alkalis, and simple earths. Berthollet reduced the systematics of affinities
to three, depending on the degrees of the same power: simple attraction,
elective attraction, and complex attraction. These species of attraction
were "all the same force, considered in different aspects and under different circumstances." Simple affinity or attraction worked between two
principles or two substances that, although composed of different principles, acted only by a collective force. When their affinity was weak, the
two substances preserved much of their individual properties, as in the
solution of sugar in water. When the affinity was strong, as in the case
of acids and alkalis, the properties of these substances disappeared in
forming a new combination. Elective affinity applied to the cases when
one body chased off another from a combination, as Geoffroy had shown
in his table of affinities. Berthollet thus limited elective affinity to the case
of displacement reactions, rather than characterizing it as a general
notion of selectivity, as historians have tended to do. Accordingly,
complex affinity referred to the cases where more than three forces were
involved. Bergman's pictorial table and Kirwan's notions of divellent and
quiescent affinities addressed these cases.

In the second lecture,[73] Berthollet took up the "real action of affinities, and the causes that can modify their effects," or the "anomalies of
chemical attraction." He was well aware of the pre-revolutionary dis-

cussions of the phenomena of reversibility and partition by Cornette, Kirwan, and Lavoisier, which came back to life in his experience with the large-scale industrial production of niter.[74] First, he pointed out as a major error the confounding of elective attraction with complex attraction. Many operations that yielded results apparently in contradiction with the affinity table were effected by more than three affinities in action. Second, the phenomena often were attributed to a wrong principle in the estimation of affinities. In the mutual precipitation of metals from their dissolution in acids, for example, it was a mistake to invoke the affinity of metals for acids, rather than for oxygen, as the principal cause of these precipitations. It was an even more serious error to suppose that the attraction of metals for a "phantastic being, phlogiston" caused their precipitation. Berthollet warned of the dangers of hypotheses and the seductions of systems that hampered the progress of knowledge. Third, the union of several principles offered another anomaly. If one poured a dissolution of ammoniac on the dissolution of corrosive muriate of mercury, there formed a precipitate composed of oxide of mercury, muriatic acid and ammoniac. Muriates of ammoniac and mercury remained in solution. That is, the ammoniac was divided between the solution and the precipitate. This illustrated that combinations that differed only in the proportion of principles could possess quite different properties. Students should pay attention, Berthollet advised, not only to the principles contained in compounds but also to their proportions. Most of the principles that showed a unique force in their action could nevertheless be composed of several principles which sometimes exercised particular actions. Fourth, the degree of affinity a substance exhibited seemed to change with the degree of saturation. In general, the tendency to combination diminished as the substance approached the state of saturation. This weakening of action in the process of combination explained a number of odd precipitations:

Sulphuric acid has more affinity with alkali than muriatic acid and nitric acid; nevertheless, these last two decompose a part of alkaline sulphate; it is because sulphuric acid tends to combine with alkali by a force which diminishes considerably from a certain proportion, so that the two other acids seize what exceeds this quantity; but they cannot remove the rest from it and it forms after this a sulphate with excess acid.[75]

In the third lecture,[76] Berthollet dealt with heat as a part of chemical constitution. Castigating the earlier, abstract conception of heat as belonging to the "physics of imagination," Berthollet applauded the birth of "experimental physics," which dealt with heat as an agent of nature. Boerhaave, Stahl, Cullen, and particularly Black deserved the credit for this turn of investigation. Berthollet accepted Lavoisier's definition of "caloric" as the "principle that produced heat" to distinguish it from the common-sense, confused uses of "heat." Furthermore, caloric was a "material principle," a chemical substance whose combinations obeyed the laws of affinity. As an "eminently elastic fluid that penetrates between the molecules of bodies," caloric dilated bodies, separated their molecules by opposing the force of attraction, and changed them from solid to liquid, then to elastic fluid. Caloric acted primarily as a solvent. Much like water, it distributed itself evenly among bodies. The attractive force of caloric allowed it to combine with other molecules of matter and to lose its elasticity. The "union of caloric with bodies" occurred in successive stages, just like a solid body placed in water. A solid body became liquid by absorbing a certain quantity of caloric, some of which became latent in the process. In sum, the phase change of bodies was a process governed by the laws of affinity between caloric and other bodies. If the air was dry, it removed water from the bodies it came into contact with. If the air was saturated with water, it abandoned a part. Likewise, when a liquid became a vapor, it removed a quantity of caloric from the adjacent bodies that had less affinity for it, producing cold. When dry air removed water from the bodies, it did not take equal quantities from each body; it removed more or less water according to the degree of affinity these bodies had with water. Likewise, a body combining with caloric took an unequal quantity from each body surrounding it until it reached an equilibrium of uniform temperature. The specific heat of a body was the comparative quantity of heat this body could give to other bodies.

Berthollet then proceeded to discuss the "application of the laws of chemical attraction to the properties of caloric." Just as the effects of attraction diminished with the increasing degree of saturation in chemical combination, the bodies weakened their grip on caloric with the increasing content of caloric. Specific heat was the first portion of caloric that was released easily with varying forces of affinity; combined caloric was the portion deployed only with difficulty. Combined caloric did not influence the distribution of heat at different temperatures, just as the water combined in a body did not influence hydrometric phenomena. If the existing combinations were ruptured by the action of complex affinities, however, the combined caloric could be released, producing heat:

Caloric can therefore be considered as an elastic fluid which is combined with the molecules of bodies; by being combined thus, it loses more or less of its elastic properties; by the elasticity which remains to it, it tends to separate the molecules of bodies, and it produces the physical phenomena that depend on this action. As a principle subject to chemical attraction, it is combined in more or less great quantity; it adheres more or less strongly; its dimensions are tightened more or less. By approaching the state of saturation, it can be removed by other bodies which are found proportionately less saturated; it is this portion which can be removed or accumulated in the lowering or elevation of temperature, which produces the phenomena of specific heat; the portion which is more strongly combined and which we call combined caloric produces sensible effects only in the changes of combination. The part which is released in the changes of temperature is placed more or less promptly in equilibrium with the affinities of other bodies, according to their more or less conducting nature: when it does not encounter sufficient obstacles, it resumes the elastic state and forms radiant heat.[77]

Berthollet's conception of caloric as a "principle subject to chemical attraction" or a fluid that interacted with other chemical bodies through its affinities encapsulated the Arsenal Group's vision of integrating heat as a part of chemical constitution. He shared this conception with the other authors of the Chemical Revolution.

8.5 The Laws of Chemical Affinity

As a chemical expert, Berthollet became as valuable to Napoleon as he had been to the revolutionary government. Between May 1796 and November 1797, he was in Italy to collect valuable art objects. Between May 1798 and October 1799, he accompanied Napoleon on the Egyptian expedition.[78] While away, he lost the commanding position in French chemistry to Fourcroy and Guyton, who now effectively controlled the *Annales de chimie* and the École polytechnique. With

Napoleon's remuneration for his services, Berthollet purchased a mansion at Arcueil and established a laboratory, which served as the new center of French physical sciences until the Restoration. As the master of a small, private research group, he (along with Laplace) nourished the French tradition of physical chemistry represented by Gay-Lussac and Dulong in the ensuing decades.[79] With ample financial resources at his disposal, Berthollet devoted his time to elucidating the problem of affinities, publishing *Recherches sur les lois affinités chimiques* in 1801 and *Essai de statique chimique* in 1803. In those works he sought to bring the phenomena of reciprocal affinity and partition, discussed before the Revolution by such leading European chemists as Baumé, Macquer, Kirwan, and Lavoisier, into a more comprehensive framework of affinity chemistry. Though historians have labeled him the last Newtonian chemist on the basis of the more speculative tone of the *Essai*, a much different insight into Berthollet's program can be gained from the *Recherches sur les lois de l'affinité*,[80] written in the days of the Institute of Egypt. They show much more clearly Berthollet's immediate concerns as a chemist, or the analytic core of his speculative model of chemical processes elaborated in the *Essai*.

In order to assess Berthollet's goal in the *Recherches*, it is necessary to pay attention to various meanings of "affinity" in the text. Berthollet did not use a rigid definition of affinity as Newtonian force throughout the text; rather, he used the term variably depending on the context. His usage can be differentiated into three different levels. At the philosophical level, he maintained the notion of affinity as the cause of chemical action. A solidly established theory of chemical affinities had to contain all the principles that caused chemical phenomena.[81] Berthollet used the Newtonian language of attraction at this philosophical level:

13. I have pointed out in this Essay, the constant and uniform cause of chemical action; I have considered all the forces which contribute to the production of combinations, and chemical phenomena arising from them; and have endeavored to ascertain the influence of each force in different circumstances.[82]

At the classificatory or theoretical level, various affinities corresponded to different categories of chemical actions or combinations. "Elective affinity" referred to the displacement reaction of a salt by an acid; "complex affinity" referred to the double displacement reactions of two salts. Berthollet regarded this distinction as no more than a convenient classification that should be eliminated. Nevertheless, he devised yet another classificatory scheme with the terms "resulting affinity" and "elementary affinity," which referred respectively to the affinity of a compound and that of a pure substance. The persistence of these naming schemes speaks for their heuristic power in organizing chemical experience. At the level of practice, affinity was closely linked to the art of chemical analysis. Berthollet objected to "elective affinity" because he could not separate the meaning of the term from the procedure of obtaining it through displacement reactions.

Though these layers of meaning seem somewhat artificial, they offer entrée to the central goal of Berthollet's *Recherches*: to refute the notion of elective affinity. It was precisely the fusion of the theoretical, philosophical, and analytical meanings "elective affinity" afforded that Berthollet sought to dismantle. "Elective affinity" referred simultaneously to the category of displacement reactions, the cause thereof, and the analytical means of ascertaining the relative affinities of substances through displacement reactions. Berthollet endeavored to put some distance between these three layers so that he could devise a more adequate analytic method for determining affinities. Such a new method would filter out contingent physical factors that modified more permanent effects of chemical affinities. The fact that historians have often characterized his efforts as a refutation of the notion of determinate selectivity in chemical actions indicates that he failed in his efforts.

The whole text of the *Recherches* was directed to proving the inadequacy of elective affinity as it was conceived and measured by Bergman. Berthollet thus began the first part with a thorough critique of Bergman's theory. He did not believe that displacement reactions provided an adequate measure of substance-specific forces that determined chemical combinations. His strongest and most consistent criticism of elective affinity focused on its method of determination via the *complete* displacement of one acid by another from its combination with a base. The method ignored reversible action, partition of a base between two acids and vice versa, and the role of the quantities of participating substances:

3. Let us suppose, says Bergman, the substance A completely saturated with the substance C, and that the combination be termed AC; if the addition of another

substance, B, to this combination, removes C, there will result the combination AB, instead of AC. He prescribes then, for determining the elective affinity of two substances, to try if one of them can remove the other from its combination with a third, and vice versa. He takes it for granted, that that body which has removed another from its combination, cannot, in like manner, be expelled by that other, and that both experiments will concur to prove that the first has a greater elective affinity than the second. He adds at the same time, that it may be necessary to employ six times as much of the decomposing substance as would be necessary to saturate immediately the substance with which it tends to combine.[83]

In particular, Berthollet wished to demolish the conception that elective affinity could be measured by complete displacement reactions without taking into account the quantities of the respective substances. Displacement reactions necessarily produced a partition of the acid between the two competing bases, the proportion of which was controlled by their respective quantities as well as by their affinities. That is, a weaker degree of affinity could be compensated by an increase of quantity:

6. I shall prove, therefore, in resuming the supposition of No. 3, that, in opposing the body A to the combination BC, the combination AC can never take place; but that the body C will be divided between the bodies A and B, proportionally to the affinity and quantity of each; that is, to their respective masses.[84]

That Berthollet opposed the method of determining the affinity specific to each pair of substances rather than the notion of such absolute affinity can be seen in the following argument. He maintained that it was impossible to determine elective affinity, even if one reinterpreted the concept according to his idea, acknowledging the fact of partition and the role of quantities:

To ascertain the elective affinity of two substances for a third, according to *our* idea of that term, it will be necessary to determine in what proportion that third substance divides its action between the two former, and the degree of saturation which each of these two can attain *when they act in competition*; their respective affinities will be commensurate with the degree of their saturation, which degree of saturation must be still further influenced by the quantity of each which acts; so that if these quantities were equal, the degree of saturation would be the exact measure of the affinity.[85]

In order to make the measurement of elective affinities meaningful, Berthollet deemed it necessary to keep the quantities of two substances equal not only at the beginning but also at each successive stage of reaction, which was practically impossible. Maintaining the same proportion throughout the course of the reaction was but one of the practical difficulties. The degree of saturation at each stage had to be determined by separating out the substances, which could be effected only by crystallization, elasticity, precipitation, or dissolution. These different means introduced "extraneous forces, which change the results, and occasion combinations, which are formed in such a manner as that the influence of these forces cannot be measured, nor their effects distinguished exactly from those of elective affinity." It was manifest, then, that "the elective affinity of two substances, with respect to a third, cannot be determined by a direct experiment," because new forces were created "at the moment that two substances are put to combine respectively with a third, which forces not only influence the results, but change even the constitution of the bodies."[86] It was this incompatibility between the meaning of elective affinity and the method of evaluating it—displacement reactions—that incurred Berthollet's strong objection:

The very term, *elective affinity* must lead into error, as it supposes the union of the whole of one substance with another, in preference to a third; whereas there is only a partition of action, which is itself subordinate to other chemical circumstances.[87]

Not only was elective affinity inadequate as a means of comparing the affinities of different substances; it also led to critical and pervasive errors in chemical analysis by compromising the purity of chemical substances obtained in this manner. In other words, the supposition of complete displacement reactions encouraged the use of impure substances as pure ones in chemical analysis, which seriously compromised the accuracy of chemical analysis. Chemists were "deceived," for example, in believing that they could obtain pure magnesia from displacement reactions.[88] Berthollet's pointed attack on the notion of elective affinity is understandable in light of the fact that it caused a serious problem for the validity of basic chemical analysis. His attack was quite successful in undermining chemists' naive confidence in the absolute order of chemical affinities obtained from displacement reactions and in the analytic purity of the substances thus obtained. He discredited affinity tables as "mere memorandums of barren facts"[89]:

5. As all tables of affinity have been formed on the supposition that, substances are endued with different degrees of affinity, from which originate all the

combinations and decompositions that take place, independently of proportion and other circumstances which influence the results, they must give an erroneous idea of the degrees of the chemical action of substances.[90]

Berthollet followed Kirwan's initiative in choosing saturation capacity as the alternative measure of the relative affinities of acids and bases. Since the affinity of a substance was compensated by its quantity, relative affinities had to be inversely proportional to "the quantities necessary to produce an equal degree of saturation"[91]:

It would appear from this consideration, that, in order to ascertain the relative affinity of acids and bases, it is sufficient to know their respective capacities of saturation, that is, the capacity of different acids for one base, or of different bases for one acid; for, considered under this point of view, the relative affinities ought to be inversely, as the quantities necessary to produce an equal degree of saturation.[92]

In other words, relative affinities were determined by neutralization reactions between an acid and a base, not by displacement reactions between an acid and a salt. Berthollet rationalized his method of evaluating relative affinities with the notion of "chemical mass," which was a compounded ratio of affinity and quantity:

These consequences are, that substances act in the compound proportion of the quantity in the sphere of action, and of the affinity with which they are endued; that the latter may be compensated by the former, and that the chemical action of each is proportional to the degree of saturation which it produces. I have distinguished by the term *chemical mass*, or *mass*, the quantities as determined by an equal degree of saturation, which quantities are consequently relative to their capacity of saturation.[93]

As Holmes and Le Grand have pointed out, Berthollet did not abandon the notion of selective affinity specific to each pair of substances in proposing the notion of chemical mass.[94] In the above quotation, he used "affinity" in the sense of a selective force between different substances. More often, however, he conveyed this sense by using the term "intensity of affinity" or "energy of affinity." It was the general law "to which chemical action is subjugated that the substances exercise in the ratio of the energy of their affinity and of their quantity."[95] Berthollet did not object, then, to the sense of selectivity; he objected to the sense of false absoluteness that the notion of elective affinity conveyed. He criticized the perception that elective affinity conceived and measured via displacement reactions was a determinate, absolute, uniform force that always determined the outcome of displacement reactions:

4. The doctrine of Bergman is founded entirely on the supposition that elective affinity is an invariable force, and of such a nature, that a body which expels another from its combination, cannot possibly be separated from the same by the body which it eliminated. Such was the certainty with which elective affinity has been considered as an uniform force, that celebrated chemists have endeavored to represent by numbers, the comparative elective affinities of different substances, independently of any difference in the proportion of their quantities.[96]

In other words, Berthollet did not deny the notion of affinity as a force of specific magnitude for the pair of substances under consideration, a notion he often emphasized with the phrase "energy of affinity" or "intensity of affinity." His crucial insight, as is well known, was the effect of quantity on chemical action. This was not an entirely new idea; Kirwan had already pointed out the effect of quantity on affinities. Berthollet's alternative delineated how a "chemical action" was caused by a combination of a definite affinity between substances and their respective quantities, or chemical mass:

5. It is my purpose to prove in the following sheets, that elective affinity, in general, does not act as a determinate force, by which one body separates completely another from a combination; but that, in all the compositions and decompositions produced by elective affinity, there takes place a partition of the base, or subject of the combination, between the two bodies whose actions are opposed; and that the proportions of this partition are determined, not solely by the difference of energy in the affinities, but also by the difference of the quantities of the bodies; so that an excess of quantity of the body whose affinity is the weaker, compensates for the weakness of affinity.

If I can prove that a weaker degree of affinity can be compensated by an increase of quantity, it will follow, that the action of any body is proportionate to the quantity of it which is necessary to produce a certain degree of saturation. This quantity, which is the measure of the capacity of saturation of different bodies, I shall call *mass*.[97]

This was a far cry from "variable" affinities, as some historians have called it.

Berthollet's critique effectively destroyed the loosely correlated usage of elective affinity as a category of reaction, the cause thereof, and an analytic method of determining relative affinities of acids and bases. One could still use it as a category of reaction in which an acid reacted with a salt. The cause of this action was now much more thoroughly analyzed in terms of the chemical affinities in action and a number of other physical factors. Berthollet applied the same rigor to another category of reaction that had been used in a similar way, that of double affinity. Bergman's idea was as follows, according to Berthollet:

> In order to give an idea of the action of four affinities, Bergman refers to the effects produced by the mixture of the solutions of the sulfate of potash, and muriate of lime: the effect is the same, he says, as if the same proportions of the acids and bases which exist in these salts were put together into water; the two bases act by their affinities on the two acids; and although the affinity of potash for sulphuric acid be greater than for the muriatic acid, yet the affinity of muriatic acid for potash, joined with that of sulphuric acid for lime, gives a sum of force which exceeds that of sulphuric acid for potash, and of muriatic acid for lime. This difference between the two concurring forces, determines an exchange of bases, so that instead of muriate of lime, and sulphate of potash, the combinations are sulphate of lime and muriate of potash. This explanation is also founded on the supposition that affinity is a constant and uniform force, independent of quantity, and of the state of saturation.[98]

Berthollet reasoned that such an exchange of bases did not take place because the partition of acids and bases was in effect. When the exchange did occur, it was due not to the affinities in action, but to other "real" causes such as crystallization and precipitation. Chemists had inadvertently ascribed the strongest affinities to the substances that precipitated easily. Other causes, such as heat and efflorescence, also contributed to the outcome of complex affinities, as was the case for natron. In short, elective affinity involved an unsaturated substance, while complex affinities worked between saturated substances. The notion of elective affinity was tightly bound to the analytic procedure in which two acids, one combined in the salt and the other free, had to compete for the common base.

If saturation capacities seemed to promise an easy way of measuring relative affinities, Berthollet tagged on other considerations to complicate the scheme further. He considered liquid as the ideal medium for chemical action, since it allowed the substances to exercise their full power.[99] The quantity of the substances that constituted chemical mass did not necessarily coincide with the amount that was put into the solution, but referred to the effective quantity, or the quantity within the "sphere of action." Technically, concentration was more likely to fit

Berthollet's conception than the total quantity of the substance. This observation offered an important starting point for the later formulation of chemical dynamics in the 1860s:

It follows from what has been said (No.5), that when a liquid substance acts on another which is solid, or when a precipitate has been formed during the operation, it is not the absolute weight of the liquid that determines the degree of quantity of its action, but the degree of concentration, or mass, within the sphere of action.[100]

In fact, Berthollet built a highly speculative model of chemical action to accommodate the effect of quantities on chemical action. In general, the reciprocal affinity between the molecules of the reacting substances was responsible for chemical combination. The "intensity" or "energy" of chemical action depended, however, on the quantity of reacting substances present in the "sphere of activity," in addition to their "energy" of affinity. Other forces, including the action of solvent, the force of cohesion, elasticity, efflorescence, and gravitation, increased or decreased the quantity of molecules in the sphere of activity, thereby strengthening or diminishing the energy of chemical action. In other words, the interplay of these forces determined the "degree of concentration" to which reacting substances could be brought.

Insolubility, elasticity, heat, and efflorescence all affected chemical action by changing the effective quantities of the substances in solution. *Insolubility* of a reacting substance doubly hampered chemical action, limiting the contact between the two substances and also opposing the reciprocal affinity with the force of cohesion that held the molecules of the insoluble salt together. The difference of specific gravity between an insoluble substance and the solvent sometimes amplified this problem, since it "tended continually to separate the insoluble substance and to withdraw it from the action of the other."[101] *Elasticity*, or the gaseous state of a substance, also hindered chemical action by removing it from the sphere of activity. The insolubility or elasticity of a product helped chemical action, however, by removing it from the sphere. *Caloric* had little effect on chemical action because the optimal quantity of caloric, which was specific to each chemical combination at a given temperature, was maintained throughout most chemical changes. When the temperature change was great, however, it influenced chemical action significantly because of the difference of the volatility of the substances involved. Although heat generally favored chemical action by diminishing the force of cohesion, it had the same effect as elasticity when one of the substances was much more volatile than the other. The application of heat may in some cases "derange the effect intended to be produced, when the substances acted on differ in volatility, and may, from the same circumstances, deceive as to the force of affinities." *Efflorescence* was the phenomenon in which a certain substance raised itself to the surface, "withdrawing itself thereby from the sphere of chemical action." A similar reasoning as in the case of insolubility and elasticity thus applied to this case.[102] The action of *solvent* generally helped chemical action through dissolution, which multiplied the points of contact by overcoming cohesion or elasticity. The action of solvent was "an alien force, influencing the action of two or more substances: by surmounting the resistance of cohesion, or of elasticity, and by multiplying the points of contact." A solvent might hinder the intended chemical action by its own "immediate action" with the reacting substances, although this was usually weaker than its action of dissolution. In sum, solvents acted on bodies "by their affinity and quantity, like all substances which tend to combine; and whatever has been said of combination in general is applicable to them."[103]

Berthollet endeavored to explain the influence of all chemical and physical factors on chemical action with a single model according to which the quantity of the substances present in the sphere of action regulated the intensity of chemical action. Despite its speculative tone, his model generalized chemical experience rather than stipulate an a priori ontology of matter and force. For example, Berthollet took it for granted that different molecules exhibited different energies of affinity toward one another, but he did not speculate as to what brought about such differences. Although he cautiously conjectured that affinity was probably similar to gravity in nature and in origin, he did not assume that the energy of affinity changed in inverse proportion to the distance, as Kapoor has argued.[104] The increase of distance only reduced the quantities of substances in the sphere of activity. Also, the action of various forces did not influence the energies of affinity, but it influenced the energy of intended chemical action by changing the quantities of molecules that actually existed within the sphere of action. The general law of chemical action was modified only by "the circumstances which limit the quantity of liquid that can simultaneously exercise its action."[105] More important, the forces that affected chemical action did not always exist or possess any inherent magnitude. Rather, they came into existence at particular moments of chemical action, and they interacted differently depending on the circumstances. Berthollet's effort to match the experimental evidence to a heuristic model appealed to later generations of physical chemists.

8.6. Constitution

Berthollet further developed his model of chemical action in the *Essai de statique chimique* (1803), offering a speculative chemistry or "la chimie flottante."[106] Whereas the *Recherches* were directed primarily at discrediting the notion of elective affinity, the *Essai* articulated a more systematic understanding of chemical affinity and constitution. The *Essai* is a difficult text even for the initiated, owing to Berthollet's multifaceted interpretations of concrete chemical reactions and meandering generalizations thereof without a nexus of classification or an adequate vocabulary. The most valuable part of the book contains his thoughts on the various categories of chemical action and his attempts to sort out the chemical and physical conditions that modified them. This worked for the benefit of advanced researchers, however, rather than for a broader audience. The introduction does provide a précis of his understanding of chemical combination, his view of chemistry in relation to other sciences, and an overview of the book. The main body of the text is divided into two parts, each devoted to the discussion of "chemical action in general" and to a variety of concrete chemical actions. Since the core of Berthollet's analytic program has been analyzed with the *Recherches*, I will concentrate in this section on his general conception of chemical "constitution," given in the introduction.

The *Essai* is well known for its Newtonian language. Berthollet envisioned a unified "physics" of which chemistry would become a part when it was transformed into a true science. Constructing the true science of chemistry depended on subjecting chemical affinity to the laws of mechanics, since chemical affinity differed from general attraction only in its origin. Articulating the general laws of chemical affinity analogous to those of mechanics by way of observation was the goal of chemical theorizing. Despite his grand rhetoric, however, Berthollet emphasized the difference between chemical attraction and astronomical attraction. The name "affinity" was given to "the powers which produce chemical phenomena," he concurred with Bergman, "to distinguish it from astronomical attraction." Although chemical attraction or affinity probably had the same property as astronomical attraction, it depended on the shapes of molecules, their intervals, and their particular affections. Its effects were "so changed by particular and frequently indeterminate conditions" that they could "not be deduced from any general principle" such as the inverse square law. The dependence of affinity on particular conditions made deductions from general principles inadequate and mandated particular observations.

After a two-page discussion of the nature of chemical attraction, Berthollet moved on to the notion of chemical combination and constitution. The "immediate effect" of chemical affinity was always a "combination," although it was modified by the quantity of the bodies under consideration. Berthollet did not abandon the sense of specificity assigned to each affinity:

Every substance which has a tendency to enter into combination, acts in the ratio of its affinity and of its quantity. These facts are the ultimate end of every chemical observation.[107]

The result of chemical action depended not only on "the affinity peculiar to its component parts" and their quantities but also on a variety of conditions that shaped the "constitution" of a body:

2d. The chemical action of a body does not depend solely upon the affinity peculiar to its component parts and upon the quantity; it also depends upon the state in which those parts are found, whether it be that of actual combination, which causes a larger or smaller portion of their affinity to disappear, or that, by their dilatation or condensation, their reciprocal distances are varied: it is these conditions which, by modifying the properties of the elementary parts of a substance, form that which I call its constitution: to obtain the analysis of chemical action, not only each of these conditions must be appreciated, but also every circumstance with which they have any connection.[108]

In other words, Berthollet regarded chemical combination not as a permanent species, but as the result of a transient equilibrium of several forces acting on the system. These included physical as well as chemical forces. Consequently, the chemical and the physical properties of a body were often related, which implied an intimate relationship among the various sciences that dealt with these forces. The analysis of chemical action required a coordination of various sciences, the totality of which constituted "la physique." As yet, a general theory that comprised all the processes, results, and causes of chemical action was not available. The *Essai* was devoted precisely to this end of accounting for all the causes of chemical action and their interdependence.

In general, two opposing affinities determined the outcome of chemical action between a liquid and a solid: the force of cohesion and that of solution. The force of cohesion united the particles of a body as in crystallization. The force of solution opposed the force of cohesion, dissolved the solid in a liquid, and produced a different chemical combination. Dissolution itself was, therefore, "a combination considered with respect to the force of cohesion." In other words, the force of solution was not a physical force different from affinity, but a kind of affinity acting in opposition to the force of cohesion. It thus followed the law of chemical action:

Since the immediate effect of every chemical action is a combination, dissolution itself is only a combination considered with respect to the force of cohesion; because in every combination it will be observed that the action of any substance is always in proportion to the quantity contained within the sphere of its activity: an immediate consequence of this law is that the action of a substance diminishes in proportion to the saturation it experiences.[109]

Berthollet modified this general conception of chemical action with the notion of "predominant affinity," which retained his earlier principalist approach to the problem. He argued that in practice there existed some affinities which prevailed over others. These predominant and energetic affinities gave "distinctive properties" to the substances and thus were useful for their classification. Acids and alkalis exhibited the "most striking examples" of predominant affinity; they imparted opposing properties to bodies, which consequently strove for mutual saturation. For this reason, the reciprocal saturation of acids and alkalis could be regarded as "the measure of their affinity, if the respective quantities which are necessary to produce this effect were taken into consideration." That is, the action of an acid on an alkali was determined by the combined effect of its capacity for saturation and its quantity. When several acids worked on one alkali, they shared it in proportion to their chemical masses. This contradicted the behavior prescribed by the doctrine of elective attraction, which predicted the exclusive formation of one salt:

> ... I maintain that whenever several acids act upon one alkaline base, the action of one of the acids does not overpower that of the others, so as to form an insulated combination, but each of the acids has a share in the action proportionate to its capacity for saturation and to its quantity: I call this compound result by the denomination of a *chemical mass*.[110]

The general law of chemical action "that substances act in the ratio of the energy of their affinity and their quantity" was modified not only by the force of cohesion but also by caloric, "the cause of heat," or the "principle of expansibility" (section III). The concrete effect of caloric on chemical combination differed depending on the circumstances, although it always effected the separation of particles: "The effect of caloric may, according to the circumstances, contribute to the combination of that substance with others, or it may be an obstacle to it."[111] Elastic fluids deserved particular attention because their actions could point to the quantitative relation between caloric and the force of cohesion (section IV). Phase changes took place mostly at a constant temperature, indicating that the elevation of temperature depended "only on the resistance which the force of cohesion opposes to that of the caloric."[112] However, the force of cohesion seemed to have no sensible effect on gases, probably because of the distance between their molecules. In other words, gases could exhibit isolated instances of powerful affinities. It was important, therefore, to distinguish permanently elastic fluids from vapors and to study further the changing capacities of gases with temperature.[113] The interplay of multiple causes impinging on chemical action produced a chemical combination that did not necessarily have a determinate composition (section V):

> The result of the different causes, which interpose during chemical action, is sometimes a combination whose proportions are constant; sometimes, on the contrary, the proportions of the combinations which are formed, are not fixed,

and vary according to the circumstances under which they are produced: in the first case, it requires an accumulation of powers to change the proportions equal to those which tend to maintain their state of combination; this obstacle overcome, chemical action continues to produce its effects in the ratio of the energy of the affinities and the quantity of the bodies which exercise it. I have endeavored to ascertain the conditions which thus limit the proportions, in some combinations, and which appear to place an obstacle to the progress of chemical action.[114]

Berthollet explained dissolution and crystallization as reversible chemical processes. Two chemical forces—cohesion and solution—determined the resulting constitution of the solution, modified by the action of caloric and other contingent factors. He argued that solution was "a true combination," taking pains to establish this somewhat contentious point in his system. His argument rested squarely on the observation that dissolution observed the same general law of chemical affinity as other combinations. It decreased with increasing saturation, as in the case of the prototypical chemical system—acids and alkalis. Berthollet thus sought to invalidate the common distinction between solution and chemical combination. He believed that this distinction, based on the rule of determinate proportions, confounded the action of affinity with various forces that affected the separation of combinations:

39. Some chemists, influenced by having found determinate proportions in several combinations, have frequently considered it as a general law that combinations should be formed in invariable proportions; so that, according to them, when a neutral salt acquires an excess of acid or alkali, the homogeneous substance resulting from it is a solution of the neutral salt in a portion of the free acid or alkali.

This is an hypothesis which has no foundation but a distinction between solution and combination, and in which the properties which cause a separation are confounded with the affinity which produces the combination; but those circumstances must not be overlooked which can determine the separation of combinations in a certain state, and which, by that means, limit the effects of the general law of affinity.[115]

If it was a matter of principle not to distinguish between solution and neutralization, it was a practical necessity to do so. Just after he annulled the distinction, Berthollet introduced a differentiation of the "two species of saturation," one in neutralization and the other in solution. He admitted that the latter was "very analogous to mechanical effects." The neutralization of acids and alkalis served as the prototype of chemical actions in which the distinctive properties of substances disappeared or became latent. Needless to say, much of Berthollet's conception of affinity was shaped by this category of reaction, which exhibited the most powerful display of chemical affinities. He used it extensively to model his discussion of the contingent forces on chemical action in general. His discussion is rendered almost unintelligible, however, by his inconsistent use of terminology. Most critical was his obliteration of the distinction between the affinity of aggregation and that of composition under the umbrella term "force of cohesion." Not only did he use the term inconsistently; he often substituted it by a variety of descriptive terms and phrases, leaving it to the reader to figure out what he meant. Sometimes, the force of cohesion applied equally to the affinity of aggregation and to that of composition. Other times, it applied exclusively to the affinity of aggregation. As far as I can discern, Berthollet wished to sort out chemical forces from other forces related to volatility, fixity, cohesion, or specific gravity. Chemical forces would include the affinities of composition and aggregation in Macquer's terminology.[116]

A measuring scheme ensued from the conception of neutralization. Berthollet proposed to estimate the "power" of an acid from how much of it was required to saturate a given quantity of the same alkali, a scheme earlier used by Homberg and revived by Kirwan. It was this "capacity for saturation" that provided the "comparative force" of the affinity in question:

> It may therefore be said, that the affinity of the different acids for an alkaline base, is in the inverse ratio of the ponderable quantity of each of them which is necessary to neutralize an equal quantity of the same alkaline base; but by proportioning the quantities to the affinity, they produce the same effect; so that the force put in action depends on the affinity and on the quantity, and one can supply the place of the other.
>
> 48. I have designated by the name of *chemical mass* this faculty of producing a saturation, this power, which is compounded of the ponderable quantity of an acid and of its affinity.[117]

The saturation capacity could not be determined with the necessary exactitude, however, until the "real quantity" of acids and alkalis was known. This knowledge constituted the foundation of all other chemical analysis. Despite his refutation of definite proportions as a matter of principle, Berthollet was keen on improving precision in chemical analysis. He took Kirwan's work as a starting point:

The determination of the proportions of a substance which can be made to act, or which exist in a combination, is therefore, the foundation of all chemical enquiries: the object of all the methods, of all the processes, is to arrive at it, and this object ought never to be lost sight of by chemists.[118]

With a new conception of chemical affinity and constitution, Berthollet moved away from his earlier principalist approach to composition. This move had consequences for his theory of acidity.[119] He could no longer define acidity or alkalinity in terms of quality-bearing principles such as oxygen and hydrogen. Instead, he invoked the "predominating" or "characteristic" affinities to account for their peculiar behavior. Acids and alkalis were now defined by their "operational" characteristics, which were similar to Fourcroy's earlier conception of affinities:

37. Among the affinities of a substance there is frequently one which predominates, and which stamps it with a peculiar character, and, upon which the principal part of its properties are dependent. It is these predominating affinities which are more particularly serviceable in the classification of the chemical properties of different substances, and of the chemical phenomena derived from them: thus, affinity for oxygen distinguishes inflammable substances; the reciprocal affinity of acids and alkalis constitute acidity and alkalinity; for this reason, these affinities and their effects are the principal object of chemical consideration.[120]

8.7   Divergent Paths

Contemporary chemists much admired Berthollet's sophisticated treatment of chemical affinity, but they did not subscribe to his system.[121] The destructive impact of his work on the existing structure of chemical theory became clear within a few years. In particular, the line of battle became sharply drawn over the issue of constant and definite proportions. Berthollet's conception of chemical "constitution" as an interplay of chemical and physical forces called for indeterminate proportions in chemical combinations, although he allowed for certain cases of constant proportions under particular physical circumstances. This practical implication, brought into focus by the debate with Proust, undermined the feasibility of Berthollet's elaborate program.[122] Although Proust was no match for Berthollet in experimental skill or in theoretical sophistication, he could draw on the centuries-old common-sense assumption that chemical combination hinged on the identity and the quantity of its constituents. Furthermore, if Berthollet dealt a devastating blow to the paradigm of elective affinity and affinity tables, he did not offer a radically new alternative. The determination of saturation capacities, a program already growing in chemical practice, did not yield values sensitive to various chemical and physical factors. In any case, he undermined the validity of such determinations by pointing to the changing quantities of substances during the course of chemical actions. His sophisticated yet impractical vision left other chemists perplexed and disoriented. Thomas Thomson professed ignorance:

The truth is, and it is not full time to declare it, that we are profoundly ignorant of everything regarding the strength of affinity. Berthollet has succeeded in overturning all our preconceived opinions on the subject, but he has not been so successful in establishing his own.[123]

The circulation of Dalton's ideas on the chemical atom also reinforced the cause of constant and definite proportions, thereby rapidly diminishing the appeal of Berthollet's complicated notions.[124] Though Dalton did not offer a new investigative program, his atoms as a calculating scheme helped such leading chemists as Thomas Thomson and Jacob Berzelius to simplify chemical theory. Thomson's and Berzelius's stoichiometric atomism appealed to a broader range of chemical practitioners. Although it was possible to embrace Berthollet's vision in general terms and to proceed with the empirical determinations for specific cases, as Gay-Lussac did in developing his law of combining volumes, the majority of European chemists found it difficult to translate Berthollet's vision into viable investigative programs.[125] Even in France, only the chemists trained at Arcueil under Berthollet's close supervision and patronage followed his program. In addition to Gay-Lussac, Pierre Louis Dulong sought to characterize partial decomposition of double salts.[126]

Although the Arcueil chemists eventually came to control French elite chemistry, they had to contend with serious opposition from Fourcroy and his protégés who nurtured a very different approach to chemistry. In the absence of better terms, we might characterize the different outlooks of Berthollet and Fourcroy as physicalist vs. naturalist. If Berthollet sought to integrate physical forces into chemical "constitution," Fourcroy approached the problem of chemical affinity and composition with the classificatory language of natural history. After his "conversion" to Lavoisier's oxygen theory, Fourcroy worked steadily to sort out the implications of the new chemistry for his operational understanding of affinities. The nomenclature reform of 1787 contained a significant compromise between the principalist and the affinity approach to composition. While it embraced the principalist understanding by according a dominant role to one of the two constituents of a compound, such as the oxygen in acids, the proportions of the two components were determined by their affinity relations. Fourcroy generalized this compromise position in his *Tableaux Synoptique de Chimie* (1800) by classifying chemical substances according to their complexity in composition. Though Macquer advocated this "geometrical" approach to chemical composition, he could not implement it, because the analytically pure substances were not identified. Fourcroy drew on the new analytic definition of elements and implemented the geometrical order of exposition in his new textbook.[127] Such a reordering was possible because the affinity relations determined the complexity of chemical bodies. In other words, Fourcroy implemented the affinity approach to chemical composition down to the level of simple substances by incorporating Lavoisier's accomplishments and the principalist understanding. His binary chemical systematics synthesized a variety of the "two-component" theories[128] by combining the operationally (affinity-) determined binary composition of salts with the principalist stipulation of dominant principles.

Berthollet's and Fourcroy's divergent approaches to chemical composition and affinity were informed by their respective expertise in industrial and pharmaceutical chemistry. Berthollet and his associates appreciated the fickle nature of the chemical processes used in industrial production and tried to rationalize their erratic behavior by characterizing the physical conditions that affected outcomes. Jean-Antoine-Claude Chaptal (1756–1832), who had a serious interest in industrial chemistry, advocated Berthollet's system.[129] If affinity provided the guiding light for the younger authors of the Chemical Revolution, then, their conceptions of affinity differed markedly. Fourcroy argued that Berthollet's consideration of mass action did not alter the doctrine of elective affinities, despite the reactions that did not follow the order prescribed in the existing affinity tables.[130] It was indeed possible to characterize Berthollet's work as a modification of the existing theory, since Berthollet had questioned not the existence of absolute affinities specific to each pair of substances but the ability of chemists to measure them accurately. Insofar as they shared the objective of determining the order of affinities specific to each pair of substances, they could presumably accommodate any changes by modifying the affinity table. Fourcroy stuck by the doctrine of elective affinities despite the changes he introduced to the empirical laws of affinity in the various editions of his textbooks. Fourcroy's stature as the leading Parisian chemical teacher did not help Berthollet's cause.

The intersections between the theory domains of composition and affinity created by Lavoisier's unique instrumentalist approach to chemistry complicated the French chemical tradition, making it less accessible to other European chemists. The most productive line of chemical investigation in the aftermath of the Chemical Revolution was the determination of combining weights, a natural extension of the measurement of saturation capacities. The study of combining ratios, or stoichiometry, became the new analytic frontier of European chemistry. The concomitant theoretical development led to the articulation of chemical atomism. Viewed from this angle, it was Richard Kirwan and other British chemists rather than French academicians who foreshadowed the nineteenth-century development of chemical atomism. To be more precise, it was the affinity approach to composition, rather than the principalist approach to composition, that called for the laws of constant and definite proportions. If the principles contained in bodies determined their properties, including affinities, it should be possible to conjure up an infinite variety of compounds with minute shades of difference in their composition. The specificity of affinities curtailed this realm of speculative possibilities. In order to elucidate the relationship between the Chemical Revolution and the chemical atomism, then, we must trace the revolutionaries' articulation of the affinity program rather than their antiphlogistic path.

Berzelius led European chemistry during the first half of the nineteenth century by drawing on the native Swedish analytic tradition as well as on the French nomenclature reform. He reified the concept of constitution to "signify the composition of a complex substance portrayed as a higher order compound of preexisting lower order constituents." A compound, in his scheme, was composed of a pair of electrochemically opposite constituents. These binary constituents were made up, in turn, of two simpler electrochemically opposite constituents, and this procedure could continue until one reached the level of elements.[131] Berzelius thus made explicit the affinity-determined composition of a compound, while incorporating the principalist conception of dominant principles. In Berzelius's definition, "constitution" lost the reference to contingent physical factors in the composition of chemical bodies. His notion of constitution, which continued the compromise of the 1787 nomenclature reform, became the main theoretical objective of organic chemistry. In other words, the theoretical or classificatory sense of affinity as relationship guided the constitutional studies in organic chemistry throughout the first half of the nineteenth century. Organic chemists sought to minimize random efforts in their rapidly expanding field by devising a rational system of classification that would organize their experience and guide the investigation in a predictive manner. The connection between constitution and affinity became implicit, barely visible only in their shared investigative agenda—a more refined resolution of chemical compounds and their reactions. In other words, the "constitution of organic bodies" as the dominant theoretical paradigm in organic chemistry carried on the investigative agenda of the affinity table.[132]

If the theoretical agenda of affinity survived in the classification of organic bodies, its philosophical meaning became formalized in textbooks. Throughout the first half of the nineteenth century, chemical affinity was invariably defined as attraction at insensible distances, maintaining its Newtonian heritage, albeit in a compromised form.[133] Despite the stability of the formal discourse, however, chemical affinity defined as attraction remained completely inoperative in shaping the investigative programs of chemistry. Although Humphry Davy and Michael Faraday actively pursued the connection between chemical and electrical attraction, this speculative trend did not open a viable path toward the quantification of affinities.[134] The versatility and productivity of chemical affinity in the nineteenth century can only be understood in its practical meaning, as an investigative agenda—a better understanding of chemical combination and reaction. In its most consistent usage, the concept was responsible for chemical combination and chemical reaction. This investigative agenda of chemical affinity continued to expand the horizon of nineteenth-century chemical enterprise, be it in organic chemistry or in chemical mechanics.

Chemical affinity reemerged as the explicit investigative goal of chemical mechanics only in the second half of the nineteenth century. Decades of stoichiometric investigations stabilized the chemical molecule, allowing chemists to tackle "Berthollet's problem" with a new set of analytic procedures.[135] Although early attempts (such as Williamson's and Wilhelmy's) could be found in the 1850s, it was Marcellin Berthelot who explicitly drew on Berthollet's authority to formulate a viable investigative program of chemical dynamics.[136] The treatment of chemical equilibria during the second half of the nineteenth century differed fundamentally from Berthollet's. Whereas Berthollet's mass action called for indefinite proportions, later chemists usually characterized it as the partition between stoichiometrically well-defined chemical species. The concept of chemical affinity went through various metamorphoses depending on what these chemists measured and how. By the 1860s, they had developed two clearly different notions of affinity: the notion of absolute valency for each atom of a substance in organic chemistry and the notion of variable affinity subject to different physical conditions.[137] These multiple meanings of chemical affinity have made the concept elusive for historians.[138]

Epilogue
A Tale of Three Fathers

Modern chemistry has three fathers: Robert Boyle, Antoine-Laurent Lavoisier, and John Dalton. Their respective share of paternity is determined mainly by their relevance to the nineteenth-century development of chemical atomism. Boyle's claim to paternity hinges on his analytic definition of chemical elements and his corpuscular characterization of them. In the conventional historiography, which located the origin of modern chemistry in Lavoisier's operational definition of chemical elements and in its subsequent translation into a corpuscular picture by Dalton, Boyle emerged as a strikingly modern precursor who linked the operational definition of chemical elements to a corpuscular model, preparing the ground for the "modern" chemical atomism.[1] Marie Boas has pointed out, however, that Boyle's definition of elements differed substantially from Lavoisier's in that "the same few elements are found in *all* bodies, that *all* bodies should be resolvable into the same number of elements, so that every body consists of different arrangement of the same elements." Furthermore, Boyle did not endorse this commonly accepted definition of elements and principles, but intended to "question" it. He lifted it from French didactic discourse, albeit in a more concise form, to point out its shortcomings. In *The Sceptical Chymist*, he explicitly expressed the doubt that there existed such simple bodies that made up all others. Boas concluded, therefore, that Boyle's contribution to modern chemistry lay more in the destruction of the existing discursive system of chemistry than in the construction of a new, viable alternative.[2] Denying Boyle's paternity for modern chemistry has made it difficult for historians to establish a connection between the Scientific Revolution and the Chemical Revolution.[3] Donovan has suggested that the Chemical Revolution yielded a positive science by displacing chemistry from Newtonianism or the reigning tradition of natural philosophy.[4]

Lavoisier still "reigns supreme as the father of modern chemistry," thanks to the elaborate historical reconstruction over the course of two centuries.[5] The legacy of his revolution has been ambiguous, however, leading to Holmes's recent suggestion that Lavoisier's central role in the episode of the Chemical Revolution does not automatically make him the father of modern chemistry.[6] Although historians have pointed out the logical connection between Lavoisier's analytic definition of elements and Dalton's chemical atom, the investigative tracks of European chemists did not proceed from one to the other. Siegfried and Dobbs discerned "widespread confusion and uncertainty" during the twenty years separating Lavoisier's *Traité* and Dalton's *New System of Chemical Philosophy*. They attributed this state of uncertainty to "the changes the new system produced in the concepts of composition," or "the nature of the chemical element."[7] Siegfried has recently shown how Fourcroy worked out a new classification of chemical bodies according to the "number and proportions of their constituent principles," thereby arranging them in a sequence "from the greatest simplicity to the utmost complication in their composition."[8] What was new in Fourcroy's post-revolutionary system was not the idea of such a "geometrical" arrangement, however, but the way in which the order of complexity was determined in practice. The development of pneumatic chemistry and subsequent refinements in analytic standards allowed chemists to resolve chemical bodies to their elementary level and to effect transformations of analytically defined elements. Their experience in the chemistry of salts guided their efforts to determine the composition of bodies in definite proportions. In other words, chemists could implement the affinity approach to composition at the level of elements. The classificatory language of natural history helped them to systematize the nomenclature. Although Lavoisier shared this goal of rationalizing chemical systematics, which paved the way for nineteenth-century stoichiometric investigations, it was not the primary aim of his theoretical project. His comprehensive vision consisted less in chemical systematics than in the dynamic understanding of chemical constitution as an interplay of heat and affinities. Berthollet pushed this line of thought further in his *Essai*, but failed to enlist the European chemical community in the project. In simplistic terms, the Arsenal Group chemists merged the traditional theory domains of composition and affinity through a comprehensive nomenclature and a more sophisticated understanding of chemical constitution. The classificatory effort and the nomenclature reform, further elaborated by Fourcroy, met with a certain degree of success by providing pathways for nineteenth-century stoichiometric investigations. The second project, continued by Berthollet and the Arcueil Group, remained an elite affair until it resurfaced during the 1860s in physical chemistry.

Dalton's fame rests squarely on his authorship of the chemical atom. He provided a corpuscular ontology for chemical elements, thereby enabling chemists to represent their practice in a logical manner.[9] If chemical atomism occupies a canonical status in the historiography of nineteenth-century chemistry, however, Dalton was not the central figure in its historical evolution. The tortuous path of chemical atomism, which required the intensive labor of many chemists to be stabilized, indicates that the connection between chemical elements and Dalton's atoms was by no means obvious to nineteenth-century chemists. It was not the ontology of the atom but the calculating schemes of stoichiometry that fostered chemical atomism.[10] Chemists used "atom" in their practice to designate the unit weight not of chemical elements but of elements or compounds. The "stoichiometric atom," which referred to the "definite weight of the chemical element or compound," reflected chemists' effort to implement the affinity approach to composition.[11] Berzelius, not Dalton, pushed the frontier of stoichiometry with a meticulous determination of combining ratios. He adopted and modified Dalton's innovative system of calculating chemical composition to stabilize chemical atomism.[12]

The tale of three fathers, which has fueled intensive research on Boyle, Lavoisier, and Dalton, has produced a rather distorted picture of chemistry in history. These three reputed fathers did not represent the chemists of their times. None of them was regarded as a skilled chemist by the contemporary chemists. Du Clos, the chief chemist of the Académie, criticized Boyle for his deficient knowledge of chemical literature and for his lack of manipulative capacities.[13] Boyle's project saw a brief resurgence in the Académie (largely as a result of Homberg's efforts), but the main lineage of French elite chemistry ran from Nicolas Lemery to Etienne-François Geoffroy, Jean Hellot, Pierre-Joseph Macquer, Antoine François de Fourcroy, and Claude-Louis Berthollet, who primarily pursued pharmaceutical and industrial research. Macquer and his contemporaries considered Lavoisier as a physicien who introduced physical methods of investigation to chemical practice. Dalton's "chemical" atom emerged in a context far removed from contemporary chemical theories and debates.[14] His ontology of the atom was designed to refute the understanding of the atmosphere as a chemical compound governed by affinity relations, which had become a standard interpretation since Lavoisier. The leading chemists, including Berzelius and Thomson, had to tailor Dalton's scheme to their stoichiometric investigation, thereby securing its status as a chemical theory.

In fact, the three reputed fathers of modern chemistry differed significantly from the ordinary chemical practitioners of their times. Boyle and Lavoisier pioneered special branches of chemistry that fostered theoretical sophistication for the field at large, precisely because of their critical distance from apothecaries and metallurgists. Boyle wished to make alchemical and pharmaceutical literature intelligible to natural philosophers and lay audience. To this end, he sought to bring instrumental precision to chemical practice and philosophical logic to chemical discourse. He seeded the genre of philosophical chemistry that enjoyed its heyday during the Enlightenment. Most eighteenth-century chemists whose names are familiar to us—Herman Boerhaave, William Cullen, Joseph Black, Joseph Priestley, Pierre-Joseph Macquer—were philosophical chemists who read extensively in chemistry and experimental philosophy and sought to establish chemistry's relevance for the Enlightenment audience. They were favorably inclined toward physical instruments such as the air-pump and the thermometer. Richard Kirwan, Henry Cavendish, Guyton de Morveau, and Antoine-Laurent Lavoisier inherited this tradition of philosophical chemistry and pushed it one step further with quantitative measurements and algebraic reasoning. Lavoisier stands out among this later generation of philosophical chemists for his extensive reliance on metric measurements. He endeavored to establish a much more rigorous algebra of chemical operations by deploying the uniform scales of various meters: the barometer, the thermometer, the hydrometer, the gasometer, the calorimeter. As the master of the Arsenal Group, he instituted the tradition of physical chemistry in French elite chemistry, which was continued by Berthollet and the Arcueil Group. Berthollet's *Essai de statique chimique* became the locus classicus for the development of physical chemistry during the second half of the nineteenth century.

The century-long activity of philosophical chemists introduced instrumental and linguistic precision to the practice of chemistry, transforming it from an apothecaries' trade to a philosophical knowledge. Seventeenth-century chemistry already possessed a sophisticated material culture of the laboratory, an organized teaching practice and textbook tradition, and a discourse of legitimation in the form of Chemical Philosophy. The culture of chemistry came under serious attack from natural philosophers, however, for its radical political associations, its artisanal social status, its hermetic discourse, and its "sooty" practice. Boyle and later generations of philosophical chemists sought to rehabilitate chemistry by translating its vast material knowledge into a public language. To this end, they introduced instruments that depended less on the operator's skills. In contrast to chemists' fire which varied from furnace to furnace and from operator to operator, Homberg could provide a more objective measure of the intensity of fire the burning glass commanded in terms of the distance at which the object of investigation was placed from the focus of the burning glass. This powerful furnace also delivered "pure" fire unadulterated by other materials and operators, giving rise to the French phlogiston theories of the mid eighteenth century. The thermometer promised an even more objective measure of chemists' fire for Boerhaave because of its supposed ability to detect the fire in all its gradations. Metric measurements fostered algebraic reasoning, as Lavoisier demonstrated. In addition to introducing precision to the existing theory and problem domains, philosophical chemists also established a different set of priorities in chemical practice. They were interested less in producing new medicaments than in forging a true representation of nature by utilizing chemical knowledge. They scrutinized the existing literature of chemistry more for its logical coherence than for its potential application. Through their efforts, chemistry emerged as an organized discourse with a stable systematics and an algebraic language by the end of the eighteenth century.

If the philosophical chemists dominated the public scene, however, they did not dictate the practice of chemistry. When we take a closer look at the development of chemistry in history, we are drawn to more likely father figures: Etienne-François Geoffroy, Pierre-Joseph Macquer, Jacob Berzelius. Ursula Klein has asserted Geoffroy's paternity for modern chemistry, for example, by pointing out that the affinity table embedded the modern conception of the "chemical compound."[15] As an heir to a thriving pharmaceutical dynasty, Geoffroy was thoroughly immersed in the contemporary pharmaceutical practice, but became a member of the Académie and took a medical degree at the Faculté. This unique combination of practical training and exposure to an elite literary environment enabled him to invent the affinity table which became a lasting monument of eighteenth-century chemistry. The table offered guidance for the laboratory practice of apothecaries and provided a legitimating rationale for chemistry. The discerning elites of the Académie wished for a predictive science. By skillfully exploiting the multiple meanings of "rapport" as mathematical ratio and relationship, Geoffroy projected the affinity table as an instrument that would transform chemistry into a predictive science based on definite laws.

Macquer's contribution to modern chemistry has been largely ignored by historians because of his allegiance to the phlogiston theory. Nevertheless, he played a crucial role in consolidating affinity chemistry as the theoretical frontier of European chemistry. He translated Stahl's principalist approach to composition into an operational chemistry of affinities that classified chemical experience in a systematic manner. He thus presented a theoretical discourse of affinity closer to the practice of apothecaries. His effort to join the theory and the practice of chemistry was no doubt facilitated by his partnership with the apothecary Antoine Baumé.[16] At the same time, Macquer adapted the Newtonian language of attraction to affinity chemistry, thereby linking apothecaries' knowledge to the prevailing discourse of natural philosophy. He also fulfilled manifold functions as a writer, teacher, correspondent, industrial chemist, and statesman, securing the social and literary space of chemistry in the public. In short, Macquer presided over the transformation of chemistry from an apothecaries' trade to a public science in France, which enticed young lawyers such as Guyton and Lavoisier to distinguished careers in chemistry.

Berzelius played a central role in the development of European chemistry during the first half of the nineteenth century. In addition to his exact analyses and comprehensive theoretical structure, which guided chemical practitioners for decades, he compiled detailed annual reports that consolidated the European chemical community.[17] Although Dalton provided a clear representational model for the emerging field of stoichiometry, chemists did not embrace his mechanistic ontology wholeheartedly. Berzelius and Thomson translated Dalton's ideas into useful calculating schemes for stoichiometry, initiating a spectacular yet "quiet" revolution in organic chemistry. Unlike the three reputed fathers of chemistry, Geoffroy, Macquer, and Berzelius represented ordinary chemists of their times. Deeply immersed in the contemporary practice of chemistry and well versed in the current investigations, they introduced significant analytic and/or theoretical innovations and sought to systematize the current sum of chemical knowledge. Their proximity to contemporary chemical practice has obscured their extraordinary contributions to the development of modern chemistry as a sophisticated material culture. The list of these more realistic fathers is bound to grow as we enrich our understanding of the historical development of chemistry. In order to understand the family tradition, we need a complete genealogy.

Our choice of Boyle, Lavoisier, and Dalton as the founding fathers of modern chemistry is rooted in our understanding of chemistry as a "physical" science. We have looked purposely at the figures who contributed to bringing chemistry closer to physics as the heroes in the evolution of chemistry as a science. Our historiographical bias has been informed by the rhetorical warfare that natural philosophers conducted from the seventeenth century onward. The condemnation of alchemists by Mersenne and Gassendi in the 1620s, Boyle's critique of "vulgar" chemists, and Boerhaave's condescension toward chemists all served to establish an unfavorable image of chemists as unintelligible mystics or sooty empirics who needed philosophical discourse and physical experimentation to discipline themselves into veritable savants. This biased characterization of ordinary chemists by the literary/philosophical "chemists" has become entrenched in historiography. Historians as readers occupy a symmetric position with these philosophical chemists relative to the ordinary chemists of pre-Lavoisian era. We possess a cultivated sympathy toward these philosophical/literary chemists.

Chemistry has not always been a physical science, however, in the way we understand the category. Although chemistry was included in the section of "physical sciences" in the Academy of Sciences, this category referred to a group of sciences related to medicine rather than to physics in our sense of the word. This medico-pharmaceutical orientation of chemistry was particularly pronounced in France, where metallurgical chemistry lagged behind other Continental countries such as Germany and Sweden. If we wish to understand chemistry's place in French history, we have to trace its changing associations with medicine. Eighteenth-century histories of chemistry are useful in this aspect. Although Macquer condemned the practice of alchemists, he acknowledged the role of later alchemists who pursued universal medicine in preparing the ground for eighteenth-century chemistry, of Paracelsus in fostering interest in chemical medicine, and of French textbook writers in inventing a new art of chemistry. These "true citizens," along with various authors who wrote on metallurgy and other arts with a philosophical mind, delivered chemistry from the disease of alchemy by encoding chemical operations in a positive and accessible manner.[18] Macquer's emphasis on medicine as the nurturing matrix for chemistry was just as whiggish as our emphasis on physics. Precisely for this reason, his characterization conveys the importance of medicine for chemistry in Enlightenment society and culture. After all, the leading European chemical authorities, such as Boerhaave and Stahl, were renowned professors of medicine.

Medicine provided the main institutional context for the practice of chemistry throughout the Enlightenment, although the particular social paths chemists took differed somewhat in different countries. In Britain, Boyle's gentlemanly pursuit of chemistry, exemplified by Boerhaave, became a tradition of philosophical chemistry practiced by university medical professors and lay gentlemen.[19] The British elite chemistry of the Enlightenment differed substantially from its Continental counterpart for this reason, focusing on heat and air rather than on salts.[20] In Germany, the nexus of mining interests intersected with the medical establishment of the universities to produce first-rate analytic chemists, who converged on Berlin after Stahl's move.[21] Despite their sophisticated analytic practice, German professors such as Stahl aimed primarily at teaching the university audience and did not translate their knowledge into an easily accessible language. In contrast, French elite institutions provided exceptional opportunities for upward social mobility for apothecaries. We can discern the contour of their social evolution in their relationship with the established medical and royal institutions.[22] During the seventeenth century, Paracelsian physicians and apothecaries struggled to gain acceptance within the medical community controlled by the Faculté de médecine, which provided them with a singular identity as iatrochemists. The establishment of the Jardin du roi helped their cause, although it maintained the social distinction between physicians and apothecaries in their respective positions of professors and demonstrators. The Académie royale des Sciences ameliorated this distinction to a certain degree as well as providing a more upscale intellectual context for vocational chemists. If chemists of the first generation such as Duclos and Bourdelin did not adapt well to this new social environment, the generation of Wilhelm Homberg and Nicolas Lemery fared much better. They consciously cultivated corpuscular language and imagery as the public discourse of chemistry, thereby adapting chemical theory to the prevailing philosophical discourse. Louis Lemery and Etienne-François Geoffroy, academicians of the next generation, were both educated at the Faculté; this signaled a social transformation of the elite apothecaries into physicians. In other words, the elite core of French Enlightenment chemistry was produced by the assimilation over generations of Huguenot physicians and apothecaries into the existing medical establishment of the Faculté.

If medicine as a literary and social practice provided a positive social identity for chemists, pharmacy maintained the disciplinary identity of chemistry as a material culture. Our knowledge of this culture is woefully inadequate, largely because of the long-standing historiographical bias that labeled pharmacy as the business of "sooty empirics." Even a brief glance at the contemporary pharmaceutical and metallurgical literature reveals a plethora of analytic procedures, which were condensed to a few lines in chemical textbooks. Nicolas Lemery's pharmaceutical

treatises include a long list of minutely controlled distillational procedures, which were classified into four "degrees of Fire" in his chemical
textbook. Sophisticated procedures designed for the preparation of individual remedies gave way to the "easiest manner of performing those
operations" in the chemical textbook.[23] The evolution of Parisian elite
chemistry depended heavily on the medico-pharmaceutical culture of
Montpellier and other Protestant cities, although the cluster of royal
institutions and public lectures in Paris offered certain advantages in formulating theoretical chemistry. The expanding list of salts used in pharmaceutical preparations and in industrial applications fueled academic
chemical investigations throughout the eighteenth century. The theoretical debates conducted among the philosophical chemists hinged on this
advancement in the knowledge of salts. The material, social, and political culture of Huguenot apothecaries in France deserves much more
extensive study.

To understand chemistry's place in Enlightenment France we must
discern the relationship between philosophical chemistry, pharmacy, and
medicine. The apothecaries' vast and sophisticated material culture
allowed philosophical chemists to delve into chemistry with relative ease.
Except in France, however, chemistry was represented and taught in
public primarily by physicians who could converse in diverse genres of
literature. If the public literature of chemistry—textbooks, journal literature, dictionaries, encyclopedias—is inadequate for unpacking the
complex material culture of apothecaries, therefore, it is important in
understanding the transformation of chemistry into a public science.
Though this relatively public literature did not capture the entire expanse
of contemporary chemical practice, it played a disproportionately large
role in shaping the theoretical discourse and the disciplinary identity of
chemistry. In a "series of debates" that was the Enlightenment,[24] the
public literature of chemistry shaped the contour of thought on what
chemistry was and should be. The French Chemical Revolution led by
Guyton and Lavoisier constituted a culmination of these debates that
transformed an apothecaries' trade into a public science.

I have not taken issue in this book with the historiographical designation of the Chemical Revolution as a singular event flowing from
Lavoisier's accomplishments. For the moment, my aim is limited to relat-

ing the Parisian chemical tradition to Lavoisier's revolution. I wish to make heard, however, multiple voices that shaped this event—Lavoisier's vision for the Revolution was not the same as Guyton's or Berthollet's. The number of voices should continue to grow, of course, as we cast an increasingly wide net to include other participants in the Revolution, such as apothecaries, physicians, mineralogists, and industrial chemists.[25] The ease with which philosophical chemists acquired the art of chemistry point to the presence of a stable and transferable body of practical knowledge in the possession of the local apothecaries, physicians, and metallurgists. Philosophical chemists supplemented this practical knowledge of the material world with literary knowledge drawn from books of various genres, past and present. The "applied" fields of chemistry also legitimated chemistry as a useful knowledge. If mineralogy seemed a mere classificatory endeavor that required the assistance of chemists to gain a solid footing as a science, it brought chemistry into the drawing rooms and salons of the Enlightenment and to the attention of powerful patrons, private and public. Budding chemical industries provided a new social niche for the elite chemists in which they could practice chemistry without the obligations of a personal physician. Large-scale productions of relatively simple substances, rather than delicate preparations of complex medicaments, transformed the problem and theory domains of chemistry. When these studies on various "applied" fields of chemistry are fully phased in, therefore, the very notion of the Chemical Revolution will be altered beyond recognition. The debate over the Scientific Revolution as a historiographical term is a case in point.[26] Revolutions in science entail not only an intellectual shift but also an institutional and social transformation.

In tracing the development of chemistry in history, it might be useful to speak of "chemical revolutions" rather than "the Chemical Revolution." Eighteenth-century authors often called significant events or acts that changed the status quo "revolutions." These included establishing new empires, contriving new religions, and advancing new schemes in philosophy. In using "revolutions" in the plural, however, they retained the "overtones of cycles of regeneration" alongside those of violent, progressive changes. This ambiguity seems to indicate that "revolutions," when used in regard to empires and philosophies, referred to the occurrence of radical changes rather than to the particular outcomes of such changes. Revolutions in political regimes and belief systems often brought back the older regimes and systems. Revolutions in sciences and arts took on the meaning of linear progress from early on, however, because they were deemed irreversible. Fontenelle thus described the development of calculus as constituting "the epoch of an almost total revolution" that would alter the state of mathematics completely. Rousseau characterized the transition from the primitive to the organized state of society as a revolution caused by the invention of metallurgy and agriculture. According to d'Alembert, Newton laid the foundation for a revolution that was completed by the next generation. Diderot wrote of "a great revolution in the sciences" that entailed a complete rejection of geometry.[27]

The prophecies of a revolution in chemistry should be understood in this wider context of usage.[28] Venel, who called for a new Paracelsus to bring about a revolution in chemistry, was a close associate of Diderot. Within a few years, Macquer spoke of "revolutions"—recurring periods of intense, rapid development that allowed chemistry to acquire its modern characteristics. The sense of linear progress through these revolutions, though hampered by the epidemic of alchemy, comes through clearly in various historical epochs that Macquer identified.[29] Lavoisier's utterance of the word "revolution" early in his career must be situated in this context of usage. Historians have long puzzled over the audacity of a young, inexperienced academician's predicting "a revolution in physics and chemistry."[30] When he wrote these words, however, Lavoisier did not foresee a complete rejection of the phlogiston theory. He was excited by the promise the subject of fixed air held. He expected that it might open up a new epoch in chemistry, and he spoke publicly of "an almost complete revolution." The anticipation of a revolution had to do with the uncharted territory of airs, rather than with the particular theory of phlogiston. Baumé thus referred to the existence of some physiciens who spoke of "a total revolution" in chemistry resulting from the properties of fixed air.[31] Others followed the suit: Bucquet anticipated a great revolution in chemistry resulting from the studies of gases in 1778. Fourcroy observed in 1782 that Macquer was "quite convinced of the great revolution" Lavoisier's pneumatic studies would occasion. These whispers of a revolution could simply indicate that chemists appreciated the potential consequences of pneumatic studies for chemistry. If we follow Macquer's conception of "revolutions," chemistry was about to enter a new revolutionary period that would alter its foundation fundamentally. This was exactly the way in which Fourcroy understood the significance of Lavoisier's work. As early as 1782, he assigned a new epoch to pneumatic studies in his brief history of chemistry. In subsequent years, he gave much more detailed expositions of chemical revolutions in his article "Chimie" in the *Encyclopédie méthodique* (1797) and in his *Système de connaissances chimique* (1800).[32]

Radical changes in chemical practice and theory often came about through the cultivation of a new subject area or new analytic methods rather than through the "overthrow" of the existing theory. As we have seen, the infusion of solution methods into pharmaceutical practice destabilized the doctrine of five principles, leading to its demise in the 1720s. The concomitant change in model systems from vegetables to minerals shaped the new theory domain of affinities. To the degree that the "epistemic things" of chemical investigation were constituted by the material system and by the analytic methods of chemistry, any radical change in each of these factors could destabilize the "experimental system" and lead to fundamental changes in theory and practice.[33] The premonitions of an imminent revolution in chemistry during the last quarter of the eighteenth century owed much to chemists' appreciation of the vast new territory that pneumatic studies promised to open up for chemical investigation. The concomitant changes in analytic methods and precision transformed chemical theory and practice in a profound manner. Theoretical changes often trailed these shifts in analytic methods and material systems of chemistry. Understanding chemistry's evolution as a scientific discipline requires, then, a much more sophisticated conception of scientific change than has been available.

The dynamics of change in chemical tradition was further complicated by the social hierarchy in which early modern chemists operated. Eighteenth-century chemists functioned within a clearly demarcated social hierarchy that allowed philosophical chemists to represent the apothecaries' art in public and to regulate the traffic between their literary and material practices. If apothecaries controlled the material culture of the laboratory, philosophical chemists commanded the written texts of Enlightenment sciences and philosophies.[34] If the apothecaries' art gave rise to the main theory domains of chemistry, the philosophical chemists dwelled upon them to articulate visions of theoretical chemistry in tune with the reigning philosophical paradigms. Their critical distance from the mundane practice of ordinary apothecaries allowed the philosophical chemists to produce a more systematic discourse on the main theory domains of chemistry. Wilhelm Homberg and Louis Bernard Guyton de Morveau produced the most systematic discussions of composition and affinity, respectively, which constituted the two theoretical moments in French Enlightenment chemistry. However, the same distance limited their capacity for innovation relevant to the material practice of chemistry. While Homberg took up the long-standing issue of composition, he did not introduce significant innovation in the art of distillation or dispute the chemical doctrine of five principles. Instead, he sought to reconcile these chemical practices and theories with the new analytic method of the burning glass as well as with the corpuscular conception of bodies. While Guyton was acutely aware of the need to reformulate the discourse of affinities and to quantify them, he produced not a universally applicable method but a montage of existing literature on the subject. The fact that Lavoisier did not undertake a literary exercise on the existing theory domains similar to those of Homberg and Guyton, but instead undertook a full-scale investigative program that charted the new territory of airs for chemical theory, renders credit to the Paris Academy as a leading scientific institution. As the Enlightenment progressed, academicians increasingly distanced themselves from the popular expositions of natural philosophy and metaphysical speculations fit for polite consumption and endeavored to set the standards for a disciplined analysis of nature.[35] Lavoisier heeded these tacit standards of the Academy in his deliberate analysis of nature with meticulous instrumentation, although he did mobilize the British philosophical chemistry to spotlight the importance of pneumatic and thermometric studies for chemistry.

Chemistry did not become a "science" through the singular event of the Chemical Revolution. Understanding the continuity and discontinuity throughout the Chemical Revolution requires a much more complex account of cognitive, methodological, epistemological, and institutional transitions.[36] Although historians have long been haunted by the question as to whether the Chemical Revolution transformed an art into a science or whether it revolutionized an existing science, such a question is necessarily vitiated by the historically variable definition of "science."[37] Historians of early modern chemistry have shown, for example, how fluid the boundary between alchemy and chemistry had been until the 1720s.[38] The demarcation of chemistry as the more scientific practice, separate from the mystical, secretive, and gold-making pursuit of alchemy, was drawn in the particular socio-political context that imbued alchemy with connotations of dangerous political ideologies.[39] If natural philosophers and philosophical chemists shunned alchemy in public and regarded the practice of ordinary apothecaries as beneath the standards of science for their lack of linguistic and instrumental precision, twentieth-century historians of science find these philosophical chemists equally unscientific for their reliance on speculative languages. The positive material contributions of the apothecaries seem more scientific to us than the literary exercises of the philosophical chemists. In order to pose historically sensitive questions, we should perhaps move away from our obsession with the scientific status of chemistry and characterize it instead as an evolving material culture that acquired various philosophical languages, theoretical structures, and social niches over the course of time. The identity of chemistry and chemists kept changing depending on the particular configuration of these factors. We could speak of many revolutions that shaped this transformation, as eighteenth-century chemists did in their own time.

If chemistry as a material culture continued to evolve gradually over time, chemistry as a discursive system of knowledge experienced a distinct cultural moment that transformed it from an apothecaries' art into a public science during the second half of the eighteenth century.[40] The popularity of Rouelle's lectures and Macquer's textbooks attests to the existence of a broader public for chemistry—a public that clamored for a rational representation of chemical art.[41] Many factors contributed to shaping this memorable moment in the evolution of chemistry. The establishment of chemistry in the medical curriculum and the emergent public sphere of the Enlightenment provided a positive profile for "useful" sciences, enticing a new class of young men into the previously denigrated practice of "sooty empirics." The emergence of industrial chemistry offered a new social space for chemists somewhat independent of medicine and its problem domains. The chemistry of salts, which provided model systems and a main theory domain for chemistry, had reached a phase of explosive growth. Pneumatic studies allowed detection of new elements as well as a new level of analytic precision in chemical practice. The expanding material, literary, and social territory of chemistry helped transform chemistry from an esoteric laboratory practice into a public culture.

Most important, in the middle of the eighteenth-century chemistry acquired a focal heuristic concept or a disciplinary metaphor that organized its laboratory practice as well as securing its theoretical potential and public appeal—that of chemical affinity. Aside from Goethe who studied the affinity theories of Macquer, Bergman, and Berthollet minutely to unite literature and the sciences,[42] a telling testimony for the centrality of affinity for chemistry comes from Charles Darwin in *The Origin of Species*. Embattled by the critics who objected to personifying Nature, he sought to defend natural selection as a legitimate cause for the transformation of species and as a binding paradigm for the studies of organic nature by pointing to the legitimate historical usages of such metaphorical expressions:

In the literal sense of the word, no doubt, natural selection is a false term; but who ever objected to chemists speaking of the elective affinities of the various elements?—and yet an acid cannot strictly be said to elect the base with which it in preference combines. It has been said that I speak of natural selection as an active power or Deity; but who objects to an author speaking of the attraction of gravity as ruling the movements of the planets? Every one knows what is meant and is implied by such metaphorical expressions; and they are almost necessary for brevity.[43]

A history of rhetorical reasoning underlies Darwin's argument. In order to establish the autonomy of mathematical natural philosophy, Newton had proposed the concept of gravity as the unifying cause of celestial and terrestrial mechanics regardless of its ontological status in the conventional discourse of philosophy and theology. He had further generalized the concept of gravity into that of attraction, which encompassed a range of experimental fields (electricity, magnetism, heat, light), thereby circumscribing the scope of what later became experimental physics. Chemists in the next century invoked the very same rhetoric to assert the autonomy of chemistry from natural philosophy, as did Darwin nearly another century later. Gravity or attraction, chemical affinity, and natural selection enjoyed privileged status as the focal heuristic concepts or disciplinary metaphors in the speciation of mathematical natural philosophy, chemistry, and evolutionary biology. Other examples can be found in the concept of vital force for physiology, that of energy for physics, and that of gene for genetics.[44] These disciplinary metaphors deserve particular attention from historians of science because they serve to establish the autonomy of the subject by defining the object of investigation for the discipline in general terms. Nevertheless, they have ambiguous ontological references, and they change constantly depending on the methods of investigation. If their constant metamorphoses in the laboratory or in the field make them elusive targets for historical studies, they function as productive metaphors precisely because of their malleability in practice. When they can no longer adapt to ongoing practices, they cease to be useful and usually are abandoned. The life cycles of these metaphors have much to teach us about the growth of scientific disciplines.[45]

Appendix

Chemistry Teachers at the Jardin du roi (source: J.-P. Contant, *L'enseignement de la chimie au jardin royal des plantes de Paris*, A. Coueslant, 1952)

Professors	Date of appointment	Demonstrators
	1648	William Davidson
	1651	Nicaise Lefebvre
	1660	Christopher Glaser
G. C. Fagon	1672	Moyse Charas
A. de Saint-Yon	1695	S. Boulduc
E. F. Geoffroy	1712	
	1729	G. F. Boulduc
L. Lemery	1730	
L. C. Bourdelin	1743	G. F. Rouelle
	1768	H. M. Rouelle
P. J. Macquer	(1771) 1777	
	1779	A. L. Brongniart
A. F. de Fourcroy	1784	
	1793	

Chemists at the Acadèmie royale des sciences, 1666–1785 (source: *Index biographique de l'Académie des sciences, 1666–1978,* Gauthier-Villars, 1979)

Académicien chemiste		
1666	S.C.Du Clos	C.Bourdelin
1674		J. Borelly
1691	W. Homberg	
1692		M. Charas
1694		S. Boulduc
1698			Langlade

	Pensionnaire			**Associé**	
1699		N. Lemery			N. Lemery
					E. F. Geoffroy
1712				L. Lemery	
1715	E.F. Geoffroy	L.Lemery		J. Lemery	C. J. Geoffroy
1721				(d'Isnard)	
1722				(F. Petit)	
1723			C.J. Geoffroy		
1724					C. F. Dufay
1727				G. F. Boulduc	
1731	C.F. Dufay				L. C. Bourdelin
1739	La Condamine	(Hellot)			
1743		Hellot			
1744				Malouin	
1752			L.C. Bourdelin		G. F. Rouelle
1766		Malouin		Macquer	
1770					Cadet
1772	Macquer			Lavoisier	Sage
1777			Cadet		
1778		Lavoisier			
1784	Sage				

Élève			(de Tournefort)	
E. F. Geoffroy	Tuillier	G. F. Boulduc	C. Berger	
C. Berger			L. Lemery	1700
	L. Lemery		Chomel	1702
			C. J. Geoffroy	1707
Imbert	J. Lemery			1712
Adjoint				1715
				1716
C. F. Dufay				1723
(F. J. Hunauld)				1725
		L. C. Bourdelin		1727
(Duhamel)				1730
La Condamine		Grosse		1731
Hellot				1735
Malouin				1742
G. F. Rouelle				1744
		Macquer		1745
Baron d'Henouville				1752
G. Jars	(Lavoisier)	Cadet		1766
		Sage		1770
		Bucquet		1778
		Berthollet		1780

Life Dates and Positions

Berger, Claude (1679–1712); (élève de Tournefort 1699), éléve 1700

Berthollet, Claude Louis (1748–1822); adjoint 1780, associé 1785, pensionnaire 1792

Borelly, Jacques (?–1689); académicien chimiste 1674

Boulduc, Gilles-François (1675–1742); élève 1699, adjoint 1716, associé 1727

Boulduc, Simon (1652–1729); académicien chimiste 1694, pensionnaire 1699

Bourdelin, Claude (1621–1699); académicien chimiste 1666, pensionnaire 1699

Bourdelin, Louis-Claude (1696–1777); adjoint 1727, associé 1731, pensionnaire 1752

Bucquet, Jean-Baptiste-Marie (1746–1780); adjoint 1778

Cadet de Gassicourt, Louis-Claude (1731–1799); adjoint 1766, associé 1770, pensionnaire 1777

Charas, Moyse (1619–1698); académicien chimiste 1692

Du Clos, Samuel Cottereau (1598–1685); académicien chimiste 1666

Dufay, Charles-François de Cisternay (1698–1739); adjoint 1723, associé 1724, pensionnaire 1731

Geoffroy, Claude-Joseph (1685–1752); (élève de Tournefort 1707, associé botaniste 1711), associé 1715, pensionnaire 1723

Geoffroy, Etienne-François (1672–1731); élève 1699, associé 1699, pensionnaire 1715

Grosse, Jean (?–1744); adjoint 1731

Hellot, Jean (1685–1766); adjoint, 1735, pensionnaire surnuméraire 1739, pensionnaire 1743

Homberg, Wilhelm (1652–1715); académicien chimiste 1691; pensionnaire 1699

La Condamine, Charles-Marie de (1701–1774); adjoint 1730, (associé géomètre 1735), pensionnaire 1739

Langlade (?–1717); académicien chimiste 1698; associé 1699

Lavoisier, Antoine Laurent (1743–1794); adjoint surnuméraire 1768, associé 1772, pensionnaire 1778

Lemery, Jacques (1678–1721); élève 1712, associé 1715

Lemery, Louis (1677–1743); (élève de Tournefort 1700) élève 1702, associé 1712, pensionnaire 1715

Lemery, Nicolas (1645–1715); associé January 1699, pensionnaire November 1699

Macquer, Pierre Joseph (1718–1784); adjoint 1745, associé 1766, pensionnaire 1772

Malouin, Paul Jacques (1701–1777); adjoint 1742, associé 1744, pensionnaire 1766

Rouelle, Guillaume François (1703–1770); adjoint 1744, associé 1752

Sage, Baltazar-Georges (1740–1824); adjoint 1770, associé 1777, pensionnaire 1784

Tuillier, Adrien (1674–1702); élève 1699

Lavoisier's Formula for Metallic Dissolutions   (source: "Considérations générales sur la dissolution des métaux dans les acides," *Oeuvres* II, 509–527, at 515–517. Parentheses express the manner in which the molecules were grouped in solution. Alchemical symbols are replaced here by Latin letters.)

Metallic substance	*SM*
Oxygen principle	*O*
Nitric acid	*Nr*
Nitrous acid	*N*
Acid	*A*
Iron	*F*
Water	*W*

I. The general formula for all metallic dissolutions before the mixture

$$\text{e.g. iron in nitric acid } = (F)\ (W\ Nr) \tag{1}$$
$$= (F)\ (W\ O\ N). \tag{2}$$

Since a given quantity of iron would require a determinate quantity of acid to dissolve it, if

a = the quantity of iron
b = the ratio between the quantity of acid and that of iron
ab = the quantity of acid necessary to dissolve a of iron
ab/r = the portion of water in the acid
ab/s = the portion of oxygen principle in the acid
ab/t = the portion of nitrous air in the acid.

Adding two portions of water, (2) becomes

$$(a\ F) + (2ab\ W + ab/r\ W) + (ab/s\ O + ab/t\ N). \tag{3}$$

II. After the mixture,

a/p = the quantity of oxygen necessary to saturate a of iron.

Subtracting this quantity of oxygen from nitric acid and adding it to iron, (3) becomes

$$(a\ F + a/p\ O) + (2ab\ W + ab/r\ W) + (ab/s\ O - a/p\ O + ab/t\ N).$$

Subtracting the quantity of nitrous air, nearly equal to the weigth of oxgyen principle:

$(a F + a/p \, O) + (2ab \, W + ab/r \, W) + (ab/s \, O - a/p \, O + ab/t \, N - a/p \, N)$.

Assuming that the quantity of acid employed is 1 livre,

$(a F + a/p \, O) + (2W + 1/q \, W) + (1/s \, O - a/p \, O + 1/t \, N - a/p \, N)$.

Notes

Introduction

1. Jeremy Adler, "Goethe's Use of Chemical Theory in His *Elective Affinities*," in *Romanticism and the Sciences*, ed. A. Cunningham and N. Jardin (Cambridge University Press, 1990).

2. David Knight, *Humphry Davy* (Cambridge University Press, 1992).

3. A unique exception is Trevor Levere, *Affinity and Matter* (Clarendon, 1971).

4. On the notion of cultural memory, see the introduction to *Acts of Memory*, ed. M. Bal et al. (University Press of New England, 1999).

5. Bernadette Bensaude-Vincent, *Lavoisier* (Flammarion, 1993), 58; Mary Jo Nye, *From Chemical Philosophy to Theoretical Chemistry* (University of California Press, 1993), 32–55.

6. On the importance of these locales, see Mi Gyung Kim, "Labor and Mirage: Writing the History of Chemistry," *Studies in History and Philosophy of Science* 26, 1995, 155–165.

7. Arthur Donovan, *Philosophical Chemistry in the Scottish Enlightenment* (Edinburgh University Press, 1975).

8. See B. Bensaude-Vincent's introduction to L. B. Guyton de Morveau, A. L. Lavoisier, C. L. Berthollet, and A. F. de Fourcroy, *Méthode de nomenclature chimique* (Seuil, 1994). See also W. R. Albury, The Logic of Condillac and the Structure of French Chemical and Biological Theory, 1780–1801, Ph.D. dissertation, Johns Hopkins University, 1972; John G. McEvoy, "The Enlightenment and the Chemical Revolution," in *Metaphysics and Philosophy of Science in the Seventeenth and Eighteenth Centuries*, ed. R. Woolhouse (Kluwer, 1988).

9. J. R. R. Christie and J. V. Golinski, "The Spreading of the Word: New Directions in the Historiography of Chemistry, 1600–1800," *History of Science* 20, 1982, 235–266; Jan Golinski, *Science as Public Culture* (Cambridge University Press, 1992).

10. Arthur Donovan, "Lavoisier as Chemist and Experimental Physicist: A Reply to Perrin," *Isis* 81, 1990, 270–272.

11. In medicine, "simple" referred to a plant that was used as a remedy by itself. More often, apothecaries concocted a "composition" from various plants.

12. On the notion of "experimental system," see Hans-Jörg Rheinberger, *Toward a History of Epistemic Things* (Stanford University Press, 1997). For the related concept of "model systems," see Robert E. Kohler, *Lords of the Fly* (University of Chicago Press, 1994); Angela N. H. Creager, *The Life of a Virus* (University of Chicago Press, 2002).

13. This vision was carried on by Berthollet with his Arcueil group. See Maurice Crosland, *The Society of Arcueil* (Harvard University Press, 1967).

14. Alan Rocke (*The Quiet Revolution*, University of California Press, 1993) portrays the development of organic chemistry in the second half of the nineteenth century as a "quiet revolution" for its mundane yet explosive nature. A similar characterization would apply, by the same token, to the development of stoichiometry during the first half of the century. These developments seem "quiet" to us, however, only because we do not recognize the noise they created in transforming the material, analytic basis of chemistry. Refinements in analytic methods and apparatus often bring about a gigantic stride in chemical practice. Good examples are Dumas's and Liebig's apparatus for organic analysis in the 1830s and the inventions of Pyrex and nuclear magnetic resonance in the twentieth century.

15. For interesting perspectives on the use of history in the practice of law, instead of historical circumstances determining the perimeters of law, see the editors' introduction to *History, Memory, and the Law*, ed. A. Sarat and T. Kearns (University of Michigan Press, 1999).

16. Steven Shapin, "Pump and Circumstance: Robert Boyle's Literary Technology," *Social Studies of Science* 14, 1984, 481–520; Steven Shapin and Simon Schaffer, *Leviathan and the Air-Pump* (Princeton University Press, 1985).

17. I am drawing on Foucault's notion of *genealogy*, which focuses on the techniques which sustain or change the discursive structure of knowledge. The fluctuations in power relations that are intimately bound with the techniques of social control could alter these techniques radically to bring about a rupture in the discursive system of knowledge. In contrast, Foucault's *archeology* seeks to unearth the field of contemporaneous meanings and perceptions or *epistemes* underlying the representational structures of different historical periods. See Michel Foucault, *The Order of Things* (Vintage Books, 1973); idem, *The Archaeology of Knowledge and The Discourse on Language* (Pantheon Books, 1972); idem, *Discipline and Punish* (Vintage Books, 1979); Arnold I. Davidson, "Archaeology, Genealogy, Ethics," in *Foucault: A Critical Reader*, ed. D. Hoy (Blackwell, 1986).

18. Mario Biagioli, "The Anthropology of Incommensurability," *Studies in History and Philosophy of Science* 21, 1990, 183–209; idem, "From Relativism to Contingentism," in *The Disunity of Science*, ed. P. Galison and D. Stump (Stanford University Press, 1996).

19. Dorinda Outram, *Georges Cuvier* (Manchester University Press, 1984); E. C. Spary, *Utopia's Garden* (University of Chicago Press, 2000).

20. Robert Fox, "The Rise and Fall of Laplacian Physics," *Historical Studies in the Physical Sciences* 4, 1974, 89–136.

21. Mi Gyung Kim, "Constructing Symbolic Spaces: Chemical Molecules in the Académie des Sciences," *Ambix* 43, 1996, 1–31.

22. Jan Golinski, "Chemistry in the Scientific Revolution: Problems of Language and Communication," in *Reappraisals of the Scientific Revolution*, ed. D. Lindberg and R. Westman (Cambridge University Press, 1990); idem, *Science as Public Culture*.

23. For modified uses/critique of Habermas's notion of "public sphere" for the French Enlightenment, see Benjamin Nathans, "Habermas's 'Public Sphere' in the Era of the French Revolution," *French Historical Studies* 16, 1990, 620–644; Daniel A. Bell, "The 'Public Sphere,' the State, and the World of Law in Eighteenth-Century France," *French Historical Studies* 17, 1992, 912–934; Dena Goodman, *The Republic of Letters* (Cornell University Press, 1994); Keith Michael Baker, "Defining the Public Sphere in Eighteenth-Century France: Variations on a Theme by Habermas," in *Habermas and the Public Sphere*, ed. C. Calhoun (MIT Press, 1994).

24. On the relevance of science to the public sphere, see Thomas Broman, "The Habermasian Public Sphere and 'Science in the Enlightenment,'" *History of Science* 36, 1998, 123–149. Although Habermas did not consider science as a part of the public sphere, chemistry and medicine offer particularly suitable cases for conceptualizing the public sphere since they possessed both the commercial and the literary aspects (Colin Jones, "The Great Chain of Buying: Medical Advertisement, the Bourgeois Public Sphere, and the Origins of the French Revolution," *American Historical Review* 101, 1996, 13–40).

25. Jan Golinski, "Utility and Audience in Eighteenth-Century Chemistry: Case Studies of William Cullen and Joseph Priestley," British *Journal for the History of Science* 21, 1988, 1–31; Roy Porter, "Science, Provincial Culture and Public Opinion in Enlightenment England," *British Journal for Eighteenth-Century Studies* 3, 1980, 20–46.

26. Awen A. M. Coley, "The Theme of Experiment in French Literature of the Earlier Eighteenth Century (1715–1761)," *Studies on Voltaire and the Eighteenth Century* 332, 1995, 213–333; Christian Licoppe, *La formation de la pratique scientifique* (Découverte, 1996).

27. Clifford Geertz, *Local Knowledge* (Basic Books, 1983). Excellent examples are Crosbie Smith and M. Norton Wise, *Energy and Empire* (Cambridge University Press, 1989) and William Clark, Jan Golinski, and Simon Schaffer, eds., *The Sciences in Enlightened Europe* (University of Chicago Press, 1999).

28. Owen Hannaway, *The Chemists and the Word* (Johns Hopkins University Press, 1975), ix. For a critique of Hannaway's approach, see Christie and Golinski, "The Spreading of the Word."

29. This ideal provided the prototype of what historians have discerned in the Chemical Revolution as the "principalist" approach to chemical composition. See H. E. LeGrand, "The 'Conversion' of C.-L. Berthollet to Lavoisier's Chemistry," *Ambix* 22, 1975, 58–70, at 58.

30. Mi Gyung Kim, "The Layers of Chemical Language I: Constitution of Bodies v. Structure of Matter," *History of Science* 30, 1992, 69–96.

31. F. L. Holmes, *Eighteenth Century Chemistry as an Investigative Enterprise* (Berkeley: Office for History of Science and Technology, 1989).

32. Martin Fichman, "French Stahlism and Chemical Studies of Air, 1750–1770," *Ambix* 18, 1971, 94–122. In contrast, Jon Eklund (Chemical Analysis and the Phlogiston Theory, 1738–1772, Ph.D. dissertation, Yale University, 1971) identifies the Stahlist period between 1737 and 1772, focusing on the uses of phlogiston theory; I thank William Newman for this reference.

33. Betty Jo Teeter Dobbs and Margaret C. Jacob, *Newton and the Culture of Newtonianism* (Humanities Press, 1995); Margaret C. Jacob, *Scientific Culture and the Making of the Industrial West* (Oxford University Press, 1997).

34. Guyton de Morveau, "Essai physico-chymique sur la dissolution et la crystallisation, pour parvenir à l'explication des affinités par la figure des parties constituantes des corps," in *Digressions academiques* (Dijon: L. N. Frantin, 1772).

35. Mi Gyung Kim, Practice and Representation: Investigative Programs of Chemical Affinity in the Nineteenth Century, Ph.D. dissertation, University of California, Los Angeles, 1990.

Chapter 1

1. Owen Hannaway, *The Chemists and the Word* (Johns Hopkins University Press, 1975), ix.

2. A. Kent and O. Hannaway, "Some New Considerations on Beguin and Libavius," *Annals of Science* 16, 1960, 241–250.

3. Hannaway, *The Chemists*; Bruce T. Moran, "Medicine, Alchemy, and the Control of Language: Andreas Libavius versus the Neoparacelsians," in *Paracelsus*, ed. O. Grell (Brill, 1998); idem, "Libavius the Paracelsian? Monstrous Novelties, Institutions, and the Norms of Social Virtue," in *Reading the Book of Nature*, ed. A. Debus and M. Walton (Thomas Jefferson University Press, 1998).

4. Allen G. Debus, *The Chemical Philosophy* (Science History Publications, 1977); idem, *The French Paracelsians* (Cambridge University Press, 1991).

5. Hugh Trevor-Roper, "The Paracelsian Movement," in *Renaissance Essays* (University of Chicago Press, 1985), 149–199; idem, "Paracelsianism Made Political, 1600–1650," in *Paracelsus*, ed. Grell.

6. A. G. Chevalier, "The 'Antimony-War'—A Dispute between Montpellier and Paris," *Ciba Symposia* 2, 1940, 418–423; Howard M. Solomon, *Public Welfare, Science, and Propaganda in Seventeenth Century France* (Princeton University Press, 1972); Edouard Brygoo, "Les médecins de Montpellier et le Jardin du Roi à Paris," *Histoire et Nature* 14, 1979, 3–29.

7. Laurence Brockliss and Colin Jones, *The Medical World of Early Modern France* (Clarendon, 1997), 119–128. I thank Caroline Hannaway for this reference.

8. Jean Béguin, *Les Elemens de chymie*, second French edition (Geneva: I. Celerier, 1624), 3.

9. Christopher Glaser, *The Compleat Chymist* (London: printed for John Starkey at the Miter, 1677), 3.

10. Nicolas Lemery, *Pharmacopée universelle* (Paris: Laurent d'Houry, 1697), preface.

11. Hélène Metzger, *Les Doctrines chimiques en France du début du XVIIe à la fin du XVIIIe siècle*, new edition (Blanchard, 1969), 17–97.

12. Trevor-Roper, "The Paracelsian Movement," 166–169.

13. Hugh Trevor-Roper, "The Sieur de la Rivière, Paracelsian Physician of Henri IV," in *Science, Medicine and Society in the Renaissance*, ed. A. Debus (Science History Publications, 1972), volume 2.

14. Du Chesne, *Liber de priscorum philosophorum verae medicinae materiae* (1603) (Leipzig: T. Schürer and B. Voight, 1613); idem, *Ad veritatem hermeticae medicinae ex Hoppocratis veterumque decretis ac therapeusi* (1604) (Frankfurt: W. Richter and C. Nebenius, 1605). For an English translation containing large sections of both works, see *The Practice of Chymicall, and Hermeticall Physicke, for the Preservation of Health* (London: T. Creede, 1605).

15. Debus, *Chemical Philosophy*, 159–173, and *French Paracelsians*, 48–65; Moran, "Medicine, Alchemy, and the Control of Language."

16. Thomas Gibson, "A Sketch of the Career of Theodore Turquet de Mayerne," *Annals of Medical History* 5, 1933, 315–326.

17. A. Germain, *L'Apothicairerie à Montpellier sous l'ancien régime universitaire* (Montpellier: J. Martel, 1882).

18. Solomon, *Public Welfare*, 162–200; Harcourt Brown, *Scientific Organizations in Seventeenth Century France (1620–1680)* (Russell & Russell, 1934), 17–40.

19. Clara de Milt, "Early Chemistry at le Jardin du Roi," *Journal of Chemical Education* 18, 1941, 503–510.

20. Henry Guerlac, "Guy de La Brosse: Botanist, Chemist and Libertine," in Guerlac, *Essays and Papers in the History of Modern Science* (Johns Hopkins University Press, 1977); Rio C. Howard, "Guy de La Brosse: Botanique et chimie au début de la révolution scientifique," *Revue d'histoire des sciences et leurs*

applications 31, 1978, 301–326; idem, "Guy de la Brosse and the Jardin des Plantes in Paris," in *The Analytic Spirit*, ed. H. Woolf (Cornell University Press, 1981); Yves Laissus, "Le Jardin du Roi," in *Enseignement et diffusion des sciences en France au XVIIIe siècle*, ed. R. Taton (Hermann, 1964).

21. For the complete text of the 1635 edict, see Jean-Paul Contant, *L'Enseignement de la Chimie au Jardin Royal des Plantes de Paris* (Cahors: A. Coueslant, 1952), 13–18.

22. Henry Guerlac, "Guy de La Brosse and the French Paracelsians," in *Science, Medicine, and Society in the Renaissance*, ed. A. Debus, volume 1; Howard, "Guy de La Brosse: Botanique et chimie."

23. Brown, *Scientific Organizations*, 17–40; Solomon, *Public Welfare*; Geoffrey V. Sutton, *Science for a Polite Society* (Westview, 1995), 19–41.

24. The most comprehensive study of chemistry teaching at the Jardin with a genealogy of teachers is Contant, *L'Enseignement de la Chimie*. See also John Read, *Humour and Humanism in Chemistry* (Bell, 1947), 81–123.

25. J. Read, "William Davidson of Aberdeen: The First British Professor of Chemistry," *Ambix* 9, 1961, 70–101.

26. Owen Hannaway, "Nicaise Le Febvre," in *Dictionary of Scientific Biography*.

27. Roger Goulard, "A propos de l'Affaire des poisons, le célèbre édit de 1682," *Bulletin de la Société française d'histoire de la médecine* 13, 1914, 260–268. On the intrigue-ridden atmosphere of the French court in which chemists' knowledge of poison was feared, see Christine Pevitt, *The Man Who Would Be King* (Morrow, 1997), passim.

28. Ferenc Szabadváry, *History of Analytic Chemistry* (Pergamon, 1966).

29. Lefebvre, *A Compleat Body of Chymistry* (London: T. Ratcliffe, 1664), 1–20.

30. William Newman ("Art, Nature, and Experiment among some Aristotelian Alchemists," in *Texts and Contexts in Ancient and Medieval Science*, ed. E. Sylla and M. McVaugh, Brill, 1997, 308) argues that the Latin Medieval alchemy was "inveterately" Aristotelian.

31. R. Hooykaas, "Die Elementenlehre des Paracelsus," *Janus* 39, 1935, 175–188; Walter Pagel, *Paracelsus*, second edition (Karger, 1982), 82–104.

32. L. W. B. Brockliss, "Aristotle, Descartes and the New Science: Natural Philosophy at the University of Paris, 1600–1740," *Annals of Science* 38, 1986, 33–69; idem, *French Higher Education in the Seventeenth and Eighteenth Centuries* (Clarendon, 1987); Peter Dear, *Discipline and Experience* (University of Chicago Press, 1995). For a survey of philosophy curriculum in French colleges and universities, see various articles in *L'Enseignement et diffusion*, ed. Taton.

33. J. S. Spink, *French Free-Thought from Gassendi to Voltaire* (Greenwood, 1969).

34. R. Hooykaas, "Die Elementenlehre der Iatrochemiker," *Janus* 41, 1937, 1–28.

35. Debus, *The English Paracelsians* (Franklin Watts, 1966), 90 and 128 (n. 7); idem, *The French Paracelsians*, 51.

36. Joseph Duchesne, *Le Grand miroir du monde* (Lyon: pour Barthelem Honorat, 1587; second edition, Lyon: pour les Heritiers d'Eustache Vignon, 1593), 529 and 532; quoted in Debus, *French Paracelsians*, 52.

37. Debus, *Chemical Philosophy*, 159–166.

38. Debus, *French Paracelsians*, 54.

39. It appeared in more than forty Latin, French, and English editions. See T. S. Patterson, "Jean Beguin and His *Tyrocinium Chymicum*," *Annals of Science* 2, 1937, 243–298.

40. Jean Béguin, *Les Elemens de chymie*, "Au Lecteur."

41. Natural bodies such as plants, minerals, and animals were regarded as "mixts" or "mixt bodies" because they could be resolved into simpler "principles" through chemical analysis, mostly distillation.

42. This was also Libavius's conclusion in the aftermath of the Paris debate (Debus, *French Paracelsians*, 60–62).

43. With "physics" Béguin seems to refer mostly to mechanics, although in the seventeenth and eighteenth centuries *la physique* in French was equivalent to "natural philosophy" in English.

44. Béguin, *Les Elemens*, 39–40.

45. The same example was given in Oswald Croll, *Basilica Chymica* (1609); see Hannaway, *The Chemists*, 37.

46. Béguin, *Les Elemens*, 33.

47. Spink, *French Free-Thought*.

48. Didier Kahn, "Entre atomisme, alchimie et théologie: la réception des thèses d'Antoine de Villon et Étienne de Clave contre Aristote, Paracelsus et les 'cabalistes,'" *Annals of Science* 58, 2001, 241–286.

49. Debus, *Chemical Philosophy*, 205–293; W. L. Hine, "Mersenne and Alchemy," in *Alchemy Revisited*, ed. Z. von Martels (Brill, 1990).

50. Etienne de Clave, *Nouvelle lumière philosophique des vrais principes et elemens de nature, & qualité d'iceux* (Paris: O. de Varennes, 1641), "Au Lecteur."

51. Ibid., 159.

52. De Clave, *Cours de chimie* (Paris, 1646), 4, quoted in Metzger, *Les Doctrines*, 54–55.

53. Ibid.

54. De Clave, *Nouvelle lumière*, 260.

55. These elements were "ideal substances, real in the sense that they existed in matter, but not real in the sense that they could be handled and observed" (Marie Boas, *Robert Boyle and Seventeenth-Century Chemistry*, Cambridge University Press, 1958, 86).

56. John Rudolph Glauber, *Furni novi philosophici über Beschreibung einer Neu-erfundenen Distillir-Kunst* (Amsterdam: Johann Fabeln, 1646), translated as *A Description of New Philosophical Furnaces, or Art of Distilling* (London: R. Coats, 1651); John French, *The Art of Distillation* (London: R. Coats, 1650).

57. Nicaise Lefebvre, *A Compleat Body of Chymistry*, 1–4.

58. Ibid., 107.

59. Ibid., "To the Apothecaries of England," first page.

60. Ibid., 107–112.

61. Ibid., 18.

62. Ibid., 15.

63. Ibid., 19–20.

64. Ibid., 20.

65. Clara de Milt, "Christopher Glaser," *Journal of Chemical Education* 19, 1942, 53–60; Roy G. Neville, "Christophle Glaser and the *Traité de la chmie*, 1663," *Chymia* 10, 1965, 25–52. Michel Bougard (*La chimie de Nicolas Lemery*, Turnhout: Brepols, 1999), 24–26) presents evidence that Moyse Charas claimed the authorship for Glaser's book.

66. Christopher Glaser, *The Compleat Chymist*, 2–4.

67. Ibid., 5.

68. Allen G. Debus, "Fire Analysis and the Elements in the Sixteenth and the Seventeenth Centuries," *Annals of Science* 23, 1967, 127–147; F. L. Holmes, "Analysis by Fire and Solvent Extractions: The Metamorphosis of a Tradition," *Isis* 62, 1971, 129–148.

69. Allen G. Debus, "The Paracelsian Compromise in Elizabethan England," *Ambix* 8, 1960, 71–97; idem, *English Paracelsians*.

70. P. M. Rattansi, "Paracelsus and the Puritan Revolution," *Ambix* 11, 1963, 24–32; idem, "The Helmontian-Galenist Controversy in Restoration England," *Ambix* 12, 1964, 1–23; C. Hill, *The World Turned Upside Down* (Temple Smith, 1972).

71. J. Andrew Mendelsohn, "Alchemy and Politics in England, 1649–1665," *Past and Present* 135, 1992, 30–78.

72. On the political overtones of natural philosophy, see Stephen Shapin and Simon Schaffer, *Leviathan and the Air-Pump* (Princeton University Press, 1985).

73. On the delicate liaisons Boyle had to establish between the scholar and the craftsman, see Malcolm Oster, "The Scholar and the Craftsman revisited: Robert Boyle as Aristocrat and Artisan," *Annals of Science* 49, 1992, 255–276; idem,

"Biography, Culture, and Science: The Formative Years of Robert Boyle," *History of Science* 31, 1993, 177–226.

74. Walter Pagel, *Jan Baptista van Helmont* (Cambridge University Press, 1982); Charles Webster, *The Great Instauration* (Holmes & Meier, 1975), 273–282.

75. On the Hartlib circle's interest in chemistry and alchemy, and particularly in van Helmont and Glauber, see R. S. Wilkinson, "The Hartlib Papers and Seventeenth-Century Chemistry," *Ambix* 15, 1968, 54–69; 17, 1970, 85–110; Stephen Clucas, "The Correspondence of a XVII-Century 'Chymicall Gentleman': Sir Cheney Culpeper and the Chemical Interests of the Hartlib Circle," *Ambix* 40, 1993, 147–170; William R. Newman, "George Starkey and the Selling of Secrets," in *Samuel Hartlib and Universal Reformation*, ed. M. Greengrass et al. (Cambridge University Press, 1994).

76. Ladislao Reti, "Van Helmont, Boyle and the alkahest," in *Some Aspects of Seventeenth-Century Medicine and Science* (Los Angeles: William Andrews Clark Memorial Library, 1969); Antonio Clericuzio, "Robert Boyle and the English Helmontians," in *Alchemy Revisited*, ed. Martels; idem, "From van Helmont to Boyle. A Study of the Transmission of Helmontian Chemical and Medical Theories in Seventeenth-Century England," *British Journal for the History of Science* 26, 1993, 303–334; William Newman, *Gehennical Fire* (Harvard University Press, 1994).

77. Rattansi, "The Helmontian-Galenist Controversy"; Mendelsohn, "Alchemy and Politics."

78. The classic account is Marie Boas, *Robert Boyle*. See also Boas's introductory essay in *Robert Boyle on Natural Philosophy* (Indiana University Press, 1965), 81–93. On the role of chemistry in shaping Boyle's experimental philosophy, see Rose-Mary Sargent, "Learning from Experience: Boyle's Construction of an Experimental Philosophy," in *Robert Boyle Reconsidered*, ed. M. Hunter (Cambridge University Press, 1994); idem, *The Diffident Naturalist* (University of Chicago Press, 1995).

79. Lawrence M. Principe, *The Aspiring Adept* (Princeton University Press, 1998), 27–62.

80. *The Works of the Honourable Robert Boyle*, new edition (London, 1772) [hereafter cited as *Works*], volume 1, 298–443.

81. Boyle, "Some Specimens of an Attempt to Make Chymical Experiments Useful to Illustrate the Notions of the Corpuscular Philosophy," *Works*, volume 1, 354–359.

82. Boyle, "A Proëmical Essay, Wherein, with Some Considerations Touching Experimental Essays in General, Is Interwoven Such an Introduction to All Those Written by the Author, as Is Necessary to Be Perused for the Better Understanding of Them," *Works*, volume 1, 299–318, 300.

83. Robert Boyle, *The Sceptical Cymist* (1661), facsimile edition (London: Dawsons of Pall Mall, 1965), 202–204. This passage has invited conflicting interpretations concerning the status of alchemy. Clericuzio ("Carneades and the

Chemists: A Study of *The Sceptical Chymist* and Its Impact on Seventeenth-Century Chemistry," in *Robert Boyle Reconsidered*, ed. Hunter) discerns Boyle's willingness to leave alchemy in the realm of darkness while trying to forge a respectable discourse of chemistry, while Principe pitches for Boyle's sincere affection for alchemical adepti and their secrets which stood above the "vulgar" doctrines of textbook writers; see also Principe, *The Aspiring Adept*, 49.

84. On the evolution of Boyle's thought, see Marie Boas, "An Early Version of Boyle's *Sceptical Chymist*," *Isis* 45, 1954, 153–168.

85. Boyle had in his possession one of De Clave's books: Hartlib's Ephemerides, November 1649, HP 28/1/32A. I thank Lawrence Principe for this reference.

86. Boyle, *Sceptical Chymist*, 46.

87. Ibid., 16.

88. Ibid., 298–299.

89. Ibid., 37–46.

90. Ibid., 171.

91. For further elaboration of the relationship between the corpuscles of various sizes and shapes and chemists' principles, see Robert Boyle, "The Origin of Forms and Qualities, according to the Corpuscular Philosophy," *Works*, volume 3, 1–137; idem, "Experiments, Notes, &c. about the Mechanical Origin or Production of diverse particular Qualities," *Works*, volume 4, 230–354. See also Clericuzio, "A Redefinition of Boyle's Chemistry and Corpuscular Philosophy," *Annals of Science* 47, 1990, 561–589.

92. Boyle, *Sceptical Chymist*, 152–153.

93. Thomas S. Kuhn, "Robert Boyle and Structural Chemistry in the Seventeenth Century," *Isis* 43, 1952, 12–36.

94. Boyle, "The Origin of Forms and Qualities," *Works*, volume 3, 105.

95. Yung Sik Kim, "Another Look at Robert Boyle's Acceptance of the Mechanical Philosophy: Its Limits and its Chemical and Social Contexts," *Ambix* 38, 1991, 1–10; Alan Chalmers, "The Lack of Excellency of Boyle's Mechanical Philosophy," *Studies in History and Philosophy of Science* 24, 1993, 341–364; Sargent, *The Diffident Naturalist*.

96. Clericuzio, "From van Helmont to Boyle"; idem, "A Redefinition."

97. See Muriel West, "Notes on the Importance of Alchemy to Modern Science in the Writings of Francis Bacon and Robert Boyle," *Ambix* 9, 1961, 102–114; William R. Newman, "Boyle's Debt to Corpuscular Alchemy," in *Robert Boyle Reconsidered*, ed. Hunter; Lawrence M. Principe, "Robert Boyle's Alchemical Secrecy: Codes, Ciphers, and Concealments," *Ambix* 39, 1992, 63–74; idem, "Boyle's alchemical pursuits," ibid., 91–105; idem, *The Aspiring Adept*; Michael Hunter, "Alchemy, Magic and Moralism in the Thought of Robert Boyle," *British Journal for the History of Science* 23, 1990, 387–410.

98. Clericuzio, "A Redefinition."

99. Oster, "The Scholar and the Craftsman"; idem, "Biography, Culture, and Science"; idem, "Virtue, Providence and Political Neutralism: Boyle and Interregnum Politics," in *Robert Boyle Reconsidered*, ed. Hunter.

100. Yung-Sik Kim, "Another Look."

101. Jan Golinski, "Scepticism and Authority in Seventeenth-century Chemical Discourse," in *The Figural and the Literal*, ed. A. Benjamin et al. (Manchester University Press, 1987).

102. Boyle, *Sceptical Chymist*, 299–308.

103. On the transformation of Newton's "private passion" for chemistry/alchemy into the "public doctrine" of Newtonian chemistry, see Jan Golinski, "The Secret Life of an Alchemist," in *Let Newton Be! A New Perspective on His Life and Works*, ed. J. Fauvel et al. (Oxford University Press, 1988).

104. Thomas Sprat, *The History of the Royal Society of London* (London: printed by T. R. for J. Martyn at the Bell, 1667; new edition, St. Louis: Washington University, 1958), 37. For a careful analysis of Sprat's rhetorical stance in writing this official "apology," see Brian Vickers, "The Royal Society and English Prose Style: A Reassessment," in *Rhetoric and the Pursuit of Truth* (Los Angeles: William Andrews Clark Memorial Library, 1985).

105. J. R. Jacob, "Robert Boyle and Subversive Religion in the Early Restoration," *Albion* 6, 1974, 275–293; idem, "Restoration, Reformation and the Origins of the Royal Society," *History of Science* 13, 1975, 155–176; idem, *Robert Boyle and the English Revolution* (Franklin, 1977); idem, "Boyle's Atomism and the Restoration Assault on Pagan Naturalism," *Social Studies of Science* 8, 1978, 211–233; idem, "Restoration Ideologies and the Royal Society," *History of Science* 17, 1980, 25–38; Simon Schaffer, "Godly Men and Mechanical Philosophers: Souls and Spirits in Restoration Mechancial Philosophy," *Science in Context* 1, 1987, 55–85.

106. Michael Hunter, *Robert Boyle by Himself and His Friends* (Pickering, 1994), 29.

107. Boyle, *Sceptical Chymist*, 286–297 and 290.

108. Robert Boyle, "Experiments and Notes about the Producibleness of *Chymical Principles*, being Parts of an Appendix designed to be added to *The Sceptical Chymist*," *Works*, volume 1, 587–661. Boyle's continuing preoccupation with the problem of principles can be discerned in "Of the Imperfection of the Chemists' Doctrine of Qualities (1675)," in *Selected Philosophical Papers of Robert Boyle*, ed. M. Stewart (Manchester University Press, 1979). This work recapitulated more succinctly and powerfully his earlier criticisms in *The Sceptical Chymist*.

109. Ibid., 589–599.

110. Ibid., 609.

111. Ibid., 652.

112. Ibid., 655.

113. Ibid., 629–630.

114. Ibid., 635–636.

115. Ibid., 595.

116. The text of this proposal is printed in *Oeuvres complètes de Christiaan Huygens* (Société hollandaise des sciences, 1888–1950), volume 4, 325–329; Trevor McClaughlin, "Sur les rapports entre la Compagnie de Thévenot et l'Académire royale des Sciences," *Revue d'histoire des sciences et leurs applications* 28, 1975, 235–242.

117. On the prehistory and organization of the Paris Academy, see M. G. Bigourdan, "Le premières sociétés scientifiques de Paris au XVIIe siècle. Les réunions du P. Mersenne et l'Académie de Montmor," *Comptes Rendus* 164, 1917, 129–134, 159–162, and 216–220; Harcourt Brown, "The Utilitarian Motive in the Age of Descartes," *Annals of Science* 1, 1936, 182–192; idem, *Scientific Organizations*; John Milton Hirschfield, *The Académie Royale des Sciences, 1666–1683* (Arno, 1981), 1–78; René Taton, *Les Origines de l'Académie Royale des Sciences* (Université de Paris, Palais de la Decouverte, 1965); Roger Hahn, *The Anatomy of a Scientific Institution* (University of California Press, 1971), 1–19; Wendy Perkins, "The Uses of Science. The Montmor Academy, Samuel Sorbière and Francis Bacon," *Seventeenth-Century French Studies* 7, 1985, 155–162; David S. Lux, "Colbert's Plan for the Grande Academie: Royal Policy toward Science, 1663–67," *Seventeenth-Century French Studies* 12, 1990, 177–188; David J. Sturdy, *Science and Social Status* (Boydell, 1995), 1–142. Sturdy also offers an insightful synopsis of scientific and political context as well as of the context of patronage. His book is an excellent source of biographical information on academcians during the stated period.

118. Alice Stroup, *A Company of Scientists* (University of California Press, 1990).

119. Charles-Joseph-Etienne Wolf, *Histoire de l'Observatoire de Paris de sa Fondation à 1793* (Gauthier-Villars, 1902), 1–112.

120. Hirschfield, *The Académie*, 97–110.

121. Stroup, *A Company of Scientists*, 55.

122. E. C. Watson, "The Early Days of the Académie des Sciences as Portrayed in the Engravings of Sébastien Le Clerc," *Osiris* 7, 1939, 556–587.

123. For example, Mlle. de Chetaignaires's interest in chemistry was not viewed kindly; see Sutton, *Science for a Polite Society*, 104–105.

124. Alice Stroup, "Censure ou querelles scientifiques: l'affaire Duclos (1675–1685)" (manuscript). There exist records of four Samuel Du Closes, all of them physicians, chemists, and Protestants (Doru Todériciu, "Sur la vraie biographie de Samuel Duclos (Du Clos) Cortreau," *Revue d'histoire des sciences et leurs applications* 27, 1974, 64–67; Sturdy, *Science and Social Status*, 107–109).

125. P. Dorveaux, "Apothicaires Membres de l'Académie Royale des Sciences," *Bulletin de la Société d'Histoire de la Pharmacie*, 1929, 4–12; Sturdy, *Science and Social Status*, 112–115.

126. Sturdy, *Science and Social Status*, 80–142.

127. "Projet d'Exercitations Physiques, proposé a l'assemblée par le Mr. Du Clos," *procès-verbaux de l'Académie royale des sciences*, Paris [hereafter cited as *PV*] 1, 1666, 1–16 (December 31). For a published version, see "Dissertation sur les principes des Mixtes naturels (1781)," in *Mémoires de l'Académie royale des sciences, 1666–1699* (Paris, 1731), volume 4.

128. Clericuzio, "From van Helmont to Boyle," 304. Du Clos's student Lefebvre cited van Helmont and Glauber as the two greatest chemists of his time (Lefebvre, *A Compleat Body of Chymistry*, "To the apothecaries of England").

129. Sturdy, *Science and Social Status*, 108.

130. *PV* 4, 1668, 58r–66r (June 16th and 23rd).

131. Samuel Cottereau Du Clos, *Observations sur les Eaux Minérales de Plusieurs Provinces de France, Faites en l'Académie Royale des Sciences en l'Année 1670 et 1671*(Paris, 1675). For a survey of analyses of mineral water in Europe, including Du Clos's, see Jon Eklund, Chemical Analysis and the Phlogiston Theory, 1738–1772, Ph.D. dissertation, Yale University, 1971, 231–291.

132. Brown, *Scientific Organizations*; Marie Boas Hall, "The Royal Society's Role in the Diffusion of Information in the Seventeenth Century," *Notes and Records of the Royal Society of London* 29, 1974–1975, 173–192; Barrie Walters, "The *Journal des savants* and the Dissemination of News of English Scientific Activity in Late Seventeenth-Century France," *Studies on Voltaire and the Eighteenth Century* 314, 1993, 133–166.

133. R. E. W. Maddison, "Studies in the Life of Robert Boyle. Part I. Robert Boyle and Some of His Foreign Visitors," *Notes and Records of the Royal Society of London* 9, 1951, 1–33.

134. A. Rupert Hall, "Henry Oldenburg et les relations scientifiques au XVIIe siècle," *Revue d'histoire des sciences et leurs applications* 23, 1970, 285–303.

135. Alice Stroup, "Duclos on Boyle: A French Academician Criticizes *Certain Physiological Essays*," presentation at the History of Science Society Annual meeting (Vancouver, 2000); idem, Censure ou querelles scientifiques; idem, *A Company of Scientists*.

136. *Histoire de l'Académie royale des sciences, avec mémoires, Histoire* [hereafter cited as *Histoire*], 1666–1686, volume 1, 79–81. See the discussion in John C. Powers, " 'Ars sine arte': Nicolas Lemery and the End of Alchemy in Eighteenth-Century France," *Ambix* 45, 1998, 163–189, esp. 166.

137. *PV* 1, 1666, 30–38 (December 31); Stroup, *A Company of Scientists*, 70–71.

138. *PV* 4, 1668, 48r–54v (June 9); Stroup, *A Company of Scientists*, 73–74.

139. Hirschfield, *The Académie*, 146–168; Stroup, *A Company of Scientists*, 89–102; Yves Laissus and Anne-Marie Monseigny, "*Les Plantes du Roi*. Note sur un grand ouvrage de botanique préparé au XVIIe siècle par l'Académie royale des sciences," *Revue d'histoire des sciences et leurs applications* 22, 1969, 193–236.

140. Stroup, *A Company of Scientists*, 75–116. Du Clos renounced alchemy publicly and converted to Catholicism on his death bed to ensure the career of his nephew (Stroup, "Censure").

141. Holmes, *Eighteenth-Century chemistry as an Investigative Enterprise* (Berkeley: Office for History of Science and Technology, 1989), 68–73.

142. For a bibliography of Lemery's works, see A. J. J. Vandevelde, "L'oeuvre bibliographique de Nicolas Lemery," *Bulletin de la société chimique de Belgique* 30, 1921, 153–166. See also Bougard, *La chimie de Nicolas Lemery*, 393–437.

143. The standard story, relayed by Fontenelle ("Eloge de Monsieur Lemery," in *Eloges des Academiciens, avec l'histoire de l'Acaddemie royale des sciences*, La Haye: Isaac vander Kloot, 1740, I), is that Lemery spent barely two months with Glaser in 1666, but Bougard (*La chimie de Nicolas Lemery*, 25) estimates that he must have stayed at least three years.

144. Owen Hannaway, "Nicolas Lemery," *Dictionary of Scientific Biography*; Paul-Antoine Cap, "Nicolas Lémery" (Paris: Fain, 1839); D. Joseph Tonnet, "Notice sur Nicolas Lémery, chimiste" (Niort: Robin, 1840); Paul Dorveaux, "Apothicaires membres de l'Académie Royale des Sciences. VI. Nicolas Lemery," *Revue d'histoire de la pharmacie* 75, 1931, 208–219; Bougard, *La chimie de Nicolas Lemery*.

145. Jean-Claude Guedon, "Protestantisme et Chimie: Le Milieu Intellectuel de Nicolas Lémery," *Isis* 65, 1974, 212–228.

146. Clara de Milt, "Christopher Glaser," *Journal of Chemical Education* 19, 1942, 53–60.

147. Metzger, *Les Doctrines*, 281–338.

148. Powers, "'Ars sine arte'"; Sutton, *Science for a Polite Society*.

149. Lefebvre, *A Compleat Body of Chymistry*, 119–120 and 123.

150. Ibid., 237–238.

151. Ibid., 292–293.

152. Lemery, *A Course of Chemistry* (London, 1677), 229–230.

153. Ibid., 26 and 35.

154. Ibid., 26–27.

155. Ibid., 2.

156. Ibid., 2–4.

157. Ibid., 4.

158. Ibid., 5.

159. Guedon, "Protestantisme et Chimie."

160. Sturdy, *Science and Social Status*, 228.

161. Nicolas Lemery, *An Appendix to a Course of Chymistry* (London, 1680), 5–11. This English edition was extracted from the second and the third French edition.

162. Ibid., 11.

163. Ibid., 12.

164. Ibid., 13–14.

165. Ibid., 14–15.

166. On Boyle's connection to Huguenots, see Margaret E. Rowbottom, "Some Huguenot friends and acquaintances of Robert Boyle (1627–1691)," *Proceedings of the Huguenot Society of London* 20, 1959–60, 177–194.

167. For Lemery's 1682 and 1687 letters, see *The Correspondence of Robert Boyle*, ed. M. Hunter et al. (Pickering and Chatto, 2001), V, 322 and volume VI, 210–212.

168. Nicolas Lemery, *A Course of Chymistry*, second English edition (London, 1686), 5.

Chapter 2

1. F. L. Holmes, "The Communal Context for Etienne-François Geoffroy's 'Table des rapports,'" *Science in Context* 9, 1996, 289–311.

2. *Histoire*, 1702, 45–46.

3. F. L. Holmes, *Eighteenth-Century Chemistry as an Investigative Enterprise* (Berkeley: Office for History of Science and Technology, 1989), 64–68.

4. Boyle, *The Sceptical Chymist* (1661), facsimile edition (London: Dawsons of Pall Mall, 1965), 37–46.

5. Although there exists an extensive historiography on Robert Boyle and on the Chemical Revolution, the relationship between these two important foundational moments of modern chemistry has not been established in a satisfactory manner. Rather, Thomas Kuhn and Ursula Klein have argued that Boyle's mechanical philosophy had little influence on the development of modern chemistry (Kuhn, "Robert Boyle and Structural Chemistry in the Seventeenth Century," *Isis* 43, 1952, 12–36; Klein, *Verbindung und Affinität*, Birkhäuser, 1994; idem, "Robert Boyle—Der Begründer der neuzeitlichen Chemie?" *Philosophia Naturalis* 31, 1994, 63–106).

6. Stroup, *A Company of Scientists*, 46–61 and 103–116; David J. Sturdy, *Science and Social Status* (Boydell, 1995), 214–238; Jack A. Clarke, "Abbé Jean-Paul Bignon 'Moderator of the Academies' and Royal Librarian," *French Historical Studies* 8, 1973, 213–235.

7. The secret of making phosphorus was much sought after by the court circle of chemists and natural philosophers. See Pamela H. Smith, *The Business of Alchemy* (Princeton University Press, 1994).

8. Sturdy, *Science and Social Status*, 226–228 and 230–233; Fontenelle, "Éloge de Monsieur Homberg (1715)," reprinted in *Eloges des academiciens avec l'histoire de l'académie royale des sciences* (La Haye: Isaac vander Kloot, 1740), volume 1, 350–368. Alice Stroup ("Wilhelm Homberg and the Search for the Constituents of Plants at the 17th-century Académie royale des sciences," *Ambix* 26, 1979, 184–201) has disputed some of the biographical details offered by Fontenelle.

9. Leonard M. Marsak, introduction to *The Achievement of Bernard le Bovier de Fontenelle* (Johnson Reprint, 1970); S. Delorme, "Tableau chronologique de la vie et des oeuvre des Fontenelle," *Revue d'histoire des sciences et leurs applications* 10, 1957, 289–309; A. Niderst, *Fontenelle* (Plon, 1991); Erica Harth, *Cartesian Women* (Cornell University Press, 1992); Sturdy, *Science and Social Status*, 239–241; Geoffroy V. Sutton, *Science for a Polite Society* (Westview, 1995); Aram Vartanian, *Science and Humanism in the French Enlightenment* (Rookwood, 1999), 5–44.

10. Marsak, "Bernard de Fontenelle: The Idea of Science in the French Enlightenment," *Transactions of the American Philosophical Society* 49, 1959, pt. 7, p 33, n. 3.

11. Fontenelle, "Preface to the History of the Academy of Sciences, from 1666 to 1699," translated in *The Achievement of Bernard le Bovier de Fontenelle*, part IV, 17–23, at 18.

12. Marsak, "Bernard de Fontenelle"; Charles B. Paul, *Science and Immortality* (University of California Press, 1980).

13. Sturdy, *Science and Social Status*, 281–292; Holmes, "The Communal Context."

14. Lemery, *Traité de l'Antimoine* (Paris: Jean Boudot, 1707).

15. *PV* 18, 1699, 142r–144r (February 28).

16. Stroup, "Wilhelm Homberg"; John Milton Hirschfield, *The Académie royale des sciences, 1666–1683* (Arno, 1981), 117–120 and 146–162; Yves Laissus et Anne-Marie Monseigny, "Les Plantes du Roi. Note sur un grand ouvrage de botanique préparé au XVIIe siècle par l'Académie royale des Sciences," *Revue d'histoire des sciences et leurs applications* 12, 1969, 193–236.

17. *PV* 13, 1692, 81v and 85v (February 27 and March 18).

18. *PV* 13, 1692, 115v–117v (September 3); *Histoire* (1666–1699), volume 2, 148–150.

19. *PV* 14, 1695, 122r–125r (June 8); *Histoire* (1666–1699), volume 2, 246–247.

20. "Sur les huiles des plantes," *Histoire*, 1700, 56–57; Homberg, "Observations sur les huiles des plantes," *Histoire de l'Académie royale des sciences, avec mémoires, Mémoires* [hereafter cited as *Mémoires*], 1700, 206–211.

21. Dodart, *Mémoire pour servir à l'histoire des plants* (Paris, 1731); Laissus and Monseigny, "Les Plantes du Roi"; Stroup, *A Company of Scientists*, 65–116.

22. One removed a large amount of phlegm from wine to prepare *eau de vie*, and still more phlegm to obtain *esprit de vin* ("Sur les epreuves de l'eau de vie & de l'esprit de vin," *Histoire*, 1718, 33–35).

23. Homberg, "Nouvelle maniere de distiller sans aucune chaleur," *PV* 16, 1697, 126v–129v (May 15).

24. Homberg, "Essays sur l'analyse des sels des plantes," *PV* 16 bis, 1697, 194r–198v (July 17).

25. Homberg, "Observations sur les sels fixes des Plantes," *PV* 18, 1698, 37v–39v (December 3).

26. Homberg, "Observations sur le sel urineux des plants," *PV* 17, 1697, 40r–43r (November 27).

27. Within the Academy, Charas took up the subject in 1695. See *PV* 14, 1695, 98v–101r (April 17); "Sur la Nature des sels," *Histoire de l'Académie royale des sciences* (1666–1699), volume 2, 252–255.

28. "Sur les fermentations," *Histoire*, 1701, 66–68; Homberg, "Observations sur quelques effets des Fermentations," *Mémoires*, 1701, 95–99.

29. Stroup ("Wilhelm Homberg," 188) mentions that Homberg's procedure here was reminiscent of Dodart's in 1674 and 1676 in his *Mémoire pour servir à l'histoire des plants* (Paris, 1731). Apparently, Dodart hoped to ascertain precise degrees of volatility and fixity in salts by determining their specific gravity, testing them in common water saturated with fixed salts.

30. "Sur les esprits acides," *Histoire* (1666–1699), volume 2, 250–252, at 251.

31. Homberg seems to have introduced a new areometer a couple of years before this since he says in 1699 that he did so six years ago. On Boyle's use of specific gravity in the analysis and identification of compounds, see Marie Boas, *Robert Boyle and Seventeenth-Century Chemistry* (Cambridge University Press, 1958), 132.

32. "Mesure des sels volatils acides contenus dans les esprits acides," *Histoire*, 1699, 52–53; Homberg, "Observation sur la quantité exacte des sels volatiles acides contenus dans les differens esprits acides," *Mémoires*, 1699, 44–51.

33. Homberg, "Observations sur la quantité d'acides absorbés par les alcalis terreux," *Mémoires*, 1700, 64–71.

34. Camphor was an "aromatic extract from the sap of certain trees found in Brazil and the Far East" (Jon Eklund, *The Incompleat Chymist*, Smithsonian Institution Press, 1975, 23).

35. "Sur les fermentations," *Histoire*, 1701, 66–68.

36. "Sur une dissolution d'argent," *Histoire*, 1706, 30–32, at 30.

37. Holmes, *Eighteenth Century Chemistry*, 68–72.

38. "Analyse de l'ipecacuanha," PV, 19, 1700, 1r–6r (January 9); *Histoire*, 1700, 46–50; Boulduc, "Analyse de l'ipecacuanha," *Mémoires*, 1700, 1–5.

39. "Analises de la Coloquinte, du Jalap, de la Gomme Gutte," *Histoire*, 1701, 58–62; *PV* 20, 1701, 162r–167v (Jalap, April 30) and 276r–280v (Gomme Gutte, July 30); Boulduc, "Observations Analytiques du Jalap," *Mémoires*, 1701, 106–109 and "Remarques sur la nature de la Gomme-gutte, & ses differentes Analyses," ibid., 131–135.

40. See preceding note.

41. Holmes, *Eighteenth-Century Chemistry*, 66–68.

42. Volatility and fixity were among the "diverse qualities" Boyle tried to explain in mechanical terms. They became dominant classificatory criteria in Homberg's chemistry; Boyle, "Experiments, Notes, &c. about the Mechanical Origin or Production of Diverse Particular Qualities," *Works*, volume 4, 230–354.

43. Homberg, "Observations sur les analyses des plantes," *PV* 20, 1701, 209r–215v (June 18); *Mémoires*, 1701, 113–117.

44. "Sur les analyses des plantes," *Histoire*, 1701, 68–70.

45. Homberg began to present the first portion before February 4, continuing through February 15 (*PV* 21, 1702, 61r, February 15).

46. *Histoire*, 1702, 45–46.

47. I originally developed this layering technique in dealing with the literature of organic chemistry in the nineteenth century. The construction of multiple realities, less visible in the nineteenth century because of the strong positivist rhetoric then prevalent, is quite evident in Homberg's works. See Mi Gyung Kim, "The Layers of Chemical Language I: Constitution of Bodies v. Structure of Matter," *History of Science* 30, 1992, 69–96.

48. Homberg mostly uses the word '*matières*' for chemical substances. This idomatic French usage remained in place well into the nineteenth century.

49. W. D. Hackmann, "Scientific Instruments: Models of Brass and Aids to Discovery," in *The Uses of Experiment*, ed. D. Gooding et al. (Cambridge University Press, 1989)

50. Andrew Pickering's pragmatic realism calls for an "interactive stabilization" of microspic entities such as quarks. See Pickering, "Living in the Material World: On Realism and Experimental Practice," in *The Uses of Experiment*, ed. Gooding et al.

51. Wilhelm Homberg, "Essays d'Elemens de Chimie,"*PV* 21, 1702, 61r–73v (February 15); "Essays de Chimie. Article Premier: Des Principes de la Chimie en general," *Mémoires*, 1702, 33–52.

52. *La Physique Chimique*, meaning the chemical part of physics or natural philosophy.

53. "Sur les sels volatils des plantes," *Histoire*, 1701, 70–71, at 70 (emphasis added).

54. Though Homberg did not explicitly designate these compounds as middle salts, they would fit his definition of middle salt as being partly fixed and partly volatile (Holmes, *Eighteenth-Century Chemistry*, 35–36).

55. Nicolas Lemery, *An Appendix to a Course of Chymistry* (London, 1680), 14–15.

56. Homberg, "Essays de chimie. Article Premier," 41–44.

57. "Sur les analyses des plantes (1692)," *Histoire* (1666–1699), volume 2, 148–150.

58. "Sur les huiles des plantes," *Histoire*, 1700, 56–57; Homberg, "Observations sur les huiles des plantes," *Mémoires*, 1700, 206–210.

59. The instrument must have arrived at Paris near the end of 1701 since Homberg reports on June 3, 1702 that it came about six months ago (*PV* 21, 1702, 227r–231r, June 3; Homberg, "Observations faites par le moyen du Verre ardant," *Mémoires*, 1702, 141–149). Fontenelle (*Histoire*, 1702, 46) mentions that Homberg's previous treatise on salt was mostly based on the work done without the burning glass.

60. "Sur des experiences faites a un miroir ardent convex," *Histoire*, 1702, 34–38. We should take Fontenelle's remarks with a grain of salt, since he obviously wished to complement the duke on his generosity as a patron of science.

61. Homberg, "Observations sur la quantité d'acides absorbés par les alcalis terreux," *Mémoires*, 1700, 64–71, at 69.

62. *PV* 21, 1702, 227r–231r (June 3), 341r–343r (August 9 and 12) and 399v–401v (November 15); Homberg, "Observations faites par le moyen du Verre ardant," *Mémoires*, 1702, 141–149.

63. *Histoire*, 1702, 34.

64. "Sur l'analise du soufre commun," *Histoire*, 1703, 47–49; Homberg, "Analise du souffre commun," *PV* 22, 1703, 85r–88r (March 27); "Essay de l'analyse du Souffre commun," *Mémoires*, 1703, 31–40.

65. For a detailed description of the process, see Holmes, "The Communal Context," 300.

66. Fontenelle states in the *Histoire* (1703, 48) that Homberg identified this oil as "the true inflammable or sulphurous part" of sulphur, or the "sulphur principle." However, Homberg states in the *Mémoires* (1703, 38) that the sulphur principle as well as the salt principle have not yet become sensible and are "encased" in some other matter.

67. It was referred to under this title by Geoffroy and in the *Histoire* of 1705, but the actual title of presentation was "Sur le souffre Principe," *PV* 24, 1705, 125r–130v (April 22).

68. Homberg, "Suite des Essays de chimie. Article Troisième. Du Souphre Principe," *Mémoires*, 1705, 88–96, at 89.

69. Du Clos, "Expériences de l'augmentation du poids de certaines matieres en les calcinane a la chaleur ou du Soleil, ou du feu ordinaire," *PV* 1, 1667, 40–52 (January 8).

70. 'Regulus' indicated pure metal separated from the mineral substance containing the metal. "Regulus of Mars" was pure iron.

71. Homberg, "Suite de Essays de Chimie," *Mémoires*, 1705, 96.

72. Geoffroy, "Maniere de recomposer le Souffre commun par la réunion de ses principes, et d'en composer de nouveau par le mélange de semblables substances, avec quelques conjectures sur le composition des métaux," *Mémoires*, 1704, 278–286.

73. "Sur la récomposition du souffre," *Histoire*, 1704, 37–39, at 37.

74. Nicolas Lemery had earlier defined that iron was "a very porous Metal compounded of a Vitriolick Salt, of Sulphur, and Earth ill digested together" (Lemery, *A Course of Chemistry*, 1677, 68).

75. Ibid., 39.

76. Metzger, *Les Doctrines chimiques en France du début du XVIIe à la fin du XVIIIe siècle*, new edition (Paris, 1969), 399–410; Mi Gyung Kim, "Chemical Analysis and the Domains of Reality: Wilhelm Homberg's Essais de Chimie, 1702–1709," *Historical Studies in the Physical Sciences* 31, 2000, 37–69.

77. "Sur la generation du fer," *Histoire*, 1705, 64–65; Geoffroy, "Problème de chimie. Trouver des Cendres qui ne contiennent aucunes parcelles de fer," *PV* 24, 1705, 393r–395r (December 16); *Mémoires*, 1705, 362–363.

78. "Sur la nature du Miel," *Histoire*, 1706, 36–38; Lemery le fils, "Du Miel & de son Analyse Chimique," *PV* 25, 1706, 262r–272r (July 17); *Mémoires*, 1706, 272–283.

79. "Sur le fer des plantes," *Histoire*, 1706, 38–40; Lemery le fils, "Que les Plantes contiennent réellement du fer, & que ce métal entre necesairement dans leur composition naturelle," *Mémoires*, 1706, 411–417.

80. "Sur la nature du fer," *Histoire*, 1706, 32–36; Lemery le fils, "Diverses Experiences et Observations chimiques et physiques. Sur le Fer et sur l'Aimant," *PV* 25, 1706, 123v–134r (April 14); *Mémoires*, 1706, 119–135.

81. Homberg, "Observations sur le fer au verre ardent," *PV* 25, 1706, 168r–172r (May 8); *Mémoires*, 1706, 158–165.

82. Homberg, "Suite de l'article trois des Essais de chimie," *Mémoires*, 1706, 260–272.

83. "Sur la Nature du fer," *Histoire*, 1707, 43–45; Geoffroy, "Eclaircissemens sur la production artificielle du fer, & sur la composition des autres Métaux," *Mémoires*, 1707, 176–188.

84. This was an objection raised in 1707 by a Dutch philosopher to Homberg's earlier experiments with gold ("Sur la vitrification de l'or," *Histoire*, 1707, 30–32; Homberg, "Eclaircissemens touchant la vitrification de l'or au verre ardent," *Mémoires*, 1707, 40–48).

85. "Sur les metaux imparfaits exposés au verre ardent," *Histoire*, 1709, 36–38; Geoffroy, "Expériences sur les metaux, faites avec le Verre ardent du Palais Royal," *Mémoires*, 1709, 162–176.

86. Homberg, "Suite des essais de chimie. Art. IV. Du Mercure," *Mémoires*, 1709, 106–117.

87. This is the first time Homberg defines the "chemical principle" explicitly.

88. "Sur les souffres des vegetaux et des mineraux," *Histoire*, 1710, 46–48; Homberg, "Observations sur les matiers sulphureuses et sur la facilité de les changer d'une espece de souffre en une autre," *Mémoires*, 1710, 225–234. In summarizing Homberg's work in 1710, Fontenelle refers to Geoffroy's work in 1709 as Homberg's.

89. "Sur les metaux imparfaits exposés au verre ardent," *Histoire*, 1709, 36–38; Geoffroy, "Experiences sur les metaux, faites avec le Verre ardent du palais Royal," *Mémoires*, 1709, 162–176 (read on May 8)

90. Geoffroy, "Eclaircissemens sur la Table insérée dans les Mémoires de 1718, concernant les rapports observés entre différentes substances," *Mémoires*, 1720, 20–34, at 29.

91. Holmes, "The Communal Context," 306–307.

92. Lavoisier, "Réflexions sur les experiences qu'on peut tenter a l'aide du miroir ardent" (August memorandum), reprinted in Henry Guerlac, *Lavoisier—The Crucial Year* (Cornell University Press, 1961).

93. Homberg, "Mémoire touchant les Acides & les Alkalis, pour servir d'addition à l'article du Sel principe, imprimé dans nos Mémoires de l'année 1702. pag. 36," *Mémoires*, 1708, 312–323.

94. Homberg, "Observations touchant l'effet de certains Acides sur les Alcalis volatils," *Mémoires*, 1709, 354–363.

95. "Sur des expériences faites a un miroir ardent convexe," *Histoire*, 1702, 34–38, at 34.

Chapter 3

1. Ursula Klein, "The Chemical Workshop Tradition and the Experimental Practice: Discontinuities within Continuities," *Science in Context* 9, 1996, 251–287.

2. For example, M. de Fronville claimed in 1701 that he had discovered a "universal solvent, or alkaëst of Paracelsus & of van Helmont" (*Histoire*, 1701, 73), but it proved to be a weak regal water. See also Bernard Joly, "L'alkahest, dissolvant universel ou quand la théorie rend pensable une pratique impossible," *Revue d'histoire des sciences et leurs applications* 49, 1996, 305–344.

3. Marie Boas, *Robert Boyle and Seventeenth-Century Chemistry* (Cambridge University Press, 1958), 133–141. See also A. Albert Baker, "A History of Indicators," *Chymia* 9, 1964, 147–167; Peta Dewar Buchanan, J. F. Gibson, and Marie Boas Hall, "Experimental History of Science: Boyle's Colour Changes," *Ambix* 25, 1978, 208–210; William Eamon, "New Light on Robert Boyle and the Discovery of Colour Indicators," *Ambix* 27, 1980, 204–209.

4. "Examen d'eaux minerales," *Histoire*, 1699, 55–57, at 56; "Sur les eaux de Passy," *Histoire*, 1701, 62–66, at 63. For a broad survey of analyses of mineral

water in the eighteenth century, see Jon Eklund, Chemical Analysis and the Phlogiston Theory, 1738–1772: Prelude to Revolution, Ph.D. dissertation, Yale University, 1971), 231–291.

5. "Sur les huiles essentielles des plantes," *Histoire*, 1707, 37–40, at 39.

6. Homberg, "Mémoire touchant les Acides & les Alkalis, pour servir d'addition, à l'article du Sel principe, imprimé dans nos Memoires de l'année 1702. Pag. 36," *Mémoires*, 1708, 312–323, at 320.

7. Lemery, *A Course of Chemistry* (London, 1677), 5.

8. "Sur l'acide de l'antimoine," *Histoire*, 1700, 57–58.

9. The same kind of reversal in opposite direction happened in the development of organic chemistry in the 1830s. See John Hedley Brooke, "Organic Synthesis and the Unification of Chemistry—A Reappraisal," *British Journal for the History of Science 5*, 1971, 363–392.

10. "Sur une dissolution d'argent," *Histoire*, 1706, 30–32.

11. "Sur les precipitations," *Histoire*, 1711, 31–35, at 33.

12. Father Nicolas Malebranche (1638–1715) in the Academy developed a mathematically grounded metaphysics in which "rapports" were the only truth open to man's finite intellect. Relations of equality and inequality, or of magnitude, were regarded as the clearest and most distinct relations (Henry Guerlac, *Newton on the Continent*, Cornell University Press, 1981, 53–59). One can discern the contemporary mathematicians' interest in the "science of rapports" from de Lagny's long paper "La Science des Rapports ou Methode nouvelle & generale pour trouver exactement le rapport de toutes les grandeurs commensurables, & la suite de tous les nombres qui'expriment le plus exactment qu'il est possible et en plus petits fermes les rapports des grandeurs" (*PV*, 35, 1716, 11r–21v).

13. "Sur les rapports de differentes substances en Chimie," *Histoire*, 1718, 35–37, at 37.

14. Jean-Claude Guedon, "Protestantisme et Chimie: Le Milieu Intellectuel de Nicolas Lémery," *Isis 65*, 1974, 212–228.

15. David J. Sturdy, *Science and Social Status* (Boydell, 1995), 308–312.

16. Louis Lemery, "Conjectures & reflexions sur la matiere du feu ou de la lumier," *PV* 28, 1709, 445r–457v (December 20); *Mémoires*, 1709, 400–418.

17. Boerhaave's discussion of fire in his *Elements of chemistry* was often referred to as *Traité du feu*.

18. Lemery, *A Course of Chymistry* (London, 1686), translator's preface (emphasis added).

19. Homberg, "Observations sur la quantité d'acides absorbés par les alcalis terreux," *Mémoires*, 1700, 64–71, at 68–69.

20. Lemery, "Conjectures et reflexions."

21. Louis Lemery, "Memoire sur les precipitations Chimiques; ou l'on examiné par occasion la dissolution de l'Or & de l'Argent, la nature particuliere des esprits acides, & la maniere dont l'esprit de nitre agit sur celuy de Sel dans la formation de l'Eau regale ordinaire," *Mémoires*, 1711, 56–79.

22. Nicolas Lemery in his earlier work on the composition of camphor had to dissolve it in a variety of solvents before distillation. He provided an exhaustive list of solvents, divided into three categories of sulphurous liquids, acids and alkalis. See "Sur le Camphre," *Histoire*, 1705, 59–62.

23. Louis Lemery, "Memoire sur les precipitations Chimiques," 65.

24. Ibid., 66.

25. Louis Lemery, "Conjectures sur les couleurs differentes des Précipités de Mercure," *Mémoires*, 1712, 51–70.

26. Lemery, "Second Memoire sur les Couleurs differentes des Précipités du Mercure," *Mémoires*, 1714, 259–280.

27. Lemery, "Explication mechanique de quelques differences asses curieuses qui resultent de la dissolution de differents Sels dans l'Eau commune," *Mémoires*, 1716, 154–172 (read June 27).

28. Ibid., 158.

29. Louis Lemery, "Observations nouvelle et singuliére sur la dissolution successive de plusieurs Sels dans l'eau commune," *Mémoires*, 1724, 332–347; "Second Memoire ou Reflexions nouvelles sur une précipitation singuliere de plusieurs sels par un autre sel," *Mémoires*, 1727, 40–49; "Troisième Memoire ou Réflexions nouvelles sur une Précipitation singuliére de plusieurs Sels par un autre Sel," ibid., 214–228.

30. Louis Lemery, "Reflexions physiques sur le défaut & le peu d'utilité des Analises ordinaires des Plantes & des Animaux," *Mémoires*, 1719, 173–188.

31. Louis Lemery, "Second Mémoire sur les analises ordinaires de chimie," *Mémoires*, 1720, 98–107; "Troisième Mémoires sur les analises de chimie," ibid., 166–178; "Quatrième Mémoire sur les analises ordinaires des plantes et des animaux"; *Mémoires*, 1721, 22–44.

32. "Sur la volatilité des sels urineux," *Histoire*, 1721, 35–36.

33. G. Planchon, "La dynastie des Geoffroy, apothicaires de Paris," *Journal de pharmacie et de chimie* 8, 1898, 289–293 and 337–345; Olivier Lafont, "Peronnalisation des rapports individu-puissance publique ou: Geoffroy et la famille Le Tellier," *Revue d'histoire de la pharmacie* 38, 1991, 15–23.

34. I. Bernard Cohen, "Isaac Newton, Hans Sloane and the Académie royale des sciences," in *Mélanges Alexandre Koyré*, ed. I. Cohen and R. Taton (Hermann, 1964), volume 1.

35. Fontenelle, "Éloge de M. Geoffroy," *Histoire*, 1731, 93–100; P. Dorveaux, "Étienne-François Geoffroy," *Revue d'histoire de la pharmacie* 2, 1931, 118–126.

36. E. F. Geoffroy, *Tractatus de materia medica* (Paris: Johannis Desaint & Caroli Saillant, 1741); *Traité de matière médicale* (Paris: J. Desaint & C. Saillant, 1743); W. A. Smeaton, "Geoffroy, Etienne-François," *Dictionary of Scientific Biography*.

37. Laurence Brockliss, "Consultation by Letter in Early Eighteenth-Century Paris: The Medical Practice of Etienne-François Geoffroy," in *French Medical Culture in the Nineteenth Century*, ed. A. La Berge and M. Feingold (Rodopi, 1994).

38. Geoffroy's lectures on affinities began December 14, 1715 and continued through March 26, 1716.   ˌ

39. Geoffroy, "Des différents Rapports observés en Chymie entre différentes substances," PV 37, 1718, 233v–242r (August 27 and 31); *Mémoires*, 1718, 256–269; English translation: "Table of the different relations observed in chemistry between different substances," *Science in Context* 9, 1996, 313–319.

40. Various authors mentioned such an order of reactivity before Geoffroy (Geber, Paracelsus, Newton), as did Stahl slightly later. See A. M. Duncan, "Some Theoretical Aspects of Eighteenth-Century Tables of Affinity," *Annals of Science* 18, 1962, 177–196 and 217–232, at 177.

41. Geoffroy, "Des différents Rapports," 202.

42. Ibid., 203.

43. Ibid.

44. Geoffroy, "Eclaircissemens sur la Table insérée dans les Mémoires de 1718, concernant les rapports observés entre différentes substances," *Mémoires*, 1720, 200–219, at 200–201.

45. F. L. Holmes, *Eighteenth-Century Chemistry as an Investigative Enterprise* (Berkeley, 1989), 39–40.

46. Ursula Klein, "E. F. Geoffroy's Table of different 'Rapports' observed between different chemical substances—A Reinterpretation," *Ambix* 42, 1995, 251–287.

47. Nicolas Lemery, "Reflexions & Experiences sur le sublimé corrosif," *Mémoires*, 1709, 42–47.

48. Holmes, *Eighteenth Century Chemistry*, 40–41.

49. "Sur les rapports de differentes substances en Chimie," *Histoire*, 1718, 35–37, at 37.

50. Geoffroy, "Eclaircissemens," *Mémoires*, 1720, 200–219. For a summary, see Duncan, "Some Theoretical Aspects," 187.

51. "Sur les Rapports des differentes Substances en Chimie," *Histoire*, 1720, 32–36, at 32.

52. Stahl had persuaded Neumann in 1716 to resume his services for Prussia. Neumann then spent three years on a study tour of England, France, and Italy at royal expense (Karl Hufbauer, *The Formation of the German Chemical Community, 1720–1795*, University of California Press, 1982, 173).

53. In 1851, Chevreul identified the merit of Geoffroy's table as its impulse for multiple analyses (*Journal des Savants*, 1851, 104; quoted in Cohen, *Melanges Alexandre Koyré*, 80).

54. Duncan, "Some Theoretical Aspects," 181; *Laws and Order in Eighteenth-Century Chemistry* (Clarendon, 1996), 110–176.

55. Cohen, "Isaac Newton, Hans Sloane"; W. A. Smeaton, "E. F. Geoffroy Was Not a Newtonian Chemist," *Ambix* 18, 1971, 212–214.

56. Lissa Roberts, "Setting the Table: The Disciplinary Development of Eighteenth-Century Chemistry as Read through the Changing Structure of Its Tables," in *The Literary Structure of Scientific Argument*, ed. P. Dear (University of Pennsylvania Press, 1991). Roberts advances a much broader interpretative claim concerning the evolution of chemistry from an art to a science which cannot be summarized here. See also her "Filling the Space of Possibilities: Eighteenth-Century Chemistry's Transition from Art to Science," *Science in Context* 6, 1993, 511–553.

57. Evelyn Fox Keller, "Linking Organisms and Computers: Theory and Practice in Contemporary Biology," in *Connecting Creations*, ed. M. Safir (Xunta de Galicia, 2000). I thank Professor Keller for providing me with a copy of this article.

58. Holmes, *Eighteenth Century Chemistry*, 39–41; "The communal context for Etienne-François Geoffroy's 'Table des rapports,'" *Science in Context* 9, 1996, 289–311.

59. Klein, "E. F. Geoffroy's Table," 92; idem, "Origin of the Concept of Chemical Compound," *Science in Context* 7, 1994, 163–204.

60. Thomas S. Kuhn, "Robert Boyle and Structural Chemistry in the Seventeenth Century," *Isis* 43, 1952, 12–36, at 13.

61. Klein, "E. F. Geoffroy's Table."

62. Klein, "E. F. Geoffroy's Table," 84–86; Holmes, "Communal Context," 299–307.

63. Eklund, "Chemical Analysis and the Phlogiston Theory."

64. Geoffroy, "Eclaircissemens," 28.

65. Stahl, *Zymotechnia Fundamentalis* (Halle, 1697). This was Stahl's first publication in which phlogiston was discussed for the first time; see (Kevin) Ku-Ming Chang, "Fermentation, Phlogiston, and Matter Theory: Chemistry and Natural Philosophy in Georg Ernst Stahl's *Zymotechnia Fundamentalis*," *Early Science and Medicine* 7, 2002, 31–64.

66. Stahl, *Ausführliche Betrachtung und zulänglicher Beweiss von den Saltzen, dass dieselbe aus einer zarten Erde, mit Wasser innig verbunden bestehen* (Halle, 1723); anonymously translated, *Traité des sels* (Paris, 1771). See especially chapter 22.

67. Gilles-François Boulduc, "Essai d'analyse en général des nouvelles eaux minerales de Passy," *Mémoires*, 1726, 306–327; Du Hamel and Grosse, "Sur les différentes manières de rendre le Tartre soluble," *Mémoires*, 1732, 323–342.

68. Stahl, *Philsophical Principles of Universal Chemistry* (London, 1723), 2.

69. Holmes, *Eighteenth Century Chemistry*, 45–46.

70. See Robert P. Multhauf, *Neptune's Gift* (Johns Hopkins University Press, 1978).

71. See Robert P. Multhauf, "Sal Ammoniac: A Case History in Industrialization," *Technology and Culture* 6, 1965, 569–586.

72. "Sur l'origine du sel armoniac," *Histoire*, 1716, 28–30; "Sur l'origine du sel armoniac," *Histoire*, 1720, 46–50; Claude-Joseph Geoffroy, "Observations sur la nature et la composition du sel ammoniac," *Mémoires*, 1720, 189–207.

73. "La genie François n'est pas misterieux même en Chimie"; "Sur le sel armoniac," *Histoire*, 1723, 38–40, at 40; Claude-Joseph Geoffroy, "Suite des Observations sur la Fabrique du Sel Ammoniac, avec sa décomposition pour en tirer le sel volatil, que l'on nomme vulgairement sel d'Angleterre," *Mémoires*, 1723, 210–222.

74. "Sur l'origine du nitre," *Histoire*, 1717, 29–34; Louis Lemery, "Premier Memoire sur le Nitre," *Mémoires*, 1717, 31–52; "Second Memoire sur le Nitre," ibid., 122–146.

75. Allen G. Debus, "The Paracelsian Aerial Niter," *Isis* 55, 1964, 43–61; Robert G. Frank Jr., *Harvey and the Oxford Physiologists* (University of California Press, 1980), 117–128; A. Rupert Hall, "Isaac Newton and the Aerial Nitre," *Notes and Records of the Royal Society of London* 52, 1998, 51–61.

76. For cases with more direct relevance, see Charles C. Gillispie, "The Discovery of the Leblanc Process," *Isis* 48, 1957, 152–170; Holmes, *Eighteenth Century Chemistry*, 85–102; John Graham Smith, *The Origins and Early Development of the Heavy Chemical Industry in France* (Oxford University Press, 1979).

77. Noel G. Coley, "The Preparation and Uses of Artificial Mineral Waters," *Ambix* 31, 1984, 32–48; idem, "Physicians, Chemists, and the Analysis of Mineral Waters," in *The Medical History of Waters and Spas*, ed. R. Porter (Wellcome Institute for the History of Medicine, 1990).

78. Gilles-François Boulduc, "Essai d'analyse en general des nouvelles eaux minerales de Passy," *Mémoires*, 1726, 306–327; Holmes, *Eighteenth Century Chemistry*, 43–44.

79. Boulduc, "Essai d'analyse," 324.

80. "Sur un sel naturel de Dauphiné," *Histoire*, 1727, 29–31.

81. "Sur les eaux minérales chaudes de Bourbon-l'Archambaut," *Histoire*, 1729, 22–27, at 22.

82. "Sur les différents vitriols et sur l'alun," *Histoire*, 1728, 34–35; C. J. Geoffroy, "Examen des différentes Vitriols; avec quelques Essais sur la formation artificielle du Vitriol blanc & de l'Alun," *Mémoires*, 1728, 301–310.

83. "Sur le vinaigre concentré par la gelée," *Histoire*, 1729, 16–19.

84. Louis Lemery, "Expériences & Réflexions sur le Borax," *Mémoires*, 1728, 273–288; "Second Mémoire sur le Borax," *Mémoires*, 1729, 282–300.

85. "Sur le Borax et sur des expériences nouvelles de ce sel," *Histoire*, 1732, 52–54; Claude-Joseph Geoffroy, "Nouvelles Expériences sur le Borax, avec un moyen facile de faire le Sel Sédatif, & d'avoir un Sel de Glauber, par la même opération," *Mémoires*, 1732, 398–418.

86. Louis Claude Bourdelin, "Mémoire sur la Formation des Sels lixiviels," *Mémoires*, 1728, 384–400; "Mémoire sur le sel lixiviel du Gayac," *Mémoires*, 1730, 33–44.

87. Gilles-François Boulduc, "Manière de faire le sublimé corrosif en simplifiant l'opération," *Mémoires*, 1730, 357–362; "Sur un sel connu sous le nom de Polychreste de Seignette," *Mémoires*, 1731, 124–129; "Recherches du sel d'Epsom," ibid., 347–357.

88. *Histoire*, 1728, 39–43; "Sur la précipitation du sel marin dans la fabrique du salpestre," *Histoire*, 1729, 19–22.

89. Demachy, "Exposition d'une nouvelle Table des principales combinaisons chymiques, connue jusqu'à présent sous le nom de Table des Rapports ou 'Affinités,'" 82–305 and plates I–VII in his *Recueil de Dissertations physico-chimiques, présentées a différentes académie* (Paris: Nyon & Barrois, 1781), 94–100. Demachy obtained a copy, carefully guarded by his students, from Grosse's nephew in 1748.

90. *Histoire*, 1728, 38–39.

91. "Sur le Tartre soluble," *Histoire*, 1732, 47–50.

92. Du Hamel and Grosse, "Sur les différentes manières."

93. Holmes, *Eighteenth Century Chemistry*, 44.

94. "Sur le sel armoniac," *Histoire*, 1735, 23–26; Du Hamel, "Sur les sel ammoniac," *Mémoires*, 1735, 106–116, 414–434, and 483–504.

95. Du Hamel, "Sur la Base du Sel marin," *Mémoire*, 1736, 215–232; Holmes, *Eighteenth Century Chemistry*, 41–42.

96. "Sur les vitriols," *Histoire*, 1735, 26–31; Louis Lemery, "Nouvel Eclaircissement sur l'Alun, sur les Vitriols, et particulierement sur la Composition naturelle, & jusqu'à présent ignorée, du Vitriol blanc ordinaire. Premier Mémoire," *Mémoires*, 1735, 262–280; "Second Mémoire sur les Vitriols, et particuliérement sur le Vitriol blanc ordinaire," ibid., 385–402; "Supplement aux deux Mémoires que j'ai donnés en 1735, sur l'Alun et sur les Vitriols," *Mémoires*, 1736, 263–301.

97. "Sur la Terre de l'Alun," *Histoire*, 1744, 16–18.

98. Rouelle, "Mémoire sur les sels neutres, dans lequel on propose une division méthodique de ces sels, qui facilite les moyens pour parvenir à la théorie de leur crystallisation," *Mémoires*, 1744, 353–364. Rouelle presented this to the Academy in 1743 as a part of an election bid (*Histoire*, 1743, 106). See also

"Sur la crystallisation du sel marin," *Histoire*, 1745, 32–34; Rouelle, "Sur le Sel marin. Premiere partie. De la crystallisation du Sel marin," *Mémoires*, 1745, 57–79.

99. Eklund, "Chemical Analysis and the Phlogiston Theory," 41.

Chapter 4

1. Douglas McKie, "Guillaume-François Rouelle (1703–1770)," *Endeavour*, July 1953, 130–133; Rhoda Rappaport, "G.-F. Rouelle: An Eighteenth-Century Chemist and Teacher," *Chymia* 6, 1960, 68–101; Jean Mayer, "Portrait d'un chimiste: Guillaume-François Rouelle (1703–1770)," *Revue d'histoire des sciences et leurs applications* 23, 1970, 305–332.

2. Henry Guerlac, "Some French Antecedents of the Chemical Revolution," *Chymia* 5, 1959, 73–112; Theodore Porter, "The Promotion of Mining and the Advancement of Science: The Chemical Revolution of Mineralogy," *Annals of Science* 38, 1981, 543–570.

3. For a list of Rouelle's students, see Rappaport, G.-F. Rouelle, 76. Kirwan attended Rouelle's lectures during 1754–55; see Diane Elizabeth Heggarty, Richard Kirwan: The Natural Philosopher and the Chemical Revolution, Ph.D. dissertation, University of Washington, 1978, 76.

4. Venel, "Chymie," Diderot's *Encyclopédie*, volume 3, 437, quoted and translated in Rappaport, G.-F. Rouelle, 99.

5. Published for the first time in *Annales de la société Jean-Jacques Rousseau* 12–13, 1918–1921. On Rousseau's chemical education and work, see Théophile Dufour, *Les Institutions chimiques de Jean-Jacques Rousseau* (Geneva, 1905).

6. Jean Mayer, *Diderot, homme de science* (Rennes: Bretonne, 1959); Jacques Roger, *Les sciences de la vie dans la pensée française du XVIIIe siècle; la génération des animaux de Descartes à l'Encyclopédie* (Paris: A. Colin, 1963); Aram Vartanian, *Science and Humanism in the French Enlightenment* (Rookwood, 1999), 92–179.

7. These lectures were never published, however. For a full manuscript, see "Cours de chymie de M. Rouelle, rédigé par M. Diderot," mss. 564–565, Bordeaux Public Library. For a list of manuscript copies in various libraries, see Rhoda Rappaport, G.-F. Rouelle: His *Course de Chimie* and Their Significance for Eighteenth Century Chemistry, master's thesis, Cornell University, 1958; Jean Jacques, "Le 'Cours de chimie de G.-F. Rouelle recueilli par Diderot,'" *Revue d'histoire des sciences et leurs applications* 38, 1985, 43–53. On Rouelle's influence on Diderot, see Jean-Claude Guedon, "Chimie et matérialisme: La stratégie anti-Newtonienne de Diderot," *Dix-huitième siècle* 11, 1979, 185–200.

8. On Rouelle's central (albeit hidden) role in the network of chemical contributors to Diderot's *Encyclopédie*, see Jean-Claude Daniel Roger Guedon, The Still Life of a Transition: Chemistry in the Encyclopédie, Ph.D. dissertation,

University of Wisconsin, 1974. It would go too far to argue, as Guedon does, that the chemistry of the *Encyclopédie* was Rouellian since the six major chemical contributors—d'Alembert, Diderot, Baron d'Holbach, de Jaucourt, Malouin, and Venel—read widely on their own.

9. For a list of Baron d'Holbach's translation of German scientific works, see Max Pearson Cushing, *Baron d'Holbach* (New York, 1914), 27–31; Rhoda Rappaport, "Baron d'Holbach's campaign for German (and Swedish) Science," *Studies on Voltaire and the Eighteenth Century* 323, 1994, 225–246.

10. André Morellet, *Mémoires inédits de l'abbé Morellet*, second edition (Paris, 1822), volume 1, 118 and 132–134; Alan Charles Kors, *D'Holbach's Coterie* (Princeton University Press, 1976). Kors counts Darcet among the regular members.

11. Friedrich M. Grimm, *Correspondance littéraire, philosophique et critique par Grimm, Diderot, Raynal, Meister, etc.* ed. M. Tourneux (Paris, 1877–1882), volume 9, 106–109.

12. Daniel Gordon, " 'Public Opinion' and the Civilizing Process in France: The Example of Morellet," *Eighteenth-Century Studies* 22, 1989, 302–328; idem, "Beyond the Social History of Ideas: Morellet and the Enlightenment," in *André Morellet (1727–1819) in the Republic of Letters and the French Revolution*, ed. J. Merrick and D. Medlin (Lang, 1995).

13. While the essentially "bourgeois" nature of Habermas's public sphere needs some modification when applied to the French case, the concept provides a useful demarcation between the seventeenth-century and the eighteenth-century public for science in the way the notion of the "republic of science" cannot. The need to include science in characterizing the public sphere has been argued, with an excellent review of the relevant literature, by Thomas Broman,"The Habermasian Public Sphere and 'Science in the Enlightenment,'" *History of Science* 36, 1998, 123–149. On the modification of the public sphere for French Enlightenment, see Benjamin Nathans, "Habermas's 'Public Sphere' in the Era of the French Revolution," *French Historical Studies* 16, 1990, 620–644; Dena Goodman, *The Republic of Letters; A Cultural History of the French Enlightenment* (Cornell University Press, 1994), 12–15; Keith Michael Baker, "Defining the Public Sphere in Eighteenth-Century France: Variations on a Theme by Habermas," in *Habermas and the Public Sphere*, ed. C. Calhoun (MIT Press, 1994).

14. Hélène Metzger, *Newton, Stahl, Boerhaave et la Doctrine chimique* (Paris: Félix Alcan, 1930).

15. Pierre Brunet, *L'Introduction des théories de Newton en France au XVIIIe siècle avant 1738* (Paris: Albert Blanchard, 1931); Arnold Thackray, *Atoms and Powers* (Harvard University Press, 1970); A. Rupert Hall, "Newton in France: A New View," *History of Science* 13, 1975, 233–250; Henry Guerlac, *Newton on the Continent* (Cornell University Press, 1981).

16. In the history section of his textbook, Boerhaave listed the chemical references a beginner could consult and benefit from, which included the textbooks by Glaser, Le Febvre, Lemery (1716 edition), Becher, Kunkel, van Helmont, Boyle, and Paracelsus, Stahl's *Fundamenta Chymiae*, and the articles by Homberg, E. F. Geoffroy, and Louis Lemery in the *Mémoires* of the French Academy; Boerhaave, *Elements of Chemistry* (London: printed for J. and J. Pemberton, 1735), 17–18.

17. William Cullen and Joseph Black successively built on Boerhaave's accomplishments to effect a more thoroughgoing face-lift for chemistry in Scotland; Arthur Donovan, *Philosophical Chemistry in the Scottish Enlightenment* (Edinburgh University Press, 1975); Jan Golinski, *Science as Public Culture* (Cambridge University Press, 1992).

18. Ramez Bahige Maluf, Jean Antoine Nollet and Experimental Natural Philosophy in Eighteenth Century France, Ph.D. dissertation, University of Oklahoma, 1985; John Heilbron, *Electricity in the 17th & 18th Centuries* (University of California Press, 1979), 279–289 and 352–362; Simon Schaffer, "Natural Philosophy and Public Spectacle in the Eighteenth Century," *History of Science* 21, 1983, 1–43; Geoffrey V. Sutton, *Science for a Polite Society* (Westview, 1995), 223–240.

19. Pierre Lemay, "Les cours de Guillaume-François Rouelle," *Revue d'histoire de la pharmacie* 123, 1949, 434–442; idem, "Le Cours de Pharmacie de Rouelle," *Revue d'histoire de la pharmacie* 152, 1957, 17–21.

20. Diderot, "Plan d'une Université pour le Gouvernment de Russie, ou d'une éducation publique dans toutes les sciences," in *Oeuvres complètes de Diderot*, ed. J. Assézat and M. Tourneux (Paris, 1875–1877), volume 3, at 464.

21. Diderot, "Notices sur le peintre Michel Vanloo et le chimiste Rouelle," in ibid., volume 4, at 407.

22. Mayer, "Portrait d'un chimiste."

23. Martin Fichman, "French Stahlism and Chemical Studies of Air," *Ambix* 18, 1971, 94–122.

24. P. J. Macquer, *Élémens de chymie—theorique* (Paris, 1749); *Élémens de chymie—Pratique* (Paris, 1751).

25. Roy G. Neville and W. A. Smeaton, "Macquer's *Dictionnaire de Chymie*: A Bibliographical Study," *Annals of Science* 38, 1981, 613–662.

26. Chemistry texts by Macquer and Baumé were on the reading list of the young Comte de Montlosier, who was studying public law; see François Dominique, comte de Montlosier, *Mémoires* (Paris: Dufey, 1830), volume 1, 35; quoted in Pierre Serna, "The Noble," in *Enlightenment Portraits*, ed. M. Vovelle (University of Chicago Press, 1997), 38.

27. Condorcet, "Éloge de M. Macquer," *Histoire*, 1784, 20–30, at 23–24.

28. "The Stahlianism has been known in France by the lectures of M. Rouelle and by the works of M. Macquer" (Charles Henry, ed., *Introduction a la chymie*, Paris, 1887, 67).

29. Macquer entered into an equal partnership with Baumé to start a private course in chemistry. See Pierre Julien, "Sur les relations entre Macquer et Baumé," *Revue d'histoire de la pharmacie* 39, 1992, 65–77.

30. Macquer, "Affinité," in *Dictionnaire de chymie*, second edition (Paris: P. Fr. Didot jeune, 1778), volume 1, 65–66. This section was added in the second edition.

31. Baumé, *Chymie expérimentale et raisonnée* (Paris: P. F. Didot le jeune, 1773), volume 1, 22.

32. Kathleen Wellman, *La Mettrie* (Duke University Press, 1992), 23; L. W. B. Brockliss, "The Medico-Religious Universe of an Early-Eighteenth-Century Parisian Doctor: The Case of Philippe Hecquet," in *The Medical Revolution of the Seventeenth Century*, ed. R. French and A. Wear (Cambridge University Press, 1989).

33. On standpoint epistemology, see Donna Haraway, "Situated Knowledges: The Science Question in Feminism as a Site of Discourse on the Privilege of Partial Perspective," *Feminist Studies* 14, 1988, 575–599, reprinted in her *Simians, Cyborgs, and Women* (Routledge 1991); see also Sandra Harding, *The Science Question in Feminism* (Cornell University Press, 1986); idem, *Whose Science? Whose Knowledge? Thinking from Women's Lives* (Cornell University Press, 1991).

34. Johanna Geyer-Kordesch, "George Ernst Stahl's Radical Pietist Medicine and Its Influence on the German Enlightenment," in *The Medical Enlightenment of the Eighteenth Century*, ed. A. Cunningham and R. French (Cambridge University Press, 1990); Roger French, "Sickness and the Soul: Stahl, Hoffmann and Sauvages on Pathology," ibid., 88–110; Jürgen Konert, "Academic and Practical Medicine in Halle during the Era of Stahl, Hoffmann, and Juncker," *Caduceus* 13, 1997, 23–38.

35. Irene Strube, *Georg Ernst Stahl* (Teubner, 1984).

36. Karl Hufbauer, *The Formation of the German Chemical Community (1720–1795)* (University of California Press, 1982).

37. Hans Aarsleff, "The Berlin Academy under Frederick the Great," *History of the Human Sciences* 2, 1989, 193–206.

38. For a detailed exposition of these Stahlian texts, see Metzger, *Newton, Stahl, Boerhaave.*

39. See chapter 3 for textual references to Stahl's works during the 1720s and the 1730s by French academicians.

40. David Oldroyd, "An Examination of G. E. Stahl's *Philosophical Principles of Universal Chemistry*," *Ambix* 20, 1973, 36–52.

41. Stahl, *Zufällige Gedancken und nützliche Bedencken über den Streit von dem sogennanten Sulphure, und zwar sowohl dem gemeinen verbrennlichen oder flüchtigen als unverbrennlichen oder fixen* (Halle: Verlegung des Maysenhauses, 1718); translated by Baron d'Holbach as *Traité du Soufre, tant commun, combustible ou volatil, que fixe, &c.* (Paris: P. F. Didot, 1766) and *Ausführliche*

Betrachtung und zulänglicher Beweiss von den Saltzen, dass dieselbe aus einer Zarten Erde, mit Wasser innig verbunden bestehen (Halle, 1723); translated by Baron d'Holbach as *Traité des Sels, dans lequel on démonstre qu'ils sont composés d'une terre subtile intimément combinée avec de l'eau* (Paris: Vincent, 1771). For a list of Stahl's works, see J. R. Partington, *A History of Chemistry* (London: MacMillan, 1961–1972), volume 2, 659–662.

42. Hélène Metzger, "La Philosophie de la matière chez Stahl et ses Disciples," *Isis* 8, 1926, 427–464; idem, "La Théorie de la composition des sels et la théorie de la combustion d'après Stahl et ses Disciples," *Isis* 9, 1927, 294–325; idem, *Newton, Stahl, Boerhaave, et la Doctrine chimique* (Paris: Félix Alcan, 1930).

43. Geoffroy, *Treatise of the Fossil, Vegetable, and Animal Substances That Are Made Use of in Physick* (London, 1736), 8–12; discussed in Smeaton, "E. F. Geoffroy Was Not a Newtonian Chemist,"*Ambix* 18, 1971, 212–214.

44. Stahl, *Philosophical Principles of Universal Chemistry* (London, 1730), 5.

45. Ibid., 3.

46. Ibid., 4.

47. Ibid., 12.

48. Fourcroy, *Leçons élémentaire d'Histoire naturelle et de Chimie* (Paris, 1782), 35.

49. Stahl, *Traité des sels* (Paris: Vincent, 1783), 308 and 342.

50. Stahl, *Philosophical Principles*, 23–24; "23. An *Aggregate* is distinguish'd from an *Atom*, in that an *Atom* is one numerical individual; but an *Aggregate* several *Atoms* combined together by contiguity" (ibid., 11).

51. Ibid., 43, 24.

52. Ibid., 25–26.

53. Arnold Thackray, *Atoms and Powers* (Harvard University Press, 1970), 94–95; A. M. Duncan, *Laws and Order in Eighteenth-Century Chemistry* (Clarendon, 1996), 78–82.

54. Thackray, *Atoms and Powers*, 94.

55. Senac, *Nouveau cours de chimie, suivant les principes de Newton & de Sthall* (Paris: Jacques Vincent, 1723), 2.

56. Ibid., 76.

57. Ibid., 125.

58. Ibid., 1.

59. Ibid., 29–30.

60. G. A. Lindeboom, *Herman Boerhaave* (Methuen, 1968).

61. The subject of his philosophical disputation in 1686 was on cohesion. On the intellectual environment of Leyden, see F. L. R. Sassen, "The Intellectual Climate in Leiden in Boerhaave's Time," in *Boerhaave and His Time*, ed. Lindeboom; H. A. M. Snelders, "Professors, Amateurs, and Learned Societies:

The Organization of the Natural Sciences," in *The Dutch Republic in the Eighteenth Century*, ed. M. Jacob and W. Mijnhardt (Cornell University Press, 1992); Lissa Roberts, "Going Dutch: Situating Science in the Dutch Enlightenment," in *The Sciences in Enlightened Europe*, ed. W. Clark et al. (University of Chicago Press, 1999).

62. Thackray, *Atoms and Powers*, 46; G. H. Lindeboom, "Pitcairne's Leyden Interlude described from the Documents," *Annals of Science* 19, 1963, 273–284. On Pitcairne's "qualified Newtonianism," which did not endorse attraction, see Anita Guerrini, "Isaac Newton, George Cheyne and the 'Principia Medicinae,'" in *The Medical Revolution of the Seventeenth Century*, ed. French and Wear.

63. On Boerhaave's influence on the British medical students, see Edgar Ashworth Underwood, *Boerhaave's Men at Leyden and After* (Edinburgh University Press, 1977); Andrew Cunningham, "Medicine To Calm the Mind: Boerhaave's Medical System, and Why It Was Adopted in Edinburgh," in *The Medical Enlightenment of the Eighteenth Century*, ed. Cunningham and French; F. Greenaway, "Boerhaave's Influence on Some 18th Century Chemists," in *Boerhaave and His Time*, ed. Lindeboom.

64. Jacob Bickerstaff, ed., *Dr Houston's Memoirs of His Own Life-Time* (London, 1747), 56–57, quoted in Andrew Cunningham, "Medicine to Calm the Mind," 41.

65. Donovan, *Philosophical Chemistry in the Scottish Enlightenment*; Golinski, *Science as Public Culture*.

66. Vicq d'Azyr acknowledged Boerhaave's prominence in the European medical scene in his eulogy of Boerhaave's French students; Daniel Roche, "Talent, Reason, and Sacrifice: The Physician during the Enlightenment," in *Medicine and Society in France*, ed. R. Forster and O. Ranum (Johns Hopkins University Press, 1980), 76. On French medical students, see Kathleen Wellman, *La Mettrie* (Duke University Press, 1992), 60–134; L. W. B. Brockliss, "The Medico-Religious Universe of an Early Eighteenth-Century Parisian Doctor: The Case of Philippe Hecquet," in *The Medical Revolution of the Seventeenth Century*, ed. French and Wear, at 197–198.

67. A pirate version of his chemical lectures titled *Institutiones et Experimenta Chemiae* was published in 1724, possibly in Leyden, although the title page bears Paris as the place of publication. This prompted Boerhaave to publish an authenticated Latin version: *Elementa Chemiae* (Leyden, 1732); English translation: *Elements of Chemistry* (London: printed for J. and J. Pemberton, 1735). A French translation appeared only in segments until the full translation of his theory in 1748; see Tenney L. Davis, "The Vicissitudes of Boerhaave's Textbook of Chemistry," *Isis* 10, 1928, 33–46; F. W. Gibbs, "Boerhaave's Chemical Writings," *Ambix* 6, 1958, 117–135.

68. Bernard Joly, "Voltaire chimiste: L'influence des théories de Boerhaave sur sa doctrine du feu," *Revue du nord* 77, 1995, 817–843; Robert L. Walters, "Chemistry at Cirey," *Studies on Voltaire* 58, 1967, 1807–1827.

69. Boerhaave, "Oration on the usefulness of the mechanical method in medicine (1703)," in *Boerhaave's Orations* (Brill, 1983).

70. R. W. Home, "Nollet and Boerhaave: A Note on Eighteenth-Century Ideas about Electricity and Fire," *Annals of Science* 36, 1979, 171–176.

71. *Boerhaave's Orations*, 193–213. Peter Shaw followed Boerhaave's themes in this oration faithfully in his account of Boyle's works; see F. W. Gibbs, "Peter Shaw and the Revival of Chemistry," *Annals of Science* 7, 1951, 211–237.

72. Herman Boerhaave, *Elements of Chemistry* (London, 1735), 1.

73. In the Latin edition, the first part occupies 29 pages and the second part, the remainder of 867 pages of the first volume. The third part, containing descriptions of 227 operations, is relegated to the second volume; see Lindeboom, *Herman Boerhaave*, 344–347.

74. For a detailed discussion, see J. R. R. Christie, "Historiography of Chemistry in the Eighteenth Century: Hermann Boerhaave and William Cullen," *Ambix* 41, 1994, 4–19.

75. Boerhaave, *Elements of Chemistry*, 2.

76. Ibid., 3.

77. Ibid., 19.

78. Ibid., 44–45.

79. Ibid., 46–47.

80. Ibid., 50.

81. Ibid., 51.

82. Ibid., 19.

83. Ibid., 78.

84. Metzger, *Newton, Stahl, Boerhaave*, 227–228.

85. Boerhaave, *Elements*, 83–85.

86. Pieter van der Star, ed., *Fahrenheit's Letters to Leibniz and Boerhaave* (Rodopi, 1983).

87. Boerhaave, *Elements*, 88.

88. Ibid., 88–89.

89. Ibid., 92.

90. Ibid., 114 (emphasis added).

91. Ibid., 199.

92. Rosaleen Love, "Some Sources of Herman Boerhaave's Concept of Fire," *Ambix* 19, 1972, 157–174.

93. Boerhaave, *Elements*, 168–169.

94. Ibid., 212–213.

95. Milton Kerker, "Herman Boerhaave and the Development of Pneumatic Chemistry," *Isis* 46, 1955, 36–49.

96. Boerhaave, *Elements*, 247–255.

97. Ibid., 314–315.

98. Ibid., 314.

99. Robert L. Walters, "Chemistry at Cirey"; Joly, "Voltaire chimiste"; Wellman, *La Mettrie*; Rousseau, *Institutions chimique*.

100. De Fouchy, "Éloge de M. Rouelle," *Histoire*, 1770, 137–152.

101. P. Dorveaux, "Apothicaires membres de l'Académie Royale des Sciences. IX. Guillaume-François Rouelle," *Revue d'histoire de la pharmacie* 4, 1933, 169–186.

102. Rappaport (G.-F. Rouelle, 9) conjectures that Rouelle began teaching in 1742.

103. Jacques Roger, *Buffon* (Cornell University Press, 1997), translated from *Buffon, un philosophe au Jardin du Roi* (Paris: Arthème Fayard, 1989); E. C. Spray, *Utopia's Garden* (University of Chicago Press, 2000).

104. Rouelle and Cadet, "Analyses d'une eau minérale" (Paris, 1755).

105. Rouelle, "Mémoire sur les sels neutres, dans lequel on propose une division méthodique de ces sels, qui facilite les moyens pour parvenir à la théorie de leur crystallization," *Mémoires*, 1744, 353–364, at 353.

106. Rouelle's reference to this particular text would mean either that he read German, which is unlikely, or that French translation of this text was in circulation well before its publication in 1766.

107. "Sur la crystallization du sel marin," *Histoire*, 1745, 32–34, at 34.

108. "Sur la surabondance d'acide qu'on observe en quelques sels neutres," *Histoire*, 1754, 79–86, at 79; This recapitulates more succinctly what Rouelle states in his 1744 memoire.

109. Rouelle, "Sur le Sel marin. Premiére Partie. De la crystallization du Sel marin," *Mémoires*, 1745, 57–79.

110. Rouelle, "Mémoire sur les Sels neutres, dans lequel on fait connoître deux nouvelles classes de sels neutres, & l'on développe le phénomème singulier de l'excès d'acide dans ces sels," *Mémoires*, 1754, 572–588.

111. "Sur la surabondance d'acide," *Histoire*, 1754, 86.

112. Lavoisier, "Sur le gypse, deuxième mémoire," *Oeuvres de Lavoisier*, volume 3, 128–144, at 128.

113. Rouelle, "Cours de Chymie, recueilli des leçons de M. Rouelle (1767)," [Ms. 1202, *Museum nationale d'histoire naturelle*], 33r. This copy seems nearly identical with the Lusanne copy analyzed by Secrétan. Both copies are missing the mineral kingdom; Claude Secrétan, "Un aspect de la chimie prélavoisienne (Le Cours de G.-F. Rouelle)," *Mémoires de la société vaudoise des sciences naturelles* 50, 1943, 220–444; idem, "Coup d'oeil sur la chimie prélavoisienne (d'après un manuscrit inédit)," *Bulletin de la société vaudoise des sciences naturelles* 61, 1941, 329–354. On the organization of Rouelle's lectures based on the comparison of several manuscripts, see Rappaport, G.-F. Rouelle, 92–99.

114. Diderot, "Notices sur Vanloo et Rouelle," 408.

115. Rouelle, "Cours de Chymie," 1r.

116. Ibid., 3v.

117. Rappaport, "Rouelle & Stahl—Phlogistic Revolution in France," *Chymia* 7, 1961, 73–102.

118. Rouelle, "Cours de Chymie," 15v.

119. Ibid., 21v–22r.

120. Ibid., 23r.

121. "Sur la Terre de l'Alun," *Histoire*, 1744, 16–18.

122. Rouelle's table was used by Venel as an illustration of his article "Chymie" in the *Encyclopédie*.

123. Antoine Fourcroy, *Système des connaissances chimiques, et de leurs applications aux phénomènes de la nature et de l'art* (Paris: Baudouin, 1800–1802), volume 7, 39; quoted and translated in Rappaport, G.-F. Rouelle, 94.

124. Stephen Hales, *Vegetable Staticks* (London: W. and J. Innys, 1727); translated by Buffon, *La Statique végétaux et l'analyse de l'air* (Paris: Debure l'aîné, 1735).

125. Bucquet, *Introduction a l'étude des corps naturels, tirés du règne minéral* (Paris, 1771), 62–63; Guerlac, "The Continental Reputation of Stephen Hales," *Archives internationale d'histoire des sciences* 15, 1951, 393–404; idem, "Lavoisier and his Biographers," *Isis* 45, 1954, 51–62, 59.

126. H.-M. Rouelle, "Tableau de l'Analyse chimique," quoted and translated in Rappaport, G.-F. Rouelle, 75.

127. Louis-Sébastien Mercier, "Tableau de Paris" (Amsterdam, 1782–1788), volume 11, 178–179; quoted and translated in Rappaport, G.-F. Rouelle, 69.

128. L. J. M. Coleby, *The Chemical Studies of P. J. Macquer* (Allen & Unwin, 1938); W. A. Smeaton, *Dictionary of Scientific Biography* and "Macquer et la médecine. Un aspect de sa vie qui nécessite des recherches," *Revue d'histoire de la pharmacie* 24, 1977, 251–254; Douglas McKie, "Macquer, the first lexicographer of chemistry," *Endeavour* 16, 1957, 133–136; Willem C. Ahlers, Un Chimiste du XVIIIe siècle: Pierre-Joseph Macquer (1718–1784), doctoral thesis, École pratique des hautes Études, 1969; Claude Viel, "Pierre-Joseph Macquer," *Janus* 73, 1986–1990, 1–27.

129. Denis I. Duveen and Roger Hahn, "Deux encyclopédistes hors de l'Encyclopédie: Philippe Macquer et l'abbé Jaubert," *Revue d'histoire des sciences et leurs applications* 12, 1959, 330–342.

130. "Sur la cause de la differente dissolubilité des huiles dans l'esprit de vin," *Histoire*, 1745, 35–38; *Mémoires*, 9–25.

131. Pierre-Joseph Macquer, *Élémens de chymie—Théorique* (Paris: J.-T. Herissant, 1749); *Élémens de chymie—Pratique* (Herissant, 1751). A second edition was published in 1756.

132. W. A. Smeaton, "P. J. Macquer's Course of Chemistry at the Jardin du roi," *Proceedings of the Tenth International Congress of the History of Science* (Paris: Hermann, 1964), 847–849.

133. On the various editions and translations of *Élémens* and *Dictionnaire*, see Douglas McKie, "Macquer, the first lexicographer of chemistry," *Endeavour* 16, 1957, 133–136; Coleby, *The Chemical Studies*, 16–17, 23–24; Roy G. Neville and W. A. Smeaton, "Macquer's *Dictionnaire de Chymie*: A Bibliographical Study," *Annals of Science* 38, 1981, 613–662.

134. Macquer, *L'Art du teinturier en soie* (Paris, 1763).

135. The main body of Macquer's papers and correspondances are preserved in the manuscript section of the *Bibiothèque nationale*, Paris [Mss. fr. 9127 to 9135, 12305 and 12306]. For guides, see M. Bouvet, "Les papiers Macquer de la Bibliothèque nationale," *Revue d'histoire de la pharmacie* 134, 1952, 1–8; Willem C. Ahlers, "La Correspondance de Macquer," *Revue de synthèse* 97, 1976, 125–127 and 137–140.

136. Macquer, *Élémens de Chymie—Théorique*, 1.

137. Ibid., 12.

138. Ibid., 15–16.

139. Ibid., 20.

140. Smeaton suggests that this use of Geoffroy's table may have been due to a purely technical reason. The posthumous publication of Geoffroy's *Traité de matière médicale* (Paris, 1743) included the affinity table. Macquer used the exact same copper plate used for his *Élémens*; see Smeaton, *Dictionary of Scientific Biography*, 622, n. 12.

141. Macquer, *Élémens de chymie—Théorique*, 256–273.

142. Macquer, "Affinité," in *Dictionnaire de chymie* (1778), 66–73.

143. Condorcet, "Éloge de M. Macquer," 23–24.

144. Macquer, "Affinité," 57. This part was unchanged from the first edition (1766).

145. Macquer, *Dictionnaire de chymie* (Paris: Lacombe, 1766), volume 1, 55.

146. Baumé, *A Manual of Chemistry* (Warrington: printed by W. Eyres for J. Johnson, 1778), 9.

147. Article "Affinité," in Macquer's *Dictionnaire de chymie* (1766).

148. For contrasting profiles of the apothecary and the physician, see Lester S. King, *The Medical World of the Eighteenth Century* (Krieger, 1958). For a social profile of physicians, see Daniel Roche, "Talent, Reason, and Sacrifice: The Physician during the Enlightenment," in *Medicine and Society in France*, ed. R. Forster and O. Ranum (Johns Hopkins University Press, 1980). On the scholarly identity of physicians in contrast to other health practitioners, see Christopher Lawrence, "Medical Minds, Surgical Bodies: Corporeality and Doctors," in *Science Incarnate*, ed. C. Lawrence and S. Shapin (University of Chicago Press,

1998); Lindsay Wilson, *Women and Medicine in the French Enlightenment* (Johns Hopkins University Press, 1993).

149. Jonathan Simon, The Alchemy of Identity: Pharmacy and the Chemical Revolution, 1777–1809, Ph.D. dissertation, University of Pittsburgh, 1997; "The Chemical Revolution and Pharmacy: A Disciplinary Perspective," *Ambix 45*, 1998, 1–13.

150. Macquer, *Élémens* (1749), xi–xii.

151. Macquer and Baumé, "Plan d'un Cours de Chymie expérimentale et raisonnée, avec un discours historique sur la chymie," (Paris, 1757), xix. It must have been written by Macquer, since it was reproduced in the *Dictionnaire* (1766) without much alteration. For an analysis of the rhetorical structure, see Wilda C. Anderson, *Between the Library and the Laboratory* (Johns Hopkins University Press, 1984), 19–34.

152. Venel used *la physique* to designate both natural philosophy and physics in a narrower sense, but he qualified the latter as "ordinary physics" in this particular context to avoid confusion; Venel, "Chymie," in Diderot's *Encyclopédie*.

Chapter 5

1. Henry Guerlac, *Newton on the Continent* (Cornell University Press, 1981), 73.

2. Pierre Brunet, *L'Introduction des théories de Newton en France au XIIIe siècle avant 1738* (Paris: Albert Blanchard, 1931); idem, *Les Physiciens hollandais et la méthode expérimental en France au XVIIIe siècle* (Paris: Albert Blanchard, 1926). For a more balanced account, see John L. Greenberg, "Mathematical Physics in Eighteenth-Century France," *Isis 77*, 1986, 59–78.

3. Voltaire, *Letters concerning the English Nation* (London, 1733); *Elémens de la philosophie de Newton, mis à la portée de tout le monde* (Amsterdam, 1738); R. L. Walters, "Voltaire, Newton and the Reading Public," in *The Triumph of Culture*, ed. P. Fritz and D. Williams (Hakkert, 1972).

4. William Jacob van 'sGravesande, *Physices elementa mathematica, experimentis confirmata* (Leyden, 1720–1721); P. van Muschenbroek, *Essai de physique* (Leyden, 1739); C. Depater, "Petrus van Musschenbroek (1692–1761): A Dutch Newtonian," *Janus 64*, 1977, 77–87.

5. Fontenelle, "Eloge de M. Newton (1727)," translated in A. Rupert Hall, *Isaac Newton* (Oxford University Press, 1999), 59–74; Michael Freyne, "L'Eloge de M. Newton dans la correspondance de Fontenelle," *Corpus 13*, 1990, 75–92. Fontenelle's letters to Newton are published in Douglas McKie, "Fontenelle et la Société Royale de Londres," *Revue d'histoire des sciences et leurs applications 10*, 1957, 334–338. On Fontenelle's position on attraction, see Charles B. Paul, *Science and Immortality* (University of California Press, 1980), 28–40.

6. Pierre Brunet, *Maupertuis* (Paris: Albert Blanchard, 1929). For a subtler account, see Mary Terrall, *The Man Who Flattened the Earth* (University of Chicago Press, 2002).

7. L. Hanks, *Buffon avant l'Histoire naturelle* (Paris, 1966); I. B. Cohen, *Franklin and Newton* (American Philosophical Society, 1956).

8. On the early "Newtonian physicians," see Anita Guerrini, "The Tory Newtonians: Gregory, Pitcairne, and Their Circle," *Journal of British Studies* 25, 1986, 266–311; idem, "Isaac Newton, George Cheyne and the 'Principia Medicinae,'" in *The Medical Revolution of the Seventeenth Century*, ed. French and Wear; Julian Martin, "Sauvage's Nosology: Medical Enlightenment in Montpellier," in *The Medical Enlightenment of the Eighteenth Century*, ed. A. Cunningham and R. French (Cambridge University Press, 1990).

9. Robert E. Schofield, *Mechanism and Materialism* (Princeton University Press, 1970); Arnold Thackray, *Atoms and Powers* (Harvard University Press, 1970). For another view of the early Newtonian group, see Christina M. Eagles, "David Gregory and Newtonian Science," *British Journal for the History of Science* 36, 1977, 216–225.

10. Stephen Hales, *La statique des végétaux et l'analyse de l'air* (Paris: Debure l'aîné, 1735); Henry Guerlac, "Stephen Hales: A Newtonian Physiologist," in his *Essays and Papers in the History of Modern Science* (Johns Hopkins University Press, 1977); reprinted from *Dictionary of Scientific Biography*.

11. On Buffon's philosophy of science, Robert Wohl, "Buffon and His Project for a New Science," *Isis* 51, 1960, 186–199.

12. Buffon, *Histoire naturelle* (Paris, 1749–1767), volume 8, xii–xiii; quoted and translated in Thackray, *Atoms and Powers*, 159.

13. Thackray, *Atoms and Powers*, 2–4.

14. Robert E. Schofield, "The Counter-Reformation in Eighteenth-Century Science—Last Phase," in *Perspectives in the History of Science and Technology*, ed. D. Roller (Oklahoma University Press, 1971).

15. Arthur Donovan, "Newton and Lavoisier—From Chemistry as a Branch of Natural Philosophy to Chemistry as a Positive Science," in *Action and Reaction*, ed. P. Theerman and A. Seeff (University of Delaware Press, 1993).

16. R. W. Home, "Out of a Newtonian Straitjacket: Alternative Approaches to Eighteenth-Century Physical Science," in *Studies in the Eighteenth Century*, ed. R. Brissenden and J. Eade (Australian National University Press, 1979); idem, "'Newtonianism' and the Theory of the Magnet," *History of Science* 15, 1977, 252–266. For other critiques and historiographical directions in dealing with Newtonianism, see P. M. Heimann, "Newtonian Natural Philosophy and the Scientific Revolution," *History of Science* 11, 1973, 1–7; A. Rupert Hall, "Newton in France: A New View," *History of Science* 13, 1975, 233–250; Robert E. Schofield, "An Evolutionary Taxonomy of Eighteenth-Century

Newtonianisms," *Studies in Eighteenth-Century Culture* 7, 1978, 175–192; Henry Guerlac, "Some Areas for Further Newtonian Studies," *History of Science* 17, 1979, 75–101.

17. Vincenzo Ferrone, *The Intellectual Roots of the Italian Enlightenment* (Humanities Press, 1995); Paula Findlen, "A Forgotten Newtonian: Women and Science in the Italian Provinces," in *The Sciences in Enlightened Europe*, ed. W. Clark et al. (University of Chicago Press, 1999); Lisa Roberts, "Going Dutch: Situating Science in the Dutch Enlightenment," Ibid., 350–388.

18. On the relative autonomy of demonstration culture in experimental physics, Geoffrey V. Sutton, *Science for a Polite Society* (Westview, 1995), 241–285.

19. Michelle Goupil, *Du Flou au Clair? Histoire de l'Affinité chimique* (Paris: C. T. H. S., 1991), 123–132 and 148–154.

20. Thackray, *Atoms and Powers*, 193–196 and 205–209.

21. On the rather exclusive social makeup of provincial academies and Freemasonry, which would have been more conducive to philosophical chemitsry, see Daniel Roche, *Le siècle des lumières en province* (Mouton, 1978).

22. A. M. Duncan, in his survey of eighteenth-century affinity tables ("Some Theoretical Aspects of Eighteenth-Century Tables of Affinity," *Annals of Science* 18, 1962, 177–196 and 217–232), was able to locate only one other table before 1749. Jean Grosse used a modified version of Geoffroy's table in his private course. For the list of affinity tables, see idem, *Laws and Order in Eighteenth-Century Chemistry* (Clarendon, 1996), 112–114.

23. C. E. Gellert, *Anfangsgründe zur metallurgischen Chymie*, second edition (Leipzig, 1776), opp. 172; J. R. Spielmann, *Institutiones Chymiae* (Strasbourg: J. G. Bauerum, 1763); French translation by Cadet le jeune, *Instituts de Chymie* (Paris: Vincente, 1770), volume 1, 24; Duncan, "Some Theoretical Aspects," 178.

24. Michelle Goupil, *Du Flou au Clair?*, 139–146; idem, "J. Ph. de Limbourg et la théorie des affinités chimiques," *Technologia* 7, 1984, 11–28; James Evans, "Gravity in the Century of Light: Sources, Construction and Reception of Le Sage's Theory of Gravitation," in *Pushing Gravity*, ed. M. Edwards (Apeiron, 2002).

25. *Dissertation de Jean philippe de Limbourg, docteur en medecine, sur les Affinités chymiques* (Liége: F. J. Desoer, 1761), 11, 30, and 48.

26. Ibid., 60.

27. Le Sage, *Essai de chymie méchanique* (1758), 10.

28. Ibid., 49–51.

29. Georges Bouchard, *Guyton-Morveau, chimiste et conventionnel* (Paris: Perrin, 1938).

30. Antoine Guyton was appointed to the faculty of law at Dijon in 1768, shortly before his death.

31. Jacques Roger, *Buffon* (Cornell University Press, 1997), 6.

32. David A. Bell, *Lawyers and Citizens* (Oxford University Press, 1994).

33. Guyton, *Mémoire sur l'éducation publique, avec le prospectus d'un college suivant les principes de cet ouvrage* (1764); Arnould de Lesseux, "L'éducation publique d'après Guyton de Morveau," *Mémoires de l'Académie des Sciences, Arts et Belles-Lettres de Dijon*, 123, 1976–1978 (pub. 1979), 207–239; W. A. Smeaton, "L. B. Guyton de Morveau (1737–1816): A Bibliographic Study," *Ambix* 6, 1957, 18–34.

34. Bouchard, *Guyton-Morveau*, 70–84; Roger Tisserand, *Au temp de l'encyclopédie* (Boivin, 1936).

35. Marcel Bouchard, *L'Académie de Dijon et le Premier Discours de Rousseau* (Paris: Société les Belles Lettres, 1950).

36. The question in the 1771 prize competition was to "determine the action of acids on oils, the mechanism of their combination and the nature of the different soap compounds that result from them." Competitions were again held in 1774 and in 1777, when the prize was divided between Voussonne and Planchon with an *accessit* for Jaubert ("Prix," *Journal de médecine, chirurgie, pharmacie, &c.* 47, 1777, 274–276). On the purchase of books, see W. A. Smeaton, "Two Books Are Added to Guyton de Morveau's Library: A Study of Personal and Academic Communications in 1785," *Ambix* 34, 1987, 140–146.

37. Guyton reported on the state of his laboratory in his letters to Macquer; W. A. Smeaton, "Guyton de Morveau's Course of Chemistry in the Dijon Academy," *Ambix* 9, 1961, 53–69; Claude Viel, "L'activité de chimiste de Guyton de Morveau à travers ses lettres à Macquer et à Picot de La Peyrouse," *Annales de Bourgogne*, 70, 1998, 55–67. According to a later report, Guyton's private laboratory was one of the finest in Europe (Arthur Young, *Travels in France during the years 1787, 1788 and 1789*, Bury St. Edmunds, 1792, volume 1, 148–154; discussed in W. A. Smeaton, "Louis Bernard Guyton de Morveau, F. R. S. (1737–1816) and His Relations with British Scientists," *Notes and Records of the Royal Society of London* 22, 1967, 113–130).

38. On Buffon's connections with the Dijon dignitaries, see Jacques Roger, *Buffon*; Henri Nadault de Buffon, ed., *Correspondance inédite de Buffon* (Hachette, 1860).

39. Emmanuel Grison, Michelle Goupil, and Patrice Bret, eds., *A Scientific Correspondence during the Chemical Revolution* (Berkeley: Office for the History of Science and Technology, 1994), 22.

40. Buffon, ed., *Correspondance inédite de Buffon*, volume 1, 144.

41. Eighteen letters from Guyton to Macquer, dated between April 14, 1762 and September 3, 1782, are among Macquer's correspondence preserved at the *Bibliothèque nationale* [Mss. f fr. 12306]. Smeaton has established the date of the first letter as 1769, however, from various textual evidences; "A Bibliographic Study," 24, n. 30.

42. Smeaton, "Guyton de Morveau and the Phlogiston Theory," in *Mélanges Alexandre Koyré*, ed. I. Cohen and R. Taton (Hermann, 1964), 522–526.

43. J. R. Partington and Douglas McKie, "Historical Studies on the Phlogiston Theory—I. The Levity of Phlogiston," *Annals of Science* 2, 1937, 361–404.

44. The best historical account is provided by Guyton's "Dissertation sur la phlogistique," in his *Digressions académique* (Dijon: L. N. Frantin, 1772).

45. Although these memoires were not published in full until 1769, a detailed abstract appeared in *Mercure de France* 2, July 1765, 127–134.

46. Ibid., 132–133.

47. Laurent Béraut, *Dissertation sur la cause de l'augmentation de poids, que certains matières acquièrent dans leur calcination* (Bordeaux, 1747). It won the prize competition at the Bordeaux Academy; Douglas McKie, "Béraut's Theory of Calcination (1747)," *Annals of Science* 1, 1936, 269–293.

48. Ribaptome, "Lettre sur l'augmentation de poids dans la calcination, adressée à Messieurs Auteurs du Journal des Sçavans," *Journal des Sçavans*, 1767, 889–894 (dated July 20, 1765).

49. "Lettre de M. Chardenon," *Journal des Sçavans*, 1768, 648–658. This piece mostly criticized Béraut's 1747 theory of calcination.

50. Ibid., 644–645; Partington and McKie, "Historical Studies, I," 379.

51. Chardenon, "Mémoire sur l'augmentation de poids des métaux calcinés," *Mémoires de l'Académie de Dijon* 1, 1769, 303–320.

52. *Registre de l'Académie de Dijon* 4, 70v–73v (January 8, 1768); ibid., 5, 41r–42r (extract of the public séance on December 11, 1768). The reference to Boerhaave appears only in the published version of his reports: "Mémoire sur les phénomènes de l'air dans la combustion," *Mémoires de l'académie de Dijon* 1, 1769, 416–438.

53. The earliest communication preserved is Guyton's letter to Macquer on April 14, 1769, but the letter suggests at least one previous contact; "Correspondance de Macquer" [*Bibliothèque nationale*, fr 12306].

54. *Registre de l'Académie de Dijon* 5, 260r–261r (December 21, 1770). Guyton continued to present successive chapters of the "Dissertation" until June 28, 1771.

55. PV 91, 1772, 31v–36r [prepared with P. J. Malouin]; Guerlac, *Lavoisier—The Crucial Year* (Cornell University Press, 1961), 127–145.

56. *Oeuvres de Condorcet* (Paris: Firmin Didot frères, 1847–1849), volume 2, 38–39; Guerlac, *The Crucial Year*, 139.

57. Smeaton, "Guyton de Morveau and the Phlogiston Theory."

58. Guyton, "Dissertation," 174.

59. Ibid., 177–178.

60. Ibid., 208.

61. "Extrait. Digressions académiques, ou Essais sur quelques sujets de Physique & d'Histoire naturelle," *Journal de médecine* 38, 1772, 195–220.

62. *Registre de l'Académie de Dijon*, VI, 189r–191r (September 27, 1772); printed as *Défense de la volatilité du phlogistique*.

63. Carleton Perrin, "Early Opposition to the Phlogiston Theory: Two Anonymous Attacks," *British Journal for the History of Science* 8, 1970, 128–144.

64. "Précis de la Doctrine de M. de Morveau, sur le phlogistique; & Observations sur cette Doctrine," *Observations sur la physique* [hereafter cited as *Observations*] 2, 1773, 281–285.

65. "Suite du discours sur le Phlogistique," *Observations* 2, 1773, 285–291.

66. "Discours sur le phlogistique et sur plusieurs points importans de Chymie," *Observations* 2, 1774, 185–200, at 193.

67. Guyton, "Essai physico-chymique sur la dissolution et la crystallisation, pour parvenir à l'explication des affinités par la figure des parties constituantes des corps," in *Digressions academique*; W. A. Smeaton, "Guyton de Morveau and Chemical Affinity," *Ambix* 11, 1963, 55–64.

68. *Registre de l'Académie de Dijon*, V, 33r–34v (June 2, 1769).

69. Article "Hépar" in *Supplément à l'Encyclopédie* (1776–1777), volume 3, 347–348 (emphasis added).

70. *Registre de l'Académie de Dijon*, V, 130r–134r (January 5 and 12, 1770).

71. *Registre de l'Académie de Dijon* 6, 71v–73r (September 15, 1771).

72. Guyton, "Essai physico-chimique," 273–275.

73. Ibid., 277.

74. Ibid., 281.

75. Ibid., 286.

76. Ibid., 284–285.

77. Ibid., 302.

78. Guyton cites Musschenbroek here as the authority on Newton's views; ibid., 310.

79. Ibid., 328.

80. Ibid., 342–343.

81. *Registre de l'Académie de Dijon* 6, 212v–214r (February 12, 1773); M. de Morveau, "Sur l'attraction ou la répulsion de l'eau & des corps huileux, pour vérifier l'exacttude de la méthode par laquelle de Docteur TAYLOR estime la force d'adhésion des surfaces, & détermine l'action du verre sur le mercure des baromêtres, faites en présence de l'Académie des Sciences, Arts & Belles-Lettres de Dijon, dans son assemblée du 12 Février 1773," *Observations*, 1, 1773, 172–173 [the article is six pages long, but it is mispaginated] and 460–461.

82. "Extract of a Letter from Dr. Brook Taylor, F. R. S. to Sir Hans Sloan, dated June 25, 1714. Giving an Account of some Experiments relating to Magnetism," *Philosophical Transactions* 31, 1720–21, 204–208.

83. D. Diderot, *Supplément à l'Encyclopédie* (Amsterdam, 1776–77). For the publication history, see Kathleen Hardesty, *The Supplément to the Encyclopédie* (Nijhoff, 1977). Guyton wrote all chemical articles for this project, beginning in early 1773. He presented the articles "air," "air fixe," and "affinité" to the Dijon Academy on April 23, 1773 and the article "phlogiston" on July 14, 1774; *Registre de l'Académie de Dijon* 6, 235; ibid., 7, 126r.

84. "Phlogiston," in *Supplément* 4, 336–340, at 336 (emphasis added).

85. Guyton conducted research in the crystallization of metals, first of iron and later, of gold, silver, platinum and other metals; "Observation de la cristallisation du fer," *Observations* 8, 1776, 348–353; "Lettre de M. de Morveau, à l'Auteur de ce Recueil, sur les crystallisations métalliques," ibid., 13, 1779, 90–92.

86. Melhado, "Oxygen," 324–325.

87. W. A. Smeaton, "Louis Bernard Guyton de Morveau, F. R. S. (1737–1816) and His Relations with British Scientists," *Notes and Records of the Royal Society of London* 22, 1967, 113–130.

88. Guyton de Morveau, *Mémoire sur l'utilité d'un cours public de chymie dans la ville de Dijon, les avantages qui en résulteroient pour la Province entiere, & les moyens de procurer à peu de frais de Établissement* (Dijon: L. N. Frantin, 1775); W. A. Smeaton, "Guyton de Morveau's Course of Chemistry in the Dijon Academy," *Ambix* 9, 1961, 53–69, 54–56; Tisserand, *Au temps de l'encyclopédie*, 610–617.

89. L. B. Guyton de Morveau, Hughes Maret, and J. F. Durande, *Elémens de chymie, théorique et pratique, rédigés dans un nouvel ordre, d'après les découvertes modernes* (Dijon: L. N. Frantin, 1777–78).

90. Ibid., 8.

91. De Fourcy, "Observations sur le tableau du produit des affinités chymiques," *Observations* 2, 1773, 197–204.

92. Antoine Baumé, *Chymie expérimentale et raisonnée* (Paris: P. F. Didot le jeune, 1773) volume 1, 22.

93. Antoine Baumé, *A Manual of Chemistry* (Warrington: printed by W. Eyres for J. Johnson, 1778), 4–6.

94. Macquer, *Dictionnaire* (1778), 65–66.

95. Bergman, "Disquisitio de Attractionibus Electivis," K. Ventenskaps Societeten i Upsala," *Nova Acta Regiae Societatis Scientiarum Upsaliensis* [2], 2, 1775, 159–248. (For an English translation, see *Dissertation on Elective Attractions*, tr. J. A. Schufle, Johnson Reprint Co., 1968.) An abstract appeared in *Observations*, 13, 1778, 298–333.

96. Bergman, *Opuscula physica et chemica* (Uppsala, 1783), volume 3; *A Dissertation on Elective Attractions* (London, 1785), facsimile reproduction with

an introduction by A. M. Duncan (Cass, 1970); *Traité des Affinités chymiques, ou Attractions electives* (Paris: Buisson, 1788). For a brief history of the French translation, see W. A. Smeaton, "F. J. Bonjour and His Translation of Bergman's 'Disquisitio de attractionibus electivis,'" *Ambix* 7, 1959, 47–50.

97. In the preface to the English translation, the translator calls the volume an "admirable manual of theoretical chemistry" which deserved a faithful translation based on his "reverence" for the author; *A Dissertation on Elective Attractions* (London, 1785).

98. J. A. Schufle, "Torbern Bergman, Earth Scientist," *Chymia* 12, 1967, 59–97; idem, *Torbern Bergman* (Cornado, 1985). The book also contains the translation of a short autobiographical fragment by Bergman written during his illness in 1782 [appendix II]. For a brief overview of Bergman's contribution, see Marco Beretta, "T. O. Bergman and the Definition of Chemistry," *Lychnos*, 1988, 37–67.

99. Torbern Bergman, ed., *Chemical Lectures of H. T. Scheffer* (Kluwer, 1992).

100. Bergman was also known for the "most excellent process for analyzing ores with the blow-pipe by the addition of different saline substances"; "XXIV. Account of the Life and Writings of Olof Torbern Bergman, Professor of Chemistry at Upsal," *Philosophical Magazine* 9, 1801, 193–200, at 197.

101. Bergman initiated the correspondence on January 19, 1768 [*Bibliothèque nationale*, Mss. fr. 12305, 88–9]. There are nine letters in this collection from Bergman to Macquer until June 12, 1782. For the letters from Macquer and Guyton to Bergman, see *Torbern Bergman's Foreign Correspondence*, ed. G. Carld and J. Nordström (Stockholm: Lychnos, 1965), 100–138, 229–255, and 437–438. On Bergman's ties with Buffon's circle in Paris, see Thackray, *Atoms and Powers*, 218–221.

102. Macquer to Bergman, July 31, 1768, in *Torbern Bergman's Foreign Correspondence*, 229–231.

103. Macquer to Bergman, May 18, 1770, in *Torbern Bergman's Foreign Correspondence*, 238–239.

104. *Opuscules chymiques et physiques de M. T. Bergman* (Dijon, 1780–1785), volume 1, viii.

105. Macquer to Bergman, July 7, 1781, in *Torbern Bergman's Foreign Correspondence*, 253–254.

106. Bergman's original dissertation, published in Latin in 1775, is nearly identical in the order and in the content with the 1783 version. I am using the English translation of the latter version in 1785 which preserves more of the contemporary wording than Schufle's translation of the 1775 version. I have taken care, however, to compare the two versions to screen out the later additions; Bergman, *A Dissertation on Elective Attractions* (London, 1785); reprinted with an introduction by A. M. Duncan (Cass, 1970).

107. Ibid., 2 (emphasis added).

108. Bergman emphasizes this difference also in his foreword to *Chemical Lectures of H. T. Scheffer*.

109. Bergman, *A Dissertation*, 3.

110. Ibid., 9.

111. Ibid., 10–12.

112. Ibid., 16.

113. Ibid., 18–24.

114. *Dissertation on Elective Attractions*, tr. Schufle, 69–71.

115. For a detailed description, see Guyton's articles "adhesion" and "affinité" in *Encyclopédie méthodique*.

116. Michael Donovan, "Biographical Account of the Late Richard Kirwan," *Proceedings of the Royal Irish Academy* 4, 1850, lxxxi–cxviii; Ernest Leonard Scott, The Life and Work of Richard Kirwan (1733–1812), Ph.D. thesis, University of London, 1979; J. Reilly and N. O'Flynn, "Richard Kirwan, an Irish chemist of the Eighteenth Century," *Isis* 13, 1930, 298–319.

117. Diane Elizabeth Heggarty, Richard Kirwan: The Natural Philosopher and the Chemical Revolution, Ph.D. dissertation, University of Washington, 1978, 76.

118. T. H. Levere and G. L. Turner, eds., *Discussing Chemistry and Steam* (Oxford University Press, 2002).

119. Kirwan, *Elements of Mineralogy* (London: printed for P. Elmsly, 1784); idem, *An Essay on the Analysis of Mineral Waters* (London, 1784).

120. Kirwan, "Experiments and Observations on the Specific Gravities and attractive Powers of various saline Substances," *Philosophical Transactions of the Royal Society of London* 71, 1781, 7–41, at 7; "Expériences sur les pesanteurs spécifiques & l'attraction des diverses substances saline," *Journal de physique* 24, 1784, 134–156.

121. Kirwan's studies on heat are examined in detail by Heggarty, Richard Kirwan; E. L. Scott, "Richard Kirwan, J. H. de Magellan, and the Early History of Specific Heat," *Annals of Science* 38, 1981, 141–153.

122. Kirwan, "Experiments and Observations," 8–9.

123. Scott, Life and Work of Richard Kirwan.

124. Kirwan, "Experiments and Observations," 9–10.

125. Ibid., 10–11.

126. Ibid., 13.

127. The term "saturation capacity" was coined later by Berthollet.

128. Ibid., 34.

129. Kirwan reports thus to Guyton on May 2, 1782. The editors of their correspondence could not locate the paper Kirwan referred to because it was not a separate paper; *A Scientific Correspondence*, 41. On Kirwan's own synthesis of

contemporary chemistry, see Seymour Mauskopf, "'I passionately desire to be able to link together all, or at least most, of chemical phenomena in a system': Richard Kirwan's Phlogiston Theory: Its Success and Fate," presented at Dexter Symposium, Chicago, 2001.

130. Richard Kirwan, "Continuation of the Experiments and Observations on the Specific Gravities and Attractive Powers of various Saline Substances," *Philosophical Transactions* 72, 1782, 179–236, at 195–196; "Expériences et observations sur les forces attractives des acides minéraux," *Journal de physique* 24, 1784, 188–199 and 356–368; 25, 1784, 13–28.

131. Kirwan, "Continuation," 210.

132. Kirwan, "Conclusion of the Experiment and Observations concerning the Attractive Powers of the Mineral Acids," *Philosophical Transactions* 73, 1783, 15–84, 34; "Expériences et observations sur les forces attractives des acides minéraux," *Journal de physique* 27, 1785, 250–261 and 321–335; "Expériences de M. Kirwan sur les affinités," *Journal de physique* 27, 1785, 447–457; 28, 1786, 94–109.

133. Ibid., 38.

134. Ibid., 39–40.

135. This reorientation has been called for by Evan M. Melhado, "Oxygen, Phlogiston, and Caloric: The Case of Guyton," *Historical Studies in the Physical Sciences* 13, 1983, 311–334.

Chapter 6

1. I. Bernard Cohen, *Revolution in Science* (Harvard University Press, 1985), 236.

2. Arthur Donovan, *Antoine Lavoisier* (Blackwell, 1993); Jean-Pierre Poirier, *Lavoisier* (University of Pennsylvania Press, 1996). For an excellent review of these books, see Evan M. Melhado, "Scientific Biography and Scientific Revolution: Lavoisier and Eighteenth-Century Chemistry," *Isis* 87, 1996, 688–694. For a guide to earlier biographies, see Henry Guerlac, "Lavoisier and His Biographers," *Isis* 45, 1954, 51–62.

3. Maurice Daumas, *Lavoisier, Théoricien et expérimentateur* (Presses universitaires de France, 1955); Henry Guerlac, *Lavoisier—The Crucial Year* (Cornell University Press, 1961); F. L. Holmes, *Lavoisier and the Chemistry of Life* (University of Wisconsin Press, 1985) and *Antoine Lavoisier—The Next Crucial Year* (Princeton University Press, 1998). For a guide to earlier works, see W. A. Smeaton, "New Light on Lavoisier: The Research of the Last Ten years," *History of Science* 2, 1963, 52–69.

4. For an excellent historiographical guide, see Bernadette Bensaude-Vincent, *Lavoisier* (Flammarion, 1994).

5. James Bryant Conant, *The Overthrow of the Phlogiston Theory* (Harvard University Press, 1957).

6. Maurice Crosland, "Chemistry and the Chemical Revolution," in *The Ferments of Knowledge*, ed. G. Rousseau and R. Porter (Cambridge University Press, 1980); A. Donovan, ed., *The Chemical Revolution, Osiris* [2], 4, 1988; C. E. Perrin, "The Chemical Revolution: Shifts in Guiding Assumptions," in *Scrutinizing Science*, ed. A. Donovan et al. (Johns Hopkins University Press, 1988); Jan Golinski, "Chemistry in the Scientific Revolution: Problems of Language and Communication," in *Reappraisals of the Scientific Revolution*, ed. D. Lindberg and R. Westman (Cambridge University Press, 1990); Lissa Roberts, ed., "The Chemical Revolution: Context and Practices," *The Eighteenth Century: Theory and Interpretation* 33, 1992; John G. McEvoy, "Positivism, Whiggism, and the Chemical Revolution: A Study in the Historiography of Chemistry," *History of Science* 35, 1997, 1–33.

7. C. E. Perrin, "Research Traditions and the Chemical Revolution," *Osiris* [2] 4, 1988, 53–81.

8. A. R. Albury, The Logic of Condillac and the Structure of French Chemical and Biological Theory, 1780–1801, Ph.D. dissertation, Johns Hopkins University, 1972; Daniel Brewer, *The Discourse of Enlightenment in Eighteenth-Century France* (Cambridge University Press, 1993).

9. Daumas, *Lavoisier*; M. Daumas and Denis Duveen, "Lavoisier's relatively unknown large-scale decomposition and synthesis of water, February 27 and 28, 1785," *Chymia* 5, 1959, 113–129; Jan Golinski, "Precision Instruments and the Demonstrative Order of Proof in Lavoisier's Chemistry," *Osiris* 9, 1994, 30–47; idem, " 'The Nicety of Experiment': Precision of Measurement and Precision of Reasoning in Late Eighteenth-Century Chemistry," in *The Values of Precision*, ed. M. Wise (Princeton University Press, 1995); Frederic L. Holmes, "The Evolution of Lavoisier's Chemical Apparatus," in *Instruments and Experimentation in the History of Chemistry*, ed. F. Holmes and T. Levere (MIT Press, 2000).

10. Trevor H. Levere, "Lavoisier: Language, Instruments and the Chemical Revolution," in *Nature, Experiment, and the Sciences*, ed. T. Levere and W. Shea (Kluwer, 1990); idem, "Balance and gasometer in Lavoisier's Chemical Revolution," in *Lavoisier et la Révolution chimique*, ed. M. Goupil (SABIX–Ecole polytechnique, 1992); Bernadette Bensaude-Vincent, "Eaux et mesures. Éclairages sur l'itinéraire intellectuel du jeune Lavoisier," *Revue d'histoire des sciences et leurs applications* 48, 1995, 49–69; idem, "The Chemist's Balance for Fluid: Hydrometers and their Multiple Identities, 1770–1810," in *Instruments and Experimentation*, ed. Holmes and Levere.

11. Tore Frängsmyr, J. L. Heilbron, and Robein E. Rider, eds., *The Quantifying Spirit in the Eighteenth Century* (University of California Press, 1990); Bernadette Bensaude-Vincent, "Between Chemistry and Politics," *The Eighteenth Century: Theory and Interpretation* 33, 1992, 217–237; Ken Alder, "A Revolution to Measure: The Political Economy of the Metric System in France," in *The Values of Precision*, ed. Wise.

12. Henry Guerlac, "Chemistry as a Branch of Physics: Laplace's Collaboration with Lavoisier," *Historical Studies in the Physical Sciences* 7, 1976, 193–276; A.

Donovan, "Lavoisier and the Origins of Modern Chemistry," in *The Chemical Revolution*, ed. Donovan; Evan M. Melhado, "Chemistry, Physics, and the Chemical Revolution," *Isis* 76, 1985, 195–211.

13. J. B. Gough, "Lavoisier and the Fulfillment of the Stahlian Revolution," in *The Chemical Revolution*, ed. Donovan; R. Siegfried, "The Chemical Revolution in the History of Chemistry," ibid., 34–50.

14. C. E. Perrin, "Chemistry as a Peer of Physics: A Response to Donovan and Melhado on Lavoisier," *Isis* 81, 1990, 259–270; Arthur Donovan, "Lavoisier as Chemist *and* Experimental Physicist: A Reply to Perrin," ibid., 270–272; Evan M. Melhado, "On the Historiography of Science: A Reply to Perrin," ibid., 273–276; idem, "Scientific Biography and Scientific Revolution."

15. A pioneering work in this direction is F. L. Holmes, *Eighteenth Century Chemistry as an Investigative Enterprise* (Berkeley: Office for the History of Science and Technology, 1989).

16. Ferdinando Abbri, "The Chemical Revolution: A Critical Assessment," *Nuncius* 4, 1989, 303–315.

17. Readers interested in a more detailed reconstruction of Lavoisier's investigative pathways should consult F. L. Holmes, *Lavoisier*; idem, *The Next Crucial Year*; Louis Yvonne Palmer, The Early Scientific Work of Antoine Laurent Lavoisier, Ph.D. dissertation, Yale University, 1998.

18. Poirier, *Lavoisier*, 4; Donovan, *Antoine Lavoisier*, 15.

19. Édouard Grimaux, *Lavoisier, 1743–1794, d'après sa correspondance, ses manuscrits, ses papiers de famille et d'autres documents inédits*, second edition (Paris: Germer Baillière, 1896), 3–4; Henry Guerlac, "A Note on Lavoisier's Scientific Education," *Isis* 47, 1956, 211–216.

20. David S. Evans, *Lacaille* (Pachart, 1992).

21. While historians have attributed Lavoisier's penchant for precision instruments to his inclinations as an experimental physicist, the instruments of experimental physics such as the thermometer and the barometer were not as precise as the astronomical instruments; Maurice Daumas, *Les Instruments scientifiques aux XVIIe et XVIIIe siècles* (Presses universitaires de France, 1953), translated as *Scientific Instruments of the Seventeenth and Eighteenth Centuries* (Praeger, 1953). The precision of the sector was a key issue, for example, in the debate between Cassini and Maupertuis; Mary Terrall, "Representing the Earth's Shape," *Isis* 83, 1992, 218–237.

22. Poirier, *Lavoisier*, 7.

23. Lavoisier, "Sur la manière d'enseigner la chimie (1792)" [Dossier Lavoisier Ms. 1259, archive of Académie des Sciences]; printed in Bernadette Bensaude-Vincent, "A view of the chemical revolution through contemporary textbooks: Lavoisier, Fourcroy and Chaptal," *British Journal for the History of Science* 23, 1990, 435–460, at 456–460.

24. Guerlac concluded thus because it seemed implausible that Lavoisier managed his law study, a variety of science lectures and the curriculum in philosophy at the Collège Mazarin all at the same time. Beretta finds it inconclusive since Lavoisier states in the above quote that he had a "bon cours de philosophie"; Guerlac, "A Note," 213 and Marco Beretta, *A New Course in Chemistry* (Firenze: L. S. Olschski, 1994), 14–15.

25. Rhoda Rappaport, "Lavoisier's Geologic Activities, 1763–1792," *Isis* 58, 1967, 375–384, at 376.

26. A. N. Meldrum, "Lavoisier's Early Work in Science, 1763–1771," *Isis* 19, 1933, 330–363; 20, 1934, 396–425. On Lavoisier's sustained interest in geology, see Rhoda Rappaport, Guettard, Lavoisier, and Monnet: Geologists in the Service of the French Monarchy, Ph.D. dissertation, Cornell University, 1964.

27. Based on Lavoisier's own statement, Perrin has suggested that Lavoisier was introduced to chemistry by La Planche at the Collège Mazarin, but Donovan thinks it more likely that he took the course at the Society of Pharmacists; C. E. Perrin, "The Lavoisier-Bucquet Collaboration: A Conjecture," *Ambix* 36, 1989, 5–40; Donovan, *Antoine Lavoisier*, 31.

28. Guerlac, "A Note," 215; Perrin, "The Lavoisier-Bucquet Collaboration."

29. For a detailed reconstruction of Lavoisier's activities in natural history during his formative years, see Palmer, "The Early Scientific Work."

30. Poirier, *Lavoisier*, 17–21; R. Rappaport, "Lavoisier's Theory of the Earth," *British Journal for the History of Science* 6, 1973, 247–260, at 252.

31. Theodore S. Feldman, "Late Enlightenment Meteorology," in *The Quantifying Spirit*, ed. T. Frängsmyr et al.; J. L. Heilbron, "The Measure of Enlightenment," Ibid., 207–242; Ken Alder, "A Revolution to Measure."

32. Donovan, *Antoine Lavoisier*, 11–24.

33. Poirier, *Lavoisier*, 13.

34. Jürgen Habermas, *The Structural Transformation of the Public Sphere* (MIT Press, 1992), 12–14.

35. Poirier, *Lavoisier*, 36–44.

36. Lavoisier, "Sur le gypse, deuxième mémoire," *Oeuvres*, volume 3, 128.

37. Meldrum, "Lavoisier's Early Works," 19, 1933, 337–340.

38. *L'Avant-Coureur*, April 21, 1766, 248; W. A. Smeaton, "L'avant-coureur. The journal in which some of Lavoisier's earliest research was reported," *Annals of Science* 13, 1957, 219–234, at 223.

39. Historians have long thought that Lavoisier drafted at this juncture two letters addressed to president and permanent secretary of the Academy, urging them to create a new section of experimental physics. These letters were unsigned, but included in the Lavoisier Dossier and printed in *Correspondances*, volume 1, 7–12. Lavoisier's authorship of these letters have been persuasively refuted by Eric Brian, "Lavoisier et le projet de classe de physique expérimen-

tale à l'Académie royale des sciences (avril 1766)," in Christiane Demeulenaere-Couyère, ed., *Il y a 200 ans Lavoisier* (Lavoisier Tec & Doc, 1995).

40. Donovan, *Antoine Lavoisier*, 40–41; Donovan also says that Lavoisier took election for granted, citing *Corresondance* 1, 104. On the practice of pre-election manuevering in the Academy, R. Rappaport, "The liberties of the Paris Academy of Sciences, 1716–1785," in *The Analytic Spirit*, ed. H. Woolf (Cornell University Press, 1981).

41. *Oeuvres de Lavoisier*, volume 3, 145–205 and 427–450.

42. Joseph Fayet, *La Révolution française et la science, 1789–1795* (Rivière, 1960); Roger Hahn, *The Anatomy of a Scientific Institution* (University of California Press, 1971).

43. It took 10 years of apprenticeship to open an apothecary's shop in Paris. Lavoisier's exposure to chemistry would have come to about a year of daily lecturing all together.

44. Lavoisier, "Sur la manière d'enseigner la chimie," quoted in Poirier, *Lavoisier*, 6.

45. Ibid., quoted in Poirier, *Lavoisier*, 11.

46. Lavoisier, "Commencement d'un traité de Chimie" [Lavoisier Dossier 380, archive of Académie des Sciences]; transcribed with an introduction in Marco Beretta, *A New Course in Chemistry*. The manuscript is "written in an unknown hand, but with several corrections and additions in the hand of Lavoisier." Having failed to trace the handwriting to any other French scientist, Beretta has concluded that it was dictated by Lavoisier, definitely after 1763 and probably before 1766.

47. Lavoisier, "Sur la manière d'enseigner la chimie," in Bernadette, "A View," 457.

48. La Planche, "Plan d'un cours de chymie, suivant les principes de Becher, de Boerhave, & de Stahl" [*Bibliothèque nationale*, R 5805]; transcribed in Marco Beretta, *A New Course*, 79–91.

49. Rouelle, "Cours de chimie" [*Museum national d'histoire naturelle*, Ms. 1202]; Macquer and Baumé, *Plan d'un cours de chymie expérimentale et raisonée avec un discours historique sur la chymie* (Paris: Jean-Thomas Herissant, 1757).

50. Pieter van Musschenbroek, *Essai de physique* (Leyden, 1739); Jean Antoine Nollet, *Leçons de physique expérimentale* (Paris: Guerin, 1743–1748). These were major texts in experimental physics at the time; John Heilbron, *Elements of Early Modern Physics* (University of California Press, 1982), 6–7.

51. Beretta, *A New Course*, 66.

52. Marco Beretta, "Lavoisier as a Reader of Chemical Literature," *Revue d'histoire des sciences et leurs applications* 48, 1995, 71–94, at 74.

53. Macquer and Baumé, *Plan d'un cours*, 7–8; Beretta, *A New Course*, 23.

54. Lavoisier, "Sur la manière d'enseigner la chimie," in Bernadette, "A View," 457.

55. Lavoisier, *Elements of Chemistry* (Dover, 1965), xix.

56. Lavoisier, "Analyse du gypse," *Oeuvres*, volume 3, 111–127; "Sur le gypse, deuxième mémoire," ibid., 128–144. The second part was published for the first time in this collection. Readers interested in this first phase of Lavoisier's chemical experiments should consult Palmer, "The Early Scientific Work of Antoine Laurent Lavoisier," 78–205.

57. Theodore Porter, "The Promotion of Mining and the Advancement of Science: The Chemical Revolution of Mineralogy," *Annals of Science* 38, 1981, 543–570; Rachel Laudan, *From Mineralogy to Geology* (University of Chicago Press, 1987); Evan M. Melhado, "Mineralogy and the Autonomy of Chemistry around 1800," *Lychnos*, 1990, 229–261.

58. Palmer, "The Early Scientific Work," 78–119.

59. Lavoisier, "Analyse du gypse," 112.

60. Beretta, *A New Course*, 21.

61. Lavoisier, "Sur la manière d'enseigner la chimie," quoted in Poirier, 6 (emphasis added).

62. Lavoisier, "Analyse du gypse," 118.

63. Ibid., 115, 117.

64. Ibid., 124.

65. J. B. Gough, "Lavoisier's Early Career in science: An Examination of Some New Evidence," *British Journal for the History of Science* 4, 1968, 52–57; Robert Siegfried, "Lavoisier's View of the Gaseous State and Its Early Application to Pneumatic Chemistry," *Isis* 63, 1972, 59–78, at 60–61.

66. Eller, "Dissertation sur les elemens ou premiers principes des corps. Dans laquelle on prouve qu'il doit y avoir des elemens et qu'il y en a effectivement; qu'ils sont sujets à souffrir divers changemens, et meme susceptibles d'une parfaite transmutation; et enfin que le feu elementaire et l'eau sont les seules choses qui meritent proprement le non d'elemens," *Histoire de l'Académie royale des sciences et des belles-lettres de Berlin* 2, 1746, 3–48.

67. Ibid., 40–41.

68. Eller gave a much more extended version of this argument in a later paper: "Sur la nature et les propriétés de l'eau commune considerée comme un dissolvant," *Histoire de l'Académie royale des sciences et des belles-lettres de Berlin* 6, 1750, 67–97.

69. Siegfried, "Lavoisier's View," 63–64.

70. For full translations of Lavoisier's two notes, see Siegfried, "Lavoisier's View," 62.

71. Siegfried, "Lavoisier's View," 63.

72. J. B. Gough, "Lavoisier's Memoires on the Nature of Water and Their Place in the Chemical Revolution," *Ambix* 30, 1983, 89–106, n. 69.

73. See the first extracted quotation in section 6.5.

74. Donovan, *Antoine Lavoisier*, 40–44.

75. Jean Baptiste Le Roy (1719–1800), a professor of medicine at Montpellier, contested in 1767 the conventional method of analyzing mineral waters; Meldrum, "Lavoisier's Early Work," 345–349.

76. On the importance of hydrometry in Lavoisier's early work, Meldrum, "Lavoisier's Early Work," 421; Bensaude-Vincent, "The Chemist's Balance."

77. Lavoisier, "Recherches sur les moyens les plus sûrs, les plus exacts et les plus commodes de déterminer la pesanteur spécifique des fluides soit pour la physique soit pour le commerce," *Oeuvres*, volume 3, 427–450.

78. Ibid., 449–450; Siegfried, "Lavoisier's View," 65–66.

79. Lavoisier, "De la nature des eaux d'une partie de la Franche-Comité de l'Alsace, de la Lorraine, de la Champagne, de la Brie et du Valois," *Oeuvres*, volume 3, 145–205; Guerlac, *Antoine-Laurent Lavoisier*, 62–63.

80. Lavoisier, "De la nature des eaux," 161–162.

81. Poirier, *Lavoisier*, 34–37.

82. Guerlac, "Chemistry as a Branch," 195–196. Brisson later published *Pesanteur spécifique des corps* (Paris: Imprimerie royale, 1787).

83. Meldrum, "Lavoisier's Work on the Nature of Water and the Supposed Transmutation of Water into Earth (1768–1773)," *Archeion* 14, 1932, 246–247; idem, "Lavoisier's Early Works in Science," *Isis* 20, 1934, 396–425; Louis Dulieu, "Un Parisien, professeur à l'Université de Médecine de Montpellier: Charles Le Roy (1726–1779)," *Revue d'histoire des sciences et leurs applications* 6, 1953, 50–59.

84. On the complicated publication history of these two papers, see J. B. Gough, "Lavoisier's Memoirs on the Nature of Water and Their Place in the Chemical Revolution," *Ambix* 30, 1983, 89–106.

85. Lavoisier, "Sur la nature de l'eau et sur les expériences par lesquelles on a prétendu prouver la possibilité de son changement en terre. Second Mémoire," *Oeuvres*, volume 2, 11–28.

86. Guerlac, *Antione-Laurent Lavoisier*, 68–69.

87. M. Berthelot, *La Révolution chimique Lavoisier* (Paris: Germer-Baillière, 1890), 45.

88. Gough, "Lavoisier's Memoires," 97.

89. First manuscript version of Lavoisier's second memoir on water, dated by de Fouchy May 10, 1769 [Lavoisier Dossier 1299, archive of Académie des Sciences]; quoted in ibid., 90.

90. It was included in the memoires of the Academy for the year 1770 as the "first memoir," along with the "second memoir" containing the experimental

part. "Sur la nature de l'eau et sur les expériences par lesquelles on a prétendu prouver la possibilité de son changement en terre. Première memoire," *Oeuvres*, volume 2, 1–11.

91. For a description of Rouelle's apparatus, see J. B. Bucquet, *Introduction a l'étude des corps naturels, tirés du règne minéral* (Paris: Jean-Th. Herrissant, 1771), 62.

92. Lavoisier, "Sur la nature de l'eau," 7.

93. Meldrum, "Lavoisier's Early Work," 420.

94. Grimaux, *Lavoisier*, 34.

95. As is well known, Guerlac has construed the year 1772 as the "crucial year" in Lavoisier's evolution as a chemist and has provided a detailed chronology of his investigations; Guerlac, *Lavoisier—The Crucial Year* (Cornell University Press, 1961). Guerlac's account should be supplemented by the works of C. E. Perrin and F. L. Holmes.

96. This is in fact what Lavoisier said in his letter to Guyton accompanying his book *Opuscules*; *Correspondance*, volume 2, 404–406.

97. *PV* 87, 1768, 90–96 (May 18).

98. "Résultat de quelques expériences faites sur le Diamant, par MM. Macquer, Cadet & Lavoisier, lu à la Séance publique de l'Académie Royale des Sciences, 29 Avril, 1772," *Introduction aux observations*, volume 2, 108–111; reprinted in Guerlac, *Lavoisier*, 199–204.

99. Cadet, "Expériences et Observations Chymiques sur le Diamant" and Mittouart, "Extrait de deux Mémoires sur les Diamans & autres Pierres Précieuses, lûs à l'Académie Royale des Sciences les 2 & 7 Mai 1772" [*Ecole de médecine*, cote 90958, t. 169, no. 4, 3–15 and 26–31].

100. D'Arcet and Rouelle, "Expériences nouvelles sur la destruction du Diamant, dans les vaisseaux fermés," *Journal de médecine, chirurgie, pharmacie, &c.* 39, 1773, 50–86.

101. Guerlac could not determine if Lavoisier read the *Vegetable Staticks* before 1772, but argued that Lavoisier should have been familiar with Hales's experiments reproduced in Rouelle's lectures; Guerlac, "Continental Reputation of Stephen Hales," 25–35. On Lavoisier's apparatus, see F. L. Holmes, "Lavoisier the Experimentalist," *Bulletin for the History of Chemistry* 5, 1989, 24–31; idem, "The Evolution of Lavoisier's Chemical Apparatus."

102. "Reflexions sur les experiences qu'on peut tenter a l'aide du miroir ardent," in Guerlac, *Lavoisier*, 208–214.

103. "Système sur les elémens," in Guerlac, *Lavoisier*, 215–223; reprinted from René Fric, "Contribution à l'étude de l'évolution des idées de Lavoisier sur la nature de l'air et sur la calcination des métaux," *Archives internationale d'histoire des sciences* 47, 1959, 137–168.

104. According to Kohler's interpretation, Lavoisier realized that Priestley had not yet exhausted the subject and thought that he could extract fixed air from

minerals with the burning glass; Robert E. Kohler, "The Origin of Lavoisier's First Experiments on Combustion," *Isis* 63, 1972, 349–355, at 350.

105. Guerlac, *Lavoisier*, 92–97; idem, "A Lost Memoir of Lavoisier," *Isis* 50, 1959, 135–139; idem, "Lavoisier's Draft Memoir of July 1772," ibid., 380–382; Robert J. Morris, "Lavoisier on Fire and Air: The Memoir of July 1772," *Isis* 60, 1969, 374–380.

106. Guerlac has argued based on the August memorandum that Lavoisier did not agree with the committee's earlier conclusion and planned to test the volatilization and decrepitation hypotheses with the burning glass. He used this evidence to argue that combustion was not the route that took Lavoisier to the *pli cacheté* of November 1772. Lavoisier did ask in the memorandum, however, what produced the differences observed by reason of the "intermediates" in which the diamonds were placed. This was the line of questioning Macquer and Mitouard took to confirm their combustion hypothesis. Besides, whether Lavoisier favored the combustion hypothesis is less important, I think, than the fact that he learned what combustion entailed; Guerlac, *Lavoisier*, 87–89; Siegfried, "Lavoisier's View," 69.

107. Lavoisier dossier 14, archive of Académie des Sciences; transcribed in Guerlac, *Lavoisier*, 144. The date of this note is uncertain, although Guerlac thinks it likely that it was written during the summer of 1772. Macquer presented a copy of Guyton's *Digressions* to the Academy on June 3; Guerlac, *Lavoisier*, 138.

108. "Experiences sur le phosphore du 10 7bre 1772"; Guerlac, *Lavoisier*, 223–224.

109. "Memoire sur l'acide du Phosphore et sur ses combinaisons avec differentes substances salines terreuses et metallique (draft memoire of October 20, 1772)" [Lavoisier dossier 1308D]; printed in Guerlac *Lavoisier*, 224–227.

110. Guerlac, *Lavoisier*, 227–228; translated in Siegfried, "Lavoisier's View," 71.

111. Andrew Norman Meldrum, *The Eighteenth Century Revolution in Science—The First Phase* (Longmans, Green, 1930); Guerlac, *The Crucial Year*; Gough, "The Origin of Lavoisier's Theory of the Gaseous State," in *Analytic Spirit*, ed. Woolf; Robert J. Morris, "Lavoisier on Fire and Air: The Memoir of July 1772," *Isis* 60, 1969, 374–380; "Lavoisier and the Caloric Theory," *British Journal for the History of Science* 6, 1972, 1–38; Maurice Crosland, "Lavoisier's Theory of Acidity," *Isis* 64, 1973, 306–325; Robert Kohler, "The Origin of Lavoisier's First Experiment on Combustion," *Isis* 63, 1972, 349–355.

112. One can see this in Lavoisier's experiments during the "next crucial year." He started out with a range of experiments that would confirm his theory and shifted constantly between them, rarely staying on the same experiment until he achieved success; F. L. Holmes, *The Next Crucial Year*.

113. C. E. Perrin, "Lavoisier's Thoughts on Calcination and Combustion, 1772–1773," *Isis* 77, 1986, 647–666.

114. Most of these works were published in *Observations* in 1773.

115. M. Berthelot, *La Révolution chimique*, 46–49; English translation in Meldrum, *The Eighteenth Century Revolution*, 8–10.

116. For a day-by-day account of Lavoisier's experiments between February and August of 1773, see Holmes, *The Next Crucial Year*.

117. Lavoisier, "Sur une nouvelle Theorie de la Calcination et de la Reduction des substances metalliques sur la cause de laugmentation de poids quelles acquierent au feu et sur differens phenomenes qui appartiennent a l'air fixe" [Fonds Lavoisier, 1303]; printed in Fric, "Contribution," 155–162. See also Holmes's discussion in *The Next Crucial year*, 30–40.

118. Holmes, *The Next Crucial Year*, 37.

119. Guyton, "Dissertation," 179–189 and 194–208.

120. Fric, "Contribution," 160–162.

121. Holmes, *The Next Crucual Year*, 86–94.

122. Lavoisier, "Essay sur la nature de l'air," printed in Fric, "Contribution," 147–151.

123. Greenway, Introduction to Lavoisier, *Essays physical and chemical* (London, 1776; second edition Cass, 1970), xxi.

124. The translator, Thomas Henry, notes that Lavoisier undertook to decide the "controversy" between Black's fixed air and Meyer's acidum pingue; ibid., xii. Bucquet discussed the rival views of Meyer and MacBride in his *Introduction à l'étude des corps naturels tirés du règne minéral*.

125. H. E. Le Grand, "A Note on Fixed Air: The Universal Acid," *Ambix* 10, 1973, 88–94.

126. Holmes, *The Next Crucial Year*, 71–75.

127. Lavoisier, *Essays*, 204.

128. Ibid., 213.

129. Ibid., 120–121.

130. Ibid., 215.

131. Ibid., 216–217.

132. Ibid., 158.

133. Ibid., 48.

134. Ibid., 30–31.

135. Ibid., 72–75.

136. Ibid., 265.

137. Ibid., 324–326 (emphasis added).

138. "Projet d'un memoire sur les differens degrés d'affinité des acides en exces avec les differentes substances (September 3, 1766)" [Lavoisier Dossier 323, archive of Académie des Sciences]. I thank Dr. Marco Beretta for a partially transcribed copy of this manuscript.

139. Ibid., 383.

140. *Correspondance de Lavoisier*, volume 2, 398–439.

141. W. C. Ahlers, "P. J. Macquer et le rapport sur les *Opuscules physiques et chimiques* de Lavoisier," *Actes de XIIe congrès international d'histoire des sciences* (Paris, 1968), 5–9; idem, Un chimiste du XVIIIe siècle: Pierre-Joseph Macquer (1718–1784), doctoral thesis, Ecole Pratique des Hautes Etudes, Paris, 1969, 138–151; C. E. Perrin, "Did Lavoisier Report to the Academy of Sciences on His Own Book?" *Isis* 75, 1984, 343–348.

142. Lavoisier, *Essays*, xxi–xxii.

143. Sidney J. French, "The Chemical Revolution—The Second Phase," *Journal of Chemical Education* 27, 1950, 83–89; Guerlac, *Antoine-Laurent Lavoisier*; Perrin, "Prelude to Lavoisier's Theory of Calcination: Some Observations on *mercurius calcinatus per se*," *Ambix* 16, 1969, 140–151; Holmes, *Lavoisier*, 41–48; Historians differ a great deal in judging the sequence of events here, but Holmes synthesizes various accounts convincingly.

144. Pierre Bayen, "Essai d'expériences chymiques, faites sur quelques précipités de mercure, dans la vue de découvrir leur nature," *Observations* 3, 1774, 280–295.

145. "Discours sur le phlogistique," *Observations* 3, 1774, 185–200.

146. Perrin, "Prelude to Lavoisier's Theory of Calcination: Some Observations on *Mercurius calcinatus per se*," *Ambix* 16, 1969, 140–151, at 144–145; William M. Sudduth, "Eighteenth-Century Identifications of Electricity with Phlogiston," *Ambix* 25, 1978, 131–147.

147. Holmes, *Lavoisier*, 43–44.

148. Ibid., 27–28.

149. Kohler and Crosland interpretes from this delay that Priestley's visit turned Lavoisier's attention not to the air produced from mercury calx, but to acid airs in general; Crosland, "Lavoisier's Theory of Acidity"; Robert E. Kohler, "Lavoisier's Rediscovery of the Air from Mercury Calx: A Reinterpretation," *Ambix* 22, 1975, 52–57.

150. Holmes, *Lavoisier*, 28.

151. Lavoisier, "Mémoire sur la nature du principe qui se combine avec les Métaux pendant leur calcination, & qui en augmente le poids," *Observations* 5, 1775, 429–433; translated in *Essays* (p. 408); published in a revised form, *Mémoires*, 1775 (1778), 520–526; *Oeuvres*, volume 2, 122–128.

152. Guerlac, *Antoine-Laurent Lavoisier*, 85–86.

153. Holmes, *Lavoisier*, 53–56.

154. Crosland, "Lavoisier's Theory of Acidity," 314.

155. Lavoisier, "Mémoire sur l'existence de l'air dans l'acide nitreux, et sur les moyens de décomposer & de recomposer cet acide" *Mémoires*, 1776 (pub. 1779), 671–680; *Oeuvres*, volume 2, 129–138; Guerlac, *Antoine-Laurent Lavoisier*, 91–92.

156. Bucquet's lectures were published as *Introduction à l'étude des corps naturels tirés du règne minéral* (Paris: Jean-Th. Herrissant, 1771) and *Introduction à l'étude des corps naturels tirés du règne végétal* (Paris, 1773). For an announcement of his public lectures on "analytic and medicinal chemistry" at the Faculté, see *Journal de medicine* 47, 1777, 276.

157. E. McDonald, "The Collaboration of Bucquet and Lavoisier," *Ambix* 13, 1966, 74–83; C. E. Perrin, "The Lavoisier-Bucquet Collaboration: A Conjecture," *Ambix* 36, 1989, 5–40.

158. *PV* 96, 1777, 527.

159. Guerlac, "Chemistry as a Branch of Physics," 196–197.

160. Lavoisier, "Mémoire sur la combustion en général," *Oeuvres*, volume 2, 225–233.

161. Ibid., 228.

162. Ibid., 230–231.

163. Guerlac, *Lavoisier*, 221–222.

164. Ibid., 222.

165. Lavoisier, "Sur la combustion du phosphore de Kunckel, et sur la nature de l'acide qui résulte de cette combustion," *Mémoires*, 1777, 65–78; *Oeuvres*, volume 2, 139–152 (presented to the Academy on April 16).

166. Bergman, "Extrait des recherches sur les attractions electives," *Observations*, 13, 1778, 298–333.

167. Crosland, "Lavoisier's Theory of Acidity," 315–316.

168. Lavoisier, "Considérations générales sur la nature des acides et sur les principes dont ils sont composés (read to the Academy, November 23, 1779)," *Oeuvres*, volume 2, 248–260, at 250.

169. An exquisite model for this type of study is Martin J. S. Rudwick, *The Great Devonian Controversy* (University of Chicago Press, 1985).

170. Lavoisier sent a copy of his *Opuscules* to Guyton on January 19, 1774; *Correspondance*, volume 2, 404.

171. Smeaton, "Guyton de Morveau and the Phlogiston Theory," in *Mélanges Alexandre Koyré*, ed. I Cohen and R. Taton (Paris: Herman, 1964), 531–532.

172. *Supplément à l'Encyclopédie* (Amsterdam, 1776–77), volume 1, 234–235.

173. Ibid., volume 2, 114.

174. Ibid., volume 4, 338–340.

175. Ibid., 339.

176. Guyton to Macquer, May 24, 1774, in "Correspondance de Macquer," volume 2 [BN f fr 12306], 134–135.

177. Guyton, "Conciliation des Principes de Sthaal avec les Expériences modernes sur l'Air fixe," *Observations* 8, 1776, 389–395. For a detailed analysis, see Evan M. Melhado, *Jacob Berzelius* (University of Wisconsin Press, 1981), 84–90.

178. Guyton, "Conciliation," 390.

179. Ibid., 395.

180. Denis I. Duveen and Herbert S. Klickstein, "A Letter from Guyton de Morveau to Macquart Relating to Lavoisier's Attack Against the *Phlogiston* Theory (1778)," *Osiris* 12, 1956, 342–367. It seems more than likely that this letter, dated on January 22, was addressed to Macquer, who had written to Guyton on January 15. It was addressed to "M. Macquart" of the Academy, but Macquart was not a member.

181. Frederic L. Holmes, "The Boundaries of Lavoisier's Chemical Revolution," in *Lavoisier et la Révolution chimique*, ed. Goupil.

182. A. F. de Fourcroy, *Leçons élémentaires d'Histoire naturelle et de Chimie* (Paris, 1782), i–ii and xxiii.

183. Ibid., 21–22.

Chapter 7

1. On the relationship between Lavoisier and the other members, see Michelle Goupil, ed., *Lavoisier et la Révolution chimique* (SABIX–Ecole polytechnique, 1992), especially Maurice Crosland, "Lavoisier, Lone Genius or 'Chef d'École'? The Testimony of Fourcroy."

2. On the notion of "opinion" during the eighteenth century, Keith Michael Baker, *Inventing the French Revolution* (Cambridge University Press, 1990), 167–199. Lavoisier regarded the Academy as the "only court" before which he should appear; Jean-Pierre Poirier, *Lavoisier* (University of Pennsylvania Press, 1996), 186.

3. Lavoisier, *Elements of Chemistry* (Dover, 1965), xxxiii–xxxiv (emphasis added).

4. Guyton stayed in Paris for about 8 months beginning in late January or early February of 1787. His "conversion" has been dated to between mid March and mid April; D. I. Duveen and H. S. Klickstein, "A Letter from Guyton de Morveau to Macquart relating to Lavoisier's attack against the *phlogiston* theory (1778); with an account of de Morveau's conversion to Lavoisier's doctrines in 1787," *Osiris* 12, 1956, 342–367.

5. C. E. Perrin, "The Triumph of the Antiphlogistians," in *The Analytic Spirit*, ed. Woolf.

6. B. Bensaude-Vincent and F. Abbri, eds., *Lavoisier in European Context* (Science History Publications, 1995).

7. Some historians have recently reconsidered the antiphlogistic interpretation of the Chemical Revolution seriously, but this is due mostly to the singular focus on Lavoisier and other philosophical chemists; Robert Siegfried, "Lavoisier and the Phlogistic Connection," *Ambix* 36, 1989, 31–40; F. L. Holmes, "The 'Revolution in Chemistry and Physics': Overthrow of a Reigning Paradigm or

Competition between Contemporary Research Programs?" *Isis* 91, 2000, 735–753.

8. Another significant genre of experiments on respiration is dealt with in F. L. Holmes, *Lavoisier and the Chemistry of Life* (University of Wisconsin Press, 1985).

9. Historians have paid due attention to Lavoisier's collaboration with Laplace; Henry Guerlac, "Chemistry as a Branch of Physics: Laplace's Collaboration with Laviosier," *Historical Studies in the Physical Sciences* 7, 1976, 193–276; Luis M. R. Saraiva, "Laplace, Lavoisier and the Quantification of Heat," *Physis* 34, 1997, 99–137; Charles Coulston Gillispie, *Pierre-Simon Laplace, 1749–1827* (Princeton University Press, 1997), 101–108. Lavoisier's work on metallic precipitation has been largely ignored. (For an exception, see Maurice Daumas, "Les conceptions de Lavoisier sur les affinités chimiques et la constitutoin de la matière," *Thalès*, 1949–50, 69–80.)

10. Lavoisier, *Essays Physical and Chemical*, second edition (Cass, 1970), xix.

11. *Supplément à l'Encyclopédie* (Amsterdam, 1776–77), volume 2, 114.

12. *Oeuvres*, volume 2, 509–527.

13. Ibid., 511–512.

14. For a detailed exposition, see Daumas, "Les conceptions de Lavoisier sur les Affinités chimiques."

15. Lavoisier, "Considération générales sur la dissolution des métaux dans les acides," *Oeuvres*, volume 2, 509–527, at 521–522.

16. Ibid., 525.

17. Lavoisier, "Mémoire sur la précipitation des substances métalliques les unes par les autres," *Oeuvres*, volume 2, 528–545.

18. Lavoisier, "Mémoire sur l'affinité du principe oxygine avec les différentes substances auxquelles il est susceptible de s'unir," *Oeuvres*, volume 2, 546–556, at 546 (emphasis added). It was deposited at the Academy on Deccember 20, 1783, along with the "Sur les dissolutions métalliques," but read on March 2 and 5, 1785.

19. Ibid., 550–551.

20. Guerlac, "Chemistry as a Branch of Physics," 225–234.

21. Magellan, "Essai sur la nouvelle théorie du feu élémentaire & de chaleur des corps," *Observations*, 17, 1781, 375–386; "Suite du mémoire de M. H. Magellan sur le feu élémentaire et la chaleur: Sommaire de l'ouvrage du docteur Crawford," ibid., 411–422. These were preceded by a prior announcement in *Observations*, 16, 1780, 62–63.

22. Crawford was a founding member of the Chapter Coffee House, along with Kirwan; Diane Elizabeth Heggarty, Richard Kirwan: The Natural Philosopher and the Chemical Revolution, Ph.D. dissertation, University of Washington, 1978; E. L. Scott, The Life and Work of Richard Kirwan (1735–1812), Ph.D.

dissertation, University of London, 1979; T. H. Levere and G. L'E Turner, *Discussing Chemistry and Steam* (Oxford University Press, 2002).

23. Lavoisier and Laplace, "Mémoire sur la chaleur (1783)," *Oeuvres*, volume 2, 283–333; translated by Henry Guerlac as *Memoir on Heat* (Neale Watson, 1982).

24. Jan Golinski, "'Fit Instruments': Thermometers in Eighteenth-Century Chemistry," in *Instruments and Experimentation in the History of Chemistry*, ed. F. Holmes and T. Levere (MIT Press, 2000), 201–206.

25. Robert J. Morris, "Lavoisier and the Caloric Theory," *British Journal for the History of Science* 6, 1972, 1–38, at 14–15.

26. *Memoir on Heat*, 28.

27. Lissa Roberts, "A Word and the World: The Significance of Naming the Calorimeter," *Isis* 82, 1991, 199–222.

28. Guerlac ("Chemistry as a Branch of Physics," 243–244) thinks that the first section is strongly influenced by Laplace, the second could have been done by either, the third is almost exclusively Laplace's, and the fourth is by Lavoisier.

29. *Memoir on Heat*, 4.

30. Ibid., Guerlac traced Lavoisier and Laplace's knowledge of this theory to J. A. Deluc, *Recherches sur les modifications de l'atmosphère* (1772). De Luc was known personally to Lavoisier since 1768 as a corresponding associate member of the Academy. He visited Paris in October 1781 and was in touch with Guettard, Lavoisier's old mentor; Guerlac, "Chemistry as a Branch of Physics," 246–249.

31. *Memoir on Heat*, 5.

32. Ibid., 6.

33. Ibid., 7–8.

34. Guerlac, "Chemistry as a Branch of Physics," 250–251.

35. *Memoir on Heat*, 15.

36. Ibid., 16.

37. A thermometer marked off in divisions proportional to the quantities of heat contained in the fluid which is used, which would allow the measurement of all possible degrees of temperature; ibid., 20.

38. Ibid., 22.

39. Ibid., 25.

40. T. H. Lodwig and W. A. Smeaton, "The Ice Calorimeter of Lavoisier and Laplace and Some of Its Critics," *Annals of Science* 31, 1974, 1–18.

41. For a symmetric account of mediations between chemistry and physics via the calorimeter, see M. Norton Wise, "Mediations: Enlightenment Balancing Acts, or the Technologies of Rationalism," in *World Changes*, ed. P. Horwich (MIT Press, 1993).

42. W. A. Smeaton, *Fourcroy* (Heffer, 1962).

43. J. B. M. Bucquet, *Mémoires sur la manière dont les animaux sont affectés par différens fluides aériformes, méphitiques; & sur les moyens de remédier aux effets de ces fluides. Précédé d'une histoire abrégée des différens fluides aériformes ou gas* (Paris: Imprimerie Royale, 1778).

44. This may have been due to his involvement in the Société royale de Médecine, which drew much opposition from the Faculté. The Société's membership included Vicq d'Azyr (permanent secretary), Bucquet, and Lavoisier (W. A. Smeaton, "Lavoisier's membership on the Société Royale de Médecine," *Annals of Science* 12, 1957, 228–244).

45. Fourcroy, *Leçons élémentaires d'histoire naturelle et de chimie* (Paris, 1782).

46. "Eloge" by Georges Cuvier, *Recueil des éloges historiques lus dans les séances publiques de l'Institut royal de France* (Strasbourg and Paris, 1819), volume 2, 17–18; quoted in Poirier, *Lavoisier*, 186.

47. Fourcroy, *Leçons*, 1.

48. Ibid., iv.

49. Ibid., xxxix.

50. Ibid., lxiii.

51. Ibid., liv.

52. Ibid., lv.

53. Ibid., lxv.

54. Ibid., 34.

55. Ibid., 37–38.

56. Ibid., 35.

57. Bernadette Bensaude-Vincent, "A View of the Chemical Revolution through Contemporary Textbooks: Lavoisier, Fourcroy, and Chaptal," *British Journal for the History of Science* 23, 1990, 435–460.

58. *Encyclopédie méthodique, chimie, pharmacie, et métallurgie*, volume 1 (Paris, 1786); The first half appeared in 1786. The second half, containing the article on air, appeared in 1789 with a new preface that made public Guyton's endorsement of Lavoisier's theory.

59. "Acide," ibid., 27–418, at 27–29.

60. "Adhérence, Adhésion," ibid., 466–490, at 466.

61. "Affinité," ibid., 535–613.

62. Ibid., 537.

63. Ibid., 541.

64. Ibid., 545.

65. Ibid., 550.

66. Ibid., 551.

67. Ibid., 552.

68. Ibid., 553–554.

69. Ibid., 555–557.

70. Ibid., 560.

71. Ibid., 565.

72. Ibid., 567–576.

73. Ibid., 577–578.

74. Ibid., 578–579.

75. Ibid., 583.

76. Ibid., 608.

77. On rhetorical moment, see Mi Gyung Kim, "Constructing Symbolic Spaces: Chemical Molecules in the Académie des sciences," *Ambix* 43, 1996, 1–31.

78. Guerlac, *Antoine-Laurent Lavoisier*, 92 and 265.

79. Seymour H. Mauskopf, "'I passionately desire to be able to link together all, or at least most, of chemical phenomena in a system.' Richard Kirwan's Phlogiston Theory: Its Success and Fate," presentation at Dexter Symposium, Chicago, 2000.

80. On the controversy over priority, see J. P. Muirhead, *Correspondance of the Late James Watt on His Discovery of the Theory of the Composition of Water* (London, 1846); Sidney M. Edelstein, "Priestley Settles the Water Controversy,' *Chymia* 1, 1948, 123–137; Robert E. Schofield, "Still More on the Water Controversy," *Chymia* 9, 1964, 71–76; W. A. Smeaton, "Is Water Converted into Air? Guyton de Morveau Acts as Arbiter between Priestley and Kirwan," *Ambix* 15, 1968, 73–83.

81. J. R. Partington, *A History of Chemistry* (Macmillan, 1961–1972), volume 3, 436–456; C. E. Perrin, "Lavoisier, Monge, and the Synthesis of Water: A Case of Pure Coincidence?" *British Journal for the History of Science* 6, 1973, 424–428.

82. "Extrait d'un mémoire lû par M. lavoisier à la séance publique de l'Académie royale des sciences, du 12 novembre, sur la nature de l'eau," *Observations* 23, 1783, 452–455.

83. Guerlac, *Antoine-Laurent Lavoisier*, 102.

84. Maurice Daumas and Denis Duveen, "Lavoisier's Relatively Unknown Large-Scale Decomposition and Synthesis of Water, February 27 and 28, 1785," *Chymia* 5, 1959, 113–129; Jan Golinski, "Precision Instruments and the Demonstrative Order of Proof in Lavoisier's Chemistry," *Osiris* 9, 1994, 30–47; idem, "'The Nicety of Experiment': Precision of Measurement and Precision of Reasoning in Late Eighteenth-Century Chemistry," in *The Values of Precision*, ed. Wise (Princeton University Press, 1995).

85. Berthollet to Guyton de Morveau, May 4, 1785 [Dossier Guyton de Morveau, #93, archive of Académie des Sciences]. As can be discerned from the

26 letters contained in this collection, Berthollet continued with the mission until Guyton's conversion in 1787, debating respective merits of phlogiston and oxygen theories in understanding the composition of acids.

86. Lavoisier, "Réflexions sur le phlogistique pour servir de développement à la théorie de la combustion & de la calcination, publiée en 1777," *Oeuvres*, volume 2, 623–655; A short description in Robert J. Morris, "Lavoisier and the Caloric Theory," *British Journal for the History of Science* 6, 1972, 1–38, at 16–21; Guerlac, *Lavoisier*, 258–261.

87. Lavoisier, "Réflexions sur le phlogistique," 629–630.

88. Ibid., 631–633.

89. Ibid., 640.

90. Ibid., 640–641.

91. Ibid., 641.

92. Ibid.

93. Ibid., 644.

94. W. A. Smeaton, "The Contributions of P.-J. Macquer, T. O. Bergman, and L. B. Guyton de Morveau to the Reform of Chemical Nomenclature," *Annals of Science* 10, 1954, 87–106. For a detailed history of the nomenclature reform and its reception, see Maurice P. Crosland, *Historical Studies in the Language of Chemistry* (Heinemann, 1962), 144–214.

95. Marco Beretta, "Torbern Bergman in France: An Unpublished Letter by Lavoisier to Guyton de Morveau," *Lychnos*, 1992, 167–170. The role mineralogy played in Bergman's communications with Kirwan, Guyton, and Macquer comes through clearly in his correspondence; see *Torbern Bergman's Foreign Correspondence*, ed. G, Carld and J, Nordström (Stockholm: Lychnos, 1965). Although Lavoisier spent a number of years in mineralogy and geology in his youth and maintained his interest long afterwards, he was more involved in instrumental measurements and in the theories of earth than in collecting specimens and classifying them; Rhoda Rappaport, Guettard, Lavoisier, and Monnet: Geologists in the Service of the French Monarchy, Ph.D. dissertation, Cornell University, 1964; Louis Yvonne Palmer, The Early Scientific Work of Antoine Laurent Laviosier, Ph.D. dissertation, Yale University, 1998.

96. Guyton de Morveau, "Mémoire sur les Terres simples, & principalement sur celles qu'on nomme absorbante; suivi d'un appendice sur une nouvelle preuve de l'existence du Phlogistique dans la chaux, & de quelques observations sur le Sel phosphorique calcaire ou substance osseuse régénérée," *Observations*, 17, 1781, 216–231, esp. 216–217.

97. Guyton de Morveau, "Mémoire sur les Dénominations Chymiques, la nécessité d'en perfectionner le systém; & les règles pour y parvenir," *Observations*, 19, 1782, 370–382, at 370–371 (emphasis added).

98. Poirier, *Lavoisier*, 182–190.

99. W. R. Albury, The Logic of Condillac and the Structure of French Chemical and Biological Theory, 1780–1801, Ph.D. dissertation, Johns Hopkins University, 1972; Lissa Roberts, "Condillac, Lavoisier, and the Instrumentalization of Science," *The Eighteenth Century: Theory and Interpretation* 33, 1992, 252–271.

100. Lavoisier, "Mémoire sur la nécessité de réformer & de perfectionner la nomenclature de la chimie, lu à l'assemblée publique de l'Académie royale des sciences du 18 avril 1787," in *Méthode de nomenclature chimique* (Seuil, 1994).

101. Guyton de Morveau, "Mémoire sur le développement des principes de la nomenclature méthodique, lu à l'Académie, le 2 mai 1787," ibid., 75–108; Fourcroy, "Mémoire pour servir à l'explication du tableau de nomenclature," ibid., 109–122.

102. Ibid., 75–76.

103. Richard Kirwan, *An Essay on Phlogiston and the Constitution of Acids* (London, 1787); translated with comments by Lavoisier, Guyton, Berthollet, and Fourcroy, *Essai sur le phlogistique et sur la constitution des acides* (Paris, 1788); second English edition with the translation of French comments, *An Essay on Phlogiston and the Constitution of Acids* (London, 1789); second edition (Cass, 1968). Keiko Kawashima, "Madame Lavoisier et la traduction française de l'Essay on Phlogiston de Kirwan," *Revue d'histoire des sciences et leurs applications* 53, 2000, 235–263.

104. The French translator's preface (supposedly by Mm. Lavoisier) mentions subtly that "the public opinion was not against them" (Kirwan, *An Essay on Phlogiston and the Constitution of Acids*, Cass, 1968, xvii).

105. Ibid., xvi.

106. Ibid., xiv–xv.

107. Ibid., 7.

108. Ibid., 5.

109. Ibid., 15 (emphasis added).

110. Ibid., 41.

111. Ibid., 52–53.

112. Ibid., 6–7.

113. Ibid., v–vi.

114. Bernadette Bensaude-Vincent, "A View of the Chemical Revolution through Contemporary Textbooks: Lavoisier, Fourcroy and Chaptal," *British Journal for the History of Science* 23, 1990, 435–460.

115. W. R. Albury, The Logic of Condillac and the Structure of French Chemical and Biological Theory, 1780–1801, Ph.D. dissertation, Johns Hopkins University, 1972.

116. Lavoisier, *Elements of Chemistry*, xix–xxiii.

117. Ibid., xxiv.

118. On other sources, see James W. Llana, "A Contribution of Natural History to the Chemical Revolution in France," *Ambix* 32, 1985, 71–91.

119. Ibid., xiv–xv.

120. Charles C. Gillispie, *The Edge of Objectivity* (Princeton University Press, 1960), 245.

121. The historical connection between the affinity tables and Lavoisier's table of simple substances has been pointed out by C. E. Perrin, "Lavoisier's Table of the Elements: A Reappraisal," *Ambix* 20, 1973, 95–105; Roberg Siegfried, "Lavoisier's Table of Simple Substances: Its Origin and Interpretation," *Ambix* 29, 1982, 29–48; Lissa Roberts, "Setting the Table: The Disciplinary Development of Eighteenth-Century Chemistry as Read through the Changing Structure of Its Tables," in *The Literary Structure of Scientific Argument*, ed. P. Dear (University of Pennsylvania Press, 1991); idem, "Filling the Space of Possibilities: Eighteenth-Century Chemistry's Transition from Art to Science," *Science in Context* 6, 1993, 511–553.

122. Lavoisier, *Elements of Chemistry*, xx–xxi.

123. Ibid., xxi.

124. Ibid., 171–172.

125. Ibid., xxi–xxii.

126. Ibid., 172.

127. Ibid., 57.

128. Ibid., 125, 141, and 241 respectively.

129. Ibid., 212–290.

130. Ibid., 78–79.

131. Ibid., 124.

132. Ibid., 4–5.

133. Ibid., 123.

134. De Morveau, "Second Avertissement pour servir d'introduction aux articles qui suivent, & pour indiquer quelques corrections à faire à ceux qui précèdent," *Encyclopédie méthodique, chimie*, volume 1, at 625.

135. "Cours de chimie expérimentale rangée suivant l'ordre naturel des idées" [Dossier Lavoisier 1260, archive of Académie des Sciences]; Maurice Daumas, "L'élaboration du Traité de Chimie de Lavoisier," *Archives internationale d'histoire des sciences* 12, 1950, 570–590.

136. Michelle Goupil, ed., *Lavoisier et la Révolution chimique*. See also Marco Beretta, "Lavoisier and His Last Printed Work: The *Mémoires de physique et de chimie* (1805)," *Annals of Science* 58, 2001, 327–356.

Chapter 8

1. H. E. Le Grand, "Genius and the Dogmatization of Error: The Failure of C. L. Berthollet's Attack upon Lavoisier's Acid Theory," *Organon* 12–13, 1976–77, 193–209; Robert Fox, *The Caloric Theory of Gases* (Clarendon, 1971).

2. Pere Grapi and Mercè Izquierdo, "Berthollet's Conception of Chemical Change in Context," *Ambix* 44, 1997, 113–130.

3. R. Siegfried and B. J. Dobbs, "Composition: A Neglected Aspect of the Chemical Revolution," *Annals of Science* 24, 1968, 275–293; Robert Siegfried, "The Chemical Revolution in the History of Chemistry," *Osiris* [2] 4, 1988, 34–50.

4. Pere Grapi, "The Marginalization of Berthollet's Chemical Affinities in the French Textbook Tradition at the Beginning of the Nineteenth Century," *Annals of Science* 58, 2001, 111–135.

5. Henry M. Leicester, "The Spread of the Theory of Lavoisier in Russia," *Chymia* 5, 1959, 138–144; Arthur Donovan, "Scottish Responses to the New Chemistry of Lavoisier," *Studies in Eighteenth-Century Culture* 9, 1979, 237–249; H. A. M. Snelders, "The New Chemistry in the Netherlands," *Osiris* [2] 4, 1988, 121–145; Anders Lundgren, "The New Chemistry in Sweden: The Debate That Wasn't," ibid., 146–168; Ramon Gago, "The New Chemistry in Spain," ibid., 169–192; Bernadette Bensaude-Vincent and Ferdinando Abbri, eds., *Lavoisier in European Context* (Science History Publications, 1995). On the contentious nature of such an authoritative claim over chemical language, see Jan Golinski, "The Chemical Revolution and the Politics of Language," *The Eighteenth Century: Theory and Interpretation* 33, 1992, 238–251.

6. Fourcroy, *Philosophie chimique* (Paris, 1792); *Séances des Écoles normales, recueillies par des sténographes, et revues par les professeurs* (Paris, 1796) [*Bibliothèque nationale*, R 22215–22220].

7. Michelle Goupil, "Claude Louis Berthollet, Collaborateur et Continuateur (?) de Lavoisier," in *Lavoisier et la Révolution chimique*, ed. Goupil.

8. Maurice Crosland, *The Society of Arcueil* (Harvard University Press, 1967); idem, *Gay-Lussac* (Cambridge University Press, 1978).

9. Kim, Practice and Representation.

10. Michel Sadoun-Goupil, *Le chimiste Claude-Louis Berthollet (1748–1822), sa vie, son oeuvre* (Vrin, 1977); Hugh Colquhoun, "On the Life and Writings of Claude-Louis Berthollet," *Annals of Philosophy* 9, 1825, 1–18, 81–96, and 161–181.

11. The Périer brothers, who introduced the steam engine into France, were staying with the Duke; Jacques Payen, *Les frères Périer et l'introduction en France de la machine à vapeur de Watt* (Paris: Palais de la Découverte, 1968). Nicholas Le Blanc also began his career as a physician to the Duke; Amédée Britsch, *La maison d'Orléans à la fin de l'ancien régime. La jeunesse de Philippe Égalité (1747–1785) d'après des documents inédits* (Paris: Payot, 1926).

12. H. E. Le Grand, "The 'Conversion' of C.-L. Berthollet to Lavoisier's Chemistry," *Ambix* 22, 1975, 58–70.

13. Berthollet, "Mémoire sur l'acid tartareux," *Journal de physique* 7, 1776, 130–148.

14. Sadoun-Goupil, *Le Chimiste*, 115–121.

15. Berthollet, *Observations sur l'air*, 22–23.

16. Ibid., 16.

17. Ibid., 45–46.

18. Buffon, *Histoire naturelle générale et particulière servant de suite à la théorie de la Terre et d'Introduction à l'Histoire des minéraux*, suppléments, volume 1 (Paris, 1774), 1–78, See F.L. Holmes, "The Boundaries of Lavoisier's Chemical Revolution," *Revue d'histoire des sciences et leurs applications* 48, 1995, 9–48.

19. The vogue for pneumatic studies owed much to their potential in determining the salubrity of air; T. H. Levere, "Measuring Gases and Measuring Goodness," in *Instruments and Experimentation in the History of Chemistry*, ed. T. Levere and F. Holmes (MIT Press, 2000).

20. "Observations sur l'Air, par M. Berthollet," *Journal de médecine, chirurgie, pharmacie, &c.* 46, 1776, 441–452 and "Suite de l'Extrait des Observations sur l'Air, par M. Berthollet," ibid., 531–536; "Lettre à M. Berthollet au sujet des ses observations sur l'air," ibid., 47, 1777, 332; In an undated letter, Berthollet thanks Macquer for his support [*Bibliothèque nationale*, Ms. fr. 12305, 109–110].

21. Berthollet, *Observations sur l'Air*, 43.

22. On the notion of experimental system, see Hans-Jörg Rheinberger, *Toward a History of Epistemic Things* (Stanford University Press, 1997).

23. Le Grand, "The 'Conversion' of C. L. Berthollet," 61–62; Colquhoun, 6–8. Historians seem to have relied on the later, published versions of Berthollet's works.

24. Berthollet, "Mémoire sur l'acide sulfureux," *PV* 96, 1777, 570r (December 20); "Suite du mémoire précédent," report by Sage and Cadet, *PV* 97, 1778, 1v–3v (January 7).

25. Berthollet, "Mémoire sur l'air fixe sulfureux tiré de l'hépatosulfuris," *PV* 97, 1778, 34v (January 24); report by Macquer and Cadet, *PV* 97, 1778, 63v–66v (February 28).

26. "Mémoire sur la décomposition de l'acide nitreux," *PV* 97, 1778, 27v (January 17).

27. Cadet and Lavoisier, "Rapport sur un mémoire sur la décomposition de l'acide nitreux," *PV* 97, 1778, 87r–89v (March 11); reprinted in *Oeuvres de Lavoisier*, volume 4, 298–300 (emphasis added).

28. Berthollet, "Mémoire sur la production de l'alcali fixe volatil, *PV* 97, 1778, 50r (February 7).

29. This was a genre of experiments extensively discussed by Wilhelm Homberg, Etienne-François Geoffroy, and Louis Lemery. See Homberg, "Memoire touchant la Volatilisation des Sels fixes des Plantes," *Mémoires*, 1714, 186–195; Geoffroy l'Aîne, "Du Changement des Sels acides en Sels alkalis volatiles urineux," *Mémoires*, 1717, 226–238; Louis Lemery, "Sur la Volatilisation vraye ou apparente des Sels fixes," *Mémoires*, 1717, 246–256.

30. Commisaires Macquer et Lavoisier, "Rapport sur un mémoire la décomposition de l'acide nitreux," PV 97, 1778, 82v–84r (March 11); printed in *Oeuvres de Lavoisier*, volume 4, 306–308.

31. Berthollet, "Mémoire sur la préparation de l'alkali caustique, sa cristallisation, et son action sur l'Esprit-de-vin," *Mémoires*, 1783, 408–415; also printed in *Journal de physique* 28, 1786, 401–406.

32. Berthollet, "Extrait d'un Mémore sur l'analyse de l'alkali volatil," *Observations* 29, 1786, 175–177; "Analyse de l'alkali volatil,"*Mémoires*, 1785, 316–326 [read June 11]. This was in line with the principalist understanding of alkalis, then in fashion; Le Grand, "Determination of the Composition of the Fixed Alkalis, 1789–1810," *Isis* 65, 1974, 59–65.

33. Berthollet, "Recherches sur la nature des substances animales, et sur leurs rapports avec les substances végétales," *Mémoires*, 1780, 120–125; "Observations sur la décomposition spontanée de quelques acides végétaux," *Mémoires*, 1782, 608–615; "Recherches sur la nature des substances animales, & sur leur rapport avec les substances végétales; ou Recherches sur l'acide du sucre," *Observations* 27, 1785, 88–91; "Suite des recherches sur la nature des substances animales et sur leurs rapports avec les substances végétales," *Mémoires*, 1785, 331–349.

34. Macquer, "Causticité," in *Dictionnaire* (1778), volume 1, 300.

35. Berthollet, "Essai sur la cause de la causticité des alcalis et des précipités métalliques," PV 98, 1779, 302r (December 1); "Essai sur la causticité des sels & des précipités métalliques (I)," *Journal de médecine, chirurgie, pharmacie,* c. 53, 1780, 50–69, at 51.

36. On Calomel, see George Urdang, "The Early Chemical and Pharmaceutical History of Calomel," *Chymia* 1, 1948, 93–108.

37. Berthollet, "Deuxième mémoire sur la causticité des sels et des précipités métalliques," PV 99, 1780, 21v–22r, 24r (February 1 and 9); "Troisième mémoire sur le même sujet," PV 99, 1780, 31r (February 19). These were published together as "Essai sur la causticité des sels métalliques," *Mémoires*, 1780, 448–470.

38. Bonjour, the French translator of Bergman's expanded dissertation on affinity, was working for Berthollet. See W. A. Smeaton, "F. J. Bonjour and His Translation of Bergman's 'Disquisitio de attractionibus electivis,'" *Ambix* 7, 1959, 47–50.

39. Lavoisier and Macquer, "Rapport sur la causticité des sels métalliques," in *Oeuvres de Lavoisier*, volume 4, 343–349, at 348–349.

40. Michelle Sadoun-Goupil, "Science pure et science appliquée dans l'oeuvre de Claude-Louis Berthollet," *Revue d'histoire des sciences et leurs applications* 27, 1974, 127–145; Barbara Whitney Keyser, "Between Science and Craft: The Case of Berthollet and Dyeing," *Annals of Science* 47, 1990, 213–260.

41. Berthollet, "Recherches sur la nature des substances animales, et sur leurs rapports avec les substances végétales," *Mémoires*, 1780, 120–125.

42. "Sur la decomposition du nitre," *Histoire*, 1781, 28–29; Berthollet, "Observations sur la décomposition de l'acide nitreux," *Mémoires*, 1781, 21–33, 228–233, and 234–242.

43. Berthollet mentions Bucquet, rather than Cadet, in his memoir.

44. Ibid., 237.

45. Le Grand, "The 'Conversion' of C. L. Berthollet," 63–64.

46. Berthollet, "Observations sur la décomposition de l'acide nitreux," 238.

47. Ibid., 241–242.

48. "Sur la decomposition du nitre," *Histoire*, 1781, 29. On Concorcet, see Keith Michael Baker, "Les debuts de Condorcet au secrétariat de l'Academie royale des sciences (1773–1776," *Revue* 20, 1967, 229–280; idem, *Concorcet* (University of Chicago Press, 1975).

49. Berthollet, "Expériences sur l'acide sulfureux," *Mémoires*, 1782, 597–601, at 600–601.

50. Berthollet, "Recherches sur l'augmentation de poids qu'éprouvent le Soufre, le Phosphore, & l'Arsenic, lorsqu'ils sont changés en Acide," *Mémoires*, 1782, 602–607.

51. Berthollet, "Mémoire sur la différence du vinaigre radical et de l'acide acéteux," *Memoires*, 1783, 403–407.

52. Le Grand, "The 'Conversion' of C. L. Berthollet,"68–69. Berthollet used "chaleur" instead of "calorique" in 1785–86, but he used "calorique" freely in 1789. See various 1785 papers on marine acid; also see "De l'Influence de la lumière," *Observations* 29, 1786, 81–86; "Extrait d'Observations sur la combinaison des oxides métalliques avec les alkalis & la Chaux," *Annales de chimie* 1, 1789, 52–63 (read to the Academy on December 19, 1788).

53. J. R. Partington, "Berthollet and the Antiphlogistic Theory," *Chymia* 5, 1959, 130–137; Duveen and Klickstein, "A Letter from Berthollet to Blagden relating to the experiments for a large-scale synthesis of water carried out by Lavoisier and Meusnier in 1785," *Annals of Science* 10, 1954, 58–62.

54. Berthollet, "Mémoire sur l'acide marin déphlogistiqué,"*Mémoires*, 1785, 276–295 (read December 21), at 276.

55. Berthollet to Guyton, May 4, 1785 [Dossier Guyton de Morveau, #93, archive of Académie des Sciences].

56. Berthollet, "Mémoire sur l'acide marin déphlogistiqué," *PV*, 1785, 55 (April 6). This memoir was not published, but an abstract appeared in the May issue of *Journal de physique* (29, 1785, 321–325).

57. Berthollet, "Mémoire sur l'acide marin déphlogistiqué," *PV* 104, 1785, 242r (December 21); "Mémoire sur l'acide marin déphlogistiqué," *Mémoires*, 1785, 276–295.

58. Berthollet, "Observations sur l'eau régale et sur quelques affinités de l'acide marin," *Mémoires*, 1785, 296–307 (read April 19).

59. Berthollet, "Mémoire sur la décomposition de l'esprit-de-vin et de l'éther, par le moyen de l'air vital," *Mémoires*, 1785, 308–315 (read April 27).

60. Berthollet, "Analyse de l'alkali volatil," *Mémoires*, 1785, 316–326 (read June 11).

61. "Lettre de M. Berthollet, à M. de La Metherie, sur la décomposition de l'eau," *Observations* 29, 1786, 138–139.

62. Berthollet, "Mémoire sur l'acide marin déphlogistiqué," 279.

63. Le Grand, "The 'Conversion' of C. L. Berthollet"; idem, "Lavoisier's Oxygen Theory of Acidity," *Annals of Science* 29, 1972, 1–18.

64. Berthollet, "Mémoire sur l'acide prussique," *Mémoires*, 1787, 148–162 (read December 15); idem, "Extrait d'un Mémoire sur l'Acide prussique, lû à l'Académie le 15 Décembre 1787," *Annales de chimie* 1, 1789, 30–39.

65. Berthollet, "Observations sur la combinaison des oxides métalliques avec les alkalis et la chaux," *Mémoires*, 1788, 728–741; idem, "Extrait d'observations sur la combinaison des oxides métalliques, avec les alkalis & la chaux, lû à l'Académi e Royale des Sciences le 19 Décembre 1788," *Annales de chimie* 1, 1789, 52–64.

66. Berthollet, "Suite des Expériences sur l'acide sulfureux," *Annales de chimie* 2, 1789, 54–72.

67. Berthollet, "Observations sur l'hydrogène sulfuré (read 11 March 1796)," *Annales de chimie* 25, 1798, 233–273.

68. Le Grand, "Genuis and the Dogmatization of error."; idem, "Berthollet's *Essai de statique chimique* and Acidity," *Isis* 67, 1976, 229–238.

69. Fourcroy, *Philosophie chimique*, 11 and 13.

70. Craig S. Zwerling, *The Emergence of the Ecole Normale Supérieure as a Center of Scientific Education in Nineteenth-Century France* (Garland, 1990), 30–38.

71. *Séances des Écoles normales, recueillies par des sténographes, et revues par les professeurs* (Paris, 1796) [Bibliothèque nationale, R22215–22220], I, 'Avertissement'.

72. *Séances des Écoles normales*, volume 1, 205–229.

73. Ibid., 311–321.

74. Grapi and Izquierdo, "Berthollet's Conception of Chemical Change."

75. *Séances*, volume 1, 320.

76. Ibid., 432–448.

77. Ibid., 446.

78. Sadoun-Goupil, *Le Chimiste*, 29–49.

79. Crosland, *The Society of Arcueil*; idem, *Gay-Lussac* (Cambridge University Press, 1978).

80. *Annales de chimie* 36, 1801, 302–317; 37, 1801, 151–181 and 221–253; 38, 1801, 3–29 and 113–134; *Mémoires de la Classe des Sciences Mathématiques et Physiques* 3, 1803, 1–96, 207–245; collected as *Recherches sur les Lois de l'Affinité* (Paris: Badouin, 1801).

81. Berthollet, *Researches into the Laws of Chemical Affinity* (Da Capo, 1966), facsimile reproduction of the American edition (Baltimore: P. H. Nicklin, 1809), 1–2.

82. Ibid., 154.

83. Ibid., 3–4.

84. Ibid., 6.

85. Ibid., 80 (emphasis added).

86. Ibid., 81–87.

87. Ibid., 146.

88. Ibid., 24–25.

89. Ibid., 154–155.

90. Ibid., 146.

91. Ibid., 86.

92. Ibid.

93. Ibid., 143–144.

94. Holmes, "Chemical Equilibria"; Le Grand, "The Conversion."

95. Berthollet, *An Essay on Chemical Statics* (London, 1804), volume 1, 17.

96. Berthollet, *Researches*, 4.

97. Ibid., 4–5.

98. Ibid., 103–104.

99. Ibid., 31.

100. Ibid., 36.

101. Ibid., 35–39.

102. Ibid., 55 and 59.

103. Ibid., 63–65.

104. Kapoor, "Berthollet," 62–73.

105. Berthollet, *An Essay*, volume 1, 36.

106. Ibid., volume 2, 4; W. A. Smeaton, "Berthollet's *Essai de statique chimique* and Its Translations: A Bibliographical Note and a Daltonian Doubt," *Ambix* 24, 1977, 149–158; "Berthollet's *Essai de statique chimique*: A Supplementary Note," *Ambix* 25, 1978, 211–212.

107. Berthollet, *An Essay*, volume 1, ix.

108. Ibid., ix–x.

109. Ibid., xxi–xxii.

110. Ibid., xxiv.

111. Ibid., xxiv–xxv.

112. Ibid., 111.

113. This provided a starting point for Gay-Lussac, an early member of the Arcueil group.

114. Ibid., xxix–xxx.

115. Ibid., 36–37.

116. Ibid., 43–48.

117. Ibid., 45.

118. Ibid., 77–78.

119. Le Grand, "Berthollet's *Essai de statique chimique*."

120. Berthollet, *An Essay*, volume 1, 34–35.

121. Grapi, "The Marginalization."

122. Satish C. Kapoor, "Berthollet, Proust, and Proportions," *Chymia* 10, 1965, 53–110; Kiyohisa Fujii, "The Berthollet-Proust Controversy and Dalton's Chemical Atomic Theory, 1800–1820," *British Journal for the History of Science* 19, 1986, 177–200.

123. Thomas Thomson, *A System of Chemistry*, third edition (Edinburgh: printed for Bell & Bradfute and E. Balfour, 1807), volume 4, 4.

124. Alan J. Rocke, "Atoms and Equivalents: The Early Development of the Chemical Atomic Theory," *Historical Studies in the Physical Sciences* 9, 1978, 225–263; idem, *Chemical Atomism in the Nineteenth Century* (Ohio State University Press, 1984).

125. J. L. Gay-Lussac, "Mémoire sur la combinaison des substances gazeuses, les unes avec les autres," *Mémoires de la Société d'Arcueil* 2, 1809, 207–234; translated in *Foundations of the Molecular Theory*, Alembic Club Reprints, no. 4 (Edinburgh, 1950), 8–24; M. P. Crosland, "The Origins of Gay-Lussac's Law of Combining Volumes of Gases," *Annals of Science* 17, 1961, 1–26.

126. Pierre-Louis Dulong, "Recherches sur la décomposition mutuelle des sels insolubles et des sels solubles," *Annales de chimie* 82, 1812, 275–308; Crosland, *The Society of Arcueil*.

127. Siegfried and Dobbs, "Composition"; Siegfried, "The Chemical Revolution in the History of Chemistry."

128. Melhado, *Jacob Berzelius*, 61–99.

129. H. E. Le Grand, "Chemistry in a Provincial Context," *Ambix* 29, 1982, 88–105; idem, "Theory and Application," *British Journal for the History of*

Science 17, 1984, 31–46; Carleton E. Perrin, "Of Theory Shifts and Industrial Innovations: The Relations of J. A. C. Chaptal and A. L. Lavoisier," *Annals of Science* 43, 1986, 511–542; Jean Dhombres, "Quelques Réflexions de et sur Chaptal. A propos de la Révolution chimique," in *Lavoisier et la Révolution chimique*, ed. Goupil; Jeff Horn and M. C. Jacob, "Jean-Antoine Chaptal and the Cultural Roots of French Industrialization," *Technology and Culture* 34, 1998, 671–698.

130. Holmes, "From Elective Affinities."

131. Melhado, *Jacob Berzelius*, 316.

132. Mi Gyung Kim, "The Layers of Chemical Language, I: Constitution of Bodies v. Structure of Matter," *History of Science* 30, 1992, 69–96; idem, "The Layers of Chemical Language, II: Stabilizing Atoms and Molecules in the Practice of Organic Chemistry," *History of Science* 30, 1992, 397–437; Alan J. Rocke, *Nationalizing Science* (MIT Press, 2001).

133. Fourcroy, *Elements of Chemistry* (1796), 36–73; M. I. A. Chaptal, *Élémens de chymie*, third edition (Paris: Deterville, 1796), 20–37; Thomas Thomson, *A System of Chemistry* (Edinburgh: Bell & Bradfute, 1810), volume 3, 414–432; Edward Turner, *Elements of Chemistry*, fourth American edition (Philadelphia: Grigg & Elliot, 1832), 109–121; L. J. Thenard, *Traité de chimie élémentaire, théorique et pratique*, second edition (Paris: Crochard, 1817), volume 1, 1–6.

134. Trevor H. Levere, *Affinity and Matter* (Clarendon, 1971).

135. Kim, Practice and Representation.

136. Ludwig Wilhelmy, "Ueber das Gesetz, nach welchem die Einwirkung der Säuren auf den Rohrzucker stattfindet," *Annalen der physik* 81, 1850, 413–526; Alexander Williamson, "Suggestions for the Dynamics of Chemistry derived from the Theory of Aetherification," *Chemical Gazette* 9, 1851, 294–298 and 334–339; Marcellin Berthelot, "Essai d'une Théorie sur la Formation des éthers," *Annales de chimie* 66, 1862, 110–128.

137. See e.g. Buff, "Einige Bemerkungen zur Affinitätslehre," *Berichte* 2, 1869, 142–147.

138. Trevor Levere, almost alone in arguing for the importance of affinity concept in nineteenth-century chemistry, called attention to its protean nature in *Affinity and Matter*.

Epilogue

1. For a brief survey of this historiography, see Thomas S. Kuhn, "Robert Boyle and Structural Chemistry in the Seventeenth Century," *Isis* 43, 1952, 12–36.

2. Marie Boas, *Robert Boyle and Seventeenth-Century Chemistry* (Cambridge University Press, 1958), 49–50.

3. Maurice Crosland, "Chemistry and the Chemical Revolution," in *Ferments of Knowledge*, ed. G. Rousseau and R. Porter (Cambridge University Press, 1980).

4. Arthur Donovan, "Newton and Lavoisier—From Chemistry as a Branch of Natural Philosophy to Chemistry as a Positive Science," in *Action and Reaction*, ed. P. Theerman and A. Seeff (University of Delaware Press, 1993).

5. Bernadette Bensaude-Vincent, "A Founder Myth in the History of Sciences? The Lavoisier Case," in *Functions and Uses of Disciplinary Histories*, ed. L. Graham et al. (Reidel, 1983); idem, "Une Mythologie Révolutionnaire dans la Chimie Française," *Annals of Science* 40, 1983, 189–196.

6. Frederic L. Holmes, "The Boundaries of Lavoisier's Chemical Revolution," in *Lavoisier et la Révolution chimique*, ed. Goupil, 13–14. These two issues have always been confounded in historiography. For a recent example, see Arthur Donovan, "Lavoisier and the Origins of Modern Chemistry," *Osiris* [2] 4, 1988, 214–231.

7. Robert Siegfried and Betty Jo Dobbs, "Composition, A Neglected Aspect of the Chemical Revolution," *Annals of Science* 24, 1968, 275–293.

8. Robert Siegfried, "The Revolution in the History of Chemistry," *Osiris* [2] 4, 1988, 34–50.

9. Henry E. Roscoe, *John Dalton and the Rise of Modern Chemistry* (Macmillan, 1895); Frank Greenway, *John Dalton and the Atom* (Cornell University Press, 1966); Henry E. Roscoe and Arthur Harden, *A New View of the Origin of Dalton's Atomic Theory* (Johnson Reprint, 1970); Elizabeth C. Patterson, *John Dalton and the Atomic Theory* (Doubleday, 1970).

10. Arnold Thackray, *John Dalton* (Harvard University Press, 1972); Alan J. Rocke, *Chemical Atomism in the Nineteenth Century* (Ohio State University Press, 1984).

11. Mi Gyung Kim, "The Layers of Chemical Language, II: Stablizing Atoms and Molecules in the Practice of Organic Chemistry," *History of Science* 30, 1992, 397–437.

12. Evan Melhado, *Jacob Berzelius* (University of Wisconsin Press, 1981).

13. Alice Stroup, "Duclos on Boyle: A French Academician Criticizes Certain Physiological Essays," presentation at annual meeting of History of Science Society, Vancouver, 2000.

14. Thackray, *John Dalton*.

15. Ursula Klein, "Origin of the Concept of Chemical Compound," *Science in Context* 7, 1994, 163–204; "E. F. Geoffroy's Table of Different 'Rapports' Observed between Different Chemical Substances—A Reinterpretation," *Ambix* 42, 1995, 78–100; "The Chemical Workshop Tradition and the Experimental Practice: Discontinuities within Continuities," *Science in Context* 9, 1996, 251–287.

16. Condorcet, "Eloge de M. Macquer," *Histoire*, 1784, 20–30.

17. Melhado, *Jacob Berzelius.*

18. Macquer and Baumé, *Plan d'un Cours de Chymie expérimentale et raisonnée, avec un discours historique sur la chymie* (Paris, 1757).

19. Arthur Donovan, *Philosophical Chemistry in the Scottish Enlightenment* (Edinburgh University Press, 1975); Jan Golinski, *Science as Public Culture* (Cambridge University Press, 1992).

20. F. L. Holmes, "The 'Revolution in Chemistry and Physics': Overthrow of a Reigning Paradigm or Competition between Contemporary Research Paradigms?" *Isis* 91, 2000, 735–753.

21. Karl Hufbauer, *The Formation of the German Chemical Community, 1720–1795* (University of California Press, 1982).

22. For a comprehensive study of French medical community, see Laurence Brockliss and Colin Jones, *The Medical World of Early Modern France* (Clarendon, 1997).

23. Nicolas Lemery, *Cours de chymie* (1675); *Pharmacopée universelle* (Paris: L. d'Houry, 1697).

24. Dorinda Outram, *The Enlightenment* (Cambridge University Press, 1995).

25. Archibald Clow and Nan L. Clow, *The Chemical Revolution* (London: Batchworth, 1952); Henry Guerlac, "Some French Antecedents of the Chemical Revolution," *Chymia* 5, 1959, 73–112; Theodore Porter, "The Promotion of Mining and the Advancement of Science: The Chemical Revolution of Mineralogy," *Annals of Science* 38, 1981, 543–570; Evan M. Melhado, "Mineralogy and the Autonomy of Chemistry around 1800," *Lychnos*, 1990, 229–261; Seymour H. Mauskopf, "Gunpowder and the Chemical Revlution," *Osiris* [2], 4, 1988, 93–118; idem, "Lavoisier and the Improvement of Gunpowder Production," *Revue d'histoire des sciences et leurs applications* 48, 1995, 95–132; Jonathan Simon, The Alchemy of Identity: Pharmacy and the Chemical Revolution, 1777–1809, Ph.D. dissertation, University of Pittsburgh, 1997; idem, "The Chemical Revolution and Pharmacy: A Disciplinary Perspective," *Ambix* 45, 1998, 1–13. For an interesting approach, see Pamela O. Long, "The Openness of Knowledge," *Technology and Culture* 32, 1991, 318–355.

26. David C. Lindberg and Robert S. Westman, *Reappraisals of the Scientific Revolution* (Cambridge University Press, 1990); Margaret J. Osler ed. *Rethinking the Scientific Revolution* (Cambridge University Press, 2000).

27. I. Bernard Cohen, *Revolution in Science* (Belknap, 1985), 197–236.

28. A. Levin, "Venel, Lavoisier, Fourcroy, Cabanis and the Idea of Scientific Revolution: The French Political Context and the General Patterns of Conceptualization of Scientific Change," *History of Science* 22, 1984, 303–320; C. E. Perrin, "Revolution or Reform: The Chemical Revolution and Eighteenth Century Concepts of Scientific Change," *History of Science* 25, 1987, 395–423.

29. Macquer and Baumé, *Plan d'un Cours de Chymie.*

30. Jerry B. Gough, "Some Early References to Revolutions in Chemistry," *Ambix* 29, 1982, 106–109; Henry Guerlac, "The Chemical Revolution: A Word from Monsieur Fourcroy," *Ambix* 23, 1976, 1–4.

31. Antoine Baumé, *Chimie expérimentale et raisonnée* (Paris: P. F. Didot le jeune, 1773), volume 3, 693.

32. Janis Langins, "Fourcroy, Historien de la Révolution chimique," in *Lavoisier et la Révolution chimique*, ed. Goupil.

33. Hans-Jörg Rheinberger, *Toward a History of Epistemic Things* (Stanford University Press, 1997).

34. On the skewed logic of representation between gentlemen and artisans, see Steven Shapin and Simon Schaffer, *Leviathan and the Air-Pump* (Princeton University Press, 1985); Steven Shapin, "The House of Experiment in Seventeenth-Century England," *Isis* 79, 1988, 373–404; idem, *A Social History of Truth* (University of Chicago Press, 1994); Malcolm Oster, "The Scholar and the Gentleman revisited: Robert Boyle as Aristocrat and Artisan," *Annals of Science* 49, 1992, 255–276.

35. Mary Terrall, "Metaphysics, Mathematics, and the Gendering of Science in Enlightenment-Century France," in *The Sciences in Enlightened Europe*, ed. W. Clark et al. (University of Chicago Press, 1999).

36. C. E. Perrin, "Research Traditions, Lavoisier, and the Chemical Revolution," *Osiris* [2] 4, 1988, 53–81; John G. McEvoy, "Continuity and Discontinuity in the Chemical Revolution," ibid., 195–213.

37. Arthur Donovan, "Lavoisier and the Origins of Modern Chemistry," *Osiris* [2] 4, 1988, 214–231; Evan M. Melhado, "Toward an Understanding of the Chemical Revolution," *Knowledge and Society* 8, 1989, 123–137; F. L. Holmes, "What Was the Chemical Revolution About?" *Bulletin for the History of Chemistry* 20, 1997, 1–9.

38. William R. Newman and Lawrence M. Principe, "Alchemy vs. Chemistry: The Etymological Origins of a Historiographic Mistake," *Early Science and Medicine* 3, 1998, 32–65.

39. J. Andrew Mendelsohn, "Alchemy and Politics in England," *Past and Present* 135, 1992, 30–78.

40. Maurice Crosland, "The Development of Chemistry in the Eighteenth Century," *Studies on Voltaire and the Eighteenth Century* 24, 1963, 369–441; idem, "Chemistry and the Chemical Revolution"; Arthur Donovan, "The Chemical Revolution and the Enlightenment—and a Proposal for the Study of Scientific Change," in *Philosophy and Science in the Scottish Enlightenment*, ed. P. Jones (Donald, 1988); John G. McEvoy, "The Enlightenment and the Chemical Revolution," in *Metaphysics and Philosophy of Science in the Seventeenth and Eighteenth Centuries*, ed. R. Woolhouse (Kluwer, 1988); idem, "The Chemical Revolution in Context," *The Eighteenth Century: Theory and Interpretation* 33, 1992, 198–216.

41. Macquer's and Baumé's chemical textbooks were among the list of books young comte de Montlosier read, along with Pluch, Fontenelle, Voltaire, Roussesau, and Diderot; Pierre Serna, "The Noble," in *Enlightenment Portraits*, ed. M. Vovelle (University of Chicago Press, 1997), 38.

42. Jeremy Adler, "Goethe's Use of Chemical Theory in His *Elective Affinities*," in *Romanticism and the Sciences*, ed. A. Cunningham and N. Jardin (Cambridge University Press, 1990).

43. Charles Darwin, *The Origin of Species by Means of Natural Selection* (Modern Library), 64.

44. John E. Lesch, *Science and Medicine in France* (Harvard University Press, 1984); M. Norton Wise, *Energy and Empire* (Cambridge University Press, 1989); Raphael Falk, "What Is a Gene?" *Studies in History and Philosophy of Science* 17, 1986, 133–173.

45. On active versus dormant metaphors of science, see N. Katherine Hayles, *Chaos Bound* (Cornell University Press, 1990), 31–60.

Bibliography

Abbreviations

AdS	Archive of Académie des Sciences
Archives	*Archives internationale d'histoire des sciences*
BJHS	*British Journal for the History of Science*
DSB	*Dictionary of Scientific Biography*
HARS	*Histoire de l'Académie royale des sciences*
Histoire	*Histoire de l'Académie royale des sciences, avec mémoires, Histoire*
HSPS	*Historical Studies in the Physical Sciences*
Journal de médecine	*Journal de médecine, chirurgie, pharmacie, &c*
Mémoires	*Histoire de l'Académie royale des sciences, avec mémoires, Mémoires*
Notes and Records	*Notes and Records of the Royal Society of London*
Observations	*Observations sur la physique*
PV	*procès-verbaux de l'Académie royale des sciences, Paris*
RAD	*Registre de l'Académie de Dijon*
Revue	*Revue d'histoire des sciences et leurs applications*
Revue pharm.	*Revue d'histoire de la pharmacie*
SHPS	*Studies in History and Philosophy of Science*
Studies on Voltaire	*Studies on Voltaire and the Eighteenth Century*

Manuscript Sources

Boulduc, Simon. "Analyse de l'ipecacuanha." *PV* 19, 1700, 1r–6r (January 9).

Boulduc, Simon, on Jalap. *PV* 20, 1701, 162r–167v (April 30).

Boulduc, Simon, on Gomme Gutte. *PV* 276r–280v (July 30).

Cadet de Gassicourt, Louis Claude, and Lavoisier, Antoine-Laurent. "Rapport sur un mémoire sur la décomposition de l'acide nitreux." *PV* 97, 1778, 87r–89v (March 11); reprinted in *Oeuvres de Lavoisier*, IV, 298–300.

Du Clos, Samuel Cottereau. "Projet d'Exercitations Physiques, proposé a l'assemblée par le Mr. Du Clos." *PV* 1, 1666, 1–16 (December 31).

Du Clos, Samuel Cottereau. "Expériences de l'augmentation du poids de certaines matieres en les calcinane a la chaleur ou du Soleil, ou du feu ordinaire." *PV* 1, 1667, 40–52 (January 8).

Geoffroy, Etienne-François. "Problème de chimie. Trouver des Cendres qui ne contiennent aucunes parcelles de fer." *PV* 24, 1705, 393r–395r (December 16).

Homberg, Wilhelm. "Nouvelle maniere de distiller sans aucune chaleur." *PV* 16, 1697, 126v–129v (May 15).

Homberg, Wilhelm. "Essays sur l'analyse des sels des plants." *PV* 16bis, 1697, 194r–198v (July 17).

Homberg, Wilhelm. "Observations sur le sel urineux des plants." *PV* 17, 1697, 40r–43r (November 27).

Homberg, Wilhelm. "Observations sur les sels fixes des Plantes." *PV* 18, 1698, 37v–39v (December 3).

Homberg, Wilhelm. "Observations sur les analyses des plantes." *PV* 20, 1701, 209r–215v (June 18).

Homberg, Wilhelm. "Analise du souffre comun." *PV* 22, 1703, 85r–88r (March 27).

Homberg, Wilhelm. "Sur le Souffre Principe." *PV* 24, 1705, 125r–130v (April 22).

Homberg, Wilhelm. "Observations sur le fer au verre ardent." *PV* 25, 1706, 168r–172r (May 8).

La Planche. "Plan d'un cours de chymie, suivant les principes de Becher, de Boerhave, & de Stahl" [Bibliothèque nationale, R 5805]. In *A New Course of Chemistry*, ed. M. Beretta (Firenze: L. S. Olschski, 1994).

Lavoisier, Antoine-Laurent. "Commencement d'un traité de Chimie (1764)" [Lavoisier Dossier 380, AdS]. In *A New Course of Chemistry*, ed. M. Beretta (Firenze: L. S. Olschski, 1994).

Lavoisier, Antoine-Laurent. "Projet d'un memoire sur les differens degrés d'affinité des acides en exces avec les differentes substances (Sept. 3, 1766)" [Lavoisier Dossier 323, AdS].

Lavoisier, Antoine-Laurent. "Système sur les elémens (July Memoir, 1772)." In Henry Guerlac, *Antoine-Laurent Lavoisier: Chemist and Revolutionary* (Scribner, 1975). Also in René Fric, "Contribution à l'étude de l'évolution des ideés de Lavoisier sur la nature de l'air et sur la calcination des métaux," *Archives* 12, 1959, 137–168.

Lavoisier, Antoine-Laurent. "Réflexions sur les experiences qu'on peu tenter a l'aide du miroir ardent (August memorandum, 1772)." In Henry Guerlac, *Antoine-Laurent Lavoisier: Chemist and Revolutionary* (Scribner, 1975).

Lavoisier, Antoine-Laurent. "Experiences sur le phosphore du 10 7bre 1772." In Henry Guerlac, *Antoine-Laurent Lavoisier: Chemist and Revolutionary* (Scribner, 1975).

Lavoisier, Antoine-Laurent. "Memoire sur l'acide du Phosphore et sur ses combinaisons avec differentes substances salines terreuses et metallique (Draft memoir of October 20, 1772) [Lavoisier Dossier, 1308D, AdS]. In Henry Guerlac, *Antoine-Laurent Lavoisier: Chemist and Revolutionary* (Scribner, 1975).

Lavoisier, Antoine-Laurent. "Essay sur la nature de l'air." In René Fric, "Contribution à l'étude de l'évolution des ideés de Lavoisier sur la nature de l'air et sur la calcination des métaux," *Archives* 12, 1959, 137–168.

Lavoisier, Antoine-Laurent. "Sur une nouvelle Theorie de la Calcination et de la Reduction des substances metalliques sur la cause de laugmentation de poids quelles acquierent au feu et sur differens phenomenes qui appartiennent a l'air fixe" [Fonds Lavoisier, 1303, AdS]. In René Fric, "Contribution à l'étude de l'évolution des ideés de Lavoisier sur la nature de l'air et sur la calcination des métaux," *Archives* 12, 1959, 137–168.

Lavoisier, Antoine-Laurent. "Sur la manière d'enseigner la chimie (1792)" [Dossier Lavoisier, 1259, AdS]. In Bernadette Bensaude-Vincent, "A View of the Chemical Revolution through Contemporary Textbooks: Lavoisier, Fourcroy and Chaptal." *BJHS* 23, 1990, 435–460.

Lavoisier, Antoine-Laurent. "Cours de chimie expérimental rangée suivant l'ordre naturel des idées" [Dossier Lavoisier 1260, AdS].

Lavoisier, Antoine-Laurent, and Cadet de Gassicourt, Louis Claude. "Rapport sur un mémoire sur la décomposition de l'acide nitreux." *PV* 97, 1778, 87r–89v (March 11).

Lemery, Louis. "Diverses Experiences et Observations chimiques et physiques. Sur le Fer et sur l'Aimant." *PV* 25, 1706, 123v–134r (April 14).

Lemery, Louis. "Du Miel & de son Analyse Chimique." *PV* 25, 1706, 262r–272r (July 17).

Lemery, Louis. "Conjectures & reflexions sur la matiere du feu ou de la lumiere." *PV* 28, 1709, 445r–457v (December 20).

Macquer, Pierre-Joseph. "Correspondance de Macquer," 2 vols. [*Bibliothèque Nationale*, manuscript section, f fr. 12305 and 12306].

Macquer, Pierre-Joseph, and Cadet de Gassicourt, Louis Claude. report on Berthollet's "Mémoire sur l'air fixe sulfureux tiré de l'hépatosulfuris." *PV* 97, 1778, 63v–66v (February 28).

Rouelle, Guillaume-François. "Cours de Chymie, recueilli des leçons de M. Rouelle (1767)" [Ms. 1202, *Museum nationale d'histoire naturelle*].

Primary Sources

Baumé, Antoine. *Chymie expérimentale et raisonnée*, 3 vols. (Paris: P. F. Didot le jeune, 1773).

Baumé, Antoine. *Manuel de Chymie, ou Exposé des opérations de la chymie et des produits d'un cours de chymie. Ouvrage utile aux personnes qui veulent suivre un cours de cette science, ou qui ont dessein de se former un cabinet de chymie* (Paris: Didot le jeune, 1763; second edition 1765).

Baumé, Antoine. *A Manual of Chemistry* (translated from second edition of *Manuel de Chymie*) (Warrington: printed by W. Eyres for J. Johnson, 1778).

Bayen, Pierre. "Essai d'expériences cymiques, faites sur quelques précipités de mercure, dans la vue de découvrir leur nature." *Observations* 3, 1774, 280–295.

Béguin, Jean. *Les Elemens de Chymie*, second French edition (Geneva: Jean Celerier, 1624).

Béraut, Laurent. *Dissertation sur la cause de l'augmentation de poids, que certains matières acquièrent dans leur calcination* (Bordeaux, 1747).

Bergman, Torbern. "Disquisitio de Attractionibus Electivis." In *K. Ventenskaps Societeten I Upsala. Nova Acta Regiae Societatis Scientiarum Upsaliensis* [2] 2, 1775, 159–248; English edition: *Dissertation on Elective Attractions* (New York and London: Johnson Reprint Corp., 1968); abstract in *Observations* 13, 1778, 298–333.

Bergman, Torbern. *Opuscula physica et chemica*, vol. III (Uppsala, 1783); *A Dissertation on Elective Attractions* (London: J. Murray, 1785), reprinted with an introduction by A. M. Duncan (Frank Cass, 1970); *Traité des Affinités chymiques, ou Attractions electives* (Paris: Buisson, 1788).

Bergman, Torbern. *Opuscules chymiques et physiques de M. T. Bergman*, 2 vols. (Dijon: L. N. Frantin, 1780 and 1785).

Bergman, Torbern, ed. *Chemical Lectures of H. T. Scheffer* (Kluwer, 1992).

Bergman, Torbern, Göte Carld, and Johan Nordström, eds. *Torbern Bergman's Foreign Correspondence* (Stockholm: Lychnos-Bibliothek, 1965).

Berthelot, Marcelin Pierre Eugène. "Essai d'une Théorie sur la Formation des éthers." *Annales de chimie* 66, 1862, 110–128.

Berthollet, Claude Louis. "Mémoire sur l'acid tartareux." *Journal de physique* 7, 1776, 130–148.

Berthollet, Claude Louis. *Observations sur l'air* (1776).

Berthollet, Claude Louis. "Essai sur la causticité des sels & des précipités métalliques (I)." *Journal de médecine* 53, 1780, 50–69.

Berthollet, Claude Louis. "Essai sur la causticité des sels métalliques." *Mémoires*, 1780, 448–470.

Berthollet, Claude Louis. "Recherches sur la nature des substances animales, et sur leurs rapports avec les substances végétales." *Mémoires*, 1780, 120–125.

Berthollet, Claude Louis. "Observations sur la décomposition de l'acide nitreux." *Mémoires*, 1781, 21–33, 228–242.

Berthollet, Claude Louis. "Expériences sur l'acide sulfureux." *Mémoires*, 1782, 597–601.

Berthollet, Claude Louis. "Recherches sur l'augmentation de poids qu'éprouvent le Soufre, le Phosphore, & l'Arsenic, lorsqu'ils sont changés en Acide." *Mémoires*, 1782, 602–607.

Berthollet, Claude Louis. "Observations sur la décomposition spontanée de quelques acides végétaux." *Mémoires*, 1782, 608–615.

Berthollet, Claude Louis. "Mémoire sur la différence du vinaigre radical et de l'acide acéteux." *Mémoires*, 1783, 403–407.

Berthollet, Claude Louis. "Mémoire sur l'acide marin déphlogistiqué." *Journal de physique* 26, 1785, 321–325 (read April 6).

Berthollet, Claude Louis. "Mémoire sur l'acide marin déphlogistiqué." *Mémoires*, 1785, 276–295.

Berthollet, Claude Louis. "Recherches sur la nature des substances animales, & sur leur rapport avec les substances végétales; ou Recherches sur l'acide du sucre." *Observations* 27, 1785, 88–91.

Berthollet, Claude Louis. "Observations sur l'eau régale et sur quelques affinités de l'acide marin." *Mémoires*, 1785, 296–307.

Berthollet, Claude Louis. "Mémoires sur la décomposition de l'esprit-de-vin et de l'éther, par le moyen de l'air vital." *Mémoires*, 1785, 308–315.

Berthollet, Claude Louis. "Analyse de l'alkali voltail." *Mémoires*, 1785, 316–326 (read June 11).

Berthollet, Claude Louis. "Suite des recherches sur la nature des substances animales et sur leurs rapports avec les substances végétales." *Mémoires*, 1785, 331–349.

Berthollet, Claude Louis. "De l'Influence de la lumière." *Observations* 29, 1786, 81–86.

Berthollet, Claude Louis. "Lettre de M. Berthollet, à M. de La Metherie, sur la décomposition de l'eau." *Observations* 29, 1786, 138–139.

Berthollet, Claude Louis. "Extrait d'un Mémoire sur l'analyse de l'alkali volatil." *Observations* 29, 1786, 175–177.

Berthollet, Claude Louis. "Mémoire sur l'acide prussique." *Mémoires*, 1787, 148–162.

Berthollet, Claude Louis. "Extrait d'un Mémoire sur l'Acide prussique, lû à l'Académie le 15 Décembre 1787." *Annales de chimie* 1, 1789, 30–39.

Berthollet, Claude Louis. "Extrait d'Observations sur la combinaison des oxides métalliques avec les alkalis & la Chaux." *Annales de chimie* 1, 1789, 52–63.

Berthollet, Claude Louis. "Suite des Expériences sur l'acide sulfureux." *Annales de chimie* 2, 1789, 54–72.

Berthollet, Claude Louis. *Eléments de l'art de la teinture*, 2 vols. (Paris: F. Didot, 1791).

Berthollet, Claude Louis. "Observations sur l'hydrogène sulfuré." *Annales de chimie* 25, 1798, 233–273.

Berthollet, Claude Louis. "Recherches sur les lois de l'affinité." *Annales de chimie* 36, 1801, 302–317; 37, 1801, 151–181 and 221–253; 38, 1801, 3–29 and 113–134; *Mémoires de la Classe des Sciences Mathématiques et Physiques* 3, 1803, 1–96 and 207–245.

Berthollet, Claude Louis. *Recherches sur les Lois de l'Affinité* (Paris: Badouin, 1801).

Berthollet, Claude Louis. "Researches respecting the Laws of Affinity." *Philosophical Magazine* 9, 1801, 146–158 and 342–352; 10, 1801, 69–74, 129–142, 197–208, and 321–330.

Berthollet, Claude Louis. *Researches into the Laws of Chemical Affinity*, facsimile reproduction of American edition (Baltimore, 1809) (Da Capo, 1966).

Berthollet, Claude Louis. *Essai de statique chimique*, 2 vols. (Paris: F. Didot, 1803); *An Essay on Chemical Statics*, 2 vols. (London: printed for J. Mawman by W. Flint, 1804).

Black, Joseph. *Experiments upon Magnesia Alba, Quick-Lime, and other Alcaline Substances* (read to Edinburgh Philosophical Society in 1755; published in Edinburgh by Wililiam Creech in 1777).

Boerhaave, Herman. *Institutiones et experimenta chemiae*, 2 vols. (Paris, 1724).

Boerhaave, Herman. *Elementa chemiae*, 2 vols. (Leyden, 1732).

Boerhaave, Herman. *Elements of Chemistry* (London: printed for J. and J. Pemberton, 1735).

Boulduc, Gilles-François. "Essai d'analyse en general des nouvelles eaux minerale de Passy." *Mémoires*, 1726, 306–327.

Boulduc, Simon. "Analyse de l'ipecacuanha." *Mémoires*, 1700, 1–5.

Boulduc, Simon. "Observations Analitiques de la Coloquinthe." *Mémoires*, 1701, 12–17.

Boulduc, Simon. "Observations Analytiques du Jalap." *Mémoires*, 1701, 106–109.

Boulduc, Simon. "Remarques sur la nature de la Gomme-Gutte, & ses differentes Analyses." *Mémoires*, 1701, 131–135.

Boulduc, Simon. "Observations sur les effets de l'Ipecacuanha." *Mémoires*, 1701, 190–195.

Bourdelin, Louis-Claude. "Mémoire sur la Formation des Sels lixiviels." *Mémoires*, 1730, 357–362.

Bourdelin, Louis-Claude. "Sur un sel connu sous le nom de Polychreste de Seignette." *Mémoires*, 1731, 124–129.

Bourdelin, Louis-Claude. "Recherches du sel d'Epsom." *Mémoires*, 1731, 347–357.

Boyle, Robert. *The Sceptical Chymist* (1661), facsimile edition (London: Dawsons of Pall Mall, 1965).

Boyle, Robert. *The Works of the Honourable Robert Boyle*, new edition, 6 vols. (London, 1772).

Boyle, Robert. "Of the Imperfection of the Chemists' Doctrine of Qualities (1675)." In *Selected Philosophical Papers of Robert Boyle*, ed. M. Stewart (Manchester University Press, 1979).

Boyle, Robert. *The Works of Robert Boyle* (ed. M. Hunter and E. Davis), 14 vols. (London: Pickering & Chatto, 1999).

Boyle, Robert. The Correspondence of Robert Boyle (ed. M. Hunter et al.), 6 vols. (London: Pickering & Chatto, 2001).

Brisson, M. J. *Pesanteur spécifique des corps. Ouvrage utile à l'histoire naturelle, à la physique, aux arts & au commerce* (Paris: Imprimerie royale, 1787).

Bucquet, Jean-Baptiste. *Introduction à l'étude des corps naturels, tirés du règne minéral*, 2 vols. (Paris: J. T. Hérissant, 1771).

Bucquet, Jean-Baptiste. *Introduction à l'étude des corps naturels tirés du règne végétal*, 2 vols. (Paris: J. T. Hérissant, 1773).

Bucquet, Jean-Baptiste. *Mémoires sur la manière don't les Animaux sont affectés par différens fluides Aériformes, Méphitiques; & sur les moyens de remédier aux effets de ces Fluides. Précédé d'une histoire abrégée des différens Fluides Aériformes ou Gas*, 8 vols. (Paris: Imprimerie royale, 1778).

Buff, Heinrich. "Einige Bemerkungen zur Affinitätslehre." *Berichte* 2, 1869, 142–147.

Buffon, H. N., ed. *Correspondance inédite de Buffon*, 2 vols. (Paris: Hachette, 1860).

Cadet, Louis Claude. "Expériences et Observations Chymiques sur le Diamant" [*Ecole de médecine*, cote 90958, t. 169, no. 4, 3–15].

Chardenon. "Lettre de M. Chardenon, docteur agrégé au Collége de Médecine de Dijon, & Membre de l'Académie de la même Villé, en réponse à celle de M. Ribaptome, insérée dans le Journal des Sçavans du mois de Décembre 1767, sur l'augmentation de poids des matiéres calcinées." *Journal des Sçavans*, 1768, 648–658.

Chardenon. "Mémoire sur l'augmentation de poids des métaux calcinés." *Mémoires de l'Académie de Dijon* 1, 1769, 303–320.

Condorcet. "Eloge de M. Macquer." *Histoire*, 1784, 20–30.

Condorcet. *Oeuvres de Condorcet*, published by A. Condorcet O'Connor and M. F. Arago, 12 vols. (Paris: Firmin Didot frères, 1847–49).

Crawford, Adair. *Experiments and Observations on Animal Heat, and the inflammation of combustible bodies: Being an attempt to resolve these phenomena into a general law of nature* (London: J. Murray, 1779).

D'Arcet and Rouelle. "Expériences nouvelles sur la destruction du Diamant dans les vaisseaux fermés." *Journal de médecine* 39, 1773, 50–86; *Observations* 1, 1773, 12–34.

De Clave, Etienne. *Nouvelle lumière philosophique des vrais principes et elemens de nature, & qualité d'iceux* (Paris: Olivier de Varennes, 1641).

De Clave, Etienne. *Cours de chimie* (1646).

De Fourcy. "Observations sur le tableau du produit des affinités chymiques." *Observations* 2, 1773, 197–204.

Deluc, J. A. *Recherches sur les modifications de l'atmosphère contenant l'histoire critique du baromètre & du thermomère, un traité sur la construction de ces instrumens, des expériences relatives à leurs usages, & principalement à la mesure des hauteurs & à la correction des réfractions moyennes*, 2 vols. (Genève, 1772).

Demachy, Jacques-François. "Exposition d'une nouvelle Table des principales combinaisons chymiques, connue jusqu'à présent sous le nom de Table des Rapports ou "Affinités." 82–305 and pl. I-VII in Demachy, *Recueil de Dissertations physico-chimiques, présentées a différentes académie* (Paris: Nyon & Barrois, 1781).

Diderot, Denis, ed. *Encyclopédie, ou, Dictonnaire raisonné des sciences, des arts et des métiers par une société des gens de lettres; mis en ordre & public par M. Diderot . . . & quant à la partie mathematique, par d'Alembert*, 17 vols. (Paris, 1751–1765).

Diderot, Denis, ed. *Supplément à l'Encyclopédie, ou dictionnaire raisonné des sciences, des arts et des métiers, par une société de gens de lettres*, 4 vols. (Amsterdam, 1776–1777).

Diderot, Denis. "Plan d'une Université pour le Gouvernment de Russie, ou d'une éducation publique dans toutes les sciences." *Oeuvres complètes de Diderot*, ed. J. Assézat and M. Tourneux (Paris, 1875–1877), III, 411–534.

Diderot, Denis. "Notices sur le peintre Michel Vanloo et le chimiste Rouelle." In *Oeuvres complètes de Diderot*, volume IV, ed. J. Assézat and M. Tourneux (Paris, 1875–1877).

Dodart, Denis. *Mémoire pour servir à l'histoire des plants* (Paris, 1731).

Duchesne, Joseph. *Le Grand Miroir du Monde* (Lyon: pour Barthelem Honorat, 1587); second edition (Lyon: pour les heritiers d'Eustache Vignon, 1593).

Duchesne, Joseph. *Liber de priscorum philosophorum verae medicinae materiae* (1603) (Leipzig: Thom. Schürer and Barthol. Voight, 1613).

Duchesne, Joseph. *Ad veritatem hermeticase medicinae ex Hoppocratis veterumque decretis ac therapeusi* (1604) (Frankfurt: Wolfgang Richter and Contrad Nebenius, 1605).

Duchesne, Joseph. *The Practice of Chymicall, and hermeticall Physicke, for the preservation of health* (London: Thomas Creede, 1605).

Du Clos, Samuel Cottereau. *Observations sur les Eaux Minérales de Plusieurs Provinces de France, Faites en l'Académie Royale des Sciences en l'Année 1670 et 1671* (Paris: l'Imprimerie royale, 1675).

Du Clos, Samuel Cottereau. *Observations on the Mineral Waters of France, made in the Royal Academy of the Sciences* (London: Henry Faithorne, 1684).

Du Clos, Samuel Cottereau. "Dissertation sur les principes des Mixtes naturels." *Mémoires de l'Académie royale des sciences* (1666–1699), 11 vols. (Paris: par la compagnie des libraires, 1729–1733), IV, 1–40.

Du Hamel du Monceau, Henri-Louis. "Sur les sel ammoniac." *Mémoires*, 1735, 106–116, 414–434, and 483–504.

Du Hamel du Monceau, Henri-Louis. "Sur la Base du Sel marin." *Mémoires*, 1736, 215–232.

Du Hamel and Grosse. "Sur les différentes manières de rendre le Tartre soluble." *Mémoires*, 1732, 323–342.

Dulong, Pierre-Louis. "Recherches sur la décomposition mutuelle des sels insolubles et des sels solubles." *Annales de chimie* 82, 1812, 275–308.

Eller, J. T. "Dissertation sur les elemens ou premiers principes des corps. Dans laquelle on prouve qu'il doit y avoir des elemens et qu'il y en a effectivement; qu'ils sont sujets à souffrir divers changemens, et meme susceptibles d'une parfaite transmutation; et enfin que le feu elementaire et l'eau sont les seules choses qui meritent proprement le non d'elemens." *Histoire de l'Académie royale des sciences et des belles-lettres de Berlin* 2, 1746, 3–48.

Eller, J. T. "Sur la nature et les propriétés de l'eau commune considerée comme un dissolvant." *Histoire de l'Académie royale des sciences et des belles-lettres de Berlin* 6, 1750, 67–97.

Encyclopédie méthodique, chimie, pharmacie, et métallurgie, vol. I (Paris, 1786).

Fontenelle, Bernard le Bovier de. "Preface to the History of the Academy of Sciences, from 1666 to 1699." In *The Achievement of Bernard le Bovier de Fontenelle* (Johnson Reprint Corp., 1970), part 4, 17–23.

Fontenelle, Bernard le Bovier de. "Eloge de M. Geoffroy." *Histoire*, 1731, 93–100.

Fontenelle, Bernard le Bovier de. *Eloges des academiciens avec l'histoire de l'académie royale des sciences en M. DC. XCIX. Avec un Discours preliminaire sur l'utilité des Mathematiques*, 2 vols. (La Haye: Isaac vander Kloot, 1740).

Fourcroy, Antoine-François de. *Leçons élémentaires d'Histoire naturelle et de Chimie* (Paris: rue et hôtel Serpente, 1782).

Fourcroy, Antoine-François de. *Mémoires et Observations de chimie* (Paris: Cuchet, 1784).

Fourcroy, Antoine-François de. "Mémoire pour servir l'explication du tableau de nomenclature (1787)." 109–122 in *Méthode de nomenclature chimique* (Paris: Seuil, 1994).

Fourcroy, Antoine-François de. *Philosophie chimique ou vérités fondamentales de la chimie moderne, disposées dans un nouvel ordre* (Paris, 1792).

Fourcroy, Antoine-François de. *Système des connaissances chimiques, et de leurs applications aux phénomènes de la nature et de l'art*, 10 vols. (Paris: Baudouin, 1800–1802).

Gay-Lussac, Louis-Joseph. "Mémoire sur la combinaison des substances gazeuses, les unes avec les autres." *Mémoires de la Société d'Arcueil* 2, 1809, 207–234; translated in *Foundations of the Molecular Theory, Alembic Club Reprints*, no. 4 (Edinburgh, 1950).

Gellert, C. E. *Anfangsgründe zur metallurgischen Chymie*, second edition (Leipzig, 1776).

Geoffroy, Claude-Joseph. "Observations sur la nature et la composition du sel ammoniac." *Mémoires*, 1720, 189–207.

Geoffroy, Claude-Joseph. "Suite des observations sur la fabrique du sel ammoniac, avec sa décomposition pour en tirer le sel volatil, que l'on nomme vulgairement sel d'Angleterre." *Mémoires*, 1723, 210–222.

Geoffroy, Claude-Joseph. "Expériences & Réflexions sur le Borax." *Mémoires*, 1728, 273–288.

Geoffroy, Claude-Joseph. "Examen des différentes Vitriols; avec quelques Essais sur la formation artificielle du Vitriol blanc & de l'Alun." *Mémoires*, 1728, 301–310.

Geoffroy, Claude-Joseph. "Nouvelles Expériences sur le Borax, avec un moyen facile de faire le Sel Sédatif, & d'avoir un Sel de Glauber, par la même opération." *Mémoires*, 1732, 398–418.

Geoffroy, Etienne-François. "Manière de recomposer le Souffre commun par la réunion de ses principes, et d'en composer de nouveau par le mélange de semblables substances, avec quelques conjectures sur le composition des métaux." *Mémoires*, 1704, 278–286.

Geoffroy, Etienne-François. "Problème de chimie. Trouver des Cendres qui ne contiennent aucunes parcelles de fer." *Mémoires*, 1705, 362–363.

Geoffroy, Etienne-François. "Eclaircissemens sur la production artificielle du fer, & sur la composition des autres Métaux." *Mémoires*, 1707, 176–188.

Geoffroy, Etienne-François. "Expériences sur les metaux, faites avec le Verre ardent du Palais Royal." *Mémoires*, 1709, 162–176.

Geoffroy, Etienne-François. "Du Changement des Sels acides en Sels alkalis volatiles urineux." *Mémoires*, 1717, 226–238.

Geoffroy, Etienne-François. "Des différents Rapports observés en Chymie entre différentes substances." *Mémoires*, 1718, 256–269; translated as "Table of the different relations observed in chemistry between different substances." *Science in Context* 9, 1996, 313–319.

Geoffroy, Etienne-François. "Eclaircissemens sur la Table insérée dans les Mémoires de 1718, concernant les rapports observés entre différentes substances." *Mémoires*, 1720, 20–34.

Geoffroy, Etienne-François. *A Treatise of the Fossil, Vegetable, and Animal Substances that are made Use of in Physick. Containing the history and description of them; with an account of their several virtues and preparations. To which is prefixed, an enquiry into the constituent principles of mixed bodies, and the proper methods of discovering the nature of medicines* (London: W. Innys and R. Manby, 1736).

Geoffroy, Etienne-François. *Tractatus de materia medica, sive de medicamentorum simplicium historia, virtute, delectu et usu*, 3 vols. (Paris: Joannis Desaint & Caroli Saillant, 1741).

Geoffroy, Etienne-François. *Traité de matière médicale, ou de l'histoire des vertus, du choix et de l'usage des remèdes simples*, 7 vols. (Paris: J. Desaint & C. Saillant, 1743).

Glaser, Christophe. *Traité de la chymie, enseignant par une briève et facile méthode toutes ses plus nécessaires préparations* (Paris; l'auteur, 1663).

Glaser, Christophe. *The Compleat Chymist, or, A New Treatise of Chymistry. Teaching by a Short and Easy Method All Its Most Necessary Preparations*, translated from fourth French edition (London: printed for John Starkey at the Miter, 1677).

Glauber, John Rudolph. *Furni Novi Philosophici oder Beschreibung einer neu-erfundenen Distillir-Kunst: Auch was für Spiritus, Olea, Flores, und andere vergleichen Vegetablische, Animalische, und Mineralische Medicamenten, damit auff eine sonderbahre Weise gantz leichtlich mit grossem Nutzen können zugericht und bereitet werden. Auch vozu solche dienen und in Medicina, Alchimia und anderen Künsten können gebraucht werden* (Amsterdam: Johann Fabeln, 1646).

Glauber, John Rudolph. *A Description of New Philosophical Furnaces, or Art of Distilling, divided into five parts. Whereunto is added a Description of the Tincture of Gold, or the true Aurum potable; also, the First part of the Mineral Work. Set forth and published for the sake of them that are studious of the TRUTH* (London: printed by Richard Coats for Tho. Williams, 1651).

Gourneux, M., ed. *Correspondance littéraire, philosophique et critique par Grimm, Diderot, Raynal, Meister, etc.* 16 vols. (Paris, 1877–1882).

'sGravesande, William Jacob van. *Physices elementa mathematica, experimentis confirmata. sive introductio ad philosophiam Newtonianam*, 2 vols. (Leyden, 1720–21).

'sGravesande, William Jacob van. *Mathematical Elements of Natural Philosophy confirmed by experiments, or an introduction to Sir Isaac Newton's Philosophy*, 2 vols. (London: printed for J. Senex and W. Taylor, 1720–21).

Grison, Emmanuel, Michelle Goupil, and Patrice Bret, eds. *A Scientific Correspondence during the Chemical Revolution: Louis-Bernard Guyton de Morveau & Richard Kirwan, 1782–1802* (Office for History of Science and Technology, University of California at Berkeley, 1994).

Guyton de Morveau, Louis-Bernard. *Mémoire sur l'éducation publique avec le prospectus d'un collège suivant les principes de cet ouvrage* (Dijon, 1764).

Guyton de Morveau, Louis-Bernard. "Mémoire sur les phénomènes de l'air dans la combustion." *Mémoires de l'Académie de Dijon* 1, 1769, 416–438.

Guyton de Morveau, Louis-Bernard. "Dissertation sur la phlogistique." In *Digressions académique ou Essais sur quelques sujets de Physique, de Chymie, & d'Histoire naturelle* (Dijon: L. N. Frantin, 1772).

Guyton de Morveau, Louis-Bernard. "Essai physico-chymique sur la Dissolution et la Crystallisation." In *Digressions académique ou Essais sur quelques sujets de Physique, de Chymie, & d'Histoire naturelle* (Dijon: L. N. Frantin, 1772).

Guyton de Morveau, Louis-Bernard. "Sur l'attraction ou la répulsion de l'eau & des corps huileux, pour vérifier l'exactitude de la méthode par laquelle de Doctreur TAYLOR estime la force d'adhésion des surfaces, & détermine l'action du verre sur le mercure des baromêtres, faites en présence de l'Académie des Sciences, Arts & Belles-Lettres de Dijon, dans son assemblée du 12 Février 1773." *Observations* 1, 1773, 172–173.

Guyton de Morveau, Louis-Bernard. *Défense de la volatilité du phlogistique.* Microfiche M12844, Bibliothèque nationale.

Guyton de Morveau, Louis-Bernard. *Mémoire sur l'utilité d'un cours public de chymie dans la ville de Dijon, les avantages qui en résulteroient pour la Province entière, & les moyens de procurer à peu de frais de Etablissement* (Dijon: L. N. Frantin, 1775).

Guyton de Morveau, Louis-Bernard. "Observation de la cristallisation du fer." *Observations* 8, 1776, 348–353.

Guyton de Morveau, Louis-Bernard. "Conciliation des Principes de Sthaal avec les Expériences modernes sur l'Air fixe." *Observations* 8, 1776, 389–395.

Guyton de Morveau, Louis-Bernard. "Lettre de M. de Morveau, à l'Auteur de ce Recueil, sur les crystallisations métalliques." *Observations* 13, 1779, 90–92.

Guyton de Morveau, Louis-Bernard. "Mémoire sur les Terres simples, & principalement sur celles qu'on nomme absorbante; suivi d'un appendice sur une nouvelle preuve de l'existence du Phlogistique dans la chaux, & de quelques observations sur le Sel phosphorique calcaire ou substance osseuse régénérée." *Observations* 17, 1781, 216–231.

Guyton de Morveau, Louis-Bernard. "Mémoire sur les Dénominations Chymiques, la nécessité d'en perfectionner le systém; & les règles pour y parvenir." *Observations* 19 1782, 370–382.

Guyton de Morveau, Louis-Bernard. "Mémoire sur le développement des principes de la nomenclature méthodique, lu à l'Académie, le 2 mai 1787." 75–108 in *Méthode de nomenclature chimique* (Seuil, 1994).

Guyton de Morveau, L. B., Hughes Maret, and J. F. Durande. *Elémens de chymie, théorique et pratique, rédigés dans un nouvel ordre, d'après les découvertes modernes*, 3 vols. (Dijon: L. N. Frantin, 1777–78).

Guyton de Morveau, L. B., A. L. Lavoisier, C. L. Berthollet, and A. F. de Fourcroy, *Méthode de nomenclature chimique* (Paris: Cuchet, 1787); reprinted with an introduction by Bernadette Bensaude-Vincent (Seuil, 1994).

Hales, Stephen. *Vegetable Staticks: or, An account of some statical experiments on the sap in vegetables: being an essay towards a natural history of vegetation. Also, a specimen of an attempt to analyse the air, by a great variety of chymostatical experiments, which were read at several meetings before the Royal Society* (London: W., and J. Innys, 1727).

Hales, Stephen. *La statique des végétaux et l'analyse de l'air. Expériences nouvelles lûes a la Société royale de Londres* (Paris: Debure l'aîne, 1735).

Histoire de l'Académie royale des sciences (1666–1699), 2 vols. (Paris: Gabriel Martin, 1733).

Histoire de l'Académie royale des sciences. Avec les Mémoires de Mathematique & de Physique, pour la même Année.

Homberg, Wilhelm. "Observation sur la quantité exacte des sels volatiles acides contenus dans les differens esprits acides." *Mémoires*, 1699, 44–51.

Homberg, Wilhelm. "Observations sur la quantité d'acides absorbés par les alcalis terreux." *Mémoires*, 1700, 64–71.

Homberg, Wilhelm. "Observations sur les huiles des plantes." *Mémoires*, 1700, 206–211.

Homberg, Wilhelm. "Observations sur quelques effets des Fermentations." *Mémoires*, 1701, 95–99.

Homberg, Wilhelm. "Observations sur les analyses des plantes." *Mémoires*, 1701, 113–117.

Homberg, Wilhelm. "Essays de Chimie. Article Premier: Des Principes de la Chimie en general." *Mémoires*, 1702, 33–52.

Homberg, Wilhelm. "Observations faites par le moyen du Verre ardant." *Mémoires*, 1702, 141–149.

Homberg, Wilhelm. "Essay de l'analyse du Souffre commun." *Mémoires*, 1703, 31–40.

Homberg, Wilhelm. "Suite de Essays de chimie. Article Troisième. Du Souphre Principe." *Mémoires*, 1705, 88–96.

Homberg, Wilhelm. "Observations sur le fer au verre ardent." *Mémoires*, 1706, 158–165.

Homberg, Wilhelm. "Suite de l'article trois des Essais de chimie." *Mémoires*, 1706, 260–272.

Homberg, Wilhelm. "Eclaircissemens touchant la vitrificaiton de l'or au verre ardent." *Mémoires*, 1707, 40–48.

Homberg, Wilhelm. "Mémoire touchant les Acide & les Alkalis, pour servir d'addition à l'article du Sel principe, imprimé dans nos Mémoires de l'année 1702. Pag. 36." *Mémoires*, 1708, 312–323.

Homberg, Wilhelm. "Suite des essais de chimie. Art. IV. Du Mercure." *Mémoires*, 1709, 106–117.

Homberg, Wilhelm. "Observations touchant l'effet de certains Acides sur les Alcalis volatils." *Mémoires*, 1709, 354–363.

Homberg, Wilhelm. "Observations sur les matieres sulphureuses et sur la facilité de les changer d'une espece de souffre en une autre." *Mémoires*, 1710, 225–234.

Homberg, Wilhelm. "Memoire touchant la Volatilisation des Sels fixes de Plantes." *Mémoires*, 1714, 186–195.

Huygens, Christiaan. *Oeuvres complètes de Christiaan Huygens*, 22 vols. (The Hague: Société hollandaise des sciences, 1888–1950).

Kirwan, Richard. "Experiments and Observations on the Specific Gravities and attractive Powers of various saline Substances." *Philosophical Transactions* 71, 1781, 7–41. Translation: *Journal de physique* 24, 1784, 134–156.

Kirwan, Richard. "Continuation of the Experiments and Observations on the Specific Gravities and Attractive Powers of various Saline Substances." *Philosophical Transactions* 72, 1782, 179–236. Translation: *Journal de physique* 24, 1784, 188–199, 356–368; 25, 1784, 13–28.

Kirwan, Richard. "Conclusion of the Experiment and Observations concerning the Attractive Powers of the Mineral Acids." *Philosophical Transactions* 73, 1783, 15–84. Translation: *Journal de physique* 27, 1785, 250–261, 321–335, 447–457; 28, 1786, 94–109.

Kirwan, Richard. *Elements of Mineralogy* (London: P. Elmsly, 1784).

Kirwan, Richard. *Essay on the Analysis of Mineral Waters* (London, 1784).

Kirwan, Richard. *An Essay on Phlogiston and the Constitution of Acids* (London: printed by J. Davis for P. Elmsly, 1787).

Kirwan, Richard. *Essai sur le phlogistique et sur la constitution des acides* (Paris: Rue et Hôtel Serpente, 1788).

Kirwan, Richard. *An Essay on Phlogiston and the Constitution of Acids* (London: J. Johnson, 1789); second edition (London: Frank & Cass Co., 1968).

Lavoisier, Antoine-Laurent. *Oeuvres de Lavoisier*, 6 vols. (Paris: Imprimerie Impériale, 1862–1893).

Lavoisier, Antoine-Laurent. "Analyse du gypse." *Mémoires de Mathématique et de Physique, présentés à l'Académie royale des Sciences, par divers Savans et lûs dans ses Assemblées* 5, 1768, 341–357; *Oeuvres* III, 111–127.

Lavoisier, Antoine-Laurent. "Sur le gypse, deuxième mémoire." *Oeuvres* III, 128–144.

Lavoisier, Antoine-Laurent. "Recherches sur les moyens les plus sûrs, les plus exacts et les plus commodes de déterminer la pesanteur spécifique des fluides soit pour la physique soit pour le commerce (read on March 23, 1768)." *Oeuvres* III, 427–450.

Lavoisier, Antoine-Laurent. "Sur la nature de l'eau et sur les expériences par lesquelles on a prétendu prouver la possibilité de son changement en terre. Première memoire." *Mémoires*, 1770 (1773), 73–82; *Oeuvres* II, 1–11.

Lavoisier, Antoine-Laurent. "Sur la nature de l'eau et sur les expériences par lesquelles on a prétendu prouver la possibilité de son changement en terre. Second Mémoire." *Mémoires*, 1770 (1773), 90–107; *Oeuvres* II, 11–28.

Lavoisier, Antoine-Laurent. "De la nature des eaux d'une partie de la Franche-Comité de l'Alsace, de la Lorraine, de la Champagne, de la Brie et du Valois." *Oeuvres* III, 145–205.

Lavoisier, Antoine-Laurent. *Opuscules physiques et chimiques* (Paris: Durand Neveu, 1774). Translation: *Essays Physical and Chemical* (London: printed for Joseph Johnson, 1776); second edition (London: Frank Cass & Co., 1970).

Lavoisier, Antoine-Laurent. "Mémoire sur la nature du principe qui se combine avec les Métaux pendant leur calcination, & qui en augmente le poids." *Observations* 5, 1775, 429–433; translated as "A Memoire on the Nature of the Principle which is combined with metals during their Calciantion, and occasions an Increase in their Weight." 407–419 in *Essays*; published in a revised form in *Mémoires*, 1775 (pub. 1778), 520–526; *Oeuvres* II, 122–128.

Lavoisier, Antoine-Laurent. "Mémoire sur l'existence de l'air dans l'acide nitreux." *Mémoires*, 1776, 671–680; *Oeuvres* II, 129–138.

Lavoisier, Antoine-Laurent. "Sur la combustion du phosphore de Kunckel, et sur la nture de l'acide qui résulte de cete combustion (Presented on April 16)." *Mémoires*, 1777, 65–78; *Oeuvres* II, 139–152.

Lavoisier, Antoine-Laurent. "Mémoire sur la Combustion en Général (read on Nov. 12, 1777)." *Mémoires*, 1777 (1780), 592–600; *Oeuvres* II, 225–233.

Lavoisier, Antoine-Laurent. "Considérations générales sur la nature des acides et sur les principes dont ils sont composés (submitted on Sept. 5, 1777 and read on Nov. 23, 1779)." *Mémoires*, 1778 (1781), 535–547; *Oeuvres* II, 248–260.

Lavoisier, Antoine-Laurent. "Extrait d'un mémoire lû par M. Lavoisier à la séance publique de l'Académie royale des sciences, du 12 novembre, sur la nature de l'eau." *Observations* 23, 1783, 452–455.

Lavoisier, Antoine-Laurent [with Laplace]. "Mémoire sur la chaleur (read on June 18, 1783)." *Mémoires*, 1780 (1784), 355–408; *Oeuvres* II, 283–333. English translation: *Memoir on Heat* (Neale Watson, 1982).

Lavoisier, Antoine-Laurent. "Considération générales sur la dissolution des métaux dans les acides." *Mémoires*, 1782 (1785), 492–511; *Oeuvres* II, 509–527.

Lavoisier, Antoine-Laurent. "Mémoires sur la précipitation des substances métalliques les unes par les autres (presented on Dec. 20, 1783)." *Mémoires*, 1782 (1785), 512–529; *Oeuvres* II, 528–545.

Lavoisier, Antoine-Laurent. "Mémoire sur l'affinité du principe oxygine avec les différentes substances auxquelles il est susceptible de s'unir (presented on Dec. 20. 1783)." *Mémoires*, 1782 (1785), 530–540; *Oeuvres* II, 546–556.

Lavoisier, Antoine-Laurent. "Réflexions sur le phlogistique, pour servir de développement à la théorie de la Combustion & de la Calcination, publiée en 1777." *Mémoires*, 1783 (1786), 505–538; *Oeuvres* II, 623–655.

Lavoisier, Antoine-Laurent. "Mémoire sur la nécessité de réformer & de perfectionner la nomenclature de la chimie, lu à l'assemblée publique de l'Académie royale des sicences, du 18 avril 1787." 63–74 in *Méthode de nomenclature chimique* (Paris: Seuil, 1994).

Lavoisier, Antoine-Laurent. *Oeuvres de Lavoisier VII: Correspondance*, 6 fascicules (1955–1997).

Lavoisier, Antoine-Laurent. *Elements of Chemistry* (Dover, 1965).

Lefebvre, Nicaise. *A Compleat Body of Chymistry*, 2 vols. (London: T. Ratcliffe, 1664).

Lemery, Louis. "Du Miel & de son Analyse Chimique." *Mémoires*, 1706, 272–283.

Lemery, Louis. "Que les Plantes contiennent réellement du fer, & que ce métal entre necessairement dans leur composition naturelle." *Mémoires*, 1706, 411–417.

Lemery, Louis. "Diverses Experiences et Observations chimiques et physiques. Sur le Fer et sur l'Aimant." *Mémoires*, 1706, 119–135.

Lemery, Louis. "Conjectures et reflexions sur la matiere du feu ou de la lumiere." *Mémoires*, 1709, 400–418.

Lemery, Louis. "Memoire sur les precipitations Chimiques; ou l'on examiné par occasion la dissolution de l'Or & de l'Argent, la nature particuliere des esprits acides, & la maniere don't l'esprit de nitre agit sur celuy de Sel dans la formation de l'Eau regale ordinaire." *Mémoires*, 1711, 56–79.

Lemery, Louis. "Conjectures sur les couleurs differentes des Précipités de Mercure." *Mémoires*, 1712, 51–70.

Lemery, Louis. "Second Memoire sur les Couleurs differentes des Précipités du Mercure." *Mémoires*, 1714, 259–280.

Lemery, Louis. "Explication mechanique de quelques differences asses curieuses qui resultent de la dissolution de different Sels dans l'Eau commune." *Mémoires*, 1716, 154–172.

Lemery, Louis. "Premiere Memoire sur le Nitre." *Mémoires*, 1717, 31–52.

Lemery, Louis. "Second Memoire sur le Nitre." *Mémoires*, 1717, 122–146.

Lemery, Louis. "Sur la Volatilisation vraye ou apparente des Sels fixes." *Mémoires*, 1717, 246–256.

Lemery, Louis. "Reflexions physiques sur le défaut & le peu d'utilité des Analises ordinaires des Plantes & des Animaux." *Mémoires*, 1719, 173–188.

Lemery, Louis. "Second Mémoire sur les analises ordinaires de chimie." *Mémoires*, 1720, 98–107.

Lemery, Louis. "Troisième Mémoire sur lse analises de chimie." *Mémoires*, 1720, 166–178.

Lemery, Louis. "Quatrième Mémoire sur les analises ordinaires des plantes et des animaux." *Mémoires*, 1721, 22–44.

Lemery, Louis. "Observations nouvelle et singuliére sur la dissolution successive de plusieurs Sels dans l'eau commune." *Mémoires*, 1724, 332–347.

Lemery, Louis. "Second Memoire ou Reflexions nouvelles sur une précipitation singuliere de plusieurs sels par un autre sel." *Mémoires*, 1727, 40–49.

Lemery, Louis. "Trosième Memoire ou Réflexions nouvelles sur une Précipitation singuliére de plusieurs Sels par un autre Sel." *Mémoires*, 1727, 214–228.

Lemery, Louis. "Expériences & Réflexions sur le Borax." *Mémoires*, 1728, 273–288.

Lemery, Louis. "Second Mémoire sur le Borax." *Mémoires*, 1729, 282–300.

Lemery, Louis. "Nouvel Eclaircissement sur l'Alun, sur les Vitriols, et particulierement sur la Composition naturelle, & jusqu'à présent ignorée, du Vitriol blanc ordinaire. Premier Mémoire." *Mémoires*, 1735, 262–280.

Lemery, Louis. "Second Mémoire sur les Vitriols, et particuliérement sur le Vitriol blanc ordinaire." *Mémoires*, 1735, 385–402.

Lemery, Louis. "Supplement aux deux Mémoires que j'ai donnés en 1735, sur l'Alun et sur les Vitriols." *Mémoires*, 1736, 263–301.

Lemery, Nicolas. *Cours de chimie* (1675).

Lemery, Nicolas. *A Course of Chemistry. containing The Easiest Manner of performing those Operations that are in Use in PHYSICK. illustrated With many Curious Remarks and Useful Discourses upon each OPERATION* (London: printed for Walter Kettilby at the Bishop's Head, 1677).

Lemery, Nicolas. *An Appendix to a Course of Chymistry, being Additional Remarks to the former Operations. Together with the Process of the Volatile Salt of Tartar, and some other useful Preparations* (London: printed for Walter Kettilby at the Bishop's Head, 1680).

Lemery, Nicolas. *A Course of Chemistry*, second English edition Translated by Walter Harris from fifth French edition (London: printed by R. N. for Walter Kettilby at the Bishop's Head, 1686).

Lemery, Nicolas. *Pharmacopée universelle, contenant toutes les compositions de pharmacie qui sont en usage dans la Medecine, tant en France que par toute l'Europe; leurs Vertus, leurs Doses, les manieres d'operer les plus simples & les meilleures.* (Paris: Laurent d'Houry, 1697).

Lemery, Nicolas. *Traité universel des drogues simples, mises en ordre alphabetique. Où l'on truove leurs differens noms, leur origine, leur choix, les principes qu'elles renferment, leurs qualitez, leur etymologie, & tout ce qu'il y a de particulier dans les Animaux, dans les Vegetaux & dans les Mineraux.* (Paris: Laurent d'Houry, 1698).

Lemery, Nicolas. *Traité de l'Antimonie, contenant l'Analyse Chymique de ce Mineral, & un recueil d'un grand nombre d'operations rapportées à l'Academie Royale des sciences, avec les raisonnemens qu'on a crus necessaires* (Paris: Jean Boudot, 1707).

Lemery, Nicolas. "Reflexions & Experiences sur le sublimé corrosif." *Mémoires*, 1709, 42–47.

Le Sage, George Louis. *Essai de chymie méchanique.*

Limbourg, Jean Philippe. *Dissertation de Jean philippe de Limbourg, docteur en medecine, sur les Affinités chymiques* (Liége: F. J. Desoer, 1761).

Macquer, Pierre-Joseph. "Sur la cause de la differente dissolubilité des huiles dans l'esprit de vin." *Mémoires*, 1745, 9–25.

Macquer, Pierre-Joseph. *Elémens de chimie—théorique* (Paris: J. T. Hérissant, 1749); second edition (Paris: P. F. Didot, 1756).

Macquer, Pierre-Joseph. *Elémens de chymie-pratique, contenant la description des opérations fondamentales de la chymie, avec des explications & des remarques sur chaque opération*, 2 vols. (Paris: J. T. Hérissant, 1751); second edition (Paris: J. T. Hérissant, 1756).

Macquer, Pierre-Joseph. *Art de la teinture en soie* (Paris: Desaint, 1763).

Macquer, Pierre-Joseph. *Dictionnaire de chymie, contenant la théorie & la pratique de cette science, son application à la physique, à l'histoire naturelle, à la médecine & à l'economie animale. Avec l'explication détaillée de la vertu & de la maniere d'agir des médicamens chymiques; et les principes fondamentaux des arts, manufactures & métiers dependans de la chymie*, 2 vols. (Paris: Lacombe, 1766); second edition 4 vols. (Paris: T. Barrois le jeune, 1778).

Macquer, Pierre-Joseph. *A Dictionary of Chemistry* (London: T. Cadell and P. Elmsly, 1771).

Macquer, Pierre-Joseph. "Résultat de quelques expériences faites sur le Diamant, par MM. Macquer, Cadet & Lavoisier, lu à la Séance pubilque de l'Académie Royale des Sciences, 29 Avril, 1772." *Introduction aux observations*, II, 108–111; reprinted in Guerlac, *Lavoisier*, 199–204.

Macquer, Pierre-Joseph, and Antoine Baumé. *Plan d'un Cours de Chymie expérimentale et raisonnée, avec un discours historique sur la chymie* (Paris: J. T. Hérissant, 1757).

Magellan. *Essai sur la nouvelle théorie du feu élémentaire, et de la chaleur des corps: avec la description des nouveaux thermomètres destinés particulierement aux observations sur ce sujet* (London: W. Richardson, 1780).

Magellan. "Essai sur la nouvelle théorie du feu élémentaire & de chaleur des corps." *Observations* 17, 1781, 375–386.

Magellan. "Suite du mémoire de M. H. Magellan sur le feu élémentaire et la chaleur: Sommaire de l'ouvrage du docteur Crawford." *Observations* 17, 1781, 411–422.

Mercier, Louis-Sébastien. *Tableaux de Paris*, 12 vols. (Amsterdam, 1782–1788).

Mitouard. "Extrait de deux Mémoires sur les Diamans & autres Pierres Précieuses, lûs à l'Académie Royale des Sciences les 2 & 7 Mai 1772, par M. Mittouart" [Ecole de médecine, cote 90958, t. 169, no. 4, 26–31].

Musschenbroek, P. van. *Essai de physique*, 2 vols. (Leyden, 1739).

Nollet, Jean Antoine. *Leçons de physique expérimentale*, 6 vols. (Paris: les frères Guérin, 1743–1748).

Ribaptome. "Lettre sur l'augmentation de poids dans la calciantion, adressée à Messieurs les Auteurs du Journal des Sçavans (dated July 20, 1765)." *Journal des Sçavans*, 1767, 889–894.

Rouelle, Guillaume-François. "Mémoire sur les sels neutres, dans lequel on propose une division méthodique de ses sels, qui facilite les moyens pour parvenir à la théorie de leur crystallisation." *Mémoires*, 1744, 353–364.

Rouelle, Guillaume-François. "Sur le Sel marin. Première partie. De la crystallization du Sel marin." *Mémoires*, 1745, 57–79.

Rouelle, Guillaume-François. "Mémoire sur les Sels neutres, dans lequel on fait connoître deux nouvelles classes de sels neutres, & l'on développe le phénomène singulier de l'excès d'acide dans ces sels." *Mémoires*, 1754, 572–588.

Rouelle et Cadet. *Analyses d'une eau minérale* (Paris, 1755) [Ecole de médecine, côte 90958, t. 64, no. 2].

Rousseau, Jean-Jacques. "Les Institutions chymiques." *Annales de la société Jean-Jacques Rousseau*, XII (1918/19)-XIII (1920/21).

Séances des Ecoles normales, recueillies par des sténographes, et revues par les professeurs, 6 vols. (Paris, 1796) [*Bibliothèque nationale*, R22215–22220].

Senac, Jean-Baptiste. *Nouveau cours de chimie, suivant les principes de Newton & de Sthall* (Paris: Jacques Vincent, 1723).

Spielmann, Jacob Reinbold. *Institutiones Chymiae* (Strasbourg: J. G. Bauerum, 1763).

Spielmann, Jacob Reinbold. *Instituts de Chymie*, 2 vols. (Paris: Vincente, 1770).

Sprat, Thomas. *The History of the Royal Society of London, for the improving of natural knowledge* (London: printed by T. R. for J. Martyn, 1667) (new edition: Washington University, St. Louis, 1958). Translation: *L'histoire de la société royale de Londre, establie pour l'enrichissement de la science naturelle* (Genève: pour Iean Herman Widerhold, 1669).

Stahl, Georg-Ernst. *Zymotechnia Fundamentalis* (Halle, 1697).

Stahl, Georg-Ernst. *Zufällige Gedancken und nützliche Bedencken über den Streit von dem sogennanten Sulphure, und zwar sowohl dem gemeinen verbrennlichen oder flüchtigen als unverbrennlichen oder fixen* (Halle: In

Verlegung des Waysenhauses, 1718). French edition: *Traité du soufre, ou remarques sur la dispute qui s'est élevée entre les Chymistes, au sujet du Soufre, tant commun, combustible ou volatil, que fixe, &c.* (Paris: P. F. Didot, 1766).

Stahl, Georg-Ernst. *Ausführliche Betrachtung und zulänglicher Beweiss von den Saltzen, dass dieselbe aus einer Zarten Erde, mit Wasser innig verbunden bestehen* (Halle, 1723). French edition: *Traité des sels, Dans lequel on démontre qu'ils sont composés d'une terre subtile, intimément combinée avec de l'eau* (Paris: Vincent, 1771); second edition (Paris, 1783).

Stahl, Georg-Ernst. *Philosophical Principles of Universal Chemistry, or The Foundation of a scientifical Manner of Inquiring into and Preparing the natural and Artificial Bodies for the Uses of Life: Both in the smaller Way of Experiment, and the larger Way of Business. Design'd as a General Introduction To the Knowledge and Practice of Artificial Philosophy: or, Genuine Chemistry in All Its Branches. Drawn from the Collegium Fenense of Dr. George Ernest Stahl. By Peter Shaw M. D.* (London: printed for John Osborn and Thomas Longman, 1730).

Star, Pieter van der, ed. *Fahrenheit's Letters to Leibniz and Boerhaave* (Rodopi, 1983).

Taylor, Brook. "Extract of a Letter from Dr. Brook Taylor, F.R.S. to Sir Hans Sloan, dated 25 June 1714. Giving an Account of some Experiments relating to Magnetism." *Philosophical Transactions* 31, 1720–21, 204–208.

Thomson, Thomas. *A System of Chemistry*, third edition, 5 vols. (Edinburgh: printed for Bell & Bradfute and E. Balfour, 1807).

Voltaire. *Letters concerning the English Nation* (London: printed for C. Davis and A. Lyon, 1733).

Voltaire. *Lettres philosophiques* (Amsterdam: E. Lucas, au Livre d'or, 1734).

Voltaire. *Elémens de la philosophie de Newton, mis à la portée de tout le monde* (Amsterdam, 1738).

Wilhelmy, Ludwig. "Ueber das Gesetz, nach welchem die Einwirkung des Säuren auf den Rohrzucker stattfindet." *Annalen der Physik* 81, 1850, 413–526.

Williamson, Alexander. "Suggestions for the Dynamics of Chemistry derived from the Theory of Aetherification." *Chemical Gazette* 9, 1851, 294–298 and 334–339.

Secondary Sources

Aarsleff, Hans. "The Berlin Academy under Frederick the Great." *History of the Human Sciences* 2, 1989, 193–206.

Abbri, Ferdinando. "The Chemical Revolution: A Critical Assessment." *Nuncius* 4, 1989, 303–315.

Adler, Jeremy. "Goethe's Use of Chemical Theory in His Elective Affinities." In *Romanticism and the Sciences*, ed. A. Cunningham and N. Jardin (Cambridge University Press, 1990).

Ahlers, Willem C. Un chimiste du XVIIIe siècle: Pierre-Joseph Macquer (1718–1784). Aspects de sa vie et de son oeuvre." Doctoral thesis, Ecole pratique es hautes Etudes, 1969.

Ahlers, Willem C. "La corresponance de Macquer." *Revue de synthèse* 97, 1976, 125–127 and 137–140.

Albury, W. R. The Logic of Condillac and the Structure of French Chemical and Biological Theory, 1780–1801. Ph.D. dissertation, Johns Hopkins University, 1972.

Alder, Ken. "A Revolution to Measure: The Political Economy of the Metric System in France." In *The Values of Precision*, ed. M. Wise (Princeton University Press, 1995).

Anderson, Wilda C. "The Rhetoric of Scientific Language: An Example from Lavoisier, *Modern Language Notes* 96, 1981, 746–770.

Anderson, Wilda C. *Between the Library and the Laboratory: The Language of Chemistry in Eighteenth-Century France* (Johns Hopkins University Press, 1984).

Ashworth, Edgar. *Boerhaave's Men at Leyden and After* (Edinburgh University Press, 1977).

Baker, A. Albert. "A History of Indicators." *Chymia* 9, 1964, 147–167.

Baker, Keith Michael. *Inventing the French Revolution* (Cambridge University Press, 1990).

Baker, Keith Michael. "Les débuts de Condorcet au secrétariat de l'Academie royale des sciences (1773–1776)." *Revue* 20, 1967, 229–280.

Baker, Keith Michael. *Condorcet: From Natural Philosophy to Social Mathematics* (University of Chicago Press, 1975).

Baker, Keith Michael. "Defining the Public Sphere in Eighteenth-Century France: Variations on a Theme by Habermas." In *Habermas and the Public Sphere*, ed. C. Calhoun (MIT Press, 1994).

Bell, David A. "The 'Public Sphere,' the State, and the World of Law in Eighteenth-Century France." *French Historical Studies* 17, 1992, 912–934.

Bell, David A. *Lawyers and Citizens: The Making of a Political Elite in Old Regime France* (Oxford University Press, 1994).

Bensaude-Vincent, Bernadette. "A Founder Myth in the History of Science? The Lavoisier Case." In *Functions and Uses of Disciplinary Histories*, ed. L. Graham et al. (Sociology of Sciences Yearbook 7) (Reidel, 1983).

Bensaude-Vincent, Bernadette. "Une mythologie révolutionaire dans la chimie française." *Annals of Science* 40, 1983, 189–196.

Bensaude-Vincent, Bernadette. "A View of the Chemical Revolution through Contemporary Textbooks: Lavoisier, Fourcroy and Chaptal." *BJHS* 23, 1990, 435–460.

Bensaude-Vincent, Bernadette. *Lavoisier: Mémoires d'une révolution* (Flammarion, 1993).

Bensaude-Vincent, Bernadette. "'The Chemist's Balance for Fluids': Hydrometers and Their Multiple Identities, 1770–1810." In *Instruments and Experimentation in the History of Chemistry*, ed. F. Holmes and T. Levere (MIT Press, 2000).

Bensaude-Vincent, Bernadette, and Ferdinando Abbri, eds. *Lavoisier in European Context: Negotiating a New Language for Chemistry* (Science History Publications, 1995).

Bensaude-Vincent, Bernadette, and Isabelle Stengers, *A History of Chemistry* (Harvard University Press, 1996).

Beretta, Marco. "T. O. Bergman and the Definition of Chemistry." *Lychnos*, 1988, 37–67.

Beretta, Marco. "Torbern Bergman in France: An unpublished letter by Lavoisier to Guyton de Morveau." *Lychnos*, 1992, 167–170.

Beretta, Marco. *The Enlightenment of Matter: The Definition of Chemistry from Agricola to Lavoisier* (Science History Publications, 1993).

Beretta, Marco, ed. *A New Course of Chemistry: Lavoisier's First Chemical Paper* (Firenze: L. S. Olschski, 1994).

Beretta, Marco. "Lavoisier as a reader of chemical literature." *Revue* 48, 1995, 71–94.

Beretta, Marco. "Humanism and Chemistry: The Spread of Georgius Agricola's Metallurgical Writings." *Nuncius* 12, 1997, 17–47.

Beretta, Marco. "Chemical Imagery and the Enlightenment of Matter." In *Science and the Visual Image in the Enlightenment*, ed. W. Shea (Science History Publications, 2000).

Beretta, Marco. "Lavoisier and His Last Printed Work: the *Mémoires de physique et de chimie* (1805)." *Annals of Science* 58, 2001, 327–356.

Berthelot, Marcelin. *La Revolution chimique—Lavoisier: Ouvrage suivi de notices et extraits des registres inédites de laboratoire de Lavoisier* (Paris: F. Alcan, 1890).

Bigourdan, M. G. "Le premières sociétés scientifiques de Paris au XVIIe siècle. Les réunions du P. Mersenne et l'Académie de Montmor." *Comptes Rendus* 164, 1917, 129–134, 159–162, and 216–220.

Boas, Marie. "An Early Version of Boyle's *Sceptical Chymist*." *Isis* 45, 1954, 153–168.

Boas, Marie. "Acid and Alkali in Seventeenth Century Chemistry." *Archives* 9, 1956, 13–28.

Boas, Marie. *Robert Boyle and Seventeenth-Century Chemistry* (Cambridge University Press, 1958).

Bouchard, Georges. *Guyton-Morveau, chimiste et conventionnel* (Paris: Perrin, 1938).

Bouchard, Marcel. *L'Académie de Dijon et le Premier Discours de Rousseau* (Paris: Société des Belles Lettres, 1950).

Bougard, Michel. *La chimie de Nicolas Lemery* (Turnhout: Brepols, 1999).

Boulad-Ayoub, Josiane. "Diderot et d'Holbach: un système matérialiste de la nature." *Dialogue* 24, 1985, 59–89.

Bouvet, M. "Les apothicaires royaux." *Revue pharm.* 1, 1930, 31–43, 76–83, 189–211, 242–262.

Bouvet, M. "Les papiers Macquer de la Bibliothèque nationale." *Revue pharm.* 134, 1952, 1–6.

Braunrot, Christabel P., and Kathleen Hardesty Doig. "The *Encyclopédie méthodique*: An Introduction." *Studies on Voltaire* 327, 1995, 1–152.

Brewer, Daniel. *The Discourse of Enlightenment in Eighteenth-Century France* (Cambridge University Press, 1993).

Briggs, Robin. "The Académie Royale des Sciences and the Pursuit of Utility." *Past and Present* 131, 1991, 38–88.

Britsch, Amédée. *La maison d'Orléans à la fin de l'ancien régime. La jeunesse de Philippe Egalité (1747–1785) d'après des documents inédits* (Paris: Payot, 1926).

Brockliss, Laurence W. B. "Aristotle, Descartes and the New Science: Natural Philosophy at the University of Paris, 1600–1740." *Annals of Science* 38, 1986, 33–69.

Brockliss, Laurence W. B. *French Higher Education in the Seventeenth and Eighteenth Centuries: A Cultural History* (Clarendon, 1987).

Brockliss, Laurence W. B. "The Medico-Religious Universe of an Early Eighteenth-Century Parisian Doctor: The Case of Philippe Hecquet." In *The Medical Revolution of the Seventeenth Century*, ed. R. French and A. Wear (Cambridge University Press, 1989).

Brockliss, Laurence W. B. "Consultation by Letter in Early Eighteenth-Century Paris: The Medical Practice of Etienne-François Geoffroy." In *French Medical Culture in the Nineteenth Century*, ed. A. La Berge and M. Feingold (Amsterdam: Rodopi, 1994).

Brockliss, Laurence, and Colin Jones. *The Medical World of Early Modern France* (Clarendon, 1997).

Broman, Thomas. "The Habermasian Public Sphere and the 'Science in the Enlightenment.'" *History of Science* 36, 1998, 123–149.

Brooke, John Hedley. "Organic Synthesis and the Unification of Chemistry—A Reappraisal." *BJHS* 5, 1971, 363–392.

Brown, Harcourt. *Scientific Organizations in Seventeenth Century France (1620–1680)* (New York: Russell & Russell, 1934).

Brown, Harcourt. "The Utilitarian Motive in the Age of Descartes." *Annals of Science* 1, 1936, 182–192.

Brunet, Pierre. *Les Physiciens hollandais et la méthode expérimentale en France au XVIIIe siècle* (Paris: Albert Blanchard, 1926).

Brunet, Pierre. *Maupertuis: Étude biographique*, 2 vols. (Paris: Albert Blanchard, 1929).

Brunet, Pierre. *L'Introduction des théories de Newton en France au XVIIIe siècle avant 1738* (Paris: Albert Blanchard, 1931).

Brygoo, Edouard. "Les médecins de Montpellier et le jardin du Roi à Paris." *Histoire et Nature* 14, 1979, 3–29.

Buchanan, Peta Dewar, J. F. Gibson, and Marie Boas Hall. "Experimental History of Science; Boyle's Colour Changes." *Ambix* 25, 1978, 208–210.

Cap, Paul-Antoine. *Nicolas Lemery* (Paris: Robin, 1839).

Chabbert, Pierre. "Jacques Borelly: Membre de l'Académie Royale des Sciences." *Revue* 23, 1970, 203–227.

Chalmers, Alan. "The Lack of Excellency of Boyle's Mechanical Philosophy." *SHPS* 24, 1993, 341–364.

Chang, (Kevin) Ku-Ming. "Fermentation, Phlogiston and Matter Theory: Chemistry and Natural Philosophy in Georg Ernst Stahl's *Zymotechnia Fundamentalis*." *Early Science and Medicine* 7, 2002, 31–64.

Chevalier, A. G. "The 'Antimony-War'—A Dispute between Montpellier and Paris." *Ciba Symposia* 2, 1940, 418–423.

Christie, J. R. R. "Narrative and Rhetoric in Hélène Metzger's Historiography of Eighteenth Century Chemistry." *History of Science* 25, 1987, 99–109.

Christie, J. R. R. "Historiography of Chemistry in the Eighteenth Century: Hermann Boerhaave and William Cullen." *Ambix* 41, 1994, 4–19.

Christie, J. R. R., and Jan V. Golinski. "The Spreading of the Word: New Directions in the Histotiography of Chemistry, 1600–1800." *History of Science* 20, 1982, 235–266.

Clarke, Jack A. "Abbé Jean-Paul Bignon, 'Moderator of the Academies' and Royal Librarian." *French Historical Studies* 8, 1973, 213–235.

Clark, William, Jan Golinski, and Simon Schaffer, eds. *The Sciences in Enlightened Europe* (University of Chicago Press, 1999).

Clericuzio, Antonio. "Robert Boyle and the English Helmontians." In *Alchemy Revisited*, ed. Z. von Martels (Brill, 1990).

Clericuzio, Antonio. "A Redefinition of Boyle's Chemistry and Corpuscular Philosophy." *Annals of Science* 47, 1990, 561–589.

Clericuzio, Antonio. "From van Helmont to Boyle: A Study of the Transmission of Helmontian Chemical and Medical Theories in Seventeenth-Century England." *BJHS* 26, 1993, 303–334.

Clericuzio, Antonio. "Carneades and the Chemists: A Study of *The Sceptical Chymist* and Its Impact on Seventeenth-Century Chemistry." In *Robert Boyle Reconsidered*, ed. M. Hunter (Cambridge University Press, 1994).

Clow, Archibald and Nan L. Clow. *The Chemical Revolution: A Contribution to Social Technology* (London: Batchworth, 1952).

Clucas, Stephen. "The Correspondence of a XVII-century 'Chymicall Gentleman': Sir Cheney Culpeper and the Chemical Interests of the Hartlib Circle." *Ambix* 40, 1993, 147–170.

Cohen, I. B. *Franklin and Newton: An Inquiry into Speculative Newtonian Experimental Science and Franklin's Work in Electricity as an Example thereof* (Philadelphia: American Philosophical Society, 1956).

Cohen, I. B. "Isaac Newton, Hans Sloane and the Académie royale des sciences." In *Mélanges Alexandre Koyré*, vol. 1, ed. I. Cohen and R. Taton (Hermann, 1964).

Cohen, I. B. *Revolution in Science* (Harvard University Press, 1985).

Coleby, L. J. M. *The Chemical Studies of P. J. Macquer* (London: George Allen & Unwin, 1938).

Coley, Awen A. M. "The Theme of Experiment in French Literature of the Earlier Eighteenth Century (1715–1761)." *Studies on Voltaire* 332, 1995, 213–333.

Coley, Noel G. "The Preparation and Uses of Artificial Mineral Waters." *Ambix* 31, 1984, 32–48.

Coley, Noel G. "Physicians, Chemists, and the Analysis of Mineral Waters: 'The Most Difficult Part of Chemistry.'" In *The Medical History of Water and Spas*, ed. R. Porter (London: Wellcome Institute for the History of Medicine, 1990).

Colquhoun, Hugh. "On the Life and Writings of Claude-Louis Berthollet." *Annals of Philosophy* 9, 1825, 1–18, 81–96 and 161–181.

Conant, James Bryant. *The Overthrow of the Phlogiston Theory: The Chemical Revolution of 1775–1789* (Harvard University Press, 1957).

Contant, Jean-Paul. *L'enseignement de la chimie au jardin royal des plantes de Paris* (Cahors: A. Coueslant, 1952).

Court, Susan and W. A. Smeaton. "Fourcroy and the *Journal de la société des pharmaciens de Paris*." *Ambix* 26, 1979, 39–55.

Creager, Angela N. H. *The Life of a Virus: Tobacco Mosaic Virus as an Experimental Model, 1930–1965* (University of Chicago Press, 2002).

Crosland, Maurice. "The Origins of Gay-Lussac's Law of Combining Volumes of Gases." *Annals of Science* 17, 1961, 1–26.

Crosland, Maurice. *Historical Studies in the Language of Chemistry* (Harvard University Press, 1962).

Crosland, Maurice. "The Development of Chemistry in the Eighteenth Century." *Studies on Voltaire* 24, 1963, 369–441.

Crosland, Maurice. *The Society of Arcueil: A View of French Science under Napoleon I* (Harvard University Press, 1967).

Crosland, Maurice. "Lavoisier's Theory of Acidity." *Isis* 64, 1973, 306–325.

Crosland, Maurice. *Gay-Lussac: Scientist and Bourgeois* (Cambridge University Press, 1978).

Crosland, Maurice. "Chemistry and the Chemical Revolution." In *The Ferment of Knowledge*, ed. G. Rousseau and R. Porter (Cambridge University Press, 1980).

Crosland, Maurice. "Lavoisier, Lone Genius or 'Chef d'Ecole'? The Testimony of Fourcroy." In *Lavoisier et la Révolution chimique*, ed. M. Goupil (SABIX—Ecole polytechnique, 1992).

Cunningham, Andrew. "Medicine to Calm the Mind: Boerhaave's Medical System, and Why It Was Adopted in Edinburgh." In *The Medical Enlightenment of the Eighteenth Century*, ed. A. Cunningham and R. French (Cambridge University Press, 1990).

Cunningham, Andrew, and Roger French, eds. *The Medical Enlightenment of the Eighteenth Century* (Cambridge University Press, 1990).

Cushing, Max Pearson. *Baron d'Holbach: A Study of Eighteenth Century Rationalism in France* (New York, 1914).

Daumas, Maurice. "Les Conceptions de Lavoisier sur les Affinités chimiques et la Constitution de la Matière." *Thalès* (1949–1950), 69–80.

Daumas, Maurice. "Les appareils d'expérimentation de Lavoisier." *Chymia* 3, 1950, 45–62.

Daumas, Maurice. "L'élaboration du *Traité de Chimie* de Lavoisier." *Archives* 12, 1950, 570–590.

Daumas, Maurice. *Les Instruments scientifiques aux XVIIe et XVIIIe siècles* (Presses universitaires de France, 1953). English edition: *Scientific Instruments of the Seventeenth and Eighteenth Centuries* (Praeger, 1953).

Daumas, Maurice. *Lavoisier, Théoricien et expérimentateur* (Presses Universitaires de France, 1955).

Daumas, Maurice, and Denis Duveen. "Lavoisier's Relatively Unknown Large-Scale Decomposition and Synthesis of Water, February 27 and 28, 1785." *Chymia* 5, 1959, 113–129.

Davidson, Arnold I. "Archaeology, Genealogy, Ethics." In *Foucault*, ed. D. Hoy (Basil Blackwell, 1986).

Davis, Tenney L. "The Vicissitudes of Boerhaave's Textbook of Chemistry." *Isis* 10, 1928, 33–46.

Dear, Peter. *Discipline and Experience: The Mathematical Way in the Scientific Revolution* (University of Chicago Press, 1995).

Debus, Allen G. "The Paracelsian Compromise in Elizabethan England." *Ambix* 8, 1960, 71–97.

Debus, Allen G. "Solution Analysis prior to Robert Boyle." *Chymia* 8, 1962, 41–62.

Debus, Allen G. "The Paracelsian Aerial Niter." *Isis* 55, 1964, 43–61.

Debus, Allen G. *The English Paracelsians* (Franklin Watts, 1966).

Debus, Allen G. "Fire Analysis and the Elements in the Sixteenth and the Seventeenth Centuries." *Annals of Science* 23, 1967, 127–147.

Debus, Allen G., ed. *Science, Medicine and Society in the Renaissance: Essays to honor Walter Pagel* (Science History Publications, 1972).

Debus, Allen G. *The Chemical Philosophy: Paracelsian Science and Medicine in the Sixteenth and Seventeenth Centuries*, 2 vols. (Science History Publications, 1977).

Debus, Allen G. "The Role of Chemistry in the Scientific Revolution." *Lias* 13, 1986, 139–150.

Debus, Allen G. "Iatrochemistry and the Chemical Revolution." In *Alchemy Revisited*, ed. Z. von Martels (Brill, 1990).

Debus, Allen G. "Chemistry and the Universities in the Seventeenth Century." *Estudos avançados* 4, 1990, 176–196.

Debus, Allen G. *The French Paracelsians: The Chemical Challenge to Medical and Scientific Tradition in Early Modern France* (Cambridge University Press, 1991).

Debus, Allen G. "Chemists, Physicians, and Changing Perspectives on the Scientific Revolution." *Isis* 89, 1998, 66–81.

Delorme, S. "Tableau chronologique de la vie et des oeuvre de Fontenelle" *Revue* 10, 1957, 289–309.

Demeulenaere-Couyère, Christiane, ed. *Il y a 200 ans Lavoisier: Actes du Colloque organisé à l'occasion du bicentenaire de la mort d'Antoine Laurent Lavoisier le 8 mai 1794* (Paris: Lavoisier Tec & Doc, 1995).

de Milt, Clara. "Early Chemistry at Le Jardin du Roi." *Journal of Chemical Education* 18, 1941, 503–510.

de Milt, Clara. "Christopher Glaser." *Journal of Chemical Education* 19, 1942, 53–60.

Departer, C. "Petrus van Musschenbroek (1692–1761): A Dutch Newtonian." *Janus* 64, 1977, 77–87.

Dhombres, Jean. "Quelques Réflexions de et sur Chaptal. A propos de la Révolution chimique." In *Lavoisier et la Révolution chimique*, ed. M. Goupil (SABIX—Ecole polytechnique, 1992).

Dobbs, Betty Jo Teeter, and Margaret C. Jacob. *Newton and the Culture of Newtonianism* (Humanities Press, 1995).

Donovan, Arthur. "James Hutton, Joseph Black and the Chemical Theory of Heat." *Ambix* 25, 1978, 176–190.

Donovan, Arthur. *Philosophical Chemistry in the Scottish Enlightenment: The Doctrines and Discoveries of William Cullen and Joseph Black* (Edinburgh University Press, 1975).

Donovan, Arthur. "Scottish Responses to the New Chemistry of Lavoisier." *Studies in Eighteenth-Century Culture* 9, 1979, 237–249.

Donovan, Arthur, ed. *The Chemical Revolution: Essays in Reinterpretation*, Osiris [2] 4, 1988.

Donovan, Arthur. "The Chemical Revolution and the Enlightenment—And a Proposal for the Study of Scientific Change." In *Philosophy and Science in the Scottish Enlightenment*, ed. P. Jones (Edinburgh: John Donald, 1988).

Donovan, Arthur. "Lavoisier as Chemist *and* Experimental Physicist: A Reply to Perrin." *Isis* 81, 1990, 270–272.

Donovan, Arthur. *Antoine Lavoisier: Science, Administration, and Revolution* (Blackwell, 1993).

Donovan, Arthur. "Newton and Lavoisier—From Chemistry as a Branch of Natural Philosophy to Chemistry as a Positive Science." In *Action and Reaction*, ed. P. Theerman and A. Seeff (University of Delaware Press, 1993).

Donovan, Arthur. "The Chemical Revolution and the Enlightenment—And a Proposal for the Study of Scientific Change." In *Philosophy and Science in the Scottish Enlightenment*, ed. P Jones (Edinburgh: John Donald, 1988).

Donovan, Michael. "Biographical Account of the late Richard Kirwan." *Proceedings of the Royal Irish Academy* 4, 1850, lxxxi–cxviii.

Dorveaux, P. "Apothicaires membres de l'Académie Royale des Sciences. III. Simon Boulduc." *Revue pharm.* 67, 1930, 5–15; "IV. Gilles-François Boulduc." ibid. 74, 1931, 113–117; "V. Etienne-François Geoffroy." ibid. 2, 1931, 118–126; "VI. Nicolas Lemery." ibid. 75, 1931, 208–219; "VII. Claude-Joseph Geoffroy." ibid. 3, 1932, 113–122; "IX. Guillaume-François Rouelle." ibid. 4, 1933, 169–186.

Dufour, Théophile, *Les Institutions chimiques de Jean-Jacques Rousseau* (Geneva: Imprimerie du Journal de Genève, 1905).

Dulieu, Louis. "Un Parisien, professeur à l'Université de Médecine de Montpellier: Charles Le Roy (1726–1779)." *Revue* 6, 1953, 50–59.

Duncan, A. M. "Some Theoretical Aspects of Eighteenth-Century Tables of Affinity." *Annals of Science* 18 (1962), 177–196 and 217–232.

Duncan, A. M. "The Function of Affinity Tables and Lavoisier's List of Elements." *Ambix* 17, 1970, 28–42.

Duncan, A. M. "Particles and Eighteenth Century Concepts of Chemical Combination." *BJHS* 21, 1988, 447–453.

Duncan, A. M. *Laws and Order in Eighteenth-Century Chemistry* (Clarendon, 1996).

Duveen, Denis I., and Roger Hahn. "Deux encyclopédistes hors de l'Encyclopédie: Philippe Macquer et l'abbé Jaubert." *Revue* 12, 1959, 330–342.

Duveen, Denis I., and Herbert S. Klickstein, *A Bibliography of the Works of Antoine Laurent Lavoisier, 1743–1794* (London: Wm. Dawson & Sons, Ltd., and E. Weil, 1954).

Duveen, Denis I., and Herbert S. Klickstein. "A Letter from Berthollet to Blagden Relating to the Experiments for a Large-Scale Synthesis of Water Carried Out by Lavoisier and Meusnier in 1785." *Annals of Science* 10, 1954, 58–62.

Duveen, Denis I., and Herbert S. Klickstein. "A Letter from Guyton de Morveau to Macquart relating to Lavoisier's Attack against the Phlogiston Theory (1778)." *Osiris* 12, 1956, 342–367.

Eagles, Christina M. "David Gregory and Newtonian Science." *BJHS* 36, 1977, 216–225.

Eamon, William. "New Light on Robert Boyle and the Discovery of Colour Indicators." *Ambix* 27, 1980, 204–209.

Edelstein, Sidney M. "Priestley Settles the Water Controversy." *Chymia* 1, 1948, 123–137.

Eklund, Jon. Chemical Analysis and the Phlogiston Theory, 1738–1772: Prelude to Revolution. Ph.D. dissertation, Yale University, 1972.

Eklund, Jon. *The Incompleat Chymist* (Smithsonian Institution Press, 1975).

Evans, David S. *Lacaille: Astronomer, Traveler* (Tucson: Pachart, 1992).

Evans, James. "Gravity in the Century of Light: Sources, Construction and Reception of Le Sage's Theory of Gravitation." In *Pushing Gravity*, ed. M. Edwards (Montreal: Apeiron, 2002).

Falk, Raphael. "What Is a Gene?" *SHPS* 17, 1986, 133–173.

Fayet, Joseph. *La Revolution française et la science, 1789–1795* (Paris: Marcel Rivière, 1960).

Feldman, Theodore S. "Late Enlightenment Meteorology." In *The Quantifying Spirit*, ed. T. Frängsmyr et al. (University of California Press, 1990).

Ferrone, Vincenzo, *The Intellectual Roots of the Italian Enlightenment: Newtonian Science, Religion, and Politics in the Early Eighteenth Century* (Humanities Press, 1995).

Fichman, Martin. "French Stahlism and Chemical Studies of Air, 1750–1770." *Ambix* 18, 1971, 94–122.

Findlen, Paula. "A Forgotten Newtonian: Women and Science in the Italian Province." In *The Sciences in Enlightened Europe*, ed. W. Clark et al. (University of Chicago Press, 1999).

Fox, Robert. *The Caloric Theory of Gases: From Lavoisier to Regnault* (Clarendon, 1971).

Fox, Robert. "The Rise and Fall of Laplacian Physics." *HSPS* 4, 1974, 89–136.

Frängsmyr, Tore, J. L. Heilbron, and Robin E. Rider, eds. *The Quantifying Spirit in the 18th Century* (University of California Press, 1990).

Frank, Robert G. Jr. *Harvey and the Oxford Physiologists* (University of California Press, 1980).

French, Roger. "Sickness and the Soul: Stahl, Hoffmann and Sauvages on Pathology." In *The Medical Enlightenment of the Eighteenth Century*, ed. A. Cunningham and R. French (Cambridge University Press, 1990).

French, Roger, and Andrew Wear, eds. *The Medical Revolution of the Seventeenth Century* (Cambridge University Press, 1989).

French, Sidney J. "The Chemical Revolution—The Second Phase." *Journal of Chemical Education* 27, 1950, 83–89.

Freyne, Michael. "L'Eloge de M. Newton dans la correspondance de Fontenelle." *Corpus* 13, 1990, 75–92.

Fric, René. "Contribution à l'étude de l'évolution des ideés de Lavoisier sur la nature de l'air et sur la calcination des métaux." *Archives* 12, 1959, 137–168.

Fujii, Kiyohisa. "The Berthollet-Proust Controversy and Dalton's Chemical Atomic Theory, 1800–1820." *BJHS* 19, 1986, 177–200.

Gago, Ramon. "The New Chemistry in Spain." *Osiris* [2] 4, 1988, 169–192.

Galison, Peter. "History, Philosophy, and the Central Metaphor." *Science in Context* 2, 1988, 197–212.

Galison, Peter. *Image and Logic: A Material Culture of Microphysics* (University of Chicago Press, 1997).

Germain, A. *L'Apothicairerie à Montpellier sous l'ancien régime universitaire* (Montpellier: J. Martel, 1882).

Geyer-Kordesch, Johanna. "Georg Ernst Stahl's Radical Pietist Medicine and Its Influence on the German Enlightenment." In *The Medical Enlightenment of the Eighteenth Century*, ed. A. Cunningham and R. French (Cambridge University Press, 1990).

Gibbs, F. W. "Peter Shaw and the Revival of Chemistry." *Annals of Science* 7, 1951, 211–237.

Gibbs, F. W. "Boerhaave's Chemical Writings." *Ambix* 6, 1958, 117–135.

Gibson, Thomas. "A Sketch of the Career of Theodore Turquet de Mayerne." *Annals of Medical History* 5, 1933, 315–326.

Gillispie, Charles Coulston. "The Discovery of the Leblanc Process." *Isis* 48, 1957, 152–170.

Gillispie, Charles Coulston. *The Edge of Objectivity: An Essay in the History of Scientific Ideas* (Princeton University Press, 1960).

Gillispie, Charles Coulston. *Science and Polity at the End of the Old Regime* (Princeton University Press, 1980).

Gillispie, Charles Coulston. *Pierre-Simon Laplace, 1749–1827: A Life in Exact Science* (Princeton University Press, 1997).

Goldgar, Anne. *Impolite Learning: Conduct and Community in the Republic of Letters, 1680–1750* (Yale University Press, 1995).

Golinski, Jan. "Hélène Metzger and the Interpretation of Seventeenth Century Chemistry." *History of Science* 25, 1987, 85–97.

Golinski, Jan. "Scepticism and Authority in Seventeenth-century Chemical Discourse." In *The Figural and the Literal*, ed. A. Benjamin et al. (Manchester University Press, 1987).

Golinski, Jan. "Utility and Audience in Eighteenth-Century Chemistry: Case Studies of William Cullen and Joseph Priestley." *BJHS* 21, 1988, 1–31.

Golinski, Jan. "The Secret Life of an Alchemist." In *Let Newton Be! A New Perspective on His Life and Works*, ed. J. Fauvel et al. (Oxford University Press, 1988).

Golinski, Jan. "Chemistry in the Scientific Revolution: Problems of Language and Communication." In *Reappraisals of the Scientific Revolution*, ed. D. Lindberg and R. Westman (Cambridge University Press, 1990).

Golinski, Jan. *Science as Public Culture: Chemistry and Enlightenment in Britain, 1760–1820* (Cambridge University Press, 1992).

Golinski, Jan. "The Chemical Revolution and the Politics of Language." *The Eighteenth Century: Theory and Interpretation* 33, 1992, 238–251.

Golinski, Jan. "Precision Instruments and the Demonstrative Order of Proof in Lavoisier's Chemistry." *Osiris* [2] 9, 1993, 30–47.

Golinski, Jan. " 'The Nicety of Experiment': Precision of Measurement and Precision of Reasoning in Late Eighteenth-Century Chemistry." In *The Values of Precision*, ed. M. Wise (Princeton University Press, 1995).

Golinski, Jan. "Fit Instruments: Thermometers in Eighteenth-Century Chemistry." In *Instruments and Experimentation in the History of Chemistry*, ed. F. Holmes and T. Levere (MIT Press, 2000).

Gooding, D., T. Pinch, and S. Schaffer, eds. *The Uses of Experiment: Studies in the Natural Sciences* (Cambridge University Press, 1989).

Goodman, Dena. *The Republic of Letters: A Cultural History of the French Enlightenment* (Cornell University Press, 1994).

Gordon, Daniel. " 'Public Opinion' and the Civilizing Process in France: The Example of Morellet." *Eighteenth-Century Studies* 22, 1989, 302–328.

Gordon, Daniel. "Beyond the Social History of Ideas: Morellet and the Enlightenment." In *André Morellet (1727–1819) in the Republic of Letters and the French Revolution*, ed. J. Merrick and D. Medlin (Peter Lang, 1995).

Gough, J. B. "Lavoisier's Early Career in Science: An Examination of Some New Evidence." *BJHS* 4, 1968, 52–57.

Gough, J. B. "Lavoisier and the Caloric Theory." *BJHS* 6, 1972, 1–38.

Gough, J. B. "The Origin of Lavoisier's Theory of the Gaseous State." In *The Analytic Spirit*, ed. H. Woolf (Cornell University Press, 1981).

Gough, J. B. "Some Early References to Revolutions in Chemistry." *Ambix* 29, 1982, 106–109.

Gough, J. B. "Lavoisier's Memoirs on the Nature of Water and Their Place in the Chemical Revolution." *Ambix* 30, 1983, 89–106.

Gough, J. B. "Lavoisier and the Fulfillment of the Stahlian Revolution." *Osiris* [2] 4, 1988, 15–33.

Goulard, Roger. "A propos de l'Affaire des poisons, le célèbre édit de 1682." *Bulletin de la Société française d'histoire de la médecine* 13, 1914, 260–268.

Goupil, Michelle. "J. Ph. de Limbourg et la théorie des affinités chimiques." *Technologia* 7, 1984, 11–28.

Goupil, Michelle. *Du flou au clair? Histoire de l'affinité chimique* (Paris: Editions du C.T.H.S, 1991).

Goupil, Michelle, ed. *Lavoisier et la Révolution chimique: Actes du colloque tenu à du bicentenaire de la publication du "Traité élémentaire de chimie"* (SABIX—Ecole polytechnique, 1992).

Goupil, Michelle. "Claude Louis Berthollet, Collaborateur et Continuateur (?) de Lavoisier." In *Lavoisier et la Révolution chimique*, ed. M. Goupil (SABIX—Ecole polytechnique, 1992).

Grapi, Pere. "The Marginalization of Berthollet's Chemical Affinities in the French Textbook Tradition at the Beginning of the Nineteenth Century." *Annals of Science* 58, 2001, 111–135.

Grapi, Pere, and Mercè Izquierdo. "Berthollet's Conception of Chemical Change in Context." *Ambix* 44, 1997, 113–130.

Greenberg, John L. "Mathematical Physics in Eighteenth-Century France." *Isis* 77, 1986, 59–78.

Greenway, Frank. *John Dalton and the Atom* (Cornell University Press, 1966).

Greenway, Frank. "Boerhaave's Influence on Some 18th Century Chemists." In *Boerhaave and His Time*, ed. G. Lindeboom (Brill, 1970).

Grell, Ole Peter, ed. *Paracelsus: The Man and His Reputation, His Ideas and Their Transformation* (Brill, 1998).

Grimaux, Edouard. *Lavoisier, 1743–1794, d'après sa correspondance, ses manuscrits, ses papiers de famille et d'autres documents inédits*, second edition (Paris: Gerner Bailière, 1896).

Guedon, Jean-Claude. The Still Life of a Transition: Chemistry in the Encyclopédie. Ph.D. dissertation, University of Wisconsin, 1974.

Guedon, Jean-Claude. "Protestantisme et Chimie: Le Milieu Intellectuel de Nicolas Lemery." *Isis* 65, 1974, 212–228.

Guedon, Jean-Claude. "Chimie et matérialisme: La stratégie anti-Newtonienne de Diderot." *Dix-huitième siècle* 11, 1979, 185–200.

Guerlac, Henry. "The Continential Reputation of Stephen Hales." *Archives* 4, 1951, 393–404.

Guerlac, Henry. "Lavoisier and His Biographers." *Isis* 45, 1954, 51–62.

Guerlac, Henry. "A Note on Lavoisier's Scientific Education." *Isis* 47, 1956, 211–216.

Guerlac, Henry. "A Lost Memoir of Lavoisier." *Isis* 50, 1959, 125–129.

Guerlac, Henry. "The Origin of Lavoisier's Work on Combustion." *Archives* 47, 1959, 113–135.

Guerlac, Henry. "Some French Antecedents of the Chemical Revolution." *Chymia* 5, 1959, 73–112.

Guerlac, Henry. "A Curious Lavoisier Episode." *Chymia* 7, 1961, 103–108.

Guerlac, Henry. *Lavoisier—The Crucial year: The Background and Origin of His First Experiments on Combustion in 1772* (Cornell University Press, 1961).

Guerlac, Henry. "Lavoisier's Draft Memoir of July 1772." *Isis* 60, 1969, 380–382.

Guerlac, Henry. "Guy de La Brosse and the French Paracelsians." In *Science, Medicine and Society in the Renaissance*, ed. A. Debus (Science History Publications, 1972).

Guerlac, Henry. *Antoine-Laurent Lavoisier: Chemist and Revolutionary* (Scribner, 1975).

Guerlac, Henry. "Chemistry as a Branch of Physics: Laplace's Collaboration with Lavoisier." *HSPS* 7, 1976, 193–276.

Guerlac, Henry. "The Chemical Revolution: A Word from Monsieur Fourcroy." *Ambix* 23, 1976, 1–4.

Guerlac, Henry. *Essays and Papers in the History of Modern Science* (Johns Hopkins University Press, 1977).

Guerlac, Henry. "The Lavoisier Papers—A Checkered History." *Archives* 29, 1979, 95–100.

Guerlac, Henry. "Some Areas for Further Newtonian Studies." *History of Science* 17, 1979, 75–101.

Guerlac, Henry. *Newton on the Continent* (Cornell University Press, 1981).

Guerrini, Anita. "The Tory Newtonians: Gregory, Pitcairne, and Their Circle." *Journal of British Studies* 25, 1986, 288–311.

Guerrini, Anita. "Isaac Newton, George Cheyne and the 'Principia Medicinae.'" In *The Medical Revolution of the Seventeenth Century*, ed. R. French and A. Wear (Cambridge University Press, 1989).

Habermas, Jürgen. *The Structural Transformation of the Public Sphere: An Inquiry into a Category of Bourgeois Society* (MIT Press, 1992).

Hackmann, W. D. "Scientific Instruments: Models of Brass and Aids to Discovery." In *The Uses of Experiment*, ed. D. Gooding et al. (Cambridge University Press, 1989).

Hahn, Roger. *The Anatomy of a Scientific Institution: The Paris Academy of Sciences, 1666–1803* (University of California Press, 1971).

Hall, A. Rupert. "Henry Oldenburg et les relations scientifiques au XVIIe siècle." *Revue* 23, 1970, 285–303.

Hall, A. Rupert. "Newton in France: A New View." *History of Science* 13, 1975, 233–250.

Hall, A. Rupert. "Isaac Newton and the Aerial Nitre." *Notes and Records* 52, 1998, 51–61.

Hall, A. Rupert. *Isaac Newton: Eighteenth-Century Perspectives* (Oxford University Press, 1999).

Hall, Marie Boas. *Robert Boyle on Natural Philosophy: An Essay with Selections from His Writings* (Indiana University Press, 1965).

Hall, Marie Boas. "The Royal Society's Role in the Diffusion of Information in the Seventeenth Century." *Notes and Records* 29, 1974–5, 173–192.

Hanks, L. *Buffon avant l'Histoire naturelle* (Presses universitaires de France, 1966).

Hannaway, Owen. *The Chemists and the Word: The Didactic Origins of Chemistry* (Johns Hopkins University, 1975).

Haraway, Donna. "Situated Knowledges: The Science Question in Feminism as a Site of Discourse on the Privilege of Partial Perspective." *Feminist Studies* 14, 1988, 575–599. Reprinted in Haraway, *Simians, Cyborgs, and Women: The Reinvention of Nature* (Routledge, 1991).

Hardesty, Kathleen. *The Supplément to the Encyclopédie* (The Hague: Nijhoff, 1977).

Harding, Sandra. *The Science Question in Feminism* (Cornell University Press, 1986).

Harding, Sandra. *Whose Science? Whose knowledge? Thinking from Women's Lives* (Cornell University Press, 1991).

Harth, Erica. *Cartesian Women: Versions and Subversions of Rational Discourse in the Old Regime* (Cornell University Press, 1992).

Hayles, N. Katherine. *Chaos Bound: Orderly Disorder in Contemporary Literature and Science* (Cornell University Press, 1990).

Heggarty, Diane Elizabeth. Richard Kirwan: The Natural Philosopher and the Chemical Revolution. Ph.D. dissertation, University of Washington, 1978.

Henry, Charles, ed. *Introduction a la chymie: Manuscrit inédit de Diderot publié avec notice sur les cours de Rouelle et tarif des produits chimiques en 1758* (Paris: E. Dentu, 1887).

Heilbron, John. *Electricity in the 17th & 18th Centuries: A Study of Early Modern Physics* (University of California Press, 1979).

Heilbron, John. *Elements of Early Modern Physics* (University of California Press, 1982).

Heilbron, John. "A Revolution to Measure." In *The Quantifying Spirit*, ed. T. Frängsmyr et al. (University of California Press, 1990).

Heimann, P. M. "Newtonian Natural Philosophy and the Scientific Revolution." *History of Science* 11, 1973, 1–7.

Hill, C. *The World Turned Upside Down: Radical Ideas during the English Revolution* (London: Temple Smith, 1972).

Hine, W. L. "Mersenne and Alchemy." In *Alchemy Revisited*, ed. Z. von Martels (Brill, 1990).

Hirschfield, John Milton. *The Académie royale des sciences, 1666–1683* (Arno, 1981).

Holmes, Frederic Lawrence. "From Elective Affinities to Chemical Equilibria: Bertholet's Law of Mass Action." *Chymia* 8, 1962, 105–145.

Holmes, Frederic Lawrence. "Analysis by Fire and Solvent Extractions: The Metamorphosis of a Tradition." *Isis* 62, 1971, 129–148.

Holmes, Frederic Lawrence. *Lavoisier and the Chemistry of Life: An Exploration of Scientific Creativity* (University Wisconsin Press, 1985).

Holmes, Frederic Lawrence. "Lavoisier's Conceptual Passage." *Osiris* [2] 4, 1988, 82–92.

Holmes, Frederic Lawrence. *Eighteenth Century Chemistry as an Investigative Enterprise* (Berkeley: Office for History of Science and Technology, 1989).

Holmes, Frederic Lawrence. "Lavoisier the Experimentalist." *Bulletin for the History of Chemistry* 5, 1989, 24–31.

Holmes, Frederic Lawrence. "The Boundaries of Lavoisier's chemical revolution." *Revue* 48, 1995, 9–48.

Holmes, Frederic Lawrence. "The Communcal Context for Etienne-François Geoffroy's 'Table des rapports.'" *Science in Context* 9, 1996, 289–311.

Holmes, Frederic Lawrence. "What was the Chemical Revolution about?" *Bulletin for the History of Chemistry* 20, 1997, 1–9.

Holmes, Frederic Lawrence. *Antoine Lavoisier—The Next Crucial Year* (Princeton University Press, 1998).

Holmes, Frederic Lawrence. "The Evolution of Lavoisier's Chemical Apparatus." In *Instruments and Experimentation in the History of Chemistry*, ed. F. Holmes and T. Levere (MIT Press, 2000).

Holmes, Frederic Lawrence. "The 'Revolution in Chemistry and Physics': Overthrow of a Reigning Paradigm or Competition between Contemporary Research Programs?" *Isis* 91, 2000, 735–753.

Holmes, Frederic L., and Trevor H. Levere, eds. *Instruments and Experimentation in the History of Chemistry* (MIT Press, 2000).

Home, R. W. "'Newtonianism' and the Theory of the Magnet." *History of Science* 15, 1979, 252–266.

Home, R. W. "Out of a Newtonian Straitjacket: Alternative Approaches to Eighteenth-Century Physical Science." In *Studies in the Eighteenth Century*, ed. R. Brissenden and J. Eade (Australian National University Press, 1979).

Home, R. W. "Nollet and Boerhaave: A Note on Eighteenth-Century Ideas about Electricity and Fire." *Annals of Science* 36, 1979, 171–176.

Hooykaas, R. "Die Elementenlehre des Paracelcus." *Janus* 39, 1935, 175–188.

Hooykaas, R. "Die Elementenlehre der Iatrochemiker." *Janus* 41, 1937, 1–28.

Hooykaas, R. "The Experimental Origin of Chemical Atomic and Molecular Theory before Boyle." *Chymia* 2, 1949, 65–80.

Horn, Jeff, and Margaret C. Jacob. "Jean-Antoine Chaptal and the Cultural Roots of French Industrialization." *Technology and Culture* 39, 1998, 671–698.

Howard, R. C. "Guy de la Brosse and the Jardin des Plantes in Paris." In *The Analytic Spirit*, ed. H. Woolf (Cornell University Press, 1981).

Howard, R. C. "Guy de La Brosse: Botanique et chimie au début de la révolution scientifique." *Revue* 31, 1978, 301–326.

Hufbauer, Karl. *The Formation of the German Chemical Community, 1720–1795* (University of California Press, 1982).

Hunter, Michael. "Alchemy, Magic and Moralism in the Thought of Robert Boyle." *BJHS* 23, 1990, 387–410.

Hunter, Michael, ed. *Robert Boyle Reconsidered* (Cambridge University Press, 1994).

Jacob, J. R. "Robert Boyle and Subversive Religion in the Early Restoration." *Albion* 6, 1974, 275–293.

Jacob, J. R. "Restoration, Reformation and the Origins of the Royal Society." *History of Science* 13, 1975, 155–176.

Jacob, J. R. *Robert Boyle and the English Revolution: A Study in Intellectual and Social Change* (New York: Burt Franklin, 1977).

Jacob, J. R. "Boyle's Atomism and the Restoration Assault on Pagan Naturalism." *Social Studies of Science* 8, 1978, 211–233.

Jacob, J. R. "Restoration Ideologies and the Royal Society." *History of Science* 17, 1980, 25–38.

Jacob, Margaret C. *Scientific Culture and the Making of the Industrial West* (Oxford University Press, 1997).

Jacques, Jean. "Le 'Cours de chimie de G.-F. Rouelle recueilli par Diderot.'" *Revue* 38, 1985, 43–53.

Jennings, Richard C. "Lavoisier's Views on Phlogiston and the Matter of Fire before about 1770." *Ambix* 27, 1981, 206–209.

Joly, Bernard. "Voltaire chimiste: L'influence des théories de Boerhaave sur sa doctrine du feu." *Revue du nord* 77, 1995, 817–843.

Joly, Bernard. "L'alkahest, dissolvant universel ou quand la théorie rend pensable une pratique impossible." *Revue* 49, 1996, 305–354.

Jones, Colin. "The Great Chain of Buying: Medical Advertisement, the Bourgeois Public Sphere, and the Origins of the French Revolution." *American Historical Review* 101, 1996, 13–40.

Julien, Pierre. "Sur les relations entre Macquer et Baumé." *Revue pharm.* 39, 1992, 65–77.

Kahn, Didier. "Entre atomisme, alchimie et théologie: la réception des thèses d'Atoine de Villon et Etienne de Clave contre Aristote, Paracelse et les 'cabalistes.'" *Annals of Science* 58, 2001, 241–286.

Kapoor, Satish C. "Berthollet, Proust, and Proportions." *Chymia* 10, 1965, 53–110.

Kawashima, Keiko. "Madame Lavoisier et la traduction française de l'Essay on Phlogiston de Kirwan." *Revue* 53, 2000, 235–263.

Kegel-Brinkgreve, E., and A. M. Luyendijk-Elshout, eds. *Boerhaave's Orations* (Brill, 1983).

Keller, Evelyn Fox. "Linking Organisms and Computers: Theory and Practice in Contemporary Biology." In *Connecting Creations*, ed. M. Safir (Xunta de Galicia, 2000).

Kent, A., and Owen Hannaway. "Some New Considerations on Beguin and Libavius." *Annals of Science* 16, 1960, 241–250.

Kerker, Milton. "Herman Boerhaave and the Development of Pneumatic Chemistry." *Isis* 46, 1955, 36–49.

Keyser, Barbara Whitney. "Between Science and Craft: The Case of Berthollet and Dyeing." *Annals of Science* 47, 1990, 213–260.

Kim, Mi Gyung. Practice and Representation: Investigative Programs of Chemical Affinity in the Nineteenth Century. Ph.D. dissertation, University of California, Los Angeles, 1990.

Kim, Mi Gyung. "The Layers of Chemical Language I: Constitution of Bodies v. Structure of matter." *History of Science* 30, 1992, 69–96.

Kim, Mi Gyung. "The Layers of Chemical Language II: Stabilizing Atoms and Molecules in the Practice of Organic Chemistry." *History of Science* 30, 1992, 397–437.

Kim, Mi Gyung. "Constructing Symbolic Spaces: Chemical Molecules in the Académie des Sciences." *Ambix* 43, 1996, 1–31.

Kim, Mi Gyung. "Labor and Mirage: Writing the History of Chemistry." *SHPS* 26, 1995, 155–165.

Kim, Mi Gyung. "Chemical Analysis and the Domains of Reality: Wilhelm Homberg's Essais de Chimie, 1702–1709." *SHPS* 31, 2000, 37–69.

Kim, Mi Gyung. "The Analytic Ideal of Chemical Elements: Robert Boyle and the French Didactic Tradition of Chemistry." *Science in Context* 14, 2001, 361–395.

Kim, Yung Sik. "Another Look at Robert Boyle's Acceptance of the Mechanical Philosophy: Its Limits and Its Chemical and Social Contexts." *Ambix* 38, 1991, 1–10.

King, Lester S. *The Medical World of the Eighteeenth Century* (Krieger, 1958).

Klein, Ursula. *Verbindung und Affinität: Die Grundlegung der neuzeitlichen Chemie an der Wende vom 17. Zum 18. Jahrhundert* (Birkhäuser, 1994).

Klein, Ursula. "Robert Boyle—Der Begründer der neuzeitlichen Chemie?" *Philosophia Naturalis* 31, 1994, 63–106.

Klein, Ursula. "Origin of the Concept of Chemical Compound." *Science in Context* 7, 1994, 163–204.

Klein, Ursula. "E.F. Geoffroy's Table of different 'Rapports' observed between different chemical substances—A Reinterpretation." *Ambix* 42, 1995, 251–287.

Klein, Ursula. "The Chemical Workshop Tradition and the Experimental Practice: Discontinuities within Continuities." *Science in Context* 9, 1996, 251–287.

Kohler, Robert E. "The Origin of Lavoisier's First Experiment on Combustion." *Isis* 63, 1972, 349–355.

Kohler, Robert E. "Lavoisier's Rediscovery of the Air from Mercury Calx: A Reinterpretation." *Ambix* 22, 1975, 52–57.

Konert, Jürgen. "Academic and Practical Medicine in Halle during the Era of Stahl, Hoffman, and Juncker." *Caduceus* 13, 1997, 23–38.

Kors, Alan Charles. *D'Holbach's Coterie: An Enlightenment in Paris* (Princeton University Press, 1976).

Kuhn, T. S. "Robert Boyle and Structural Chemistry in the Seventeenth Century." *Isis* 43, 1952, 12–36.

Lafont, Olivier. "Personnalisation des rapports individu-puissance publique ou: Geoffroy et la famille Le Tellier." *Revue pharm.* 38, 1991, 15–23.

Laissus, Yves and Anne-Marie Monseigny. "*Les Plantes du Roi*. Note sur un grand ouvrage de botanique préparé au XVIIe siècle par l'Académie royale des sciences." *Revue* 22, 1969, 193–236.

Langins, Janis. "Fourcroy, Historien de la Révolution chimique." In *Lavoisier et la Révolution chimique*, ed. M. Goupil (SABIX—Ecole polytechnique, 1992).

Laudan, Rachel. *From Mineralogy to Geology: The Foundations of a Science, 1650–1830* (University of Chicago Press, 1987).

Lawrence, Christopher. "Medical Minds, Surgical Bodies: Corporeality and the Doctors." In *Science Incarnate*, ed. C. Lawrence and S. Shapin (University of Chicago Press, 1998).

Le Grand, H. E. "Lavoisier's Oxygen Theory of Acidity." *Annals of Science* 29, 1972, 1–18.

Le Grand, H. E. "A Note on Fixed Air: The Universal Acid." *Ambix* 10, 1973, 88–94.

Le Grand, H. E. "Ideas on the Composition of Muriatic Acid and Their Relevance to the Oxygen Theory of Acidity." *Annals of Science* 31, 1974, 213–225.

Le Grand, H. E. "Determination of the Composition of the Fixed Alkalis, 1789–1810." *Isis* 65, 1974, 59–65.

Le Grand, H. E. "The 'Conversion' of C. L. Berthollet to Lavoisier's Chemistry." *Ambix* 22, 1975, 58–70.

Le Grand, H. E. "Berthollet's *Essai de statique chimique* and Acidity." *Isis* 67, 1976, 229–238.

Le Grand, H. E. "Genius and the Dogmatization of Error: The Failure of C. L. Berthollet's Attack upon Lavoisier's Acid Theory." *Organon* 12–13, 1976–1977, 193–209.

Le Grand, H. E. "Chemistry in a Provincial Context: The Montpellier Sociéte Royale des Sciences in the Eighteenth Century." *Ambix* 29, 1982, 88–105.

Le Grand, H. E. "Theory and Appication: The Early Chemical Work of J. A. C. Chaptal." *BJHS* 17, 19484, 31–46.

Leicester, Henry M. "The Spread of the Theory of Lavoisier in Russia." *Chymia* 5, 1959, 138–144.

Lemay, Pierre. "Les cours de Guillaume-François Rouelle." *Revue pharm.* 123, 1949, 434–442 .

Lemay, Pierre. "Le Cours de Pharmacie de Rouelle." *Revue pharm.* 152, 1957, 17–21.

Lesch, John E. *Science and Medicine in France: The Emergence of Experimental Physiology, 1790–1855* (Harvard University Press, 1984).

Lesseux, Arnould de. "L'éducation publique d'après Guyton de Morveau." *Académie des sciences (de Dijon), Arts et belles-lettres, Memoires,* 123, 1976–1978 (pub. 1979), 207–239.

Levere, Trevor H. *Affinity and Matter: Elements of Chemical Philosophy, 1800–1865* (Clarendon, 1971).

Levere, Trevor H. "Lavoisier: Language, Instruments, and the Chemical Revolution." In *Nature, Experiment, and the Sciences,* ed. T. Levere and W. Shea (Kluwer, 1990).

Levere, Trevor H. "Balance and Gasometer in Lavoisier's Chemical Revolution." In *Lavoisier et la Révolution chimique,* ed. M. Goupil (SABIX—Ecole polytechnique, 1992).

Levere, Trevor H. "Measuring Gases and Measuring Goodness." In *Instruments and Experimentation in the History of Chemistry*, ed. F. Holmes and T. Levere (MIT Press, 2000).

Levere, Trevor H. *A History of Chemistry from Alchemy to the Buckyball* (Johns Hopkins University Press, 2001).

Levere, T. H., and G. Turner, eds. *Discussing Chemistry and Steam: The Minutes of a Coffee-House Philosophical Society, 1780–1787* (Oxford University Press, 2002).

Levin, A. "Venel, Lavoisier, Fourcroy, Cabanis and the Idea of Scientific Revolution: The French Political Context and the General Patterns of Conceptualization of Scientific Change." *History of Science* 22, 1984, 303–320.

Licoppe, Christian. *La formation de la pratique scientifique: Le discours de l'expérience en France et en Angleterre (1630–1820)* (Découverte, 1996).

Lindeboom, G. A. "Pitcairne's Leyden Interlude Described from the Documents." *Annals of Science* 19, 1963, 273–284.

Lindeboom, G. A. *Herman Boerhaave: The Man and His Work* (Methuen, 1968).

Lindeboom, G. A., ed. *Boerhaave and His Time* (Brill, 1970).

Llana, James W. "A Contribution of Natural History to the Chemical Revolution in France." *Ambix* 32, 1985, 71–91.

Lodwig, T. H., and W. A. Smeaton. "The Ice Calorimeter of Lavoisier and Laplace and Some of Its Critics." *Annals of Science* 31, 1974, 1–18.

Long, Pamela O. "The Openness of Knowledge: An Ideal and Its Context in 16th-Century Writings in Mining and Metallurgy." *Technology and Culture* 32, 1991, 318–355.

Love, Rosaleen. "Some Sources of Herman Boerhaave's Concept of Fire." *Ambix* 19, 1972, 157–174.

Lundgren, Anders. "The New Chemistry in Sweden: The Debate That Wasn't." *Osiris* [2] 4, 1988, 146–168.

Lux, David S. *Patronage and Royal Science in Seventeenth-Century France* (Cornell University Press, 1989).

Lux, David S. "Colbert's Plan for the *Grande Académie*: Royal Policy toward Science, 1663–67." *Seventeenth-Century French Studies* 12, 1990, 177–188.

Maluf, Ramez Bahige. Jean Antoine Nollet and Experimental Natural Philosophy in Eighteenth Century France. Ph.D. dissertation, University of Oklahoma, 1985.

Marsak, Leonard M. "Bernard de Fontenelle: The Idea of Science in the French Enlightenment." *Transactions of the American Philosophical Society* 49 (1959), part 7.

Marsak, Leonard M. Introduction to *The Achievement of Bernard le Bovier de Fontenelle*, The Sources of Science, no. 76 (Johnson Reprint Corp., 1970).

Martels, Z. R. W. M. von, ed. *Alchemy Revisited: Proceedings of the International Conference on the History of Alchemy at the University of Groningen, 17–19 April 1989* (Brill, 1990).

Martin, Julian. "Sauvage's Nosology: Medical Enlightenment in Montpellier." In *The Medical Enlightenment of the Eighteenth Century*, ed. A. Cunningham and R. French (Cambridge University Press, 1990).

Mauskopf, Seymour. "Gunpowder and the Chemical Revolution." *Osiris* [2] 4, 1988, 93–118.

Mauskopf, Seymour. "Chemistry and Cannon: J.-L. Proust and Gunpowder Analysis." *Technology and Culture* 31, 1990, 398–426.

Mauskopf, Seymour. "Lavoisier and the Improvement of Gunpowder Production." *Revue* 48, 1995, 95–121.

Mauskopf, Seymour. " 'I passionately desire to be able to link together all, or at least most, of chemical phenomena in a system.' Richard Kirwan's Phlogiston Theory: Its Success and Fate." Presentation at Dexter Symposium, Chicago, 2000.

Mayer, Jean. *Diderot, homme de science* (Rennes: Bretonne, 1959).

Mayer, Jean. "Portrait d'un chimiste: Guillaume-François Rouelle (1703–1770)." *Revue* 23, 1970, 305–332.

McClaughin, Trevor. "Sur les rapports entre la Compagnie de Thévenot et l'Académie royale des Sciences." *Revue* 28, 1975, 235–242.

McDonald, E. "The Collaboration of Bucquet and Lavoisier." *Ambix* 13, 1966, 74–83.

McEvoy, John G. "The Enlightenment and the Chemical Revolution." In *Metaphysics and Philosophy of Science in the Seventeenth and Eighteenth Centuries*, ed. R. Woolhouse (Kluwer, 1988).

McEvoy, John G. "Continuity and Discontinuity in the Chemical Revolution." *Osiris* [2] 4, 1988, 195–213.

McEvoy, John G. "The Chemical Revolution in Context." *The Eighteenth Century: Theory and Interpretation* 33, 1992, 198–216.

McEvoy, John G. "Positivism, Whiggism, and the Chemical Revolution: A Study in the Historiography of Chemistry." *History of Science* 35, 1997, 1–33.

McKie, Douglas. *Antoine Laurent Lavoisier: The Father of Modern Chemistry* (Philadelphia: Lippincott, 1935).

McKie, Douglas. "Béraut's Theory of Calcination (1747)." *Annals of Science* 1, 1936, 269–293.

McKie, Douglas. "Some Early Work on Combustion, Respiration and Calcination." *Ambix* 1, 1938, 143–165.

McKie, Douglas. "Antoine Laurent Lavoisier, F.R.S. (1743–1794)." *Notes and Records* 7, 1950, 1–41.

McKie, Douglas. "Guillaume-François Rouelle (1703–170)." *Endeavour*, July 1953, 130–133.

McKie, Douglas. "Macquer, the First Lexicographer of Chemistry." *Endeavour* 16, 1957, 133–136.

McKie, Douglas. "Fontenelle et la Société Royale de Londres." *Revue* 10, 1957, 334–338.

Meinel, Christoph. "Early Seventeenth-Century Atomism: Theory, Epistemology, and the Insufficiency of Experiments." *Isis* 79 (1988), 68–103.

Meldrum, Andrew N. *The Eighteenth Century Revolution in Science—The First Phase* (Longmans, Green, 1930).

Meldrum, Andrew N. "Lavoisier's Work on the Nature of Water and the Supposed Transmutation of Water into Earth (1768–1773)." *Archeion* 14, 1932, 246–247.

Meldrum, Andrew N. "Lavoisier's Three Notes on Combustion: 1772." *Archeion* 14, 1932, 17–19.

Meldrum, Andrew N. "Lavoisier's Early Works in Science, 1763–1771." *Isis* 19, 1933, 330–363; 20, 1934, 396–425.

Melhado, Evan M. *Jacob Berzelius: The Emergence of His Chemical System* (Almqvist & Wiksell, 1981).

Melhado, Evan M. "Oxygen, Phlogiston, and Caloric: The Case of Guyton." *HSPS* 13, 1983, 311–334.

Melhado, Evan M. "Chemistry, Physics, and the Chemical Revolution." *Isis* 76, 1985, 195–211.

Melhado, Evan M. "Toward an Understanding of the Chemical Revolution." *Knowledge and Society* 8, 1989, 123–137.

Melhado, Evan M. "Mineralogy and the Autonomy of Chemistry around 1800." *Lychnos* 1990, 229–261.

Melhado, Evan M. "On the Historiography of Science: A Reply to Perrin." *Isis* 81, 1990, 273–276.

Melhado, Evan M. "Scientific Biography and Scientific Revolution: Lavoisier and Eighteenth-Century Chemistry." *Isis* 87, 1996, 688–694.

Melhado, Evan M., and Tore Frängsmyr, eds. *Enlightenment Science in the Romantic Era: The Chemistry of Berzelius and Its Cultural Setting* (Cambridge University Press, 1992).

Mendelsohn, J. Andrew. "Alchemy and Politics in England." *Past and Present* 135, 1992, 30–78.

Metzger, Hélène. "L'Evolution du règne métallique d'après les Alchimistes du XVIIe siècle." *Isis* 4, 1922, 466–482.

Metzger, Hélène. "La Philosophie de la matière chez Stahl et ses Disciples." *Isis* 8, 1926, 427–464.

Metzger, Hélène. "La Théorie de la composition des sels et la théorie de la combustion d'après Stahl et ses Disciples." *Isis* 9, 1927, 294–325.

Metzger, Hélène. *La Philosophie de la matière chez Lavoisier* (Hermann, 1935).

Metzger, Hélène. *Les Doctrines chimiques en France du début du XVIIe à la fin du XVIIIe siècle*, new edition (Paris: Albert Blanchard, 1969).

Metzger, Hélène. *Newton, Stahl, Boerhaave et la Doctrine chimique* (Paris: Félix Alcan, 1930).

Moran, Bruce T. "Paracelsus, Religion, and Dissent: The Case of Philipp Homagius and Georg Zimmermann." *Ambix* 43, 1996, 65–79.

Moran, Bruce T. "Medicine, Alchemy, and the Control of Language: Andreas Libavius versus the Neoparacelsians." In *Paracelsus*, ed. O. Grell (Brill, 1998).

Moran, Bruce T. "Libavius the Paracelsian? Mosntrous Novelties, Institutions, and the Norms of Social Virtue." In *Reading the Book of Nature*, ed. A. Debus and M. Walton (Thomas Jefferson University Press, 1998).

Morris, Robert J. "Lavoisier on Fire and Air: The Memoir of July 1772." *Isis* 60, 1969, 374–380.

Morris, Robert J. "Lavoisier and the Caloric Theory." *BJHS* 6, 1972, 1–38.

Muirhead, James Patrick. *Correspondence of the Late James Watt on His Discovery of the Theory of the Composition of Water* (London: J. Murray; Edinburgh: W. Blackwood, 1846).

Multhauf, Robert P. "Sal Ammoniac: A Case History in Industrialization." *Technology and Culture* 6, 1965, 569–586.

Multhauf, Robert P. *Neptune's Gift: A History of Common Salt* (Johns Hopkins University Press, 1978).

Nathans, Benjamin. "Habermas's 'Public Sphere' in the Era of the French Revolution." *French Historical Studies* 16, 1990, 620–644.

Neville, Roy G. "Christophle Glaser and the *Traité de la chymie*, 1663." *Chymia* 10, 1965, 25–52.

Neville, Roy G., and W. A. Smeaton. "Macquer's *Dictionnaire de Chymie*: A Bibliogrpahical Study." *Annals of Science* 38, 1981, 613–662.

Newman, William R. *Gehennical Fire: The Lives of George Starkey, an American Alchemist in the Scientific Revolution* (Harvard University Press, 1994).

Newman, William R. "Boyle's Debt to Corpuscular Alchemy." In *Robert Boyle Reconsidered*, ed. M. Hunter (Cambridge University Press, 1994).

Newman, William R. "George Starkey and the Selling of Ssecrets." In *Samuel Hartlib and Universal Reformation*, ed. M. Greengrass et al. (Cambridge University Press, 1994).

Newman, William R. "Art, Nature, and Experiment among Some Aristotelian Alchemists." In *Texts and Contexts in Ancient & Medieval Science*, ed. E. Sylla and M. McVaugh (Brill, 1997).

Newman, William R., and Lawrence M. Principe. "Alchemy vs. Chemistry: The Etymological Origins of a Historiographic Mistake." *Early Science and Medicine* 3, 1998, 32–65.

Niderst, Alain, ed. *Fontanelle: Actes du colloque tenu à Rouen du 6 au 10 Octobre, 1987* (Presses Universitaires de France, 1989).

Niderst, Alain. *Fontenelle* (Plon, 1991).

Nye, Mary Jo. *From Chemical Philosophy to Theoretical Chemistry: Dynamics of Matter and Dynamics of Discipline, 1800–1950* (University of California Press, 1993).

Oldroyd, David. "An Examination of G. E. Stahl's *Philosophical Principles of Universal Chemistry*." *Ambix* 20, 1973, 36–52.

Oldroyd, David. "From Paracelsus to Haüy: The Development of Mineralogy in Its Relation to Chemistry." *Ambix* 21, 1974, 157–178.

Oldroyd, David. "Mineralogy and the Chemical Revolution." *Centaraus* 21, 1975, 54–71.

Oster, Malcolm. "The Scholar and the Craftsman Revisited: Robert Boyle as Aristocrat and Artisan." *Annals of Science* 49, 1992, 255–276.

Oster, Malcolm. "Biography, Culture, and Science: The Formative Years of Robert Boyle." *History of Science* 31, 1993, 177–226.

Oster, Malcolm. "Virtue, Providence and Political Neutralism: Boyle and Interregnum Politics." In *Robert Boyle Reconsidered*, ed. M. Hunter (Cambridge University Press, 1994).

Outram, Dorinda. *Georges Cuvier: Vocation, Science and Authority in post-revolutionary France* (Manchester University Press, 1984).

Outram, Dorinda. *The Enlightenment* (Cambridge University Press, 1995).

Pagel, Walter. *Paracelsus: An Introduction to Philosophical Medicine in the Era of the Renaissance*, second edition (Basel: Karger, 1982).

Pagel, Walter. *Jan Baptista van Helmont: Reformer of Science and Medicine* (Cambridge University Press, 1982).

Palmer, Louis Yvonne. The Early Scientific Work of Antoine Laurent Lavoisier: In the Field and in the Laboratory, 1763–1767. Ph.D. dissertation, Yale University, 1998.

Partington, J. R. "Berthollet and the Antiphlogistic Theory." *Chymia* 5, 1959, 130–137.

Partington, J. R. *A History of Chemistry*, 4 vols. (London: MacMillan, 1961–1972).

Partington, J. R., and D. McKie. "Historical Studies on the Phlogiston Theory." *Annals of Science* 2, 1937, 361–404; 3, 1938, 1–58 and 337–371; 4, 1939, 113–149.

Patterson, Elizabeth C. *John Dalton and the Atomic Theory: The Biography of a Natural Philosopher* (Doubleday, 1970).

Patterson, T. S. "Jean Beguin and His *Tyrocinium Chymicum*." *Annals of Science* 2, 1937, 243–298.

Paul, Charles B. *Science and Immortality: The Eloges of the Paris Academy of Sciences* (1699–1791) (University of California Press, 1980).

Payen, Jacques. *Les frères Périer et l'introduction en France de la machine à vapeur de Watt* (Paris: Palais de la Decouverte, 1968).

Perkins, Wendy. "The Uses of Science. The Montmor Academy, Samuel Sorbière and Francis Bacon." *Seventeenth-Century French Studies* 7, 1985, 155–162.

Perrin, C. E. "Prelude to Lavoisier's Theory of Calcination: Some Observations on *Mercurius calcinatus per se*." *Ambix* 16, 1969, 140–151.

Perrin, C. E. "Early Opposition to the Phlogiston Theory: Two Anonymous Attacks." *BJHS* 5, 1970, 128–144.

Perrin, C. E. "Lavoisier's Table of the Elements: A Reappraisal." *Ambix* 20, 1973, 95–105.

Perrin, C. E. "Lavoisier, Monge, and the Synthesis of Water, a Case of Pure Coincidence?" *BJHS* 24, 1973, 424–428.

Perrin, C. E. "The Triumph of Antiphlogistians." In *The Analytic Spirit*, ed. H. Woolf (Cornell University Press, 1981).

Perrin, C. E. "A Reluctant Catalyst: Joseph Black and the Edinburgh Reception of Lavoisier's Chemistry." *Ambix* 29, 1982, 141–176.

Perrin, C. E. "J. B. van Mons' *Essai sur les principes de la chimie antiphlogistique*: A Mystery solved." *Ambix* 31, 1984, 1–5.

Perrin, C. E. "Did Lavoisier Report to the Academy of Sciences on His Own Book?" *Isis* 75, 1984, 343–348.

Perrin, C. E. "Lavoisier's Thoughts on Calcination and Combustion, 1772–1773." *Isis* 77, 1986, 647–666.

Perrin, C. E. "Of Theory Shifts and Industrial Innovations: The Relations of J. A. C. Chaptal and A. L. Lavoisier." *Annals of Science* 43, 1986, 511–542.

Perrin, C. E. "Revolution or Reform: The Chemical Revolution and Eighteenth Century Concept of Scientific Change." *History of Science* 25, 1987, 395–423.

Perrin, C. E. "Research Tradition, Lavoisier, and the Chemical Revolution." *Osiris* [2] 4, 1988, 53–81.

Perrin, C. E. "The Chemical Revolution: Shifts in Guiding Assumptions." In *Scrutinizing Science*, ed. A. Donovan et al. (Johns Hopkins University Press, 1988).

Perrin, C. E. "Document, Text and Myth: Lavoisier's Crucial Year Revisited." *BJHS* 22, 1989, 3–25.

Perrin, C. E. "The Lavoisier-Bucquet Collaboration: A Conjecture." *Ambix* 36, 1989, 5–40.

Perrin, C. E. "Chemistry as Peer of Physics: A Response to Donovan and Melhado on Lavoisier." *Isis* 81, 1990, 259–270.

Pevitt, Christine. *The Man Who Would Be King: The life of Philippe d'Orléans, Regent of France* (New York: Quill, 1997).

Pickering, A. "Living in the Material World: On Realism and Experimental Practice." In *The Uses of Experiment*, ed. D. Gooding et al. (Cambridge University Press, 1989).

Planchon, G. "La dynastie des Geoffroy, apothicaires de Paris." *Journal de pharmacie et de chimie* 8, 1898, 289–293 and 337–345.

Poirier, Jean-Pierre. *Lavoisier: Chemist, Biologist, Economist* (University of Pennsylvania Press, 1996).

Porter, Roy. "Science, Provincial Culture and Public Opinion in Enlightenment England." *British Journal for Eighteenth-Century Studies* 3, 1980, 20–46.

Porter, Roy, and Mikulas Teich, eds. *The Enlightenment in National Context* (Cambridge University Press, 1981).

Porter, Theodore. "The Promotion of Mining and the Advancement of Science: The Chemical Revolution of Mineralogy." *Annals of Science* 38, 1981, 543–570.

Powers, John C. " 'Ars since arte': Nicholas Lemery and the End of Alchemy in Eighteenth-Century France." *Ambix* 45, 1998, 163–189.

Principe, Lawrence M. "Robert Boyle's Alchemical Secrecy: Codes, Ciphers, and Concealments." *Ambix* 39, 1992, 63–74.

Principe, Lawrence M. "Boyle's alchemical pursuits." In *Robert Boyle Reconsidered*, ed. M. Hunter (Cambridge University Press, 1994).

Principe, Lawrence M. *The Aspiring Adept: Robert Boyle and His Alchemical Quest* (Princeton University Press, 1998).

Rappaport, Rhoda. G.-F. Rouelle, His *Cours de chimie*, and Their Significance for Eighteenth Century Chemistry. Master's thesis, Cornell University, 1958.

Rappaport, Rhoda. "G.-F. Rouelle: An Eighteenth-Century Chemist and Teacher." *Chymia* 6, 1960, 68–101.

Rappaport, Rhoda. "Rouelle and Stahl—The Phlogistic Revolution in France." *Chymia* 7, 1961, 73–102.

Rappaport, Rhoda. Guettard, Lavoisier, and Monnet: Geologists in the Service of the French Monarchy. Ph.D. dissertation, Cornell University, 1964.

Rappaport, Rhoda. "Lavoisier's Geologic Activities, 1763–1792." *Isis* 18, 1965, 375–384.

Rappaport, Rhoda. "Lavoisier's Theory of the Earth." *BJHS* 6, 1973, 247–260.

Rappaport, Rhoda. "The Liberties of the Paris Academy of Sciences, 1716–1785." In *The Analytic Spirit*, ed. H. Woolf (Cornell University Press, 1981).

Rappaport, Rhoda. "Baron d'Holbach's Campaign for German (and Swedish) Science." *Studies on Voltaire* 323, 1994, 225–246.

Rattansi, P. M. "Paracelsus and the Puritan Revolution." *Ambix* 11, 1963, 24–32.

Rattansi, P. M. "The Helmontian-Galenist Controversy in Restoration England." *Ambix* 12, 1964, 1–23.

Read, John. *Humour and Humanism in Chemistry* (London: G. Bell and Sons, 1947).

Read, John. "William Davidson of Aberdeen: The First British Professor of Chemistry." *Ambix* 9, 1961, 70–101.

Reilly, J., and N. O'Flynn. "Richard Kirwan, an Irish Chemist of the Eighteenth Century." *Isis* 13, 1930, 298–319.

Reti, L. "Van Helmont, Boyle and the Alkahest." In *Some Aspects of Seventeenth-Century Medicine and Science* (Los Angeles: William Andrews Clark Memorial Library, 1969).

Rheinberger, Hans-Jörg. *Toward a History of Epistemic Things: Synthesizing Proteins in the Test Tube* (Stanford University Press, 1997).

Roberts, Lissa. "Setting the Table: The Disciplinary Development of Eighteenth-Century Chemistry as Read Through the Changing Structure of Its Tables." In *The Literary Structure of Scientific Argument*, ed. P. Dear (University of Pennsylvania Press, 1991).

Roberts, Lissa. "A Word and the World: The Significance of Naming the Calorimeter." *Isis* 82, 1991, 199–222.

Roberts, Lissa, ed. "The Chemical Revolution: Context and Practices." *The Eighteenth Century: Theory and Interpretation* 33, 1992, 195–271.

Roberts, Lissa. "Condillac, Lavoisier, and the Instrumentalization of Science." *The Eighteenth Century: Theory and Interpretation* 33, 1992, 252–271.

Roberts, Lissa. "Filling the Space of Possibilities: Eighteenth-Century Chemistry's Transition from Art to Science." *Science in Context* 6, 1993, 511–553.

Roberts, Lissa. "Going Dutch: Situating Sciences in the Dutch Enlightenment." In *The Sciences in Enlightened Europe*, ed. W. Clark et al. (University of Chicago Press, 1999).

Roche, Daniel. *Le siècle des lumières en province: Académies et académiciens provinciaux, 1680–1789*, 2 vols. (Mouton, 1978).

Roche, Daniel. "Talent, Reason, and Sacrifice: The Physician during the Enlightenment." In *Medicine and Society in France*, ed. R. Forster and O. Ranum (Johns Hopkins University Press, 1980).

Roche, Daniel. *Les Républicains des lettres: Gens de culture et lumières au XVIIIe siècle* (Fayard, 1988).

Rocke, Alan J. "Atoms and Equivalants: The Early Development of the Chemical Atomic Theory." *HSPS* 9, 1978, 225–263.

Rocke, Alan J. *Chemical Atomism in the Nineteenth Century: From Dalton to Cannizzaro* (Ohio State University Press, 1984).

Rocke, Alan J. *The Quiet Revolution: Hermann Kolbe and the Science of Organic Chemistry* (University of California Press, 1993).

Rocke, Alan J. *Nationalizing Science: Adolphe Wurtz and the Battle for French Chemistry* (MIT Press, 2001).

Roger, Jacques. *Les sciences de la vie dans la pensée française du XVIIIe siècle* (Paris: A. Colin, 1963).

Roger, Jacques. *Buffon: A Life in Natural History* (Cornell University Press, 1997).

Roscoe, Henry E. *John Dalton and the Rise of Modern Chemistry* (Macmillan, 1895).

Roscoe, Henry E., and Arthur Harden, *A New View of the Origin of Dalton's Atomic Theory* (Johnson Reprint Corp., 1970).

Rousseau, G. S., and Roy Porter, eds. *The Ferment of Knowledge: Studies in the Historiography of Eighteenth-Century Science* (Cambridge University Press, 1980).

Rowbottom, Margaret E. "Some Huguenot Friends and Acquaintances of Robert Boyle (1627–1691)." *Proceedings of the Huguenot Society of London* 20, 1959–60, 177–194. ·

Rudwick, Martin J. S. *The Great Devonian Controversy: The Shaping of Scientific Knowledge among Gentlemanly Specialists* (University of Chicago Press, 1985).

Sadoun-Goupil, Michelle. "Science pure et science appliquée dans l'oeuvre de Claude-Louis Berthollet." *Revue* 27, 1974, 127–145.

Sadoun-Goupil, Michelle. *Le chimiste Claude-Louis Berthollet, 1748–1822: Sa vie, son oeuvre* (Vrin, 1977).

Saraiva, Luis M. R. "Laplace, Lavoisier and the Quantification of Heat." *Physis* 34, 1997, 99–137.

Sargent, R.-M. *The Diffident Naturalist: Robert Boyle and the Philosophy of Experiment* (University of Chicago Press, 1995).

Sassen, F. L. R. "The Intellectual Climate in Leiden in Boerhaave's Time." In *Boerhaave and His Time*, ed. G. Lindeboom (Brill, 1970).

Schaffer, Simon. "Natural Philosophy and Public Spectacle in the Eighteenth Century." *History of Science* 21, 1983, 1–43.

Schaffer, Simon. "Godly Men and Mechanical Philosophers: Souls and Spirits in Restoration Mechanical Philosophy." *Science in Context* 1, 1987, 55–85.

Schofield, Robert. "The Counter-Reformation in Eighteenth-Century Science— Last Phase." In *Perspectives in the History of Science and Technology*, ed. D. Roller (University of Oklahoma Press, 1971).

Schofield, Robert. "Still More on the Water Controversy." *Chymia* 9, 1964, 71–76.

Schofield, Robert. *Mechanism and Materialism: British Natural Philosophy in An Age of Reason* (Princeton University Press, 1970).

Schofield, Robert. "An Evolutionary Taxonomy of Eighteenth-Century Newtonianisms." *Studies in Eighteenth-Century Culture* 7, 1978, 175–192.

Schufle, J. A. "Torbern Bergman, Earth Scientist." *Chymia* 12, 1967, 59–97.

Schufle, J. A. *Torbern Bergman: A Man Before His Time* (Cornado, 1985).

Scott, E. L. The Life and Work of Richard Kirwan (1733–1812). Ph.D. thesis, University of London, 1979.

Secrétan, Claude. "Coup d'oeil sur la chimie prélavoisienne (d'après un manuscrit inédit), *Bulletin de la société vaudoise des sciences naturelles* 61, 1941, 329–354.

Secrétan, Claude. "Un aspect de la chimie prélavoisienne (Le Cours de G.-F. Rouelle)." *Mémoires de la société vaudoise des sciences naturelles* 50, 1943, 220–444.

Serna, Pierre. "The Noble." In *Enlightenment Portraits*, ed. M. Vovelle (University of Chicago Press, 1997).

Shapin, Steven. "Pump and Circumstance: Robert Boyle's Literary Technology." *Social Studies of Science* 14, 1984, 481–520.

Shapin, Steven. "The House of Experiment in Seventeenth-Century England." *Isis* 79, 1988, 373–404.

Shapin, Steven. *A Social History of Truth: Civility and Science in Seventeenth-Century England* (University of Chicago Press, 1994).

Shapin, Steven, and Simon Schaffer. *Leviathan and the Air-Pump: Hobbes, Boyle, and the Experimental Life* (Princeton University Press, 1985).

Siegfried, R., and B. J. Dobbs. "Composition: A Neglected Aspect of the Chemical Revolution." *Annals of Science* 24, 1968, 275–293.

Siegfried, Robert. "Lavoisier's Table of Simple Substances: Its Origin and Interpretation." *Ambix* 29, 1982, 29–48.

Siegfried, Robert. "Lavoisier's View of Gaseous State and Its Early Application to Pneumatic Chemistry." *Isis* 63, 1972, 59–78.

Siegfried, Robert. "The Chemical Revolution in the History of Chemistry." *Osiris* [2] 4, 1988, 34–50.

Siegfried, Robert. "Lavoisier and the Phlogistic Connection." *Ambix* 36, 1989, 31–40.

Simon, Jonathan. The Alchemy of Identity: Pharmacy and the Chemical Revolution, 1777–1809. Ph.D. dissertation, University of Pittsburgh, 1997.

Simon, Jonathan. "The Chemical Revolution and Pharmacy: A Disciplinary Perspective." *Ambix* 45, 1998, 1–13.

Smeaton, W. A. "The Contributions of P. J. Macquer, T. O. Bergman and L. B. Guyton de Morveau to the Reform of Chemical Nomenclature." *Annals of Science* 10, 1954, 87–106.

Smeaton, W. A. "The Early History of Laboratory Instruction in Chemistry at the Ecole Polytechnique, Paris, and Elsewhere." *Annals of Science* 10. 1954, 224–233.

Smeaton, W. A. "*L'avant-coureur*: The Journal in Which Some of Lavoisier's Earliest Research Was Reported." *Annals of Science* 13, 1957, 219–234.

Smeaton, W. A. "L. B. Guyton de Morveau (1737–1816): A Bibliographic Study." *Ambix* 6, 1957, 18–34.

Smeaton, W. A. "F. J. Bonjour and His translation of Bergman's 'Disquisitio de attractionibus electivis.'" *Ambix* 7, 1959, 47–50.

Smeaton, W. A. "Guyton de Morveau's Course of Chemistry in the Dijon Academy." *Ambix* 9, 1961, 53–69.

Smeaton, W. A. *Fourcroy: Chemist and Revolutionary* (Cambridge: Heffer, 1962).

Smeaton, W. A. "New Lights on Lavoisier; The Research of the Last Ten Years." *History of Science* 2, 1963, 51–69.

Smeaton, W. A. "Guyton de Morveau and Chemical Affinity." *Ambix* 11, 1963, 55–64.

Smeaton, W. A. "Guyton de Morveau and the Phlogiston Theory." In *Mélanges Alexandre Koyré*, 2 vols., ed. I. Cohen and R. Taton (Hermann, 1964).

Smeaton, W. A. "P. J. Macquer's Course of Chemistry at the Jardin du roi." In *Proceedings of the tenth International Congress of the History of Science* [Ithaca, 1962] (Hermann, 1964).

Smeaton, W. A. "The Portable Chemical Laboratories of Guyton de Morveau, Cronstedt and Göttling." *Ambix* 13, 1966, 84–91.

Smeaton, W. A. "Louis Bernard Guyton de Morveau, F.R.S. (1737–1816) and His Relations with Bretish Scientists." *Notes and Records of the Royal Society of London* 22, 1967, 113–130.

Smeaton, W. A. "Is Water Converted into Air? Guyton de Morveau Acts as Arbiter between Priestley and Kirwan." *Ambix* 15, 1968, 73–83.

Smeaton, W. A. "E. F. Geoffroy Was Not a Newtonian Chemist." *Ambix* 18, 1971, 212–214.

Smeaton, W. A. "Macquer et la médecine. Un aspect de sa vie qui nécessite des recherches." *Revue pharm.* 24, 1977, 251–254.

Smeaton, W. A. "Berthollet's *Essai de statique chimique* and Its translations: A Bibliographical Note and a Daltonian Doubt." *Ambix* 24, 1977, 149–158.

Smeaton, W. A. "Berthollet's *Essai de statique chimique*: A Supplementary Note." *Ambix* 25, 1978, 211–212.

Smeaton, W. A. "Two Books Are Added to Guyton de Morveau's Library: A Study of Personal and Academic Communications in 1785." *Ambix* 34, 1987, 140–146.

Smith, Crosbie, and M. Norton Wise. *Energy and Empire: A Biographical Study of Lord Kelvin* (Cambridge University Press, 1989).

Smith, John Graham. *The Origins and Early Development of the Heavy Chemical Industry in France* (Oxford, 1979).

Smith, Pamela H. *The Business of Alchemy: Science and Culture in the Holy Roman Empire* (Princeton University Press, 1994).

Snelders, H. A. M. "The New Chemistry in the Netherlands." *Osiris* [2] 4, 1988, 121–145.

Snelders, H. A. M. "Professors, Amateurs, and Learned Societies: The Organization of the Natural Sciences." In *The Dutch Republic in the Eighteenth Century*, ed. M. Jacob and W. Mijnhardt (Cornell University Press, 1992).

Solomon, Howard M. *Public Welfare, Science, and Propaganda in Seventeenth Century France: The Innovations of Théophraste Renaudot* (Princeton University Press, 1972).

Spray, E. C. *Utopia's Garden: French Natural History from Old Regime to Revolution* (University of Chicago Press, 2000).

Stengers, Isabelle. "Ambiguous Affinity: Newtonian Dream of Chemistry in the Eighteenth Century." In *A History of Scientific Thought*, ed. M. Serres (Blackwell, 1995).

Stewart, Larry. "Public Lectures and Private Patronage in Newtonian England." *Isis* 77, 1986, 47–58.

Stewart, Larry. *The Rise of Public Science: Rhetoric, Technology, and Natural Philosophy in Newtonian Britain, 1660–1750* (Cambridge University Press, 1992).

Stroup, Alice. "Wilhelm Homberg and the Search for the Constituents of Plants at the 17th-Century Académie royale des sciences." *Ambix* 26, 1979, 184–201.

Stroup, Alice. *A Company of Scientists: Botany, Patronage, and Community at the Seventeenth-Century Parisian Royal Academy of Sciences* (University of California Press, 1990).

Stroup, Alice. "Duclos on Boyle: A French Academician Criticizes *Certain Physiological Essays*." Presentation at annual meeting of History of Science Society, Vancouver, 2000.

Stroup, Alice. "Censure our querelles scientifiques: l'affaire Duclos (1675–1685)." (manuscript).

Strube, Irene. *Georg Ernst Stahl* (Leipzig: B. G. Teubner, 1984).

Sturdy, David J. *Science and Social Status: The Members of the Académie Royale des Sciences, 1666–1750* (Boydell, 1995).

Sudduth, William M. "Eighteenth-Century Identifications of Electricity with Phlogiston." *Ambix* 25, 1978, 131–147.

Sutton, Geoffrey V. *Science for a Polite Society: Gender, Culture, and the Demonstration of Enligthenment* (Westview, 1995).

Szabadváry, Ferenc. *History of Analytic Chemistry* (Pergamon, 1966).

Taton, René, ed. *Enseignement et diffusion des sciences en France au XVIIIe siècle* (Hermann, 1964).

Taton, René. *Les Origines de l'Académie Royale des Sciences* (Université de Paris and Palais de la Découverte, 1965).

Terrall, Mary. "Representing the Earth's Shape." *Isis* 83, 1992, 218–237.

Terrall, Mary. "Metaphysics, Mathematics, and the Gendering of Science in Eighteenth-Century France." In *The Sciences in Enlightened Europe*, ed. W. Clark et al. (University of Chicago Press, 1999).

Terrall, Mary. *The Man Who Flattened the Earth: Maupertuis and the Sciences in the Enlightenment* (University of Chicago Press, 2002).

Thackray, Arnold. *Atoms and Powers: An Essay on Newtonian Matter-Theory and the Development of Chemistry* (Harvard University Press, 1970).

Thackray, Arnold. *John Dalton: Critical Assessments of His Life and Science* (Harvard University Press, 1972).

Tisserand, Roger. *Au temp de l'encyclopédie: L'académie de Dijon de 1740 à 1793* (Paris: Boivin, 1936).

Tonnet, D. Joseph. *Notice sur Nicolas Lemery, chimiste* (Niort: Robin, 1840).

Trevor-Roper, Hugh. "The Sieur de la Rivière, Paracelsian Physician of Henri IV." In *Science, Medicine and Society in the Renaissance*, ed. A. Debus (Science History Publications, 1972).

Trevor-Roper, Hugh. "The Paracelsian Movement." In *Renaissance Essays*, ed. H. Trevor-Roper (University of Chicago Press, 1985).

Trevor-Roper, Hugh. "Paracelsianism Made Political, 1600–1650." In *Paracelsus*, ed. O. Grell (Brill, 1998).

Urdang, George. "The Early Chemical and Pharmaceutical History of Calomel." *Chymia* 1, 1948, 93–108.

Vartanian, Aram. *Science and Humanism in the French Enlightenment* (Charlottesville: Rookwood, 1999).

Vickers, Brian. "The Royal Society and English Prose Style: A Reassessment." 1–76 in *Rhetoric and the Pursuit of Truth: Language Change in the Seventeenth and Eighteenth Centuries* (Los Angeles: William Andrews Clark Memorial Library, 1985).

Viel, Claude. "Duhamel du Monceau, naturaliste, physicien et chimiste." *Revue* 38, 1985, 55–71.

Viel, Claude. "Pierre-Joseph Macquer." *Janus* 73, 1986–90, 1–27.

Viel, Claude. "L'activité de chimiste de Guyton de Morveau à travers ses lettres à Macquer et à Picot de La Peyrouse." *Annales de Bourgogne* 70, 1998, 55–67.

Walters, Barrie. "The *Journal des savants* and the Dissemination of News of English Scientific Activity in Late Seventeenth-Century France." *Studies on Voltaire* 314, 1993, 133–166.

Walters, R. L. "Chemistry at Cirey." *Studies on Voltaire* 58, 1967, 1807–1827.

Walton, R. L. "Voltaire, Newton & the Reading Public." In *The Triumph of Culture*, ed. P. Fritz and D. Williams (Toronto: Hakkert, 1972).

Watson, E. C. "The Early Days of the Académie des Sciences as Portrayed in the Engravings of Sébastien Le Clerc." *Osiris* 7, 1939, 556–587.

Wellman, Kathleen. *La Mettrie: Medicine, Philosophy, and Enlightenment* (Duke University Press, 1992).

West, Muriel. "Notes on the Importance of Alchemy to Modern Science in the Writings of Francis Bacon and Robert Boyle." *Ambix* 9, 1961, 102–114.

Wilson, Lindsay. *Women and Medicine in the French Enlightenment: The Debate over Maladies des Femmes* (Johns Hopkins University Press, 1993).

Wise, M. Norton. "Mediations: Enlightenment Balancing Acts, or the Technologies of Rationalism." In *World Changes*, ed. P. Horwich (MIT Press, 1993).

Wise, M. Norton, ed. *The Values of Precision* (Princeton University Press, 1995).

Wohl, Robert. "Buffon and His Project for a New Science." *Isis* 51, 1960, 186–199.

Woolf, Harry, ed. *The Analytic Spirit: Essays in the History of Science in Honor of Henry Guerlac* (Cornell University Press, 1981).

Zwerling, Craig S. *The Emergence of the Ecole Normale Supèrieure as a Center of Scientific Education in Nineteenth-Century France* (Garland, 1990).

Index